T0353346

CATIA v5

This tutorial textbook is an essential companion to using *CATIA v5* to assist with computer-aided design. Using clear CAD examples, it demonstrates the various ways through which the potential of this versatile software can be used to aid engineers in 3D modelling.

Based on 20 years of teaching experience, the authors present methods of using *CATIA v5* to model solid and surface parts, to perform parametric modelling and design of families of parts, reconstruction of surfaces, to create macros and to apply various tools and their options during 3D modelling. Importantly, this book will also help readers to discover multiple modelling solutions and approaches to solve common issues within design engineering. With a comprehensive approach, this book is suitable for both beginners and those with a good grasp of *CATIA v5*. Featuring an end chapter with questions and solutions for self-assessment, this book also includes 3D modelling practice problems, presented in the form of 2D engineering drawings of many 3D parts in both orthogonal and isometric views. Using the knowledge gained through reading the book chapters, users will learn how to approach surfaces and solids as 3D models using *CATIA v5*. This book provides detailed explanations, using clear figures, annotations and links to video tutorials.

It is an ideal companion for any student or engineer using *CATIA v5* in industries including automotive, naval, aerospace and design engineering.

Readers of this book should note that the length and distance dimensions are in millimeters and the angular dimensions are in degrees. All other parameters, such as radii, areas and volumes, also use the metric system.

CATIA v5
Advanced Parametric and Hybrid 3D Design

Ionuţ Gabriel Ghionea, Cristian Ioan Tarbă,
and Saša Ćuković

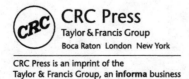

CRC Press
Taylor & Francis Group
Boca Raton London New York

CRC Press is an imprint of the
Taylor & Francis Group, an **informa** business

First edition published 2023
by CRC Press
6000 Broken Sound Parkway NW, Suite 300, Boca Raton, FL 33487-2742

and by CRC Press
4 Park Square, Milton Park, Abingdon, Oxon, OX14 4RN

CRC Press is an imprint of Taylor & Francis Group, LLC

© 2023 Ionuţ Gabriel Ghionea, Cristian Ioan Tarbă and Saša Ćuković

Reasonable efforts have been made to publish reliable data and information, but the author and publisher cannot assume responsibility for the validity of all materials or the consequences of their use. The authors and publishers have attempted to trace the copyright holders of all material reproduced in this publication and apologize to copyright holders if permission to publish in this form has not been obtained. If any copyright material has not been acknowledged please write and let us know so we may rectify in any future reprint.

Except as permitted under U.S. Copyright Law, no part of this book may be reprinted, reproduced, transmitted, or utilized in any form by any electronic, mechanical, or other means, now known or hereafter invented, including photocopying, microfilming, and recording, or in any information storage or retrieval system, without written permission from the publishers.

For permission to photocopy or use material electronically from this work, access www.copyright.com or contact the Copyright Clearance Center, Inc. (CCC), 222 Rosewood Drive, Danvers, MA 01923, 978-750-8400. For works that are not available on CCC please contact mpkbookspermissions@tandf.co.uk

Trademark notice: Product or corporate names may be trademarks or registered trademarks and are used only for identification and explanation without intent to infringe.

ISBN: 9781032250069 (hbk)
ISBN: 9781032250106 (pbk)
ISBN: 9781003281153 (ebk)

DOI: 10.1201/9781003281153

Typeset in Times
by codeMantra

Contents

Preface..vii
Authors..ix

Chapter 1   Introduction 1

      1.1   Computer-Aided Design in Conception and Development of
           Industrial Products 1
      1.2   General Aspects Regarding the Use of CATIA v5 Program Workbenches......1

Chapter 2   The Working Environment of the CATIA v5 Program 3

      2.1   Program Interface...3
      2.2   Menu Bar 3
      2.3   Specification Tree 24
      2.4   Compass 27
      2.5   Toolbars 29

Chapter 3   Hybrid 3D Modelling of Parts...33

      3.1   Methods for Creating the Working Planes 33
      3.2   Modelling of a Nut-Type Part 39
      3.3   Modelling of a Lever Part 48
      3.4   Modelling of a Stopper-Type Part 58
      3.5   Modelling of a Hinge Part...70
      3.6   Modelling of a Complex Spring 78
      3.7   Modelling of a Complex Spiral Ornament Part for Wrought Iron Fence........89
      3.8   Modelling of a Fork-Type Part 101
      3.9   Modelling of a 3D Knot 112
      3.10   Modelling of an Axle Support... 121
      3.11   Modelling of a Switch Button Part.. 130
      3.12   Modelling of a Balloon Support.. 139
      3.13   Modelling of a Rotor Part with Blades... 148
      3.14   Modelling of a Handle Knob... 163
      3.15   Modelling of a Reinforced Key Button Part 172
      3.16   Modelling of a Complex Fitting Part.. 184
      3.17   Modelling and Transformation of a Part into Two Constructive Solutions... 198
      3.18   Editing and Reconstruction of Solids Using Surfaces – Twisted Area 210
      3.19   Editing and Reconstruction of Solids Using Surfaces –
            Connected Surfaces 218
      3.20   Modelling of a Gearbox Shifter Knob 232
      3.21   Modelling of a Complex Plastic Cover...254
      3.22   Modelling of a Window Crank Handle 269
      3.23   Modelling of a Shield Using Laws 285
      3.24   Modelling of a Citrus Juicer...295
      3.25   Modelling of an Ornament Panel 326
      3.26   Modelling of Parametric Bellows in Different Constructive Solutions 349

Chapter 4 Parametric Modelling and Sheetmetal Design 361

    4.1   Introduction to Parametric Modelling of Parts and Families of Parts 361

    4.2   Parametric Modelling of a Part Using Formulas and Rules 365

    4.3   Parametric Modelling of a Connector Cover and Optimizations of
the Part 376

    4.4   Parametric Modelling of a Support Block Using Design Tables 394

    4.5   Parametric Modelling of a Hook Clamp. Creation of a Components
Catalogue 400

    4.6   Modelling of a Sheet Metal Cover 420

    4.7   Modelling of a Sheet Metal Closing Element 426

Chapter 5 Programming, Automation and Scripting 435

    5.1   Introduction to Automation and Scripting in CATIA v5 435

        5.1.1   Recording a Macro 437

        5.1.2   Getting Started with Custom Code Writing 438

        5.1.3   CATIA Automation Documentation 438

    5.2   Recording a Macro 440

    5.3   Development of VBScript Scripts 452

Chapter 6 Knowledge Assessment Tests and 2D Drawings of Parts
Proposed for Modelling 463

    6.1   Multiple-Choice Questions 463

    6.2   Answers for Multiple-Choice Questions 482

    6.3   2D Drawings of the Parts to Practise Modelling 495

Annexes

A1. Additional Online Resources: User Communities, Forums and Video Tutorials 531

A2. Video Tutorials to Support the Presented Written Tutorials 539

Bibliography 541

Index 543

Preface

This book is a part of the series of *CAD* tutorial books that presents the basic and advanced characteristics and working possibilities of modern computer-aided design software solution *CATIA v5*.

This tutorial book is intended to be not only used by students from faculties with a mechanical or industrial engineering profile but also by design engineers from various industries (automotive, aerospace, military, heavy machinery, medical technology, etc.) who need to work in this *CAD* environment. Whether they are beginners or have a good experience in using *CATIA v5*, reading all written tutorials will help them to understand, upgrade and improve their knowledge and then to apply proven 3D modelling methods, to get familiar with many new modelling operations and options, by going step by step through the solutions explained for 3D modelling problems and those proposed for practising.

Based on our 20-plus years of teaching experience, we structured and wrote this book focusing on hybrid 3D modelling of many interesting parts. Each tutorial is a challenge for the reader and gradually presents different 3D modelling techniques, carefully explained and accompanied by clear figures, with annotations for a better understanding of the context. Thus, we use numerous graphical representations, drawings, screenshots, dialog boxes, icons of the tools, etc. Text and figures support the reader in understanding the approach and highlight all important selections (geometric elements or options). By presenting all phases of the modelling in a step-by-step manner, we explain a great number of options and strategies to model complex 3D surfaces, parametric solids and family tables, macros and *Visual Basic Application* (*VBA*) scripts.

All explanations and the collection of 3D examples included at the end of this book, in all their diversity, have been carefully selected to cover additional options, and some of them are followed by video tutorials presented in the Annexes, and near most of the proposed 2D drawings. Although many theoretical aspects are briefly explained to solve the 3D modelling problems easily, this book does not present all available commands and their options. Therefore, the reader is encouraged to further explore important sub-options encountered in the dialog boxes and then to search for some new modelling solutions for solving challenging parts.

To test the reader's knowledge, the last chapter of this book contains 75 tests with their detailed answers, a scoring system (for self-assessment) and many 3D modelling problems for self-practising. Through the individual study, the reader is invited to model them in 3D because each part has an educational scope and they have different degrees of complexity, shapes and functionalities.

From students to engineers, all are advised to open and follow the pages of this book with concern and perseverance, to patiently go through all the stages of the presented tutorials, to explore the proposed 3D models and then to successfully apply the knowledge acquired in their professional activity.

The authors have made a consistent effort and passionately created all the written and video tutorials presented in this book, with great attention to detail. Several people, experts in other CAD programs, helped us create the content of this book. Our families, friends and colleagues from the university and industry supported us with patience and interest, and they proposed ideas and provided valuable observations, and we thank them for their time and feedback. The publisher (Taylor & Francis/CRC Press) also guided us with interest and professionalism to create a manuscript that will provide a great experience to our readers.

We hope that this book, by its content, will rise to the level of exigency that we set ourselves from the beginning and will be useful to all those who will have the curiosity and need to open it and learn from our knowledge.

We also challenge readers to send us other solutions for the presented 3D modelling problems and links to their own video tutorials and to contribute with their ideas and suggestions to improving the content of the future editions of this book.

Authors

Ionuț Gabriel Ghionea is an associate professor and member of the Manufacturing Engineering Department, Faculty of Industrial Engineering and Robotics, University Politehnica of Bucharest, Romania, since 2000. In 2003, he completed an internship to prepare his doctoral thesis at École Nationale Supérieure d'Arts et Métiers in Aix-en-Provence, France, and has a PhD in engineering since 2010.

Ionuț Gabriel Ghionea has published, as the first author or co-author, 11 books in the field of computer-aided design for mechanical engineering and more than 120 articles in technical journals and conference proceedings. He is a member of the editorial board and review panel of several international and national journals and conferences.

Ionuț Gabriel Ghionea participated in research and educational projects for the industry. He is one of the most known and appreciated CAD trainers in Romania, carrying out this activity since 2002.

He is also considered to be one of the main and first didactic promoters of the CATIA program in the Romanian academic environment, and he has made numerous CAD applications, video tutorials and practical works for students and CAD courses for companies in Romania and abroad.

In December 2020, he became a CATIA Champion, recognition issued by Dassault Systèmes (Paris, France) for his long-term commitment in promoting and using PLM System CATIA.

CATIA Champion
CATIA Certified Professional Part Design Specialist
Contact: ionut76@hotmail.com, http://www.fiir.pub.ro, http://www.tcm.pub.ro, http://www.catia.ro

Cristian Ioan Tarbă, PhD, is a lecturer at the University Politehnica of Bucharest, Romania. He became a member of the Manufacturing Engineering Department in 2009, after working for seven years as a CAD and CAM engineer, using different software packages. He has been using CATIA as the main tool for research and teaching activities for 13 years. He uses Part Design, Assembly Design, Generative Shape Design and Digital Mock-Up Kinematics workbenches in his lectures to both Romanian and foreign students.

In research activities, he also uses VBScripting and Python integrated with CATIA to complete his scientific work.

In December 2020, he became a CATIA Champion, recognition issued by Dassault Systèmes (Paris, France) for his long-term commitment in promoting and using PLM System CATIA.

CATIA Champion
Contact: ticris@gmail.com

Saša Ćuković is a postdoc and Marie Sklodowska Curie fellow at the Institute for Biomechanics (IfB/LMB, Prof. Dr William Taylor), ETH Zurich, Switzerland. He played a leading role in a spinal 3D modelling under the research project III41007 financed by the Ministry of Education, Science and Technological Development of the Republic of Serbia.

Saša Ćuković was a Swiss Government Excellence Scholarship holder at ETH Zurich, DAAD grant holder at Technical University of Munich (Germany) and OeAD grant holder at TU Graz and MedUni Vienna (Austria) working on non-invasive diagnosis of spinal deformities. His main research interests include CAD/CAM systems, reverse engineering and non-invasive 3D reconstruction and modelling in engineering and medicine, computational biomechanics, augmented reality and computer vision. He has valuable teaching experience and has published few books in the domain of CAD/CAM within the University of Kragujevac. He also disseminated his knowledge in various projects and in teaching abroad.

He has authored/co-authored five books and more than 80 papers. Since 2019, he is a senior scientific associate in the Institute for Information Technologies (IIT) Kragujevac, and from 2022 he is a member of the Institute for Artificial Intelligence Research and Development of Serbia, Novi Sad, Serbia. He is a group leader at ETH/IfB/LMB for computational 3D biomechanics of the spine and a senior member of IEEE.

In December 2020, he became a CATIA Champion, recognition issued by Dassault Systèmes (Paris, France) for his long-term commitment in promoting and using PLM System CATIA.

CATIA Champion
Contact: cukovic@kg.ac.rs, https://scukovic.com

1 Introduction

1.1 COMPUTER-AIDED DESIGN IN CONCEPTION AND DEVELOPMENT OF INDUSTRIAL PRODUCTS

Computer-aided design (*CAD*) is increasingly used in a wide variety of engineering fields, and it is evolving rapidly. This concerns both the general architecture and the addition of new functions and modelling tools in existing *CAD* programs, as well as their capability to create a geometry of high complexity, parametrically designed, based on the specifications imposed by the client, also allowing *FEA* simulations or machining on *NC* machine tools.

The ever-increasing complexity of products leads to some difficulties in the design and then in the manufacturing stages. The design engineer must be aware of advanced working methods, and in his work, he tries several possible variants and solutions and then optimizes the one identified as correct following the successful modelling procedure by implementing *CAD* and simulation tools. He also takes decisions related to shape, material properties and manufacturing technologies, based on the information taken from *CAD* handbooks, standards, databases, numerical analysis, the experience of the company and his intuition and knowledge. Finally, his activity leads to the validation of the projects and their transformation into successful products.

Thus, in general terms, the modern computer-aided design can be defined as the process of integrating a set of functional specifications and requirements into a complete virtual representation of the physical product or system that best meets those requirements and specifications.

The evolution of *CAD* systems has shortened the product lifecycle, increased their complexity and performance, and stimulated competition for new reliable products at attractive prices. The competition also means new jobs and opportunities for design engineers.

Companies are looking for specialists who are able to work with modern *CAD* systems, to make complex numerical programs and scripts and to use libraries with algebraic calculation methods and statistics, and who have a true vision in creating complex surfaces, which meet requirements in almost all current products. Other specialists perform specific calculations and kinematic simulations, determine moments of inertia, simulate mechanical processing on machine tools equipped with numerical control, etc.

Not only the design and development activities, but also the cost related to the design changes are intensifying as the project advances towards the launch in series production and then on the market, being much lower in the design phases when the decisions are made to establish the optimal solution.

1.2 GENERAL ASPECTS REGARDING THE USE OF CATIA v5 PROGRAM WORKBENCHES

CATIA (*computer-aided three-dimensional interactive application*), a product of *Dassault Systèmes*, is currently one of the most widely used integrated *CAD/CAM/CAE* systems worldwide, with applications in various fields, including machine manufacturing industry, aeronautics, naval industry, automotive industry, robotics, agricultural machinery and equipment, chemical industry, food industry and consumer goods. The 5th version has been available since 1999, with each new release introducing new workbenches and additional functionalities, in parallel with the improvement of the existing ones.

CATIA v5 provides a wide variety of solutions to meet all design and manufacturing issues. Among its many functionalities, advanced design of mechanical parts, interactive creation of

DOI: 10.1201/9781003281153-1

assemblies, faster 2D and assembly drawings, the ability to design in a parameterized manner, generation of complex surfaces, performing finite element analyses and simulation mechanical processing can be mentioned.

All these functionalities are possible due to the fact that the program uses virtual models, defined as the set of computer data necessary to manipulate an object created on a computer, in the same way as a real object. It is possible, thus, to test it under various mechanical and thermal stresses and their dynamic behaviour, to check whether or not an assembly is correct with all its constraints, to ensure that the moving of components, one against the other, does not generate collisions, etc.

CATIA v5 has a modular structure, which ensures great versatility and transition from one workbench to another, with the possibility of continuous editing of the project in progress. Although the number of workbenches implemented in the program is very large, some of them can be considered as principal, as the following ones:

- *CATIA Sketcher* – creates the sketch of a two-dimensional profile, being a starting point in the process of obtaining a three-dimensional feature.
- *CATIA Part Design* – is used for the three-dimensional design of parts. This workbench, along with *Sketcher*, may be considered as core workbenches of the *CATIA v5* program.
- *CATIA Assembly Design* – allows the generation of an assembly of parts using various mechanical constraints for their positioning and establishing contact between surfaces. As a scalable workbench, *Assembly Design* can be cooperatively used with other current companion products such as *Part Design* and *Generative Drafting*.
- *CATIA Drafting* – contains the necessary tools to obtain 2D drawings of the created parts and assemblies.
- *CATIA Knowledge Advisor* – is very useful in all the design stages and parameterized management of parts and assemblies, and also in the automation of certain processes of analysis with finite elements, simulation of processing on NC equipment, etc.
- *CATIA DMU Kinematics* – creates complex movement simulations based on kinematic couplings established between components of assemblies.
- *CATIA Generative Sheetmetal Design* – is used to model sheet metal parts, processed by cold plastic deformation. The user has the opportunity to obtain 2D drawings and their unfolding views.
- *CATIA Generative Structural Analysis* – is a powerful environment to perform analyses with finite elements, applied to parts and assemblies, in order to determine their behaviour under defined conditions of static or dynamic loading.
- *CATIA Generative Shape Design* – has applicability in creating objects that have complex surfaces and are difficult to design in other workbenches.
- *CATIA Prismatic Machining* – serves to simulate mechanical processing on machine tools equipped with NC devices.

2 The Working Environment of the CATIA v5 Program

2.1 PROGRAM INTERFACE

The working environment of *CATIA v5* is based on an intuitive but complex interface, and it requires the user to adapt to the toolbars and their icons. Depending on the workbench chosen for the work, *CATIA v5* offers different sets of specific toolbars, and also standard toolbars, available in the interface and/or in menus (example: tools for viewing, measuring and saving/opening files).

Also, the interface includes some tools with a very important role in the modelling process of the program, necessary for the user in the work stages (example: specification tree) and for orientation of the model in the 3D space (compass, orthogonal plane system, etc.).

As an example, Figure 2.1 shows the program interface of the *CATIA v5 Part Design* workbench. Thus, it contains some toolbars, the menu bar (*Start, File, Edit, View*, etc.), the specification tree (the tree structure at the top left corner of the interface), the compass (the structure with axes and planes in the top right corner of the graphic area), the *XY, YZ* and *ZX* planes not only in the middle of the screen but also in the specification or model tree, etc. At the bottom of the interface are the tools for saving the project in progress, those for viewing and representing 3D models in various planes, or with different visualizations (solid, wireframe, etc.), etc. All tools and interface elements will be used by the user and will be sometimes mentioned throughout this book.

Working in the *CATIA v5* program requires patience, intuition, ability to view in 3D and understand complex geometrical representations, and also a solid knowledge of technical drawing, engineering construction and *computer-aided design* (*CAD*). After becoming familiar with the interface, with the tools specific to each workbench, the user will be able to design mechanical parts and assemblies, make complex models using surfaces, perform drafting, parameterize their projects, perform simulations through finite elements, create programs for *NC* machines, etc.

2.2 MENU BAR

The *Start* menu, placed and highlighted in dark blue colour in the menu bar, provides access to the *CATIA* program workbenches. Their availability depends on the version of the licence obtained, and also on the selected user's options during the program's installation.

In the tutorials and samples presented in this book, the authors used workbenches in the *Mechanical Design, Shape* and *Knowledgeware* categories. Moving the mouse on the black arrow to the right of the category name will display the list of component workbenches. For example, Figure 2.2 partially shows the content of the *Shape* category, which contains the main workbench for defining surfaces (modelling and editing), *Generative Shape Design*.

At the bottom of the menu (Figure 2.3), a list of recently opened and working documents can be found (*Product.1*, which is checked). The user can see the extensions of these files (**.CATProduct, *.CATPart, *.CATAnalysis*, etc.), specific to the types of opened documents and, implicitly, of projects worked by the user.

Reopening such a document is very fast, and it happens by clicking on its name in the list or using the *Open* option in the *File* menu. In this second case, the file should be found on the support where it was saved and identified with a suggestive name. This name of the current file appears in the program title bar, in square brackets.

DOI: 10.1201/9781003281153-2

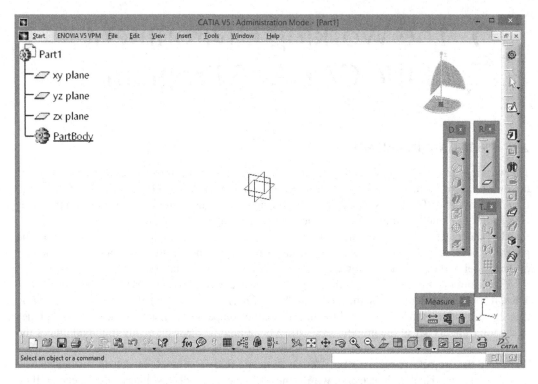

FIGURE 2.1 The program interface in the *Part Design* workbench.

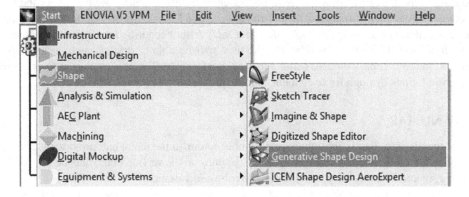

FIGURE 2.2 Accessing the *Generative Shape Design* workbench.

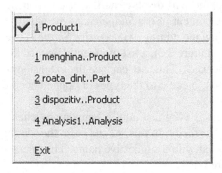

FIGURE 2.3 Different types of files opened by the user.

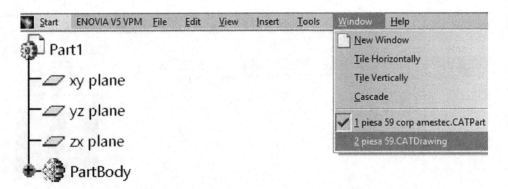

FIGURE 2.4 Opening the *Window* menu to choose what file will become current.

FIGURE 2.5 List of various types of files.

In the example in Figure 2.4, the *Window* menu is also presented, which helps the user to select which of the two open documents will become current, receiving, by selection, a check mark next to the name. At the moment, even if the user has opened multiple documents (with the same or different extensions), he can work only on one document, for example, to model, edit, or view a 3D model of a part. It is always possible to access another document/file, for example a 2D drawing of the current 3D model to insert the required dimensions.

The same figure shows the names of two different documents, the model of a part (with *CATPart* extension) and its 2D drawing/drafting (*CATDrawing*) because the user needs both files to open. The transition between documents is simple and fast, and the connection between them is also established by the program. This principle is called external association between different workbenches.

The menu bar (*File, Edit, View, Insert, Tools, Window* and *Help*) is basically a standard for *Microsoft Windows* compatible programs and contains *CATIA*-specific options. These differ depending on the workbench used, and some that are more important will be presented below, and others are used and explained in detail in the tutorials in Chapter 3.

For instance, the *File* menu contains options for saving and opening files. Selecting the *New* option leads to a selection box with the same name in which the program's file types are presented in a list. Choosing a type leads to the opening of a workbench; for example, in the case of Figure 2.5, following the *Part* selection, the *CATIA Part Design* workbench will open. Note also the *Save Management* option, which is used to save some files to different locations. The dialog box in Figure 2.6 also shows a preview of the saved entity, in this case, a part and its 2D drawing. It shows the names of the files, where they will be saved, their status, the user's rights over the respective location, etc.

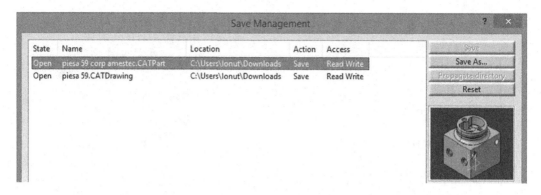

FIGURE 2.6   The management window for saving files.

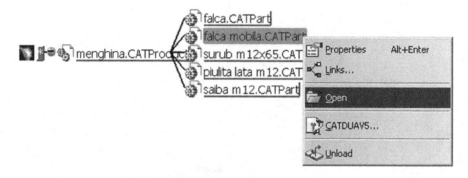

FIGURE 2.7   Tree-like structure of an assembly.

The *Close*, *Open*, *Save*, *Print*, *Document Properties* and *Exit* options are well known because they are part of any menu of a modern program in various operating systems. The *Desk* option, however, is specific to some *CAD* programs and shows, according to Figure 2.7, the links established between an assembly (*CATProduct*) and its components (*CATPart*). Also, right-clicking on a component and choosing the *Open* command has the effect of opening the document containing that component. The representation in the figure is particularly suggestive.

The *Edit* menu contains the well-known *Undo*, *Repeat*, *Cut*, *Copy*, *Paste*, etc. options. This menu also includes *Delete* (for deleting, for example, a component from an assembly) and *Update* (for updating geometrical and dimensional constraints, components' positions in the assembly, or 2D drawings after changing some dimensions of the 3D model of a part or belonging sketches, etc.).

The *Component Constraints* option involves selecting a component of an assembly and highlighting all assembly constraints in which that component is involved.

For example, Figure 2.8 shows the components of an assembly model and the list of their applied constraints. Three of them are selected, and the name of the component is observed in the name of each constraint.

Choosing the *Links* option displays the dialog box in Figure 2.9. A window with a list of components of the assembly is displayed, and the user has the possibility to select one of its parts to replace (pressing the *Replace* button) with another part.

In the selection box on the right, *Browse*, two buttons are available, as follows: *File* allows user to load a file, by default a component, from a storage medium, to replace the current selection, and *Loaded document* specifies a component already loaded and accessible from the *Window* menu. Of course, the replacement of a component must be possible and logical, and the assembly constraints are taken from the previous component to the one that replaces it.

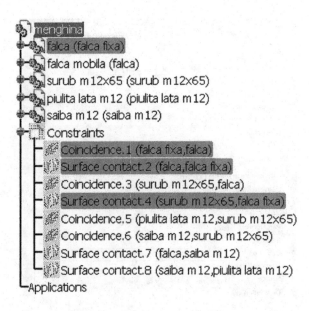

FIGURE 2.8 Structure of an assembly model with constraints applied between components.

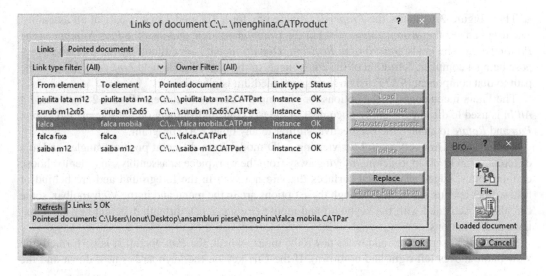

FIGURE 2.9 Replacement of components from an assembly by loading other available parts.

This method is applicable to assemblies containing parts for which several variants have been created. These variants are, usually, composed of different shapes of the same part(s), newer revisions of the same part(s) or a totally new constructive solution. Thus, different constructive solutions can be tested for the respective variants of assemblies.

It can be, also, observed, the link between the components' names in the specification trees and their files, previously saved with a certain name, and also the type or status of the links (fields *Link type* and *Status*).

The *Properties* option is used to display and edit useful information regarding a *CATIA* object; it may be an individual 3D model, a component of an assembly, a 2D drawing or a finite element analysis.

Properties   **?** **×**

Current selection : falca fixa/menghina

| Product | Graphic | Mechanical | Drafting |

Component
Instance name falca fixa
Description

☐ Visualize in the Bill Of Material
Link to Reference
falca C:\Users\Ionut\Desktop\ansambluri piese\menghina\falca.CATF
Product
Part Number   falca
Revision
Definition
Nomenclature
Source   Unknown   ∨
Description

FIGURE 2.10   Establishing the names for a component and for its instance.

Thus, Figure 2.10 shows the *Properties* dialog box for a 3D model that is part of an assembly. The user can see the *Instance name* field in the *Component* area, and also the *Part Number* in the *Product* area. The fields *Description*, *Revision*, *Definition*, etc., are editable by the user, their purpose being a complete definition of the component. In the non-editable field *Link to Reference*, the path to that component (*CATPart*) on the storage medium is placed.

The *Views* menu contains the options for viewing the current document (Figure 2.11). Thus, *Fit All In* is used to display the entire document in the working environment on the screen, and *Zoom*, *Pan* and *Rotate* to change the perspective, similar to any other *CAD* program.

These options are very useful for viewing a 3D model from different positions/angles and at a different zoom level (approaching/moving away from the workpiece or assembly), in order to choose certain points, edges, planes and surfaces that are not seen in the foreground and are behind or below other opaque surfaces. Although these options are in the menu and in the *View* toolbar, some can also be activated with the keyboard and mouse (Figure 2.12), through different key combinations and/or buttons.

Thus, if the user presses and holds down the mouse wheel, the *Pan* operation is performed (the view plane moves left-right and up-down). If the *Ctrl* key on the keyboard is held down and the

FIGURE 2.11   Options to view and navigate the workspace.

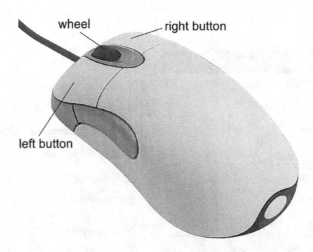

FIGURE 2.12   Identification of mouse buttons.

mouse is moved back and forth on the mouse pad while pressing the mouse wheel, *Zoom* effect is created (the view plane gets closer to the user or it's getting farther, *zoom in* or *zoom out*). This is often used to observe certain details in the working area on parts, assemblies, etc. The same *Zoom* effect can be performed only with the mouse, by holding down the wheel, then pressing and releasing its right button and moving the mouse back and forth.

Similarly, when the scroll wheel and the right mouse button are pressed simultaneously, a blue circle with a dashed line is displayed on the centre of the screen. This circle represents, in fact, a virtual sphere, and when the user moves the mouse in any direction, he will get a rotation movement (*Rotate*) of the point from which the user looks. This point is located on this sphere.

Although these visualization points give the user the impression that the 3D part and/or assembly is moved, scaled, or rotated, in fact, it is fixed in space (and in their coordinate system) and the user changes his perspective in viewing them. The orthogonal plane system retains its orientation and position relative to the visualized geometry.

The *Named Views* option (Figure 2.13) allows the visualization of the part or assembly in work in orthogonal projections (*front*, *back*, *left*, *right*, *top* and *bottom*) and isometric (*iso*) view. The meaning of the orthogonal projections is related to the system of axes and planes, as follows: for the front projection, the *YZ Plane* is orthogonally viewed from the positive direction of the *X* axis; for the top projection, the *XY Plane* is orthogonally viewed from the positive direction of the *Z* axis; etc.

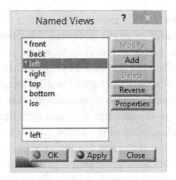

FIGURE 2.13   List with the *Named View* options.

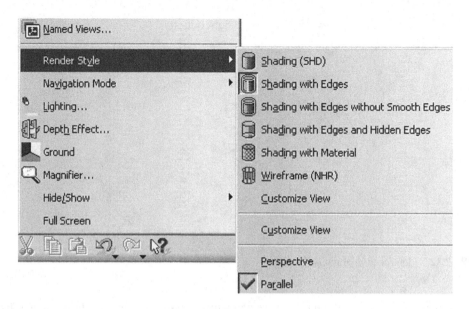

FIGURE 2.14   List with the *Render Style* view modes.

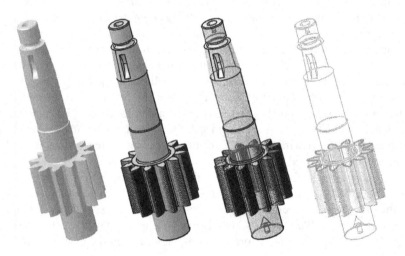

FIGURE 2.15   Pinion part displayed in four view modes.

The *Render Style* option (Figure 2.14) is used to display and visualize the models in workspace in one of the few view modes available in the program: *Shading, Shading with Edges, Shading with Edges without Smooth Edges, Shading with Edges and Hidden Edges, Shading with Material, Wireframe and Custom* (some of them are displayed in Figure 2.15).

The viewports can be displayed in *Perspective* or *Parallel* modes (Figure 2.16). The differences in the matter of visualization between the two representations are noticeable. The *CATIA* program starts by default in *Parallel* mode, so the user has a correct representation of the part or assembly to observe the parallelism or perpendicularity of the edges, planes and faces.

Generally, *CAD* uses *Parallel* representation, but the user can also choose *Perspective*, a visualization closer to reality, useful in the case of interactive presentations of assemblies.

The *View* menu contains the *Lighting* option, which can be used to add light to the workspace. According to the dialog box in Figure 2.17, three types of lights can be used (*Single Light, Two*

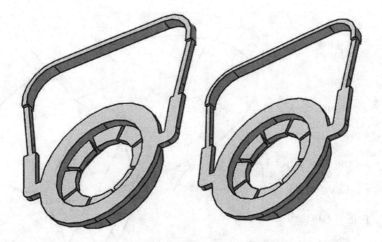

FIGURE 2.16   Part displayed in *Perspective* and *Parallel* modes.

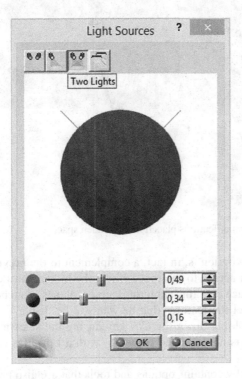

FIGURE 2.17   Different options to add lights in the workspace.

Lights and *Neon Light*). The appearance of the light (*Ambient*, *Diffuse* and *Specular*) is determined by the three sliders with their values. The *Lighting* option is recommended to be used when it is not possible to adjust the brightness of the monitor, or when it is not sufficient.

The *Hide/Show* option allows the user to select any entity (or group of entities) of the model and place them in a separate space, where they are not visible, also called 'no show space'. Those entities are 'hidden' temporarily or permanently, depending on the future requirements of the document in work.

Thus, it may be necessary at some point to hide certain curves, planes, sketches, etc., and then it may be useful to display them again.

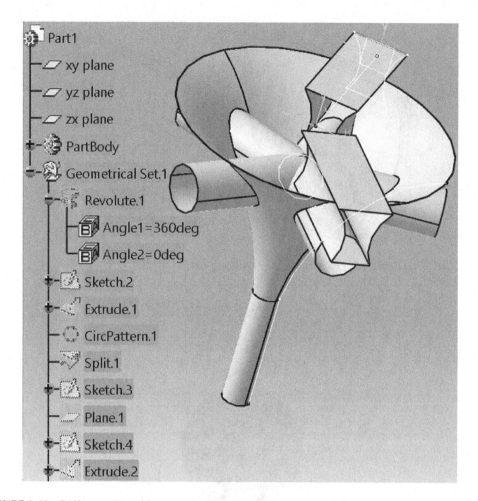

FIGURE 2.18   Different types of features placed in the hidden space.

The *Swap visible space* option is, in fact, a complement to the previous option, with the role of providing the user with access to the space where the hidden entities are placed. As seen in Figure 2.18, the hidden space contains numerous curves, planes, sketches and surfaces. The user has the possibility to select the features that he wants to be displayed again (the selection can also be a multiple one, holding down the *Ctrl* key during the selection), using the *Hide/Show* option. The hidden space is not intended for work, with all profiles being drawn by default in the visible space.

The *Insert* menu generally contains options and tools that are also present in the toolbars. It is recommended to use this menu at least in the first phases of working in *CATIA*, in the process of learning the program, because the icons and the names of the respective instruments are presented. Depending on the workbench used, the content of the menu differs.

For example, Figure 2.19 shows an excerpt from the *Insert* menu when modelling an assembly. The user can see the list of assembly constraints (icons and names), compared to the *Constraints* toolbar, specific to the *CATIA Assembly Design* workbench (shown next to it, on the left). Selecting an icon from the toolbar or choosing the same icon with its name from the menu is similar and activates a specific constraint.

The *Tools* menu contains important options related to creating mathematical relations and specific parameters for the part or assembly in work (*Formula*), to capturing an image or video sequence (*Image*), recording certain user actions to automate them (*Macro*), launching additional utilities

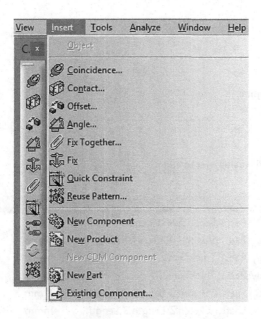

FIGURE 2.19 Content of the *Insert* menu in the *Assembly Design* workbench.

included in the program (*Utility*), displaying/hiding elements of sketches, planes or 3D models (*Show* and *Hide*), customizing and setting all the working environment parameters (*Customize* and *Options*), accessing catalogues with standardized components included in the program (*Catalog Browser*), etc.

Choosing the *Formula* option opens a dialog box (Figure 2.20), and the user has access to the parameters of the current document and, also, the possibility to establish relations between them. The complex methodology to work with parameters and their related relations is described in detail later in this book (Chapters 3 and 4).

The *Image* option, together with its options (*Capture*, *Album* and *Video*), is used to take screen-shots of static type (image) or dynamic (video), but also to manage the files thus obtained. Its options are useful for presentations and for reproducing the actions performed by the user during a work session.

Unlike *Image*, the *Macro* option records the user's actions to keep them in a file. These actions can be run at any moment, with the purpose of automating certain work steps. Recording (Figure 2.21) takes place after selecting the *Start Recording* option. The user chooses the file in which to save the recorded script, the language used and the name. The *Current macro library or document* field contains the file to which the automation applies. The language in which the automation is stored can be *Microsoft VBScript* or *CATScript*, as selected in the *Language used* field.

The *Macro* option is used to run the automation. The path and the name of the file containing the scripts (the *Available macros* field) are entered in the dialog box as shown in Figure 2.22.

Saved macros (codes for automatic actions) can be run, edited, renamed, deleted, selected, etc., using the buttons on the right. Chapter 5 of this book contains detailed information and samples related to automation and scripting.

The *Customize* option is used to personalize the *CATIA* working environment. Thus, according to the dialog box with the same name in Figure 2.23, in the *Start Menu* tab, items can be added for quick access. From the *User Workbenches* tab, it is possible to create customized workspaces.

The *Toolbars* tab presents all the toolbars of the program, and the user has the facility to create his own custom toolbars, with tools from different categories.

In addition, an important and useful option is to restore the position and contents of the standard toolbars (*Restore position* and *Restore all contents* buttons).

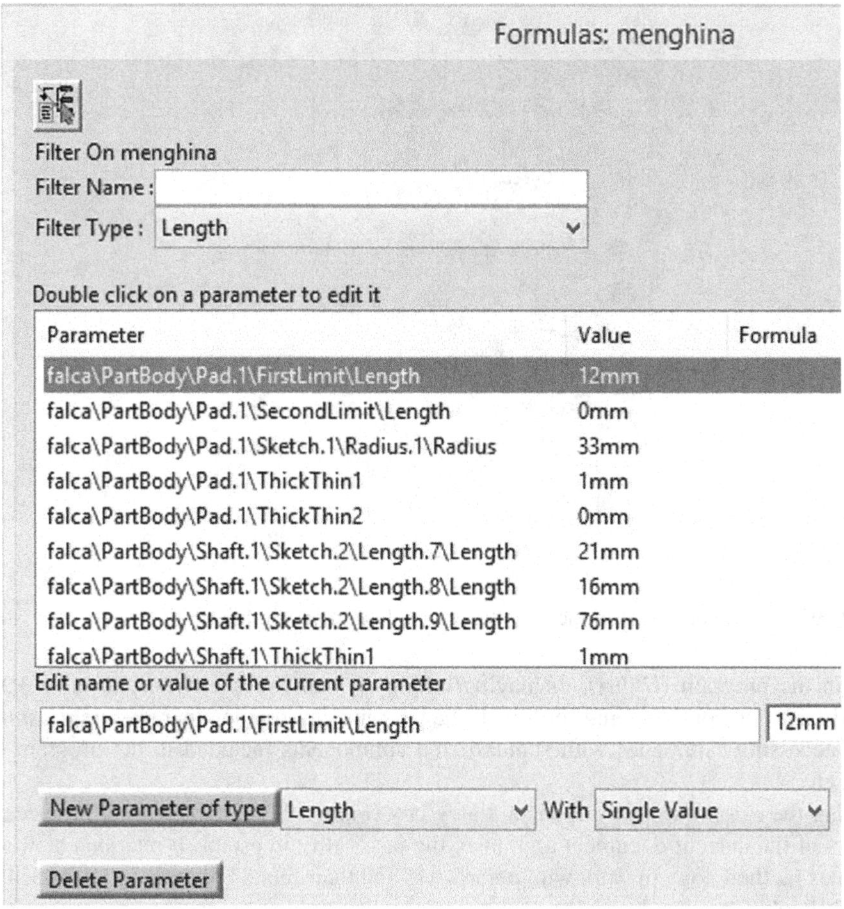

FIGURE 2.20 Content of the *Formula* dialog box showing the names and values of parameters of type *Length*.

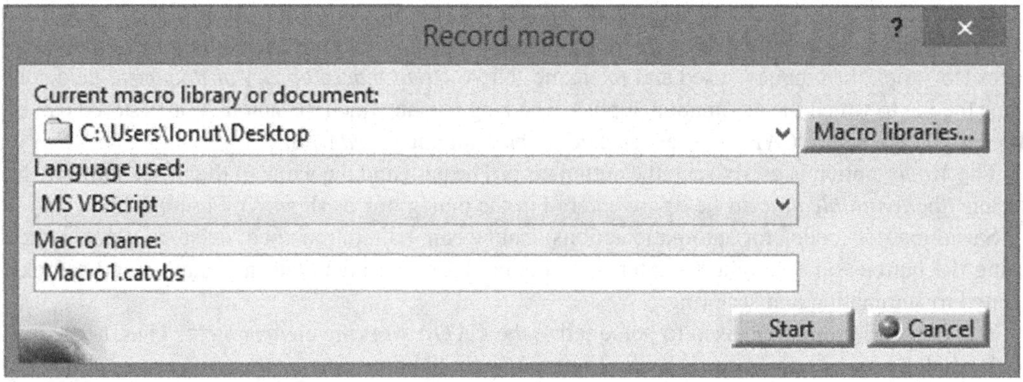

FIGURE 2.21 Preparing to record a macro.

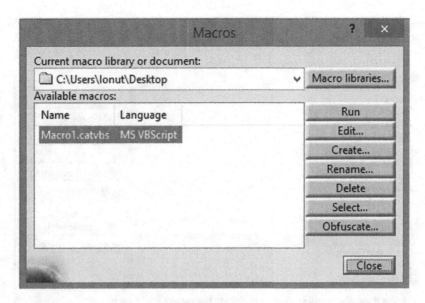

FIGURE 2.22   List of the available macros and the buttons to work with them.

The *Commands* tab of the *Customize* dialog box contains the list of menus and their options/ commands, and the user can define the icon for each, and also the shortcut in the keys for quick access.

The most important option of the *Options* tab is related to the language in which the program runs: *User Interface Language* (Figure 2.24).

Also, checking the *Tooltips* option has the role of familiarizing the user with the names of the instruments by displaying a small information box when hovering the mouse over the icon.

The establishment of a fixed position in the interface for a toolbar is done with the option *Lock Toolbar Position*. This is useful when changing the display resolution on the monitor.

All selections and options changes made by the user in the *Options* tab require *CATIA v5* to restart to take the effect of the new set-up.

In the *Tools* menu, there is *Options*; its use opens a very important and complex dialog box with the same name. The user must be aware of the role of certain selections and settings that are useful in most stages of working with the *CATIA v5* program. Throughout this book, some of the options are presented, and some important ones are mentioned below.

The choice of the background colours and of the object manipulation are made from the *General → Display* category, the *Visualization* tab (Figure 2.25). The *Performance* tab sets the 2D and 3D accuracy options, the levels of detail and the transparency of the objects/models in progress.

Performance settings are very useful when running the program on poorly equipped computing systems (low RAM, older generation microprocessor, integrated video card, etc.). Reducing the details will considerably increase the working speed.

The establishment of the measurement and, implicitly, of the working units is done from the category *General → Parameters and Measure*, *Units* tab (Figure 2.26). In the *Knowledge* tab, the user will check the *With value* and *With formula* options so that the parameters defined in the *Knowledge Advisor* workbench are displayed in the specification tree.

The *Report Generation* tab can specify the format in which *CATIA* should generate the report files, and also what it should contain.

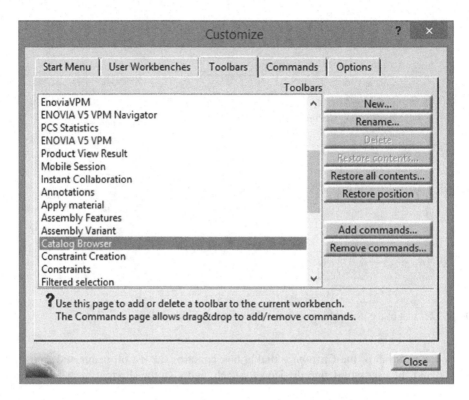

FIGURE 2.23 *Customize* dialog box with options to personalize the program.

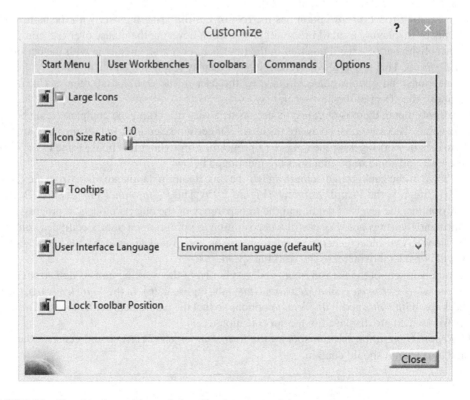

FIGURE 2.24 Showing the tooltips and choosing the program's language.

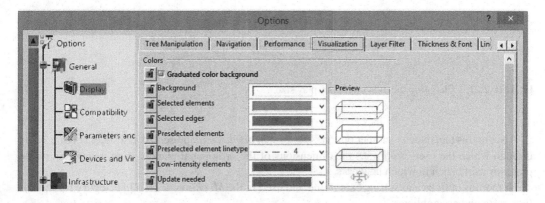

FIGURE 2.25   Choosing the colours for background, selected elements and edges.

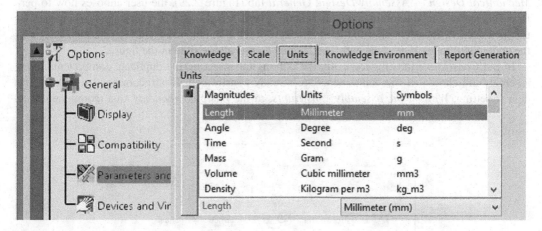

FIGURE 2.26   Choosing the measurement units.

The *Parameters Tolerance* tab sets the maximum and minimum tolerances for lengths and angles, and the *Constraints and Dimensions* tab (not shown in Figure 2.26) chooses the colours for constraints, when a profile is correctly constrained or not (over-constrained or under-constrained), etc.

Also, the constraints, parameters and relations created by the user are displayed in the specification tree if their options are checked in the *Infrastructure → Part Infrastructure* category, *Display* tab (Figure 2.27).

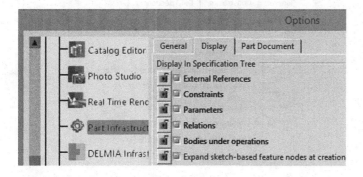

FIGURE 2.27   Checking the options to display constraints, parameters and relations in the specification tree.

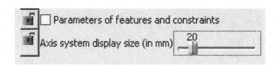

FIGURE 2.28   Changing the axis system display size.

Also, in this tab, there is another option, *Axis system display size*, presented as a slider (Figure 2.28), through which the user establishes the dimensions of the orthogonal plane system, regardless of the program workbench in which he works.

Numerous options and settings are found and used in the *Mechanical Design* category for modelling parts and assemblies.

Thus, in the case of assemblies, it is necessary to update the assembly constraints. From *Mechanical Design → Assembly Design*, *General* tab (Figure 2.29), the user chooses how to perform this update: in automatic or manual mode. The *Manual* mode is recommended because the user performs various actions with the components of the assembly (translations, rotations, etc.), and the update will take place at the end. The *Update* icon is also visible in the figure.

In the *Constraints* tab (Figure 2.30), the options for inserting the duplicate components in an assembly are chosen, if the components are already constrained. It is recommended to keep the default setting, *Without the assembly constraints*, in the *Paste components* area (components are

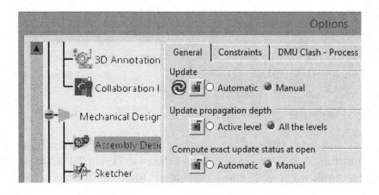

FIGURE 2.29   Setting the *Automatic* or *Manual* update mode.

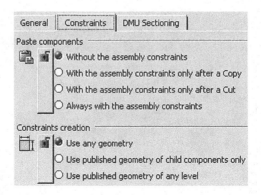

FIGURE 2.30   Setting how the constraints of an assembly will work.

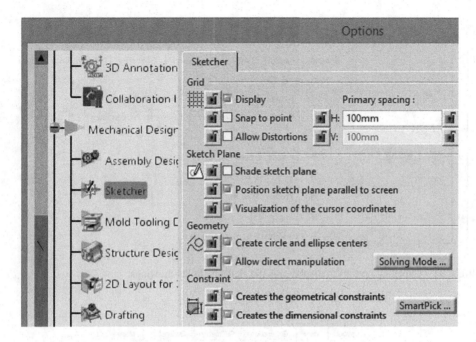

FIGURE 2.31 Setting the grid, the sketch plane, the constraints and the geometry elements.

multiplied using the *Copy-Paste* sequence) and *Use any geometry* in the *Constraints creation* area when creating new assembly constraints.

The options setting for the *CATIA Sketcher* workbench is done from the *Mechanical Design → Sketcher* category, in the tab with the same name (Figure 2.31). To display the grid (network of horizontal and vertical lines that are not considered profiles) and to use the mouse position at certain points on the grid (snap), the user will check the appropriate options in the *Grid* area. The orthogonal positioning of the sketch plane is given by the *Position sketch plane parallel to screen* option in the *Sketch Plane* area.

It is also useful to display the absolute coordinates of the cursor (right next to it), as it moves across the screen while drawing sketches (*Visualization of the cursor coordinates* option). The centers of circles and ellipses are recommended to be displayed (*Create circle and ellipse centers*) because these centers are in most cases constrained to existing geometries (in sketches or in solid modelling). Automatic creation of dimensional and geometric constraints is done using the options from the *Constraint* area of the *Sketcher* tab.

To set the *CATIA Drafting* workbench options, in which the 2D drawings are created, the user should access the *Mechanical Design → Drafting* category. The *Layout* tab selects the *View name* and *View frame* options so that the 2D drawings sheets on which the projections are created have a name and a frame.

In the *View* tab (Figure 2.32), the user will check the *Generate* options so that the 2D drawings contain by default axes, centerlines, fillets, threads and hidden lines. These elements can also be added later in each drawing and in each view from the sheets manually.

In the *Geometry* tab (Figure 2.33), the options for displaying centers of circles, arcs and ellipses (*Create circle and ellipse centers*) are selected when they are drawn by the user. The *Allow direct manipulation* option permits the selection and movement of the component elements (lines, circles, arcs, etc.) of the 2D drawings, created by the user. Of course, these drawings contain such 2D elements, obtained by projecting 3D geometries on the sheets, but the two-dimensional elements cannot be manipulated by the user.

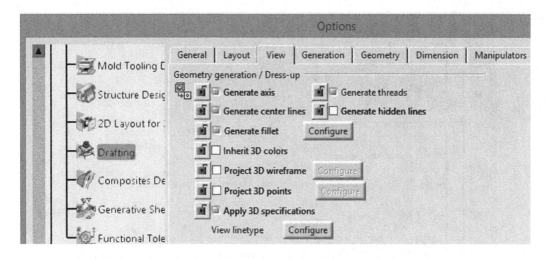

FIGURE 2.32   Checking options to display specific elements in 2D drawings.

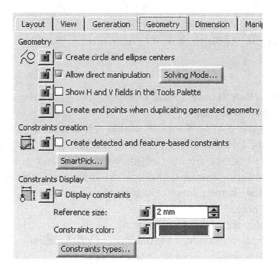

FIGURE 2.33   Checking options to display centres for circles and ellipses.

In the *Manipulators* tab (Figure 2.34), the user chooses which of the dimension's manipulators will be displayed and, thus by default, used. These can be checked in the *Creation* and *Modification* columns. Starting *CATIA* in the *Administrator mode* is very important because many options are available for selection only in this mode.

The most important manipulators are of type *Insert text* (red triangles) and *Move* (bidirectional white arrows) because they allow inserting text before/after the dimension value and, respectively, its movement along the dimension line.

The *Administration* tab (Figure 2.35) contains various restrictive administrator-type options so that the user cannot create a new set of drawings, change a set standard or update it, or have access to the drawing area for inserting the frame and the indicator.

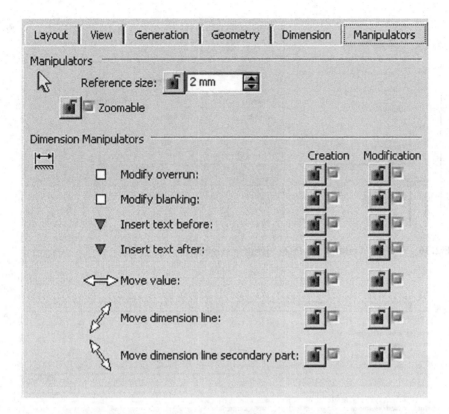

FIGURE 2.34   Checking the manipulators options is possible only in *Administrator mode*.

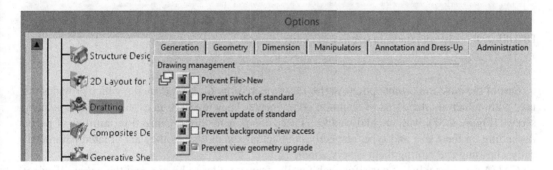

FIGURE 2.35   Specific options with restrictions in the *Administration* tab.

The *Tools* menu also contains the *Standards* option, through which various properties related to generating 2D drawings in the *CATIA Drafting* workbench can be edited in detail. Thus, the user has the possibility to intervene on the layers, the standards of the drawings, the ways of importing the *dxf* files and those of modifying/completing the projections generated by the program.

For example, Figure 2.36 shows a sequence in the process of determining how the roughness symbol is displayed for the *ISO* standard (found in the *ISO.xml* file).

All these changes are possible only if the *CATIA* program has been opened in *Administrator mode*; otherwise, the user can view the values in the fields, but cannot edit them.

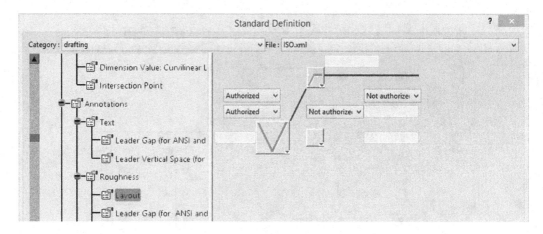

FIGURE 2.36   Sample of editing the *ISO* roughness symbol.

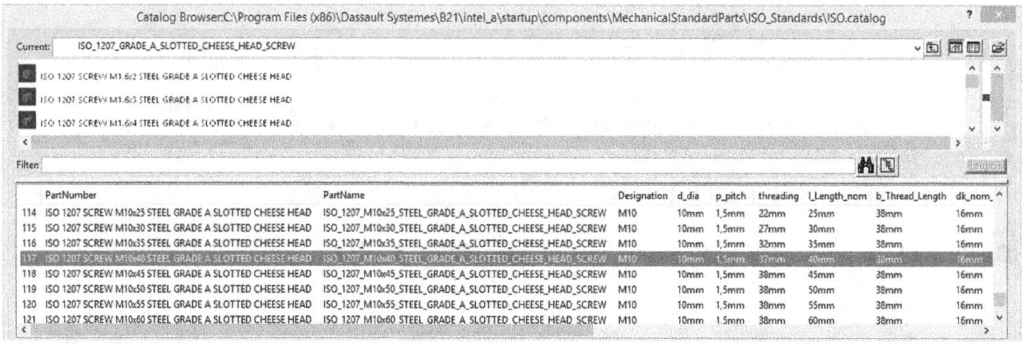

FIGURE 2.37   List with standard screws as a family of parts.

One of the most important options of the *Tools* menu is the *Catalog Browser*, which provides the user with numerous standards of common components. They are used in assemblies such as nuts, screws (Figure 2.37), washers and wedges. These components are organized in families of parts, depending on their size and type, respecting the actual standards. As they are created parametrically, choosing and editing them is very simple.

CATIA provides a list of parts for each family, from smaller to bigger, and the user will select the needed part to insert into his project. Thus, these standard parts are important due to their diversity and the easy way of working with them. *CATIA* has, by default, a few types of standard parts, stored in catalogues, available once the installation is complete, but there are companies that provide by request or from their portfolio various parts such as bushings, rivets, bearings, supports and covers.

The *Window* menu is used to organize and access different work sessions that are simultaneously opened. For example, it is possible to work on an assembly, and if it is observed that one of its components needs a change or a revision, it can be opened separately, in a new session. Thus, the user may focus only on it.

Figure 2.38 shows the *Window* menu options. The *Tile Horizontally* and *Tile Vertically* options position the windows of the working sessions horizontally and vertically, respectively, dividing the

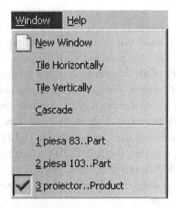

FIGURE 2.38   Various options and files in the *Window* menu.

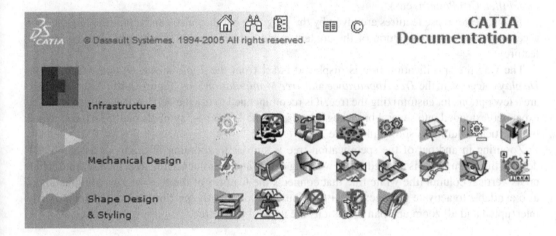

FIGURE 2.39   Accessing *CATIA Documentation* for each workbench.

screen into equal-sized areas. The *Cascade* option positions these windows in front of each other in a cascade manner.

The same figure also shows the existence of three work sessions: two parts and an assembly.

Of course, the current session is the one with a blue checkmark to the left of its name. Choosing any other session with the mouse causes the current session to be deactivated and the other to be activated.

The *Help* menu contains various options for accessing the program documentation and searching for information needed by the user in the stages of learning certain workbenches or tools.

Figure 2.39 shows a window displaying icons, grouped into categories. These icons are the program workbenches, and they are used to access and display the help content. Finding information is easy by pressing the *F1* key or accessing the *CATIA v5 Help* menu option. Browsing the original documentation is recommended for all categories of users because it is complete, with many detailed examples and explanations.

The options in the program menus change depending on the workbench chosen by the user for the current working session, but those presented above are common in most cases.

2.3 SPECIFICATION TREE

The specification tree contains the constituent elements of a part (also known as features), the components of an assembly, various information related to FEM analyses, NC simulations, various measurements, etc. It is, practically, a history of all the commands and processes applied by the user in a work session. Thus, the user has the possibility to find, in the case of a part, for example, when a threaded hole was created, which modelling tools have been applied before and which are dependent on it.

The specification tree has a very important role in the activities of selecting certain components within an assembly, visualizing its constraints, choosing the right solution in the stages of analysis with finite elements, etc. For these reasons, the specification tree has a clear and complex structure, linear and tree-like, it is expandable to the level of a graphic element (point, line, circle, spline), constraint, etc., and it can also be enlarged, reduced, moved anywhere in the workspace and even temporarily hidden.

During a working session, the specification tree is implicitly represented on the left side of the interface. The three initial work planes (*XY, YZ* and *ZX*) are easy to spot in its structure and, then, below them, various features that led, for example, to the modelling of a part (*Pad, Hole, Pocket, EdgeFillet, CircPattern*, etc.).

The names of these features are given by the program, but they also can be modified by the user accessing the *Properties* option of the context menu (which appears when right-clicking on that feature).

The way the specification tree is displayed is set from the *Tools* menu, *Options, General → Display* category, in the *Tree Appearance* and *Tree Manipulation* tabs (Figure 2.40). Although there are a few options for customizing the tree, it is recommended to use the default ones. The tree can be expanded/extended into several branches by pressing the '+' and '−' symbols on the left, connected by a vertical column (as shown in Figure 2.41).

Zooming in and out of the specification tree is similar to zooming of a part or assembly. By default, the zoom effect is applied to the working entity, and to influence the tree, the user must click on its vertical column (the white line that connects the features of the tree). A key combination is also available to activate the tree: *Shift+F3*. Thus, working on the part or assembly is temporarily interrupted and all zoom and pan operations are applied to the tree.

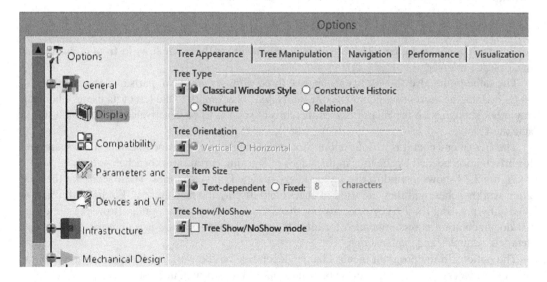

FIGURE 2.40 Setting how the specification tree will be displayed in workspaces.

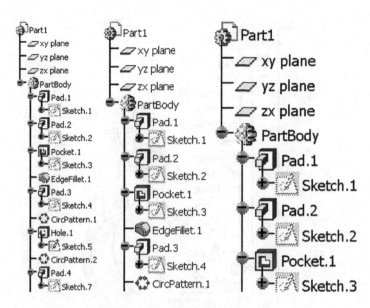

FIGURE 2.41   A specification tree displayed in three different sizes.

Figure 2.41 shows the same specification tree in three different sizes. This type of tree representation gives the user an advantage by bringing more features of the tree to the screen for complex projects (left sequence), can return to the default setting (middle sequence) or can opt for an increased representation (sequence right), useful for presentations and video tutorials.

While the specification tree is activated and various manipulations are performed on it, the part or assembly in work (the project) has a deactivated appearance, on a shade of grey, and any surface, edge, plane, etc. is not selectable.

Once the size of the specification tree is established, its linear structure (the white line) is clicked again and the user receives control over the working entities.

There are also cases in which the specification tree must be removed from the screen, an operation performed by pressing the *F3* key. Nevertheless, the user can add new features to his work session even if the tree is hidden, the display being possible also by pressing the *F3* key.

The selection of features from the specification tree is done by clicking on them. Figure 2.42 shows the model of a part and its specification tree. Note that only one item is selected: *Pocket.1*. The image shows that the shape of the pocket surface is complex; its selection is, however, very easy to choose with the mouse in the tree.

The *Pocket.1* feature contains many elements, of which the *Sketch.2* sketch is the most important due to the profile it contains and because it is the basis of the pocket cutting. Generally, sketches represent a complex combination of simple 2D graphics and dimensional and geometric constraints (radius, angle, coincidence, concentricity, etc.).

Each sketch also contains a two-dimensional *HV* (*Horizontal* and *Vertical*) axis system, because the user selects one of the *XYZ* system planes or another plane previously created.

In Figure 2.42, the specification tree has been expanded to the maximum for the *Pocket.1* feature in order to highlight its sub-elements.

In Figure 2.43, for the same part and specification tree, the user made a multiple selection of three features (*Pocket.1*, *Pocket.2* and *Pocket.4*). This selection is made by clicking with the mouse on each feature, simultaneously with the holding down of the *Ctrl* key. The selected features are highlighted in the tree, and also in the part, using the orange colour.

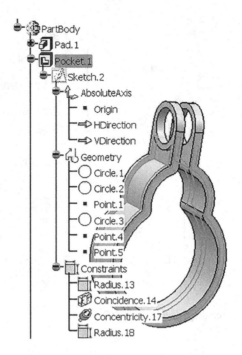

FIGURE 2.42   Specification tree displayed expanded for the feature *Pocket.1*.

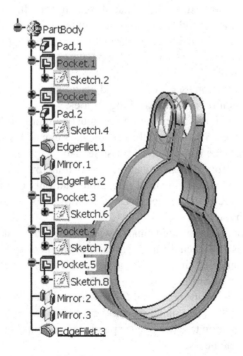

FIGURE 2.43   Full specification tree with all features of a part.

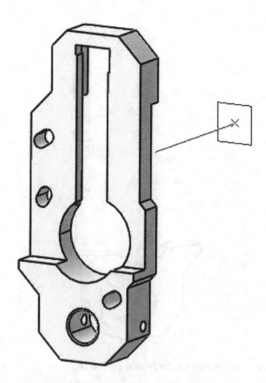

FIGURE 2.44   The compass is schematically represented near the part.

2.4   COMPASS

The compass is symbolized as a structure with axes and planes at the top right corner of the graphical interface and allows, by direct manipulation, to change the position and orientation of a three-dimensional model relative to the XYZ system coordinates or components of an assembly, if that model is part of it.

To perform direct handling, the compass must be placed on the model. Thus, the user clicks on the red square at the base of the compass, holds down the left mouse button and moves it until he can place the base of the compass on a flat surface or along an axis of the model ('drag and drop' principle).

During the movement from its default position to a surface or axis, the compass is represented schematically, simplified, by a plane and a line perpendicular to it (Figure 2.44).

Figure 2.45 shows the compass positioned on the front face of the model, and in Figure 2.46, the compass is placed on a narrow flat lateral face. If the positioning is correct (usually after zooming in on the face), the pattern is selected and the compass turns green. The user has the possibility to move and/or rotate the model using the axes, arcs and the planes of the compass.

Returning the compass to the default position is done by bringing it back by the user to that position following the procedure described above or, more simply, using the *Reset Compass* option in the *View* menu. Obviously, the compass and the possibilities of manipulating the parts of an assembly are not enough for correct positioning and assembly of them, being necessary specific geometric and dimensional constraints.

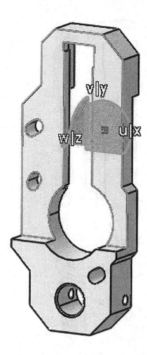

FIGURE 2.45   The compass is positioned on the frontal face of the part.

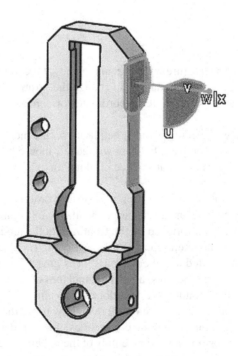

FIGURE 2.46   The compass is positioned on the lateral face of the part.

2.5   TOOLBARS

As it is already mentioned, the interface of *CATIA* contains numerous toolbars specific to each workbench. If the user switches from one workbench to another, the toolbars are updated, some of them are replaced by others, and the common ones remain unchanged.

Toolbars are very important and useful during the work stages. Although the menus contain the same tools, the toolbars provide quicker access through suggestive icons. The advantage of toolbars is that the icons are grouped and easily accessible from the interface, and the main disadvantage is that the user has to recognize these icons and associate them with certain tools needed in the work steps.

Figure 2.47 shows some toolbars specific to the *CATIA Part Design* workbench, and the respective icons can be found, with their names, in the *Insert* menu (Figure 2.48).

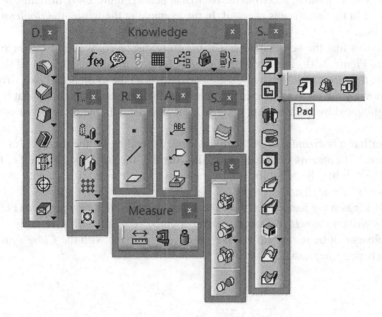

FIGURE 2.47   Toolbars from the *CATIA Part Design* workbench.

FIGURE 2.48   Icons and instruments names of *Sketch-Based Features*.

For the users who are still learning the program, it is recommended to access the tools in the menu to get used to the correspondence between the names and the icons.

It is noticeable that some icons have a small arrow pointing downwards in the toolbar. This means that under the icon, there are other tools from the same family, an example being shown in Figure 2.47 for the case of the *Pad* icon. Obviously, the tools in that toolbar can also be found in the *Insert* menu → *Sketch-Based Features* (in which case the arrow is pointing to the right).

The toolbars are movable, with the possibility of positioning them in the interface. By default, the bars are in the default places on the right and bottom of the screen. Of course, these places are insufficient for the multitude of the tools, especially if the user decides to have other toolbars displayed than the standard ones, which are available at the first run of the program.

The identification, selection and display of the additional toolbars are done by right-clicking with the mouse on any other toolbar, and from the list that appears (Figure 2.49), the name of the one that is to be displayed in the interface is checked. In the example in the figure, the *Boolean Operations* toolbar was chosen.

The user notices that the selected toolbar is not fully displayed, with only two icons available (bottom right of Figure 2.50). The reason is the lack of space in the standard area of the toolbars.

In the figure, the incompletely displayed toolbar also shows the existence of two arrows pointing downwards, showing the presence of other icons and/or toolbars not displayed at that time. The two arrows are highlighted by encirclement because their size is small and their colour is close to that of a toolbar.

Each toolbar has a horizontal or vertical grey line, depending on the direction of the string of the tools. Figure 2.51 contains some examples, showing the lines at the top and left of the toolbars, respectively. The user has the possibility to use the line of a toolbar to extract it with the mouse from the standard area of the toolbars.

Then, by clicking on the name area, the toolbar can be moved to a new position (Figure 2.52; on the name bar is written *Knowledge*) and/or can be oriented vertically or horizontally.

Unused toolbars can be removed from the *CATIA v5* interface with the *Close* symbol and displayed again whenever necessary.

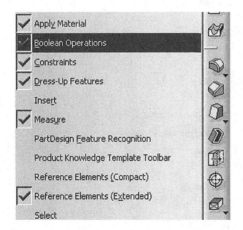
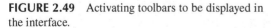

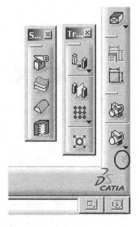

FIGURE 2.49 Activating toolbars to be displayed in the interface.

FIGURE 2.50 Arrow symbol indicates that there are more toolbars not shown.

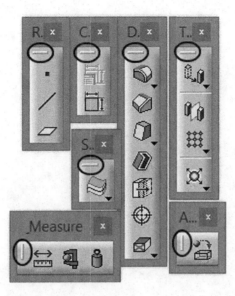

FIGURE 2.51   Line symbols used to move and position the toolbars in the interface.

FIGURE 2.52   Toolbar being repositioned in the interface.

3 Hybrid 3D Modelling of Parts

CAD applications are presented through tutorials starting from a beginner level, by creation of planes, modelling of solid parts up to an advanced level that proposes complex surface design and parametric modelling of part families. This chapter also includes creation of a sample part that evolves from one modelling solution to another and editing of solid parts with an important aid from surface features and operations.

3.1 METHODS FOR CREATING THE WORKING PLANES

This *CAD* application (tutorial) presents a few methods for creating 2D planes. In *computer-aided design* programs, planes can be used as orientation entities, as references for 2D and 3D geometric elements (features) or as supports for limiting certain operations (extrusion, revolution, repositioning, etc.). Planes can be obtained based on the other already existing planes, points, lines or curves or even by using equations.

Creating planes can be achieved by selecting the *Plane* icon in the *Reference Elements* toolbar, which will display the *Plane Definition* dialog box (Figure 3.1). From the drop-down list (*Plane type* field), the user can select different options for creating planes, most of which require, however, the previous creation of some other elements (planes, points, lines, curves and surfaces).

The first option here, *Offset from plane*, is also the easiest, and it is the default option when user activates the *Plane* icon.

To exemplify the *Offset from plane* option, in Figure 3.2, the *YZ Plane* was selected as a reference and the value of 26 mm was entered in the *Offset* field as the distance between the *YZ Plane* and the new plane, named *Plane.1*. By clicking the *Reverse Direction* button, it creates the plane on the other side of the *YZ Plane*. The appearance of a *Geometrical Set* containing the plane in question is observed in the specification tree. This geometrical set can also contain other elements such as points, sketches and surfaces.

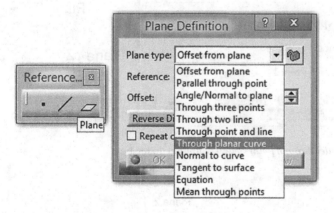

FIGURE 3.1 *Plane Definition* dialog box.

DOI: 10.1201/9781003281153-3

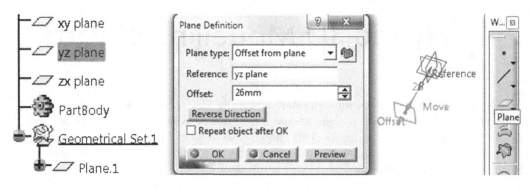

FIGURE 3.2 *Offset from plane* option definition dialog box.

Parallel through point option, which is similar to the previous one, requires a previously added *Point.1* (*Point* tool in *Reference Elements* toolbar; Figure 3.1, left) with the absolute coordinates (20, 32, 46) using the default coordinate system. When inserting the new plane, the user selects the *YZ Plane* as reference and the *Point.1* in the *Point* field. The shape of the newly created plane (*Plane.2*) is positioned to include the point (Figure 3.3) and will be parallel to the reference plane *YZ*.

For the next option for creating planes, *Angle/Normal to Plane*, in the plane *Plane.2* the *Sketch.1* is initiated and a line is drawn (Figure 3.4). In the *Plane Definition* dialog box in Figure 3.5, the user selects the line (or the sketch that contains it) in the *Rotation Axis* field, the *YZ* reference plane and the angle (35°) in the *Angle* field.

Plane.3 is displayed, and it contains the axis of rotation and an angle to the reference plane. If *Project rotation axis on reference plane* option is checked, the axis of rotation is projected on the *YZ Plane*. The projection line is not represented, but becomes the new axis of rotation.

For the *Through three points* option, three points are added to the graphic area having the following coordinates: *Point.2* (20, 32, 46), *Point.3* (37, 54, 26) and *Point.4* (28, 67, 48). From the *Plane type* drop-down list, the user chooses the *Through three points* option and selects the three

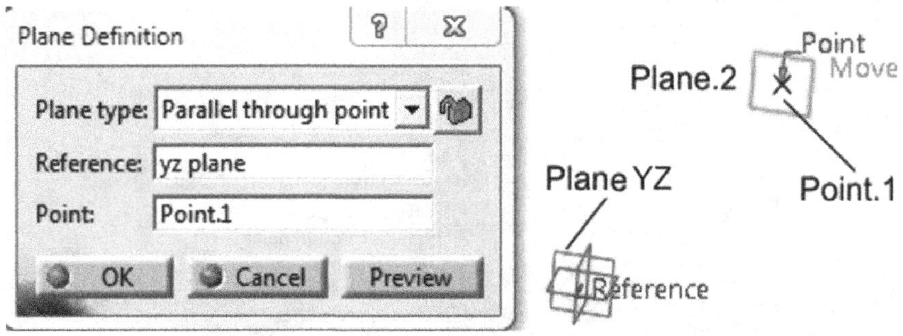

FIGURE 3.3 *Parallel through point* option definition dialog box.

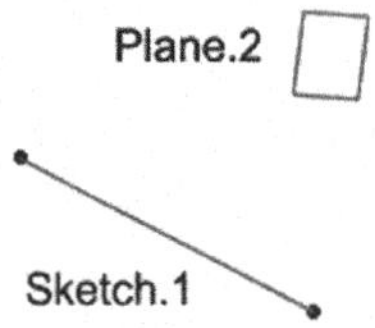

FIGURE 3.4 *Sketch.1* created in *Plane.2*.

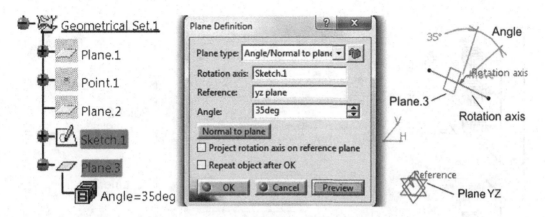

FIGURE 3.5   *Angle/Normal to Plane* option definition dialog box.

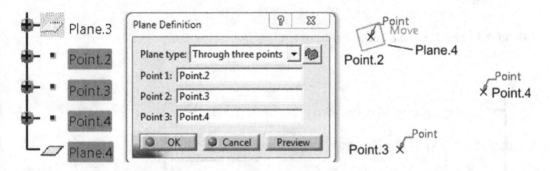

FIGURE 3.6   *Through three points* option definition dialog box.

points in the *Point* fields below. In this way, the *Plane.4* was defined by points (Figure 3.6) and it is represented in the position of the first selected point.

To illustrate the option *Through two Lines*, we will add a new point, *Point.5*, of coordinates (20, 46, 17), and draw two lines *Line.1* and *Line.2* (the tool *Line* with *Line type: Point-Point* option) between the points *Point.3* and *Point.4* and between the points *Point.2* and *Point.5*, according to Figure 3.7.

Using the option *Through two Lines* and selecting the two lines in the *Line* fields, *Plane.5* is created (Figure 3.8), its symbol being positioned on the first selected line (*Line.2*). There is also a restriction option in creating planes using this method, given by checking the *Forbid non coplanar lines* option, so that planes cannot be created unless the two lines belong to the same plane.

If the other line (*Line.1*) is selected first, *Plane.6* is created, with the plane shape attached to it. Although the lines *Line.1* and *Line.2* have different directions, the two planes *Plane.5* and *Plane.6* are parallel (the angle between the planes is equal to 0; Figure 3.9), but there is 19.98 mm between them. It is, therefore, recommended to choose the correct support for the newly created planes.

Planes can also be created by selecting a point and a line. In the *Plane Definition* dialog box, select the *Through Point and Line* option and choose *Point.2* for the point and *Line.1* for the line. *Plane.7* shape has its orientation along the line (Figure 3.10).

The *Through Planar Curve* option assumes the existence of a planar curve. For this purpose, in *Plane.7*, a new spline curve is drawn using a new sketch, *Sketch.2*, that is selected in the field *Curve*. *Plane.8* (Figure 3.11) is created, practically, in the plane of the spline curve and, in this case, coincides with *Plane.7*.

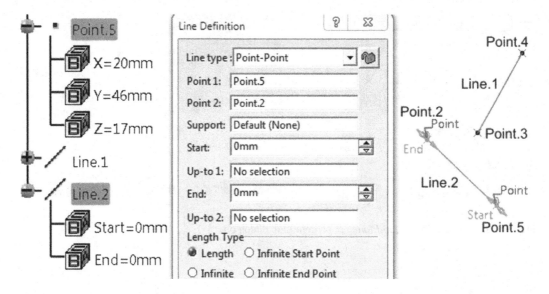

FIGURE 3.7 *Line.1* and *Line.2* definition dialog box.

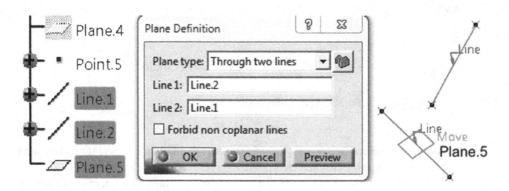

FIGURE 3.8 *Through two lines* option definition dialog box.

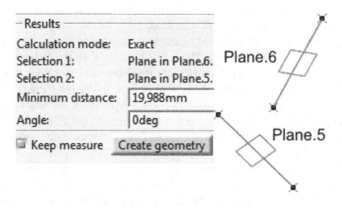

FIGURE 3.9 The angle and the distance between *Plane.5* and *Plane.6*.

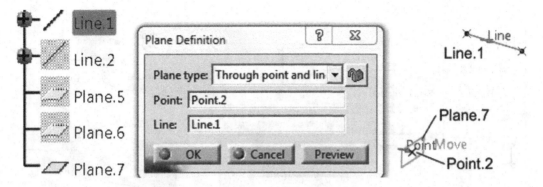

FIGURE 3.10 *Through Point and Line* option definition dialog box.

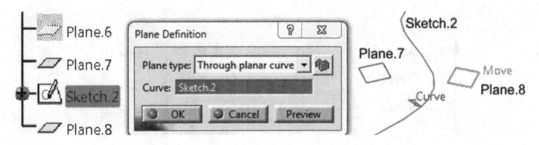

FIGURE 3.11 *Through Planar Curve* option definition dialog box.

If the user had created a *spline* curve through the points *Point.1* to *Point.5*, it could not have been selected to create *Plane.8* because it is a 3D curve.

A plane (*Plane.9*) can be perpendicular to a curve using the *Normal to curve* option. After selecting the curve from *Sketch.2*, the default proposed point is *Middle*, but usually one of its endpoints is chosen, as in the example in Figure 3.12.

To create a tangent plane to a surface, in the *Plane type* field, the user chooses the *Tangent to Surface* option, and the two fields of the dialog box require the selection of a surface and a point.

Thus, the *spline* curve in *Sketch.2* was previously extruded (the *Extrude* tool from the *CATIA Wireframe and Surface Design* workbench) to obtain the surface *Extrude.1*. The selection point does not exist yet, but the user is able to right-click in the field *Point* and a context menu appears from which *Create Midpoint* is chosen (Figure 3.13).

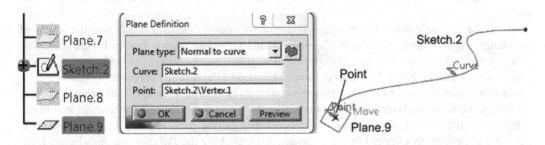

FIGURE 3.12 *Normal to Curve* option definition dialog box.

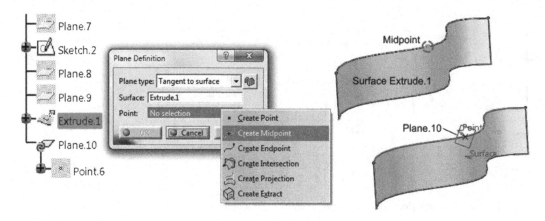

FIGURE 3.13 *Tangent to Surface* option definition dialog box.

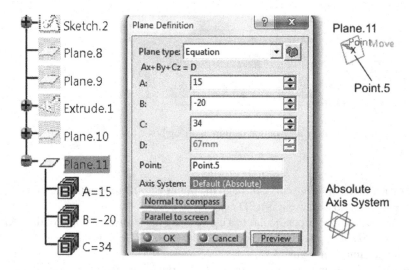

FIGURE 3.14 *Plane by equation* option definition dialog box.

By positioning the mouse on the upper curved edge of the surface, the middle of this edge becomes available. *Plane.10* becomes tangent to the surface at the specified point.

The figure shows two images of the surface *Extrude.1*, one to show how to create the midpoint of the edge using the context menu and the other (bottom) to show the plane tangent to the surface.

To create a plane defined by an equation (*Equation* option), the user must enter the values of parameters A, B, C and D in the fields of the *Plane Definition* dialog box in Figure 3.14. The user can then select a point (*Point.5*) to position the plane, in which case the field of parameter D becomes not editable. *Plane.11* is thus defined in relation to the default coordinate system.

An interesting option for creating planes by equations is *Normal to compass*; clicking the button with the same name leads to the orientation of the new plane perpendicular to the main axis of the compass (Figure 3.15). Depending on compass orientation, the parameters' values in the specific fields are automatically changed by the program. Also, clicking the *Parallel to screen* button positions the plane parallel to the current viewport.

FIGURE 3.15   Compass orientation.

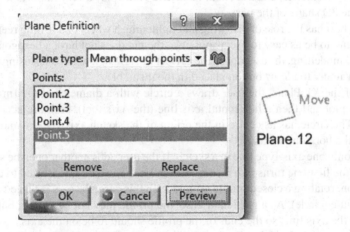

FIGURE 3.16   *Mean through points* option definition dialog box.

The last option for creating planes is *Mean through points* (Figure 3.16). At least three points must be selected in the *Points* field to insert the plane into the workspace. For three points, the new plane includes them. Selecting several points leads to a plane (*Plane.12*) interpolated between the positions of the points.

The planes inserted in the specification tree contain different parameters that can be modified by double-clicking. The user has the possibility to redefine the planes at any time. Often, planes are linked to or connected to existing geometric elements in the workspace.

Creating a plane or set of planes is a very important step, especially in the case of complex models. Almost all surface parts require planes for positioning points and curves, for changing directions, for defining sketches, etc. Planes are, also, important in defining symmetric parts, in establishing constraints between the components of assemblies.

3.2   MODELLING OF A NUT-TYPE PART

This tutorial presents the necessary stages of three-dimensional modelling of the part illustrated by the 2D drawing in Figure 3.17. It is observed that five projections are defined: four views (two

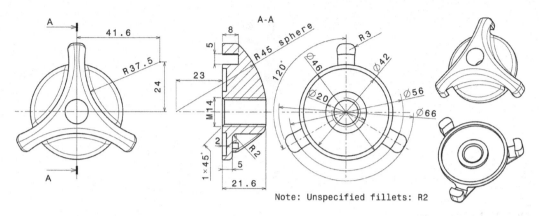

FIGURE 3.17   Drawing of the nut-type part.

orthogonal and two isometric) and a section. Isometric views play a very important role in correctly understanding the 3D shape of the part.

This nut-type part has the role of orienting and tightening a wheel of a fishing reel assembly. Due to this function and to be as easy to use as possible, the nut does not have a hexagonal shape.

To start the 3D modelling, the user opens the *CATIA Part Design* workbench using the *Mechanical Design Category* under the *Start* button placed in the menu bar.

In *Sketch.1* of the *XY* Plane, the user draws a circle with a diameter of Ø90 mm (Figure 3.18) using the *Circle* tool and then a horizontal axis line (the *Axis* tool), both are accessible from the *Profile* toolbar. The circle has its centre in the origin of the sketch axis system, and the axis line is coincident with the horizontal axis *H*.

There can be only one axis-type line in a sketch. If the user adds another one, the second becomes an axis line and the first one turns into a construction line. The axis line is used to create solid features of revolution, rotating a closed profile around it. In this case, the axis is placed on the diameter of the closed profile (circle). As a particular case, the profile may not be closed, but it must form a closed area with the axis line, so the ends of the profile should to be on the axis.

Thus, the lower semicircle of the sketched circle is removed using the *Quick Trim* tool. As shown in Figure 3.19, the axis line joins the two ends of the semicircle after the circle trimming. This trim was very simple; just half of the circle was deleted. In the case of multiple segments to be deleted, the user should double-click on the *Quick Trim* icon to keep the tool activated. So, the segments may

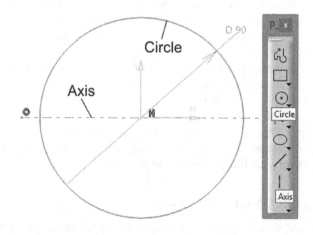

FIGURE 3.18   Circle and axis-type line inserted in the first sketch.

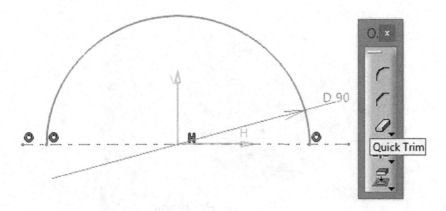

FIGURE 3.19 Half of the circle, under the axis, is trimmed.

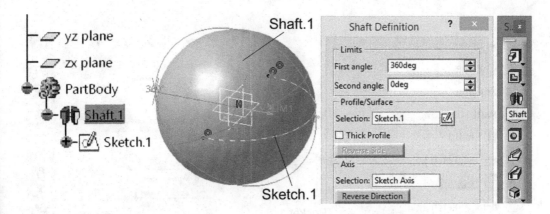

FIGURE 3.20 Rotating the semicircle around the axis to create the sphere.

be removed one by one without selecting the icon again. To end the trimming, the *Esc* key has to be pressed or the icon deselected.

The user exits the sketch using the *Exit workbench* icon. In *CATIA Part Design*, using the *Shaft* tool on the *Sketch-Based Features* toolbar, the semicircle rotates around the sketch axis, with an angle of 360° (*First Angle* in the *Shaft Definition* dialog box), thus adding volume. The result is a solid sphere (rotational feature) with a diameter of Ø90 mm highlighted in Figure 3.20 and named *Shaft.1* in the specification tree.

In the *YZ* Plane, in a new sketch, the user draws a rectangle so that its upper horizontal side is above the *H* axis at a distance of 28 mm. The other sides have some lengths, but the rectangle must enclose most of the sphere inside (Figure 3.21). This is just for a demonstration of the following operations. It is recommended that all entities of the sketch are fully dimensioned or constrained with previous features. In that case, models will be more robust and adaptable to changes.

Using the *Pocket* tool, the user intersects and cuts the sphere with the volume created by the rectangle. In Figure 3.22, in the *Pocket Definition* dialog box, he chooses the *Dimension* type, enters the value of 45 mm in the *Depth* field and checks the *Mirrored extent* option so that the *Pocket* tool cuts the sphere on either side of the plane containing its sketch.

The result of the intersection is a spherical cap with 45 mm radius, with a flat surface at a distance of 28 mm above the *XY* Plane. On this flat surface, concentric with the circular edge, a circle having a diameter of Ø46 mm is drawn in *Sketch.3* and then is extruded over a distance of 5 mm, using the *Pad* tool; the result (*Pad.1*) is presented in Figure 3.23.

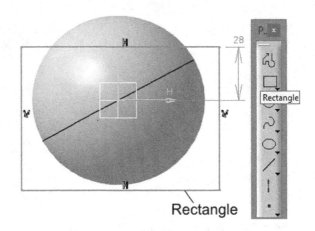

FIGURE 3.21   Drawing a rectangle to enclose most of the sphere.

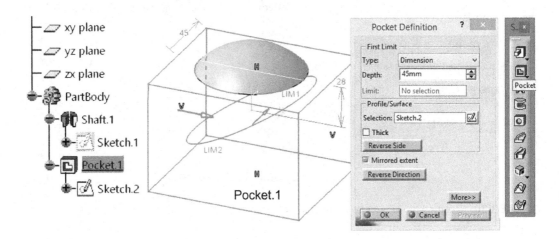

FIGURE 3.22   Cutting the sphere with the rectangle on either side of the plane *YZ*.

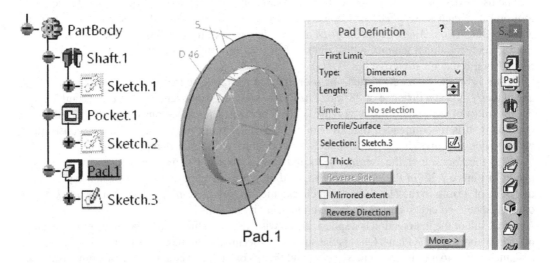

FIGURE 3.23   Circle extruded from the flat surface of the spherical cap.

The user selects the same flat surface of the spherical cap and draws in a new sketch a circle with a diameter of Ø75 mm, constrained at 41.6 mm from the *V* axis and at 24 mm from the *H* axis (according to Figure 3.24).

The circle is used by the *Pocket* tool to extract (delete) a new volume from the cap, as shown in Figure 3.25. The extraction type is *Up to next* option selected in the *Type* list from the *Pocket Definition* dialog box. *Sketch.4* and *Pocket.2* features are added to the specification tree.

The *Pocket* cut is arranged circularly on the circumference of the sphere in three instances, at angles of 120° by applying the *Circular Pattern tool* on the *Transformation Features* toolbar.

Thus, in the *Circular Pattern Definition* dialog box (Figure 3.26), in the *Parameters* field, the user chooses the *Complete crown* option, the number of instances as 3, and, as a reference element, the cylindrical surface created at the base of the cap, and then he selects in the specification tree the *Pocket.2* feature for the circular pattern. With this chosen option and the number of instances, the two angle fields (*Angular Spacing* and *Total Angle*) cannot be edited; their values are filled in automatically by the program, according to the figure.

On the flat surface at the base of the cylindrical feature *Pad.1*, two concentric circles (*Sketch.5*) are created (marked with 1 and 2). The first circle coincides with the circular edge of the face (diameter of Ø46 mm), and the other has a diameter of Ø56 mm (Figure 3.27).

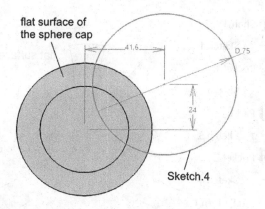

FIGURE 3.24 New circle drawn on the flat surface of the sphere cap.

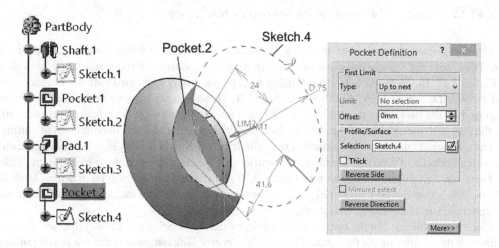

FIGURE 3.25 *Pocket* cut applied to the spherical cap.

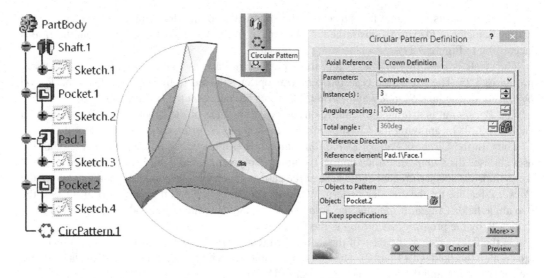

FIGURE 3.26 *Pocket* cut multiplied around the current model.

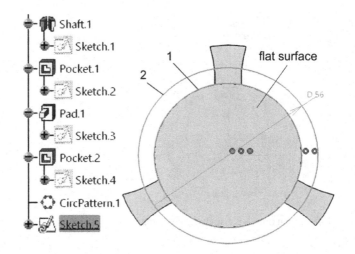

FIGURE 3.27 Two circles drawn on the flat surface at the base of the part.

The two circles are then used in a new *Pocket* cut at a depth of 8 mm towards the spherical cap. Thus, only the volume between the two circles will be extracted, the result being presented in Figure 3.28. In the *Pocket Definition* dialog box, the user may press two buttons: *Reverse Side* and *Reverse Direction* marked with letters A and B. These buttons correspond to the red arrows displayed on the part when the cutting takes place. Pressing the *Reverse Side* button, the user will cut the part with the exterior of the sketch; also, *Reverse Direction* changes the direction of cutting.

The user selects the same flat surface of the cylindrical base and initiates *Sketch.6* in which a circle with a diameter of Ø66 mm is drawn, concentric with the circular edge of the face (Figure 3.29).

The *Pocket Definition* dialog box in Figure 3.30 shows the *Dimension* type and the *Depth* of 10 mm. Any value can be considered for this cutting depth, so that the extraction volume exceeds the spherical surface of the spherical cap.

Since the profile used in the operation is a circle, by default, *Pocket* will cut a cylindrical volume defined by the contour inside the circle. Pressing the *Reverse Side* button or the arrow that is pointed towards the part's centre, the program will extract a defined volume exterior to the circle.

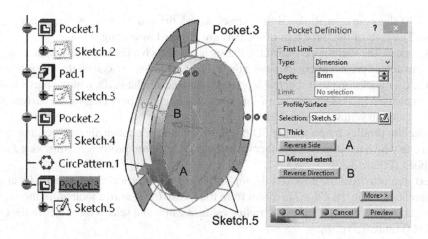

FIGURE 3.28 Cutting the part with the two concentric circles.

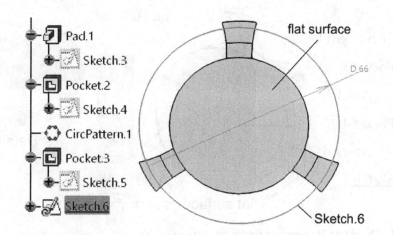

FIGURE 3.29 One circle drawn on the flat surface at the base of the part.

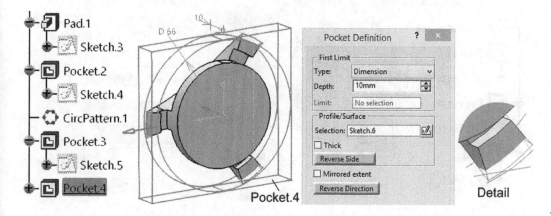

FIGURE 3.30 Cutting the part volume with the exterior of the circle.

Figure 3.30 shows in a small detail the extraction volume and the way the part's body is cut on the outside. The ends of the three legs of the part are affected by cutting.

In the next step, the user selects one of the flat surfaces of such a leg and creates a new *Sketch.7* in which it is possible to project its contour using the *Project 3D Elements* icon on the *Operation* toolbar.

Thus, the user clicks on the respective icon and then on each flat surface of the part's legs. Their geometry (which, in fact, is composed of four arcs of a circle) is projected in the current *Sketch.7* and is represented in yellow (Figure 3.31).

Obtaining the profiles by projecting 3D edges in a sketch is a much simpler and faster method than the one in which the user redraws them. On these three closed profiles, a 5 mm long *Pad* extrusion is applied (Figure 3.32).

The next step is to fillet the edges on the top surface of the spherical cap, the cylinder at its base, and the three legs, using the *Edge Fillet* tool from the *Dress-Up Features* toolbar. In the *Edge Fillet Definition* dialog box, the user enters the radius value of 2 mm and selects the edges in the *Object(s) to fillet* field (Figure 3.33).

On the flat surface at the base of the cylindrical element, a new *Sketch.8* is inserted, which contains two concentric circles with diameters of Ø20 and Ø42 mm, respectively (Figure 3.34). The two circles will be required in a *Pocket* cut operation, depth of 2 mm (Figure 3.35).

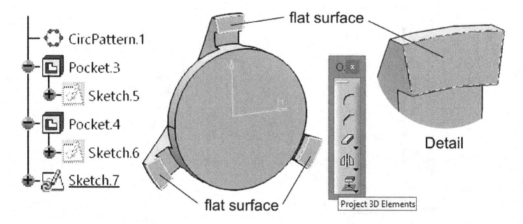

FIGURE 3.31   Projecting 3D geometry on the flat surfaces.

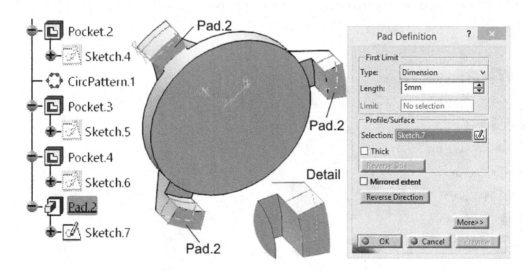

FIGURE 3.32   Extruding the projected profiles.

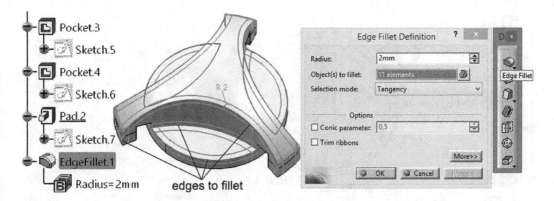

FIGURE 3.33 Applying fillets on certain edges.

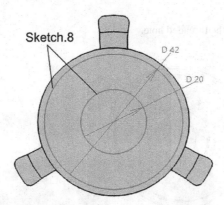

FIGURE 3.34 Two circles drawn on the flat surface at the base of the part.

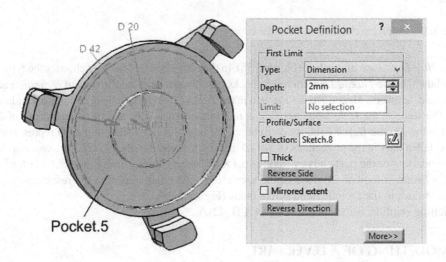

FIGURE 3.35 Pocket cut with the two concentric circles.

According to the 2D drawing, the part also has a central hole, threaded M14. The user activates the *Hole* tool, selects the flat surface on the cylindrical area left after the last *Pocket* cut and positions the centre of the hole in the centre of the circular edge.

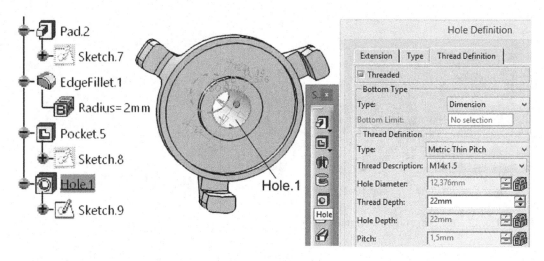

FIGURE 3.36   Parameters of the threaded hole.

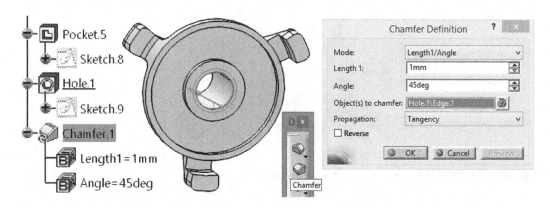

FIGURE 3.37   Applying chamfer on a circular edge.

In the *Hole Definition* dialog box (Figure 3.36), in the *Extension* tab, the user chooses the type *Up To Last*, and then, in the *Thread Definition* tab checks the *Threaded* option. In *Thread Definition* area, in the *Type* field, he selects *Metric Thick Pitch* and in the *Thread Description* field, chooses a standard M14 thread. The value of the *Hole Diameter* field is filled in automatically, and the thread depth of 22 mm is added in the *Thread Depth* field. By convention, the thread is not visible in the 3D representation in Figure 3.36, but only in the projections of the 2D drawings, created in the *CATIA Drafting* workbench.

The modelling of the part ends by making a 1×45° inner chamfer using the *Chamfer* tool on the *Dress-Up Features* toolbar. To do this, the user selects the edge, clicks on the respective icon and enters the values in the *Length 1* and *Angle* fields (Figure 3.37).

Modelling solution: https://youtu.be/_r2AXTO42vA.

3.3   MODELLING OF A LEVER PART

This tutorial presents a solution to how to model the part illustrated by the 2D drawing in Figure 3.38. It presents two orthogonal views, an isometric view and a 2D detail view. The drawing has also all the dimensions needed for the 3D modelling, from sketch to the final 3D solid.

Before starting the actual modelling procedure of the lever part, some additional settings for the working environment are required. Thus, the three default planes (*XY, YZ* and *ZX*) will be hidden by selecting them and using the *Hide/Show* option, as in Figure 3.39, from the context menu (right-click

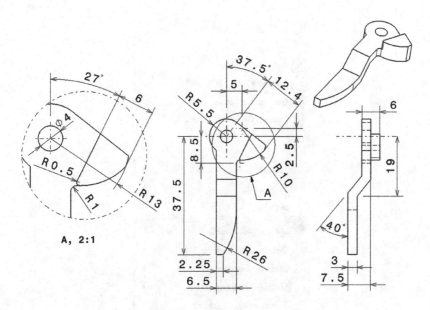

FIGURE 3.38 The 2D drawing of the part to be modelled – lever element.

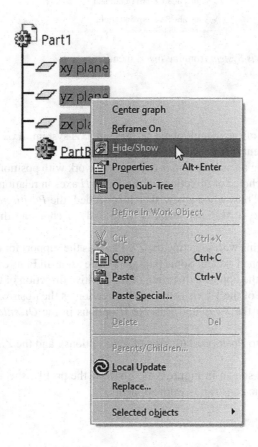

FIGURE 3.39 *Hide/Show* option on right-click.

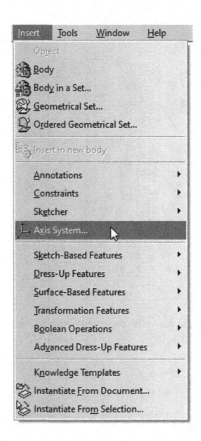

FIGURE 3.40 Insert an *Axis System* from the *Insert* menu.

on the selection). Instead of the default plane system, an axis system, *Axis System* (Figure 3.40), is inserted, accessing the menu *Insert* → *Axis System*....

The reason for inserting an axis system is that we will work with positioned sketches (*Positioned Sketch*), which allow the choice of directions of the *V* and *H* axes in relation to the *X, Y* and *Z* directions of the axis system. Thus, when a new sketch is needed, the *Positioned Sketch*... tool will be launched, from the *Insert* → *Sketcher menu* (Figure 3.41) or choosing the respective icon in the toolbar (Figure 3.42).

Modelling the part begins with selecting the *ZX Plane* as the support for drawing the basic sketch *Sketch.1*, using the positioned sketch. Some parameters are set as in Figure 3.43, so that the positive direction of the *H* axis is the opposite direction of the positive direction of the *X* axis of the system, and the positive direction of the *V* axis is the same direction as the positive direction of the *Z* axis. This is done by checking the *Swap* and *Reverse H* options in the *Orientation* area of the *Sketch Positioning* dialog box.

The sketch type is set to *Positioned* (the list of *Type* options), and the *ZX Plane* is displayed as a reference.

The sketch *Sketch.1* is shown in Figure 3.44. To draw the profile, specific drawing commands, dimensional and geometric constraints were applied.

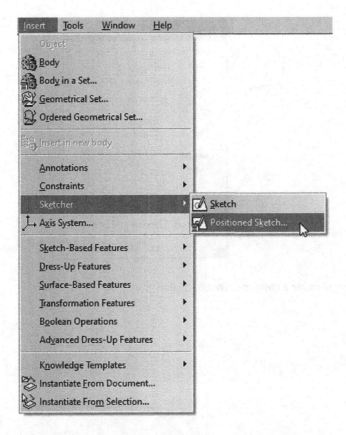

FIGURE 3.41 Launching the *Positioned Sketch* command from the *Insert* menu.

FIGURE 3.42 Launching the *Positioned Sketch* command from the *Sketch* toolbar.

The next step is to apply the *Pad* tool having the sketch *Sketch.1* as the profile to be extruded on 3.25 mm (Figure 3.45), symmetrical (check the option *Mirrored extent*) to the plane of the sketching, thus generating the feature *Pad.1*.

The modelling of the part is continued by selecting the *XY Plane* as a sketching plane for drawing *Sketch.2*, as in Figure 3.46. To create the sketch, use the *Positioned Sketch* icon again, so that

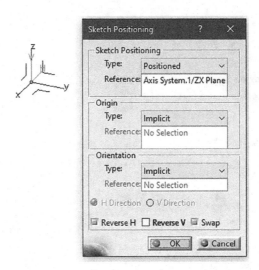

FIGURE 3.43   How to create a positioned sketch – dialog box.

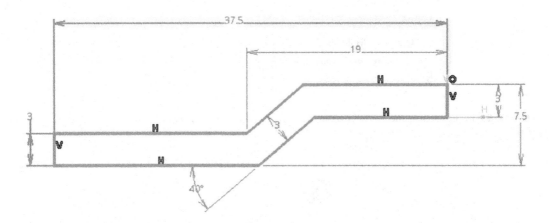

FIGURE 3.44   The first sketch drawn in the modelling process.

the positive directions of the *V* and *H* axes are as in the figure. An extrusion of the profile is then applied, using *Pad*, over a distance of 3 mm. The result is shown in Figure 3.47. Thus, the *Pad.2* feature is obtained.

To continue modelling the part, it is necessary to select a new sketching plane for *Sketch.3*. Thus, a new *Positioned Sketch* command is launched, having the parameters in Figure 3.48. The sketch needed to obtain the next feature, *Pad.3*, contains three projected edges (coloured in yellow) using the *Project 3D Elements* tool and a line, drawn at an angle of 27° to the *V* axis and 6 mm from the point representing the intersection of two circular arcs of radii R10 and R13 mm, respectively (Figure 3.49).

As a particularity, to obtain a closed profile, the *Quick Trim* tool is used on the straight lines resulting from the projection of the solid edges, but only on the left side of the line drawn at 27°, as in Figure 3.50. The result of *Quick Trim* is the transformation of projected and then deleted lines into construction lines. They will still be marked with a thin yellow line, not continuous, but interrupted. The remaining profile on the right side of the line is extruded using the *Pad* tool on a length of 3 mm.

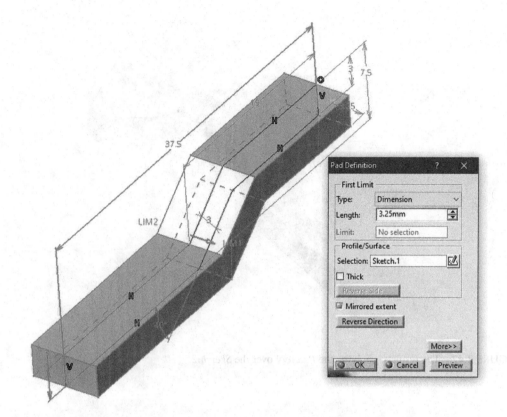

FIGURE 3.45   *Pad.1* definition dialog box.

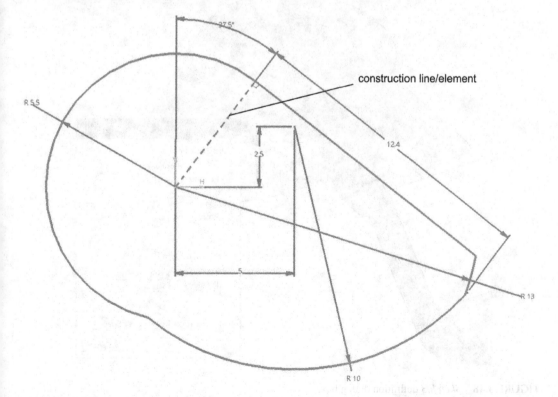

FIGURE 3.46   *Sketch.2* dimensioning.

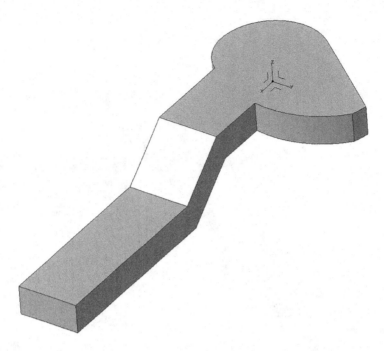

FIGURE 3.47 The result of applying the *Pad* tool over the *Sketch.2*.

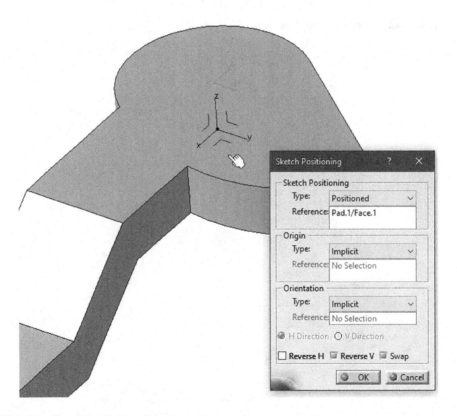

FIGURE 3.48 *Sketch.3* definition dialog box.

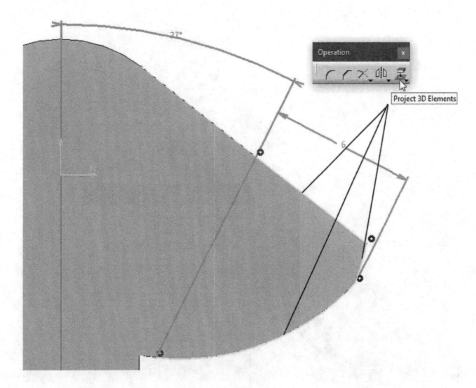

FIGURE 3.49 How to project 3D elements onto the sketch using *Project 3D Elements* tool.

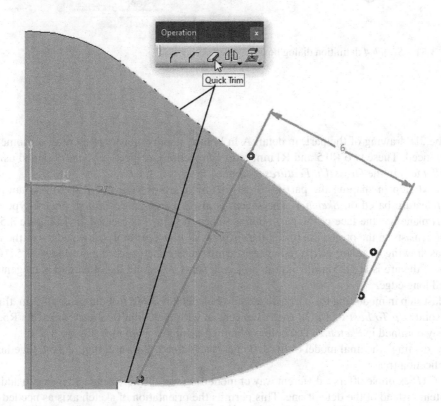

FIGURE 3.50 *Quick Trim* tool usage.

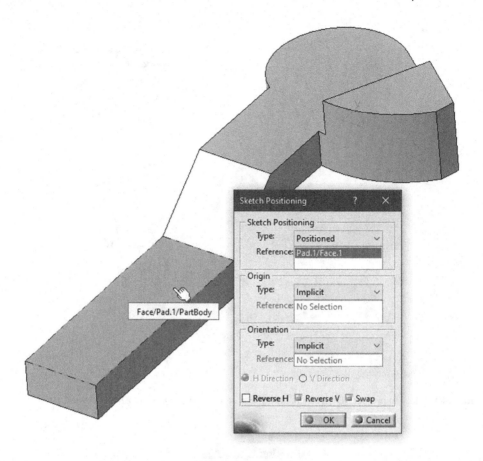

FIGURE 3.51   *Sketch.4* definition dialog box.

On the 2D drawing of the part, in detail A in Figure 3.38, two connecting radii on some edges can be noticed. These two R0.5 and R1 mm radii connections, respectively, are obtained using the *Edge Fillet* tool in the *Dress-Up Features* toolbar.

The next step in shaping the part is to cut the circle/arc of radius of R26 mm, obtaining the *Pocket.1* feature based on *Sketch.4*. The sketch is also created as *Positioned Sketch* type and as a support plane has the face of the part obtained once the *Pad.1* is created, as in Figure 3.51. The *Sketch.4* consists in the projection (using the *Project 3D Elements* tool) of two edges of the model, as well as drawing of an arc with a radius of R26 mm constrained according to Figure 3.52. One of the ends of the arc is at 2.25 mm from the projected short edge, and the other end is tangent to the projected long edge.

The last step in modelling the part is to create a hole feature *Hole.1* of diameter Ø4 mm, throughout the solid (*Up To Last* or *Up To Next*), concentric with the cylindrical face of radius R5.5 mm, previously obtained in *Sketch.2*. The hole is obtained using the *Hole* tool (Figure 3.53).

After 'drilling', the final model of the solid part is obtained, shown in Figure 3.54, together with its specification tree.

This *CAD* example offers a different way of modelling work, by creating and using an additional axis system instead of the default one. This permits the orientation of sketch axis as needed. Also, the user learned more about construction lines and their importance in a sketch.

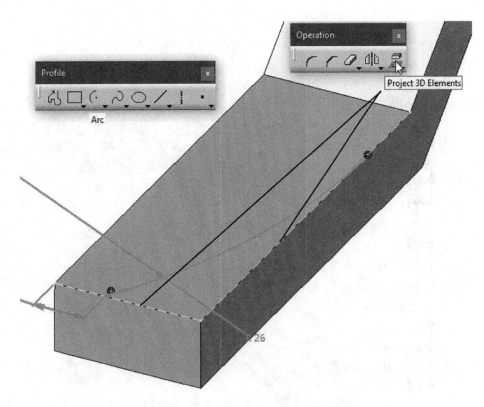

FIGURE 3.52 Dimensional and geometrical constraints of the R26 mm arc.

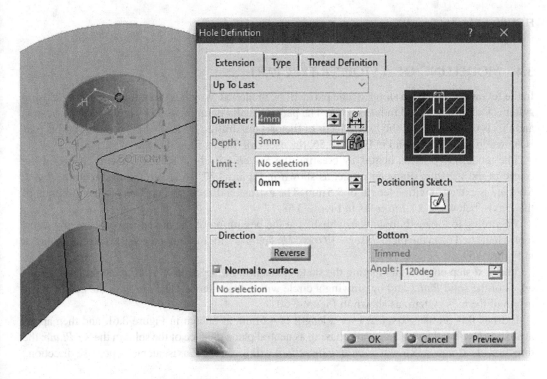

FIGURE 3.53 *Hole.1* definition dialog box.

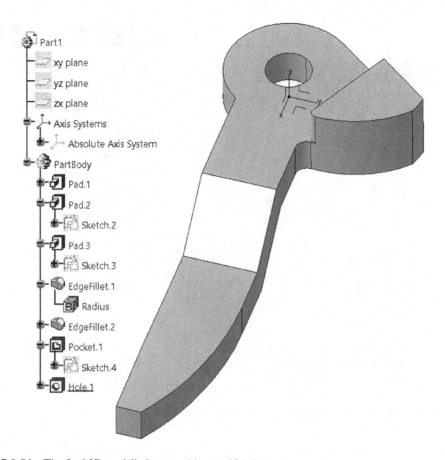

FIGURE 3.54   The final 3D modelled part and its specification tree.

3.4   MODELLING OF A STOPPER-TYPE PART

In the following tutorial, a stopper-type part will be modelled, which is often found in fixing/closing systems for household and industrial use. The part is usually manufactured by die casting processes, and this is the reason why the lateral faces of the part are drafted by 1°. The 2D drawing of this part is shown in Figure 3.55, and in Figure 3.56, the user can find its four 3D views.

The modelling process begins by creating the *Sketch.1* in the *XY Plane*, having as origin the centre of the Ø20 mm diameter circle, as shown in Figure 3.57. To create the first feature, *Pad.1*, *Sketch.1* is extruded with a length of 3.5 mm. The extrusion direction is opposite to the direction of the *Z* axis, below the *XY Plane*, as in Figure 3.58.

According to the methodology established at the beginning of the modelling process, the *Draft* tool is launched from the menu *Insert → Dress-Up Features*, for inclining the walls with 1°, as shown in Figure 3.59.

The next step consists in creating the sketch *Sketch.2*, in the same *XY Plane*, by projecting four edges of the solid *Pad.1* and by drawing a circle with a diameter of Ø20 mm with the centre at the origin of the axis system, as shown in Figure 3.60.

Extrude this profile, *Sketch.2*, over a length of 4.6 mm, as shown in Figure 3.61, and then apply, again, the *Draft* tool for all side walls, having as neutral plane the face of the solid in the *XY Plane*; the direction of extraction (known as *pulling direction*) will be on the *Z* axis, along its positive direction.

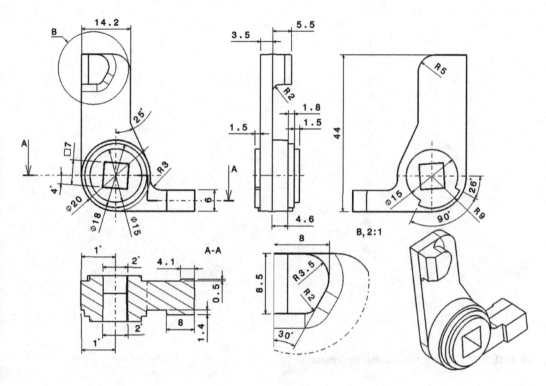

FIGURE 3.55 The 2D drawing of the part to be modelled.

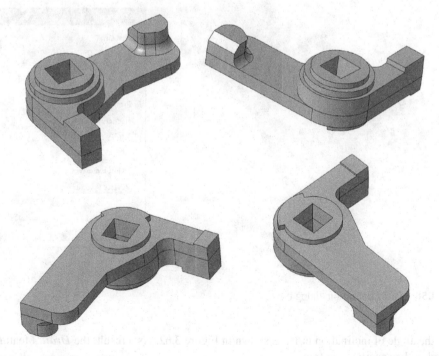

FIGURE 3.56 3D views of the part.

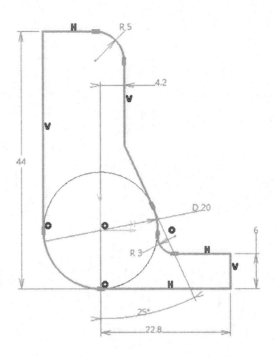

FIGURE 3.57   *Sketch.1* dimensioning constraints.

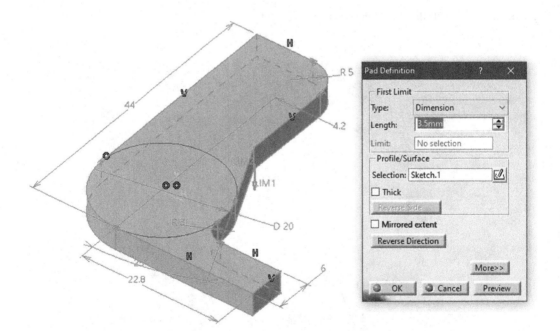

FIGURE 3.58   *Pad.1* definition dialog box.

Also, the angle of inclination is 1°, as shown in Figure 3.62. As a result, the *Draft.3* feature will be created in the specification tree.

To create the feature *Pad.3,* it is necessary to build a new sketch named *Sketch.3* (a circle of diameter Ø18 mm), on the upper face of the solid obtained in the previous step. The circle is concentric

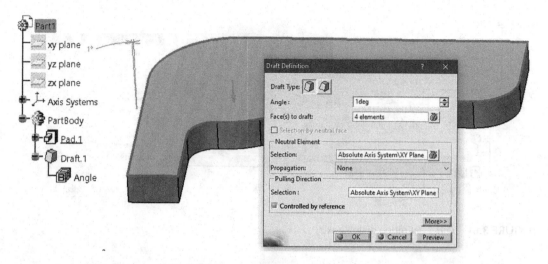

FIGURE 3.59   Drafted side walls and the *Draft* definition dialog box.

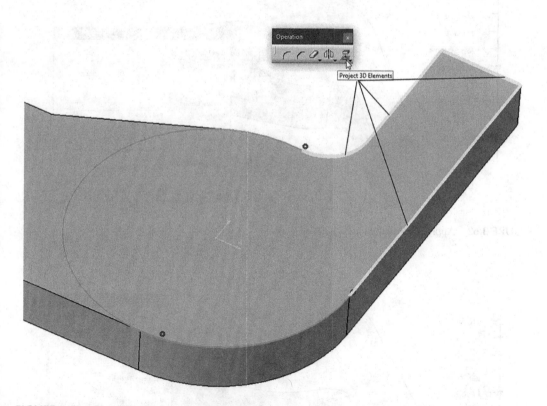

FIGURE 3.60   *Sketch.2* definition and projected edges.

to the circular edge of the face. The circle is then extruded to a height of 1.8 mm (Figure 3.63), the feature *Pad.3* is obtained, and then the *Draft* tool is applied again to the cylindrical surface, having as neutral element the surface on which *Sketch.3* was created. Thus, the feature *Draft.4* is obtained in the specification tree (Figure 3.64).

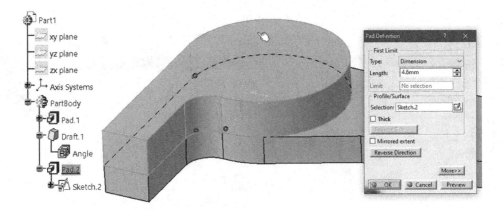

FIGURE 3.61 *Pad.2* definition dialog box.

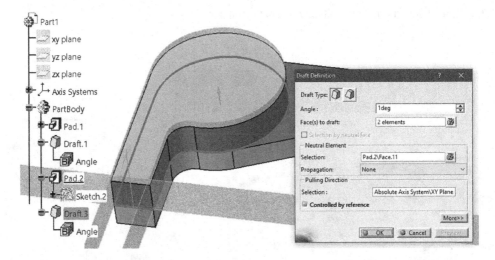

FIGURE 3.62 Applying drafting on *Pad.2*.

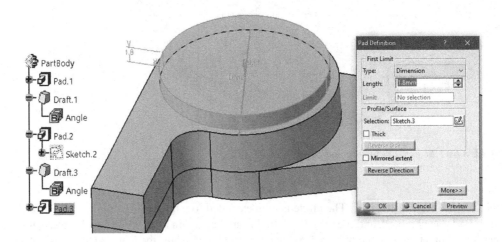

FIGURE 3.63 *Pad.3* definition dialog box.

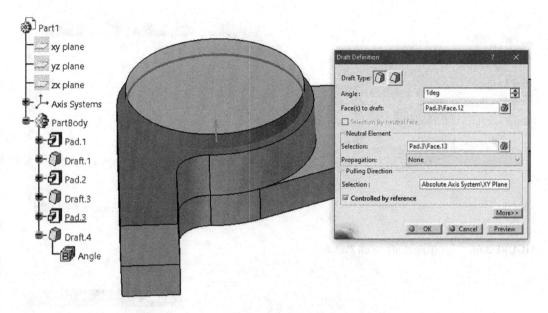

FIGURE 3.64 Applying drafting on *Pad.3*.

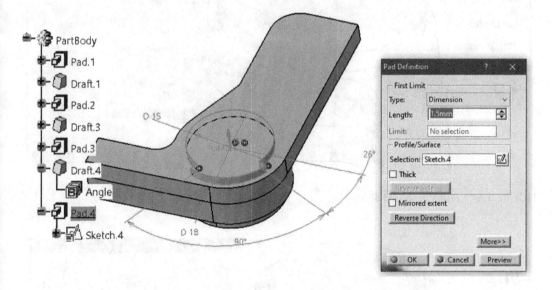

FIGURE 3.65 *Pad.4* definition dialog box.

A new sketch, *Sketch.4*, is drawn on the flat surface at the base of the part, according to the 2D drawing (right side view in Figure 3.55). The sketch is extruded at 1.5 mm, and the feature *Pad.4* represented in Figure 3.65 is obtained. The user should incline all the side walls using *Draft* tool, with an angle of 1° and having as neutral plane the plane of *Sketch.4* (Figure 3.66). Thus, the newly created feature *Draft.5* can be found in the specification tree.

Rotate the part again, and on the upper face of the feature *Pad.3* (Figure 3.63), a cylinder (*Pad.5*, Figure 3.67) will be created, with the height of 1.5 mm, which is based on a circle of diameter Ø15 mm, drawn in the *Sketch.5*. Also, the cylindrical surface will be drafted by 1°, just like the *Pad.3,* and therefore, *Draft.6* is displayed in the specification tree (Figure 3.68).

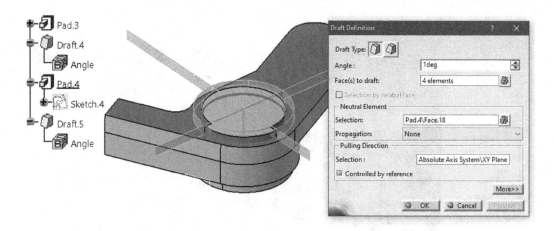

FIGURE 3.66 Applying drafting on *Pad.4*.

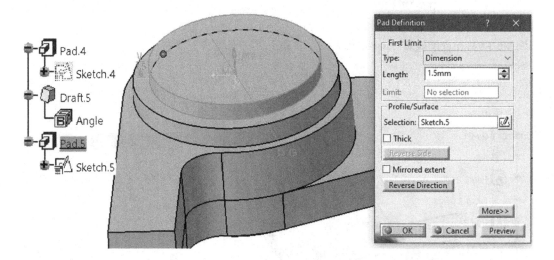

FIGURE 3.67 *Pad.5* definition dialog box.

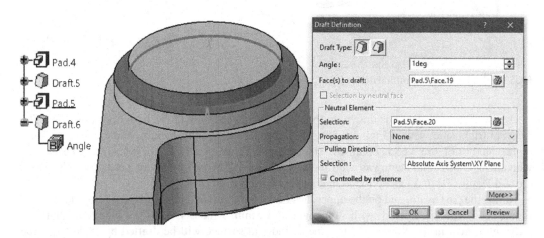

FIGURE 3.68 Applying drafting on *Pad.5*.

Next, the parallelepiped seat is made by creating in a sketch, on the upper face of the feature *Pad.5*, of a square profile with a side of 7 mm, centred in the middle of the circular face and which is rotated 4° from the *H* axis (Figure 3.69). Using this profile, the solid is cut using the *Pocket* tool, applying the *Up to last* option from the *Type* drop-down list of the *Pocket Definition* dialog box (Figure 3.70). The *Pocket.1* feature is added to the specification tree.

The walls of the *Pocket.1* are inclined, having as neutral plane a face in the *XY Plane* or even the *XY Plane*. The advanced options are also activated by clicking the *More >>* button in the *Draft Definition* dialog box (Figure 3.71). In the *Parting Element* area of the extended window, the user

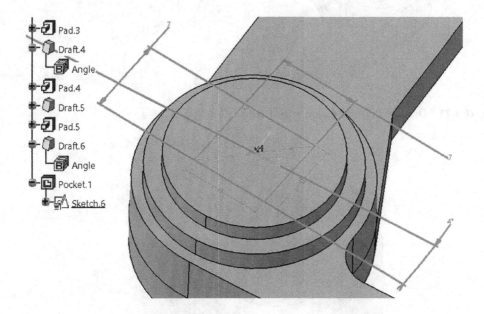

FIGURE 3.69   *Sketch.6* definition dialog box.

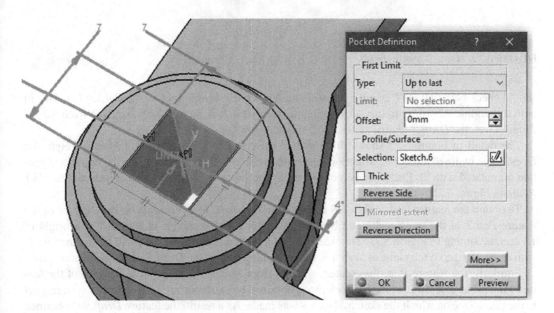

FIGURE 3.70   *Pocket.1* definition dialog box.

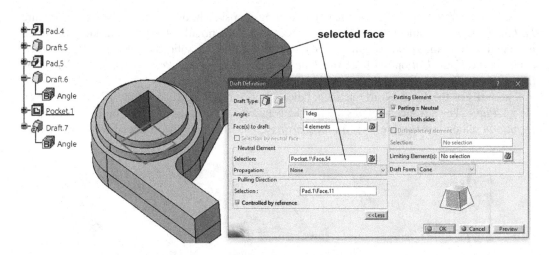

FIGURE 3.71   *Draft.7* – draft the square shape feature.

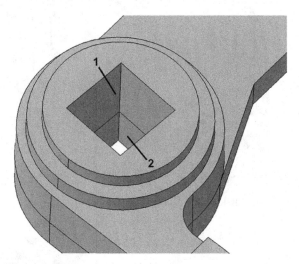

FIGURE 3.72   *Draft.7* feature result.

should check both options *Parting = Neutral* and *Draft both sides* to incline the *Pocket.1* surface in both directions, according to the 2D drawing (section view A-A, Figure 3.55), starting from the *XY Plane*. Thus, the *Draft.7* feature is obtained and added to the specification tree.

The detail in Figure 3.72 shows this inclined surface *Draft.7*, consisting, in fact, of eight flat faces, some inclined along one direction (annotated with 1) and the others along the opposite direction (annotated with 2). Each face is inclined by 1° relative to the neutral element (selected face/*XY Plane* in Figure 3.71).

To obtain the feature *Pad.7*, *Sketch.8* is created in the shape of a rectangle on the top face of the feature *Pad.2*, at the same width and 8 mm as its length. The rectangle is extruded to a height of 1.4 mm, according to Figure 3.73. The side walls of the created parallelepiped will be inclined by 1° with respect to the sketch plane of *Sketch.8*, thus obtaining the feature *Draft.8* in the specification tree.

Similarly, the feature *Pad.8* is created, starting from *Sketch.9* drawn on the flat face of the feature *Pad.1*. The side walls of the parallelepiped obtained by extrusion are inclined by 1° with respect to the plane/face on which the sketch *Sketch.9* was made. As a result, the feature *Draft.9* is obtained in the specification tree (Figure 3.74).

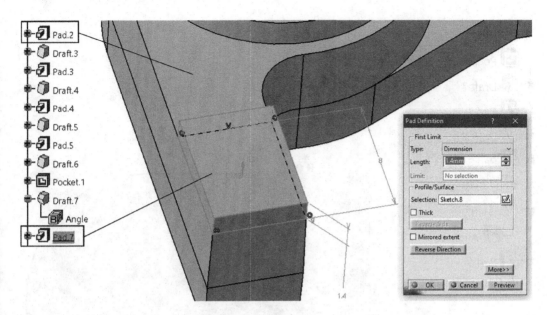

FIGURE 3.73 *Pad.7* definition dialog box.

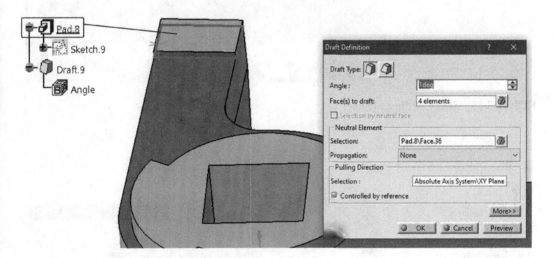

FIGURE 3.74 *Draft.9* definition dialog box.

The last part of the solid remaining to be modelled is represented in detail B in Figure 3.55, on a 2:1 scale. In the specification tree, this feature is named *Pad.9*, resulting from the extrusion on a height of 5.5 mm of the profile from *Sketch.10*, drawn in the *XY Plane* (Figures 3.75 and 3.76).

Also, the side walls of the extrusion (*Pad.9*) are also inclined by 1° to the *XY Plane*, creating the feature *Draft.10* (Figure 3.77).

The 3D model of the part is completed by applying a R2 mm radius fillet (feature *EdgeFillet.1*) and a 2×45° chamfer (feature *Chamfer.1*) on some edges, as shown in Figure 3.78.

Also, if necessary, a scaling of the 3D model with a contraction factor of 1.06 in all directions can be applied, using the *Affinity* tool (Figure 3.79). Unlike the *Scale* tool, which allows the scaling of the model with a single scaling factor along the three directions (*X*, *Y* and *Z*), *Affinity* can scale it unevenly with three different factors. In Figure 3.79, however, the scaling is uniform with the factor 1.06 (*Ratio*) in each direction.

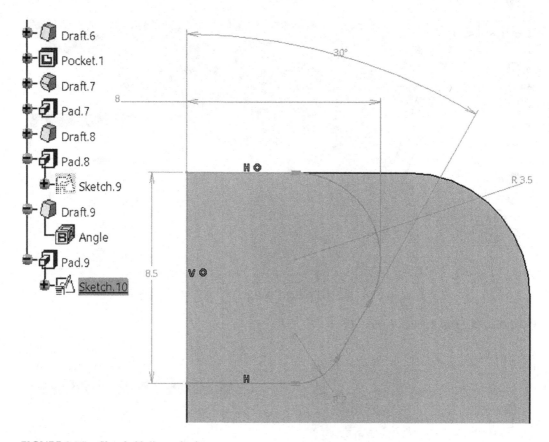

FIGURE 3.75 *Sketch.10* dimensioning.

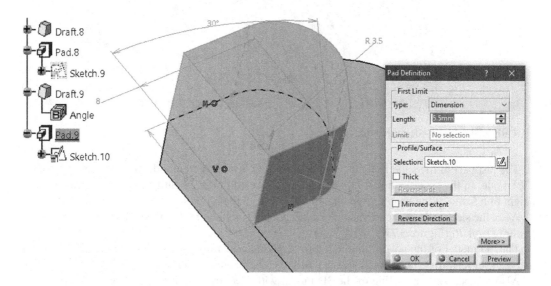

FIGURE 3.76 *Pad.9* definition dialog box.

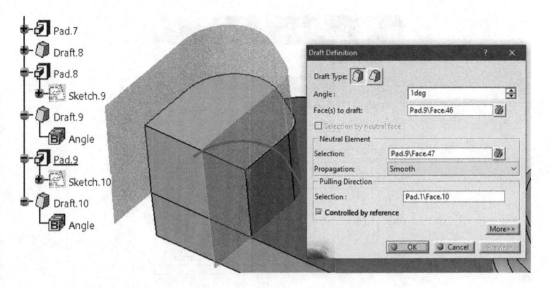

FIGURE 3.77 *Draft.10* definition dialog box.

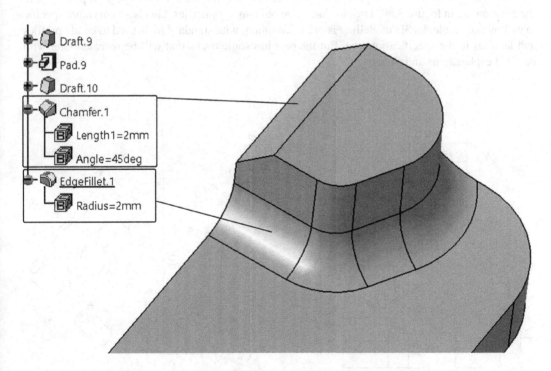

FIGURE 3.78 *EdgeFillet.1* and *Chamfer.1* applied to the 3D model.

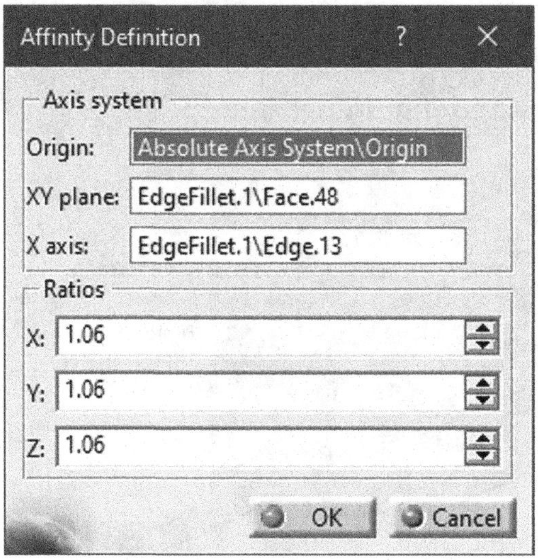

Affinity Definition
Axis system
Origin:
Absolute Axis System\Origin
XY plane:
EdgeFillet.1\Face.48
X axis:
EdgeFillet.1\Edge.13
Ratios
X:
1.06
Y:
1.06
Z:
1.06
OK
Cancel

FIGURE 3.79 *Affinity* definition dialog box.

3.5 MODELLING OF A HINGE PART

The application shows the main phases of 3D modelling process for a hinge-type part, presented in the 2D drawing in Figure 3.80. The part has some obvious symmetries, but also a curvature specific to its functional role. Its 3D modelling is pretty common, with standard tools used to create profiles and features in the specification tree. But the part has some tricks that will be revealed later with detailed explanations and figures.

40
=
=
R3
R7
R45
R35
R12
19
30
17
15
30
20
70
45°
R4.3
90°
R10
R35
45°
80
5
38
⌀12
⌀6

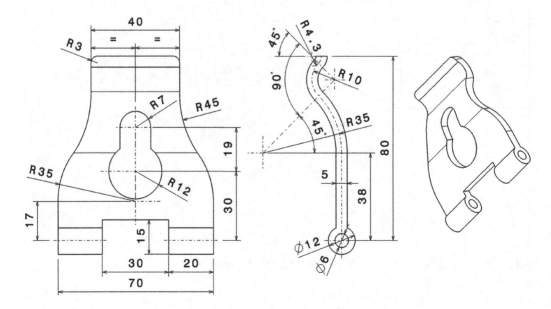

FIGURE 3.80 Drawing of the hinge part.

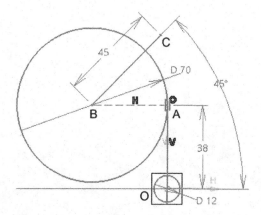

FIGURE 3.81 Drawing the first lines and a circle.

In a sketch in the *XY Plane*, the user begins the drawing of the side projection profile. First, he draws a circle with a diameter of Ø12 mm with the centre at the origin *O* of the sketch coordinate system and then the vertical line *OA* of 38 mm long, according to Figure 3.81. The circle with the diameter of Ø70 mm has its centre in point *B*, which is located 35 mm to the left of point *A* and at the same distance from the *H* axis. The circle is also tangent to the *OA* line. From point *B*, the user draws a line *BC* inclined at 45° from the *H* axis and 45 mm long.

According to Figure 3.82, the user sketches another circle with a diameter of Ø20 mm with the centre at point *C*, which is also tangent to the previous circle with the diameter of Ø70 mm (with the centre at point *B*). The *CD* line has a length of 14.3 mm and makes an angle of 90° with the *BC* line.

With a centre at point *D*, a circle with a diameter of Ø8.6 mm is created. At a distance of 80 mm from the *H* axis (passing through the origin *O*), a horizontal line is drawn, represented in Figure 3.82. The line does not contain the centre *D* of the circle with a diameter of Ø8.6 mm, and it is not tangent to the circle with a diameter of Ø20 mm.

Using the *Trim* tool, the user removes unnecessary line segments and arcs, resulting in the profile in Figure 3.83. The lines drawn between points *A-B*, *B-C* and *C-D* can be deleted or transformed into construction elements (*Construction Element* icon on the *Sketch Tools* toolbar), but it is still recommended to keep them.

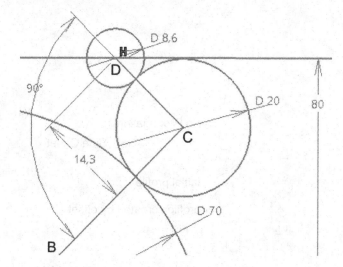

FIGURE 3.82 Drawing the next two circles and lines.

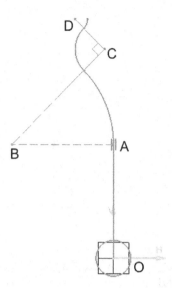

FIGURE 3.83   The hinge profile: a vertical line and three arcs.

The profile has to be multiplied on either side of it using the *Offset* tool in the *Operation* toolbar. The profiles created by this method are parallel to the initial one, similar in shape, but have other dimensions, according to the distances of 2.5 mm.

The user selects the initial profile (the line and the three circle arcs) by clicking on them and holding down the *Ctrl* key (which means multiple selection). The *Offset* tool is activated, and the *No Propagation* and *Both Side Offset* options are selected on the extended *Sketch Tools* bar (Figure 3.84). Also, 2.5 mm should be entered in the *Offset* field.

The initial profile also becomes a construction element because it no longer intervenes in the modelling process of the hinge-type part. The two profiles obtained by the offset method are possible to be created by direct drawing, after identifying the necessary dimensions, but the process is much more complicated.

A new horizontal line at a distance of 80 mm is drawn from the *H* axis, similar to the one in Figure 3.82, which will be used to limit and edit the two offset profiles.

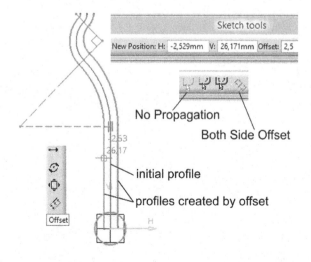

FIGURE 3.84   The hinge profile after offsetting its elements.

Figure 3.85 shows the horizontal line and two details of the areas where the offset profiles are closing to (left) and intersect (right) the line. A large zoom factor was applied for the two details. At the first glance, it seems that the horizontal line contains the ends of the offset profiles. The user is advised to zoom in on the two areas to edit the profiles.

For the left profile that does not touch the horizontal line, the *Close arc* tool is used on the *Relimitations* bar; a circle is obtained (Figure 3.86), which is then edited by the *Quick Trim* tool. It also applies to the right profile by removing the arc segment that extends beyond the horizontal line. The line is trimmed so that only the segment joining the two edited ends of the offset profiles is kept.

At the bottom of the sketch, having the centre in the origin *O*, the circle with the diameter of Ø12 mm is placed. It is intersected by the two vertical lines of the offset profiles. A trim edit is applied to remove some line and circle segments (Figure 3.87).

The *Sketch.1* results, and it is possible to be verified using the *Sketch Analysis* option in the *Tools* menu. According to Figure 3.88, an information box opens, and the status of the sketch (*All check passed*) is displayed in the *Geometry* tab. The profile is closed (condition required for *Pad* extrusion) and consists of ten curves (straight segments and arcs left after trim edits).

The sketch is extruded using the *Pad* tool on both sides of the *XY Plane* using the *Mirrored extent* option. The extrusion value in each of the two directions is 35 mm (Figure 3.89). The figure shows the profile of the sketch in the middle of the 3D solid.

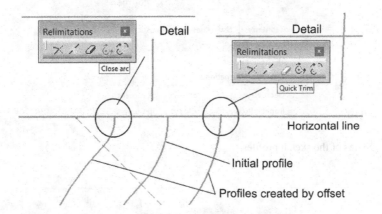

FIGURE 3.85 Details for the upper ends of the offset profiles.

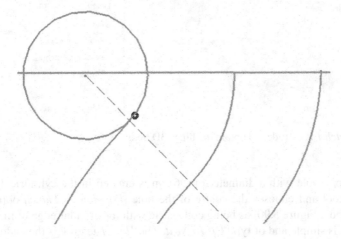

FIGURE 3.86 Circle created on the left profile – it intersects the horizontal line.

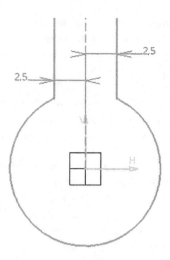

FIGURE 3.87   Details for the lower ends of the offset profiles.

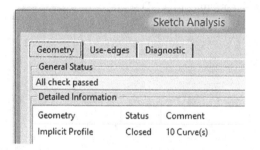

FIGURE 3.88   Status of the sketch profiles.

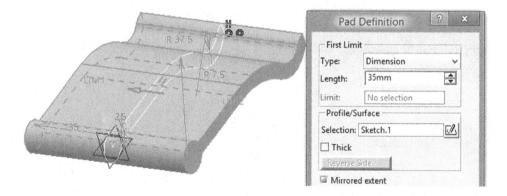

FIGURE 3.89   *Sketch.1* is extruded to obtain the hinge 3D body.

In the next step, a hole with a diameter of Ø6 mm is created in the cylindrical area. The user applies the *Hole* tool and chooses the centre of the hole (*Positioning Sketch* option in the *Hole Definition* dialog box; Figure 3.90) as being concentric with the circular edge of the flat face of the cylinder. The hole is simple and of type *Up To Next*. The *Hole.1* feature is then added to the specification tree.

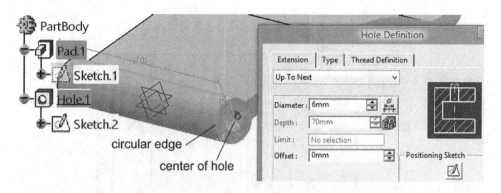

FIGURE 3.90   Parameters and elements of the *Hole.1* feature.

A rectangle is drawn in the *YZ Plane* according to the dimensions of the initial 2D drawing. The height of the rectangle is 30 mm, the right side is 9 mm from the axis of the cylindrical surface, the left side is considered outside the part, and the horizontal sides are positioned symmetrically to the *H* axis.

The representations in Figure 3.91 show the positioning of the rectangle and the result of its *Pocket* cut, on either side of the *YZ Plane*. The 2D drawing contains a dimension of 15 mm with respect to the projection of the cylinder surface generator. In the *Sketcher* workbench, this is not possible, as the generator does not materialize through a selectable line. Thus, a distance of 9 mm from the axis of the respective cylindrical surface is considered. The axis is visible and selectable only when the mouse is positioned on the surface and the dimensional constraint tool is activated.

In the same plane *YZ*, the two symmetrical profiles represented in Figure 3.92 are drawn and constrained. To do this, the user first draws a circle with a diameter of Ø70 mm having its centre at 17 mm from the axis of the cylindrical surface, as defined in Figure 3.91.

The sketch is continued by two horizontal, parallel, symmetrical lines (with respect to the *H* axis) spaced 40 mm apart. Two R45 mm *Corner* radiuses are applied between the lines and the previous circle created.

The profiles are closed on the outside of the part by the vertical and horizontal lines. The vertical lines on the left start from the tangent points (Figure 3.92) between the circle with the diameter of Ø70 mm and the horizontal edges of the part.

The *Quick Trim* tool is used to remove certain elements of sketched profiles. Being symmetrical to the *H* axis, it is possible to create only one profile and multiply it by the *Mirror* tool of the *Operation* toolbar. The condition of closing the profiles and the correctness of the drawings can be checked with the *Sketch Analysis* option in the *Tools* menu. The status of the profiles must be *Closed* (Figure 3.93).

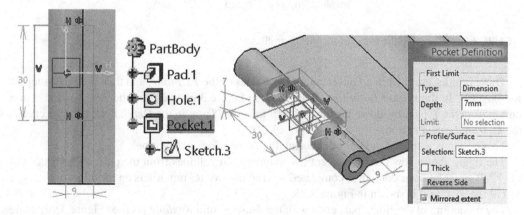

FIGURE 3.91   Drawing a rectangle and cutting the hinge with its profile.

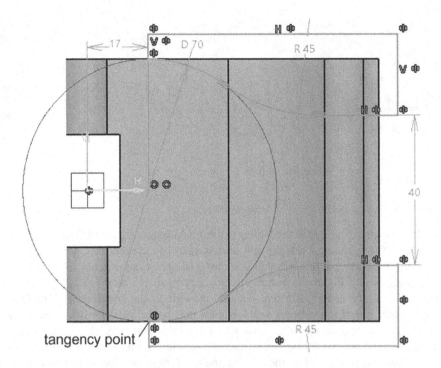

FIGURE 3.92   Drawing a circle and lines to define the hinge outer profile.

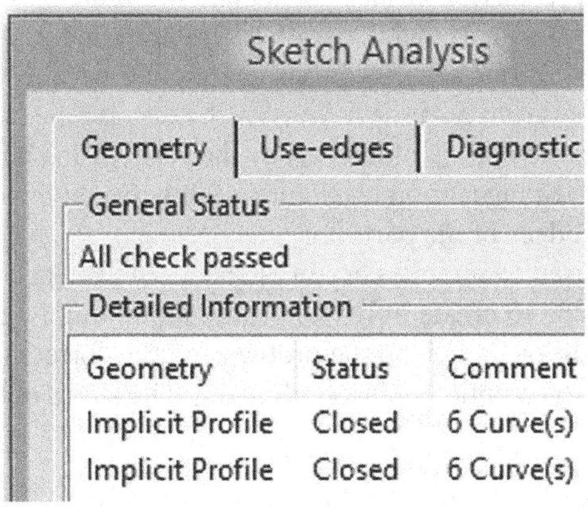

FIGURE 3.93   Checking the status of the sketch profiles.

These profiles are used in a *Pocket* cut, on both sides of the *YZ Plane*, and then used as a support for *Sketch.4*. In the *Pocket Definition* dialog box, in Figure 3.94, the user chooses the *Dimension* option from the *Type* drop-down list and enters a value of 18 mm in the *Depth* field; then, he checks the *Mirrored extent* option.

This option is activated only after choosing the *Dimension* cutting type.

The figure shows how the profiles cut two symmetrical volumes from the part, on both sides of the *YZ Plane*. The outer contour is completed by creating two R3 mm fillets on the edges at the right end of the part, as it is shown in Figure 3.95.

The modelling of the hinge part ends with the drawing of a *Keyhole* profile (Figure 3.96) using the dimensions from Figure 3.80. This profile should be placed on the flat surface or in the *YZ Plane*.

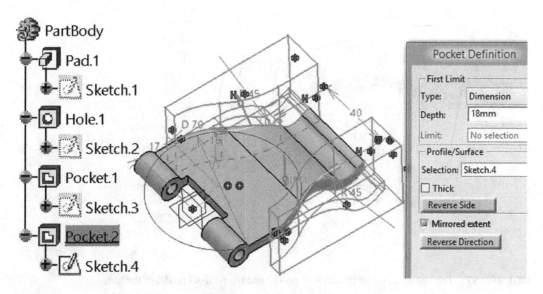

FIGURE 3.94 Cutting the part to reach the hinge outer 3D shape.

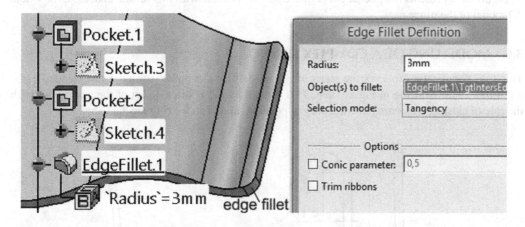

FIGURE 3.95 Edge fillets applied to the hinge 3D model.

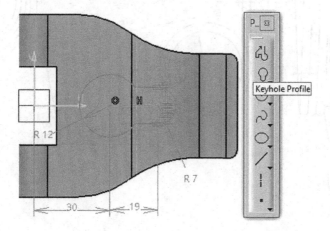

FIGURE 3.96 Drawing the keyhole profile.

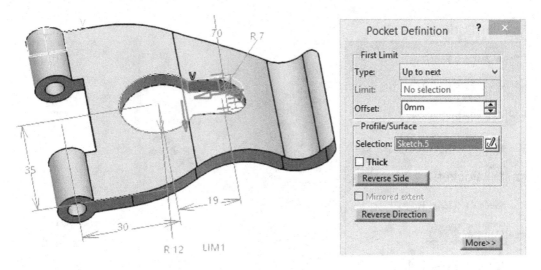

FIGURE 3.97   The last pocket with the *keyhole* profile and the final 3D model of the hinge.

Figure 3.97 presents the pocket cut (*Type: Up to next*) made with this profile and, also, the whole 3D model of the hinge part.

3.6   MODELLING OF A COMPLEX SPRING

This CAD application follows the steps of 3D modelling a complex spring with extended ends, according to the drawing in Figure 3.98. This type of spring is often a part of the assemblies with the role of closing, tensioning and keeping in position, and/or in locking mechanisms.

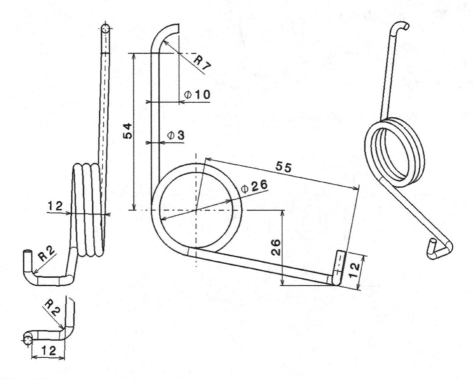

FIGURE 3.98   2D drawing of the spring part.

By analysing the 3D shape of the spring, the way it passes through different planes and has some twists in the end areas, it is necessary for the user to think and to apply a wireframe modelling approach, followed by a *Rib* extrusion of a circular profile along the 3D profile. Thus, the user should access the *CATIA Generative Shape Design* (*GSD*) workbench from the *Start → Shape* menu.

In the *Sketch.1* of the *XY Plane*, a point is created on the *H* axis at a distance of 14.5 mm from the *V* axis (Figure 3.99).

Using the *Helix* tool on the *Wireframe* bar, the parameters of the helix are established, according to Figure 3.100. The starting point is the previous one in *Sketch.1*, and the *Z* axis is chosen by right-clicking in the *Axis* field.

The helix pitch is set to 3 mm (the coils of the spring have a diameter of Ø3 mm and they are tangent to each other) and has the height of 9 mm. The winding direction is considered to be trigonometric.

At the lower end of the spring, in the *XY Plane* the point in *Sketch.1* is placed. At the other end, from the top, an extension of the coil is applied with the help of the *Extrapolate* tool (Figure 3.101) on the *Operations* bar.

In the *Boundary* field, the user selects the endpoint of the helix and this curve in the *Extrapolated* field. From the two variants available in the *Type* list (*Length* and *Up to element*), he chooses the first option and then enters the distance of 53.5 mm in the *Length* field. This value is calculated by subtracting the 1.5 mm radius of the coil from the total length of 55 mm, as it is shown in Figure 3.98. In the *Continuity* options list, the user chooses *Tangent* so that the created line is tangent to the *Helix.1* curve. The other option on the list, *Curvature*, would have extended the spring by a certain number of turns.

In the *XY Plane*, a new sketch, *Sketch.2*, is initiated to determine the position of an important point in the next stages of spring creation. Thus, a circle is drawn with the centre in the origin and a constraint is applied to coincide with the endpoint of the *Extrapol.1* line. The value of the circle diameter is not important (informative, Ø110.804 mm, just for checking).

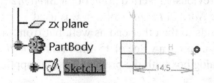

FIGURE 3.99 Inserting a point in the first sketch as the start point of a helix.

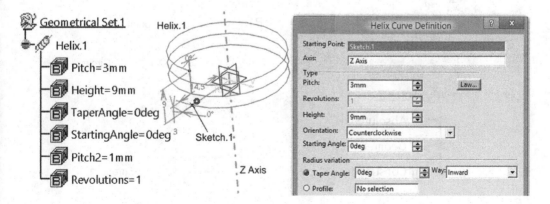

FIGURE 3.100 Creation of a helix with its parameters defined in the dialog box.

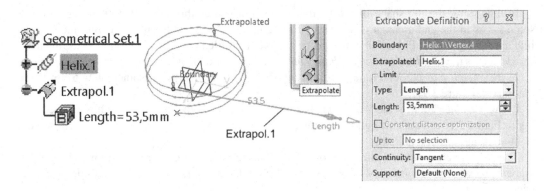

FIGURE 3.101   Extrapolation of the coil with a straight line.

A line (annotated as *Line 1*) is drawn through the same endpoint, with a random orientation. Through the origin (the point which is also the centre of the circle), a second line is created (*Line 2*), imposing a constraint of parallelism between it and *Line 1*, but also a distance of 24.5 mm (calculated by the difference 26 – 1.5 mm, the radius of the coil). The lines are drawn as construction elements (use the respective tool on the *Sketch Tools* bar to convert line type from standard to construction and vice versa). The sought point is at the intersection between the circle and *Line 2* and should be of standard to be seen and selectable after exiting the sketch. Figure 3.102 shows the elements drawn in *Sketch.2*.

Initially, the lines can be drawn with any length and orientation, but by setting up geometrical constraint of parallelism and distance, they take the correct direction, as it is shown in the figure. The ends of *Line 2* are constrained to coincide with the circle and, respectively, with the origin. The lower end of *Line 1* can be in any position as long as the two constraints are met.

The user exits the sketch and creates a plane, named *Plane.1*, perpendicular to the circle of the sketch and positioned in the previously inserted point of the *Sketch.2* (Figure 3.103). The *Plane* tool on the *Wireframe* bar with the *Normal to curve* option was used.

The *Extrapol.1* curve extends to the other end as well, but from the *Continuity* options list, the user selects *Curvature* and then enters a value of 15 mm in the *Length* field. Figure 3.104 shows the *Extrapol.2* feature and the dialog box with the previous options applied.

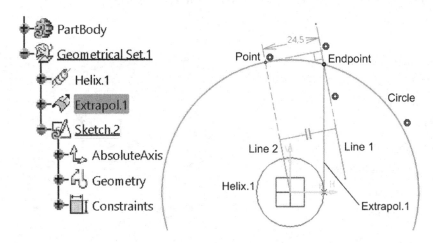

FIGURE 3.102   Auxiliary drawing to identify the position of a point.

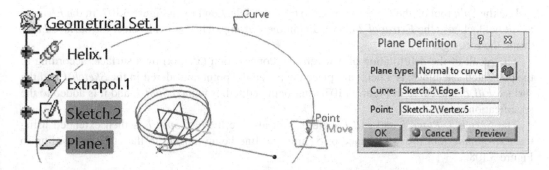

FIGURE 3.103 Plane created in the inserted point and normal to the circle.

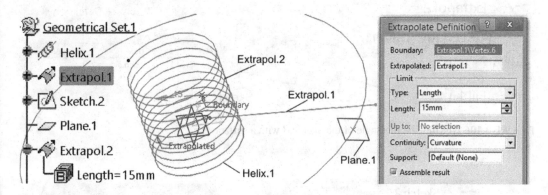

FIGURE 3.104 Extrapolation of helix while keeping its shape.

This extended curve, *Extrapol.2*, must be delimited in the *Plane.1*, previously created, by the new *Sketch.3*. Thus, a rectangle is sketched in the plane (dimensions are not important) to intersect with the curve *Extrapol.2*. The rectangle has its long sides above and below the line that symbolizes the circle from *Sketch.2* (Figure 3.105).

To activate as a boundary/delimiting element, the sketch will be filled with a surface using the *Fill* tool from the *Surfaces* toolbar. Figure 3.106 shows the closed contour of the rectangle and the surface *Fill.1*, but also the *Extrapol.2* spiral curve.

Some of the features created up to this stage will be hidden (*Hide/Show* option in their context menu). Thus, for example, the user can hide the circle, the helix, etc., the purpose being to simplify representations and explanations.

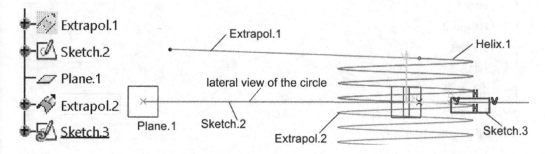

FIGURE 3.105 Drawing a rectangle in the *Plane.1* to delimit the extended helix *Extrapol.2*.

Use the *Split* tool on the *Operations* bar to open the selection box in Figure 3.107; in the *Element to cut* field, choose the *Extrapol.2* curve; and in the *Cutting elements* field, select the surface *Fill.1* inside the rectangle.

The result of the delimitation of a wireframe construction (a helix) by a surface, according to the presented selections, is a helix arc placed between the point considered in the *Sketch.1* and the surface *Fill.1*. The curve in Figure 3.107, after being edited, is named *Split.1*, and it is added to the specification tree.

The *Sketch.3*, *Plane.1* and *Fill.1* features can also be hidden. *Split.1* is then extended using the *Extrapolate* tool over a distance of 54 mm, the line being tangent to the curve, according to Figure 3.108.

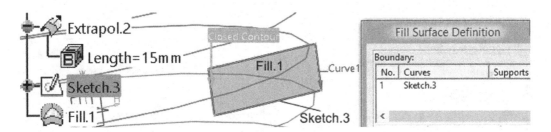

FIGURE 3.106   Filling the rectangle of the *Sketch.3* with a surface.

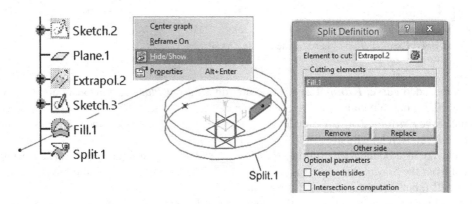

FIGURE 3.107   Splitting the curve *Extrapol.2* with the filled surface *Fill.1*.

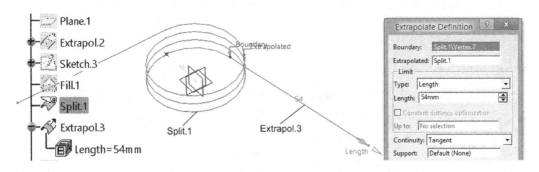

FIGURE 3.108   Extrapolation of the resulted curve, *Split.1*, over a distance of 54 mm.

The *Extrapol.3* feature is also added to the specification tree. At its free endpoint, a plane (*Plane.2*) parallel to *XY* is created (Figure 3.109). The *Plane* tool with the *Parallel through point* option from the *Plane type* list was used. Informatively, the distance between the two planes (*XY* and *Plane.2*) is about 2.435 mm.

In this plane, a new sketch is initiated, *Sketch.4*, in which a quarter of a circle, of radius R8.5 mm (7 + 1.5 mm; Figure 3.98) is sketched, according to Figure 3.110.

At the free end of the former *Extrapol.1* line (Figure 3.101), *Plane.3* is inserted perpendicular to it. From the *Plane type* list, the user chooses the option *Normal to curve* and then the curve *Extrapol.3* in the respective field (Figure 3.111). This feature now brings together all wireframe constructions: *Helix.1*, *Extrapol.1*, *Extrapol.2*, *Split.1* and *Extrapol.3*. These features are transforming from one to another, changing their names as new features or editing operations are added.

In the *Sketch.5* that belongs to the *Plane.3*, a vertical line (Figure 3.112) of length 15 mm (12 + 1.5 + 1.5 mm) is drawn. This line and the spring curve are filleted using the *Connect Curve* tool on the *Wireframe* bar.

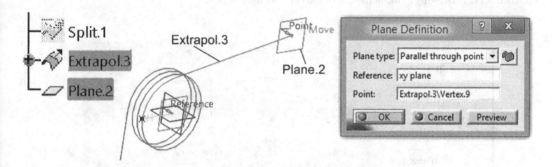

FIGURE 3.109   Creation of the *Plane.2* at the end of the curve *Extrapol.3*.

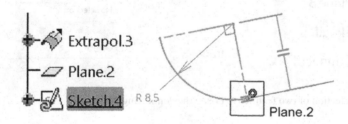

FIGURE 3.110   Quarter of circle inserted in the new sketch of the *Plane.2*.

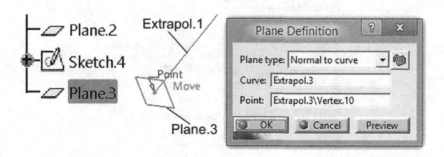

FIGURE 3.111   Definition of a new *Plane.3* at the end of the *Extrapol.1* curve.

In the *Connect Curve Definition* dialog box (Figure 3.113), the user selects the two curves *Sketch.5* and *Extrapol.3* in the *Curve* fields as the *First Curve* and the *Second Curve*, respectively.

A point must be inserted on each curve using the *Point* fields of the dialog box. Thus, after positioning the cursor in the *Point* field, the user clicks the right button on the mouse to open a context menu (Figure 3.114). He chooses the first option, *Create Point*, to open a *Point Definition* dialog box in which the position of the points on each of the two support curves is determined.

FIGURE 3.112   Drawing a line at the end of the curve in the *Plane.3*.

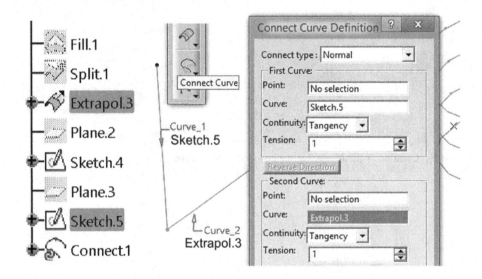

FIGURE 3.113   Selection of two features to create the connecting curve.

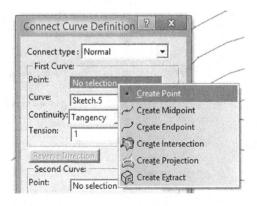

FIGURE 3.114   Available option to create points.

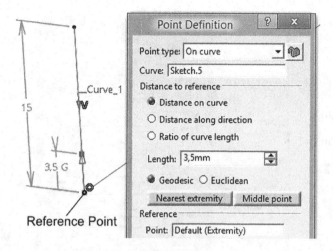

FIGURE 3.115 Inserting a new point on the line of the *Sketch.5*.

According to Figure 3.115, the user chooses a point at 3.5 mm on the vertical line (*Sketch.5*) from the intersection point (*Reference point*). Thus, from the list of *Point type* options, he selects *On curve*, then *Sketch.5* in the *Curve* field and the reference point in the *Point* field and enters the value of 3.5 mm in the *Length* field.

The selections in Figure 3.116 are similar, but the created point is on another curve, *Extrapol.3*, at a distance of 3.5 mm.

According to the dialog box in Figure 3.117, the user observes the two points, *Point.1*, respectively, *Point.2*, but also the way of creating the connection radius of 3.5 mm (2 + 1.5 mm). An arc is obtained connecting the points, tangent to the two lines/curves. The whole curve that defines this connection arc is named *Connect.1*.

Depending on how the points were inserted, it may be necessary to press the *Reverse Direction* buttons; some possible, but unwanted connection options are also shown in the figure. The user must also check the *Trim elements* option to remove the corner (the initial point of intersection of the two curves). This point is marked as *Reference Point* in Figure 3.115.

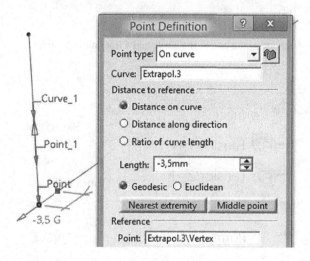

FIGURE 3.116 Inserting a new point on the curve *Extrapol.3*.

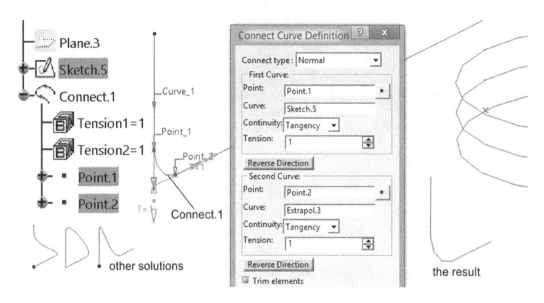

FIGURE 3.117 Connection curve between the two lines.

A new plane, *Plane.4*, is inserted at the endpoint of the profile, perpendicular to it (Figure 3.118). In the *Sketch.6*, belonging to the plane, a line of 10.5 mm (12–1.5 mm) is drawn, according to the figure.

A connection, also of R3.5 mm radius, similar to the previous one, is made between the *Connect.1* curve and this line. The name of the whole curve becomes *Connect.2* (Figure 3.119).

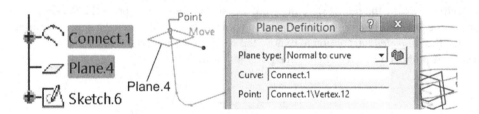

FIGURE 3.118 *Plane 4* created at the end of the curve and a line in the *Sketch.6*.

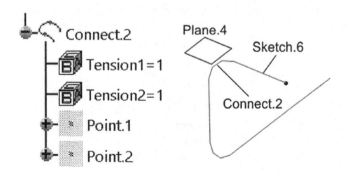

FIGURE 3.119 Connection curve between the line of *Sketch.6* and the curve.

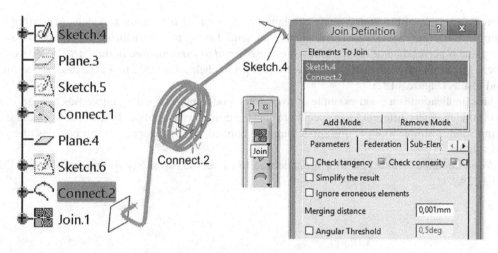

FIGURE 3.120 All the wireframe features joined to create the spring curve.

To be complete, the spring must consist of the *Connect.2* curve (which contains many other curves, created successively in the previous steps) and the arc in *Sketch.4* (Figure 3.110). Thus, the *Join* tool in the *Operations* toolbar is used. In the *Join Definition* dialog box (Figure 3.120), in the *Elements To Join* list, the user selects the two curves, and then below, in the *Parameters* tab, checks the *Check connexity* option. If all conditions are met, a single curve results for the whole spring, named *Join.1*, and it is marked in green (correct curve).

The curves and planes that led to this *Join.1* curve (added to the specification tree) can be hidden as that they are no longer needed in the 3D modelling stages.

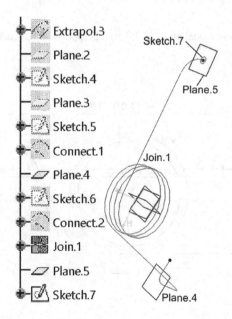

FIGURE 3.121 Drawing a circle in a plane placed at the end of the spring curve.

At either ends of the *Join.1* curve, a plane can be created to contain a sketch. According to Figure 3.121, *Plane.5* was inserted to contain *Sketch.7* having a circle with a diameter of Ø3 mm. The plane is perpendicular to the curve using the *Normal to curve* option. In the *Part Design* workbench, this circle moves along the *Join.1* curve with the help of the *Rib* tool and, thus, the spring solid results (Figure 3.122).

This application is a good example of hybrid 3D modelling. The spring curve was created in a wireframe manner; different tools specific to surface design were applied. In the end, to obtain the spring solid, a simple tool of solid modelling was applied. Thus, the spring has properties of area and volume.

As additional examples, the user is challenged to model the two safety springs in Figures 3.123 and 3.124.

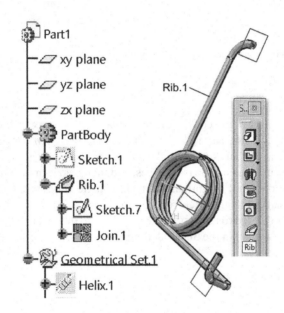

FIGURE 3.122 The spring solid model after the *Rib* tool is applied on the curve.

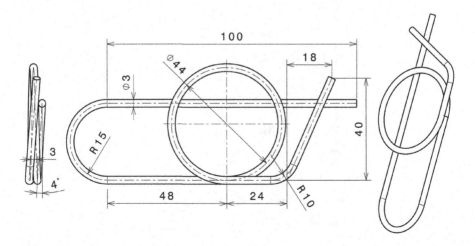

FIGURE 3.123 Drawing of the first safety spring. (Modelling solution: https://youtu.be/g09qmNWrkH4.)

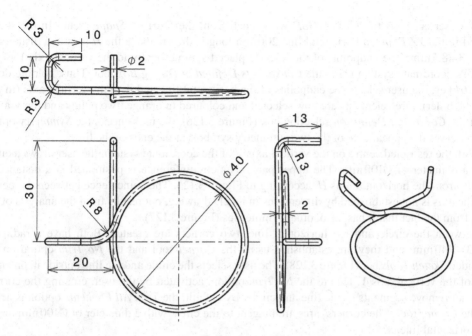

FIGURE 3.124 Drawing of the second safety spring. (Modelling solution: https://youtu.be/nrv4d0jD7r8.)

3.7 MODELLING OF A COMPLEX SPIRAL ORNAMENT PART FOR WROUGHT IRON FENCE

This *CAD* application explains how to create a spiral with a constant pitch, but with variable diameter, a part of an ornament specific to wrought iron fences. The spiral and the entire ornament are possible to be created in several ways, one of which involves wireframe, and surface modelling is further applied and explained in detail.

The 2D drawing of the whole ornament, which consists of five same spirals, is shown in Figure 3.125. Although the ornament is, in fact, an assembly, in the stages of construction of the fence, it is considered to represent a single component because the spirals are factory-welded one of another at the ends and tightened in a ring with a diameter of Ø23 mm.

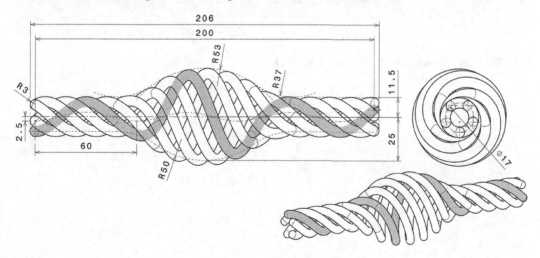

FIGURE 3.125 Drawing of the ornament part.

The user accesses the *CATIA GSD* workbench from the *Start* → *Shape* menu. In the sketch placed in the *YZ Plane*, a horizontal line 200 mm long is drawn above the *H* axis at a distance of 11.5 − 3 = 8.5 mm. The endpoints of the line are placed symmetrically to the vertical axis *V* of the sketch's coordinate system using the *Constraints Defined in Dialog Box* tool. Thus, holding down the *Ctrl* key, the user selects the endpoints of the horizontal line and then the vertical axis (in this specific order); three elements are now selected and coloured in orange (two points and the *V* axis).

In the *Constraint Definition* selection box (Figure 3.126), the user checks the *Symmetry* option and observes the appearance of the two symmetry symbols at the ends of the line.

With the restricted centre on the vertical axis *V* of the coordinate system, the user draws a circle with a diameter of Ø100 mm. The upper part of its circumference is positioned at a distance of 25 mm from the horizontal axis *H*, according to Figure 3.126. The coincidence between the centre and the axis is not established by dimensions in its 2D drawing, but results from the analysis of the part, from the fact that it has, also, other symmetries (Figure 3.127).

Between the circle and the horizontal line, two corners are created. Both have a radius of 37 + 3 = R40 mm, and they are established using the *Corner* tool and the *No Trim* option on the extended *Sketch Tools* bar (Figure 3.128). The user selects the circle and the line outside it, on either side of the vertical axis *V*. Due to the *No Trim* option, activated by the user, creating the corners does not remove segments of the line and circle. By default, the *Trim All Elements* option is active for the *Corner* tool. The corners' arcs are tangent to the circle with a diameter of Ø100 mm and to the horizontal line.

However, the line and the circle are edited by trim to remove the arc below the R40 corners and the line segment between them. Use the *Quick Trim* tool (the yellow eraser; Figure 3.129) in the *Operation* toolbar, double-*click* on its icon (to keep it activated), and select successively all segments to be removed.

The profile shown in Figure 3.129 is, obviously, open and consists of two lines and three arcs of a circle. The user can insert an axis (*Axis* element) in the *Sketch.1* to create a revolution surface in the next step. The axis, however, may be missing because the profile has been drawn with respect to the *V* and *H* axes of the sketch. The *H* axis corresponds, outside the sketch, to the *Y* axis, so the surface can be created around it.

FIGURE 3.126 Drawing the first line in the *Sketch.1*.

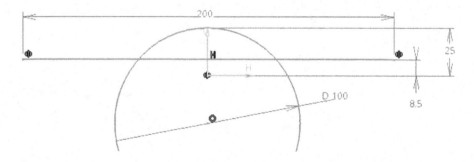

FIGURE 3.127 Drawing the circle in the *Sketch.1*.

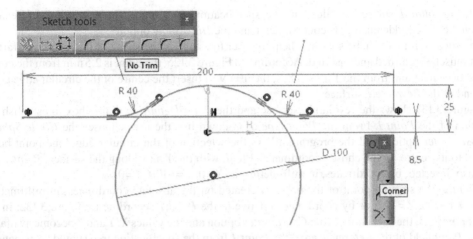

FIGURE 3.128   Creating corners between line and circle.

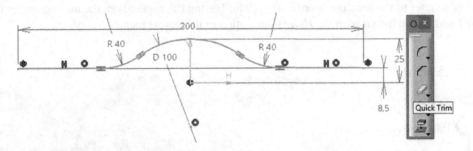

FIGURE 3.129   Profile in the *Sketch.1* with trimmed elements.

The user exits the profile's sketch (*Exit workbench* icon) and activates the *Revolve* tool on the *Surfaces* toolbar. In the *Revolution Surface Definition* dialog box, he selects *Sketch.1* in the *Profile* field, presses the right mouse button in the *Revolution Axis* field and chooses the *Y Axis* from the context menu shown in Figure 3.130. The surface is initially created at an angle of 180° (*Angle 1* field), and the value of the *Angle 2* field is 0°. The sum of the angular values of the two *Angle* fields can be a maximum of 360°; this value is now entered as *Angle 1*.

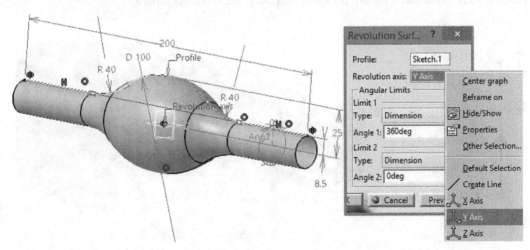

FIGURE 3.130   Surface *Revolute.1* is obtained from the *Sketch.1* profile rotated around *Y axis*.

The *Revolute.1* surface is added to the specification tree in the *Geometrical Set.1* feature. *Sketch.1* can be hidden using its context menu and the *Hide/Show* option.

In order to draw the helix that will help the user to establish the shape of the spiral, its starting point must be identified and inserted. According to Figure 3.125, the point is 2.5 mm from the centre of the ornament and from the *Y* axis. Thus, the user will insert the centre of the circular edge at the left end of the *Revolute.1* surface.

Figure 3.131 shows the revolution surface and the *Point Definition* dialog box to establish the position of the *Point.1*. From the *Point type* drop-down list, the user chooses the *Circle/Sphere/ Ellipse center* option, and the program allows the selection of the circular edge, the point being added to its centre. In the chosen coordinate system, with the *Y* axis along the surface, *Point.1* can be, also, inserted, by coordinates, in millimetres, as $X=0$, $Y=-100$, $Z=0$.

Point.2, the starting point of the helix, is created on the absolute coordinates, in millimetres, $X=0$, $Y=-100$, $Z=-2.5$ or by coordinates relative to the *Point.1* according to Figure 3.132: in the *Point type* field, the user chooses the *Coordinates* option and the values *X*, *Y* and *Z* become available. In the *Point* field of the *Reference* area, the *Point.1* from the specification tree should be selected.

Therefore, using the known position of the *Point.1*, the new *Point.2* is easier to define, without the need to refer to the absolute coordinate system. During the application, the user no longer uses *Point.1* and he can hide it with the *Hide/Show* option in its context menu.

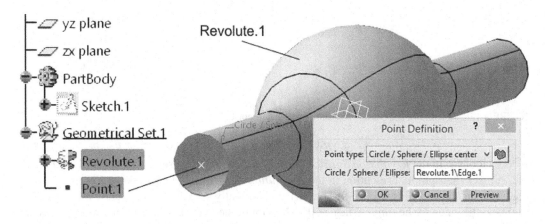

FIGURE 3.131   Insertion of the first point at the left end of the *Revolute.1* surface.

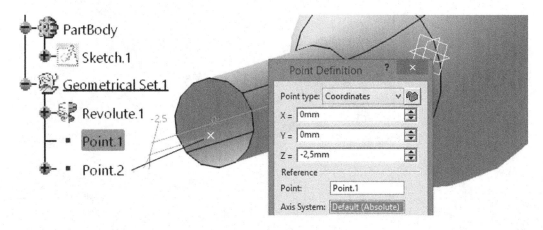

FIGURE 3.132   Insertion of the second point using absolute coordinates.

For drawing the helix, Figure 3.125 shows its dimensions of 60 mm (pitch value) and 200 mm (height). From the *Wireframe* toolbar, the user accesses the *Helix* icon, and in the *Helix Curve Definition* dialog box, he selects *Point.2* in the *Starting point* field, right-clicks on *Axis* and chooses *Y Axis* (displayed as preview) from the available context menu.

In the *Type* area, the user enters the values of 60 and 200 mm in the *Pitch* and *Height* fields, respectively. The *Orientation* list will contain the *Counterclockwise* option, and the two fields, *Starting Angle* and *Taper Angle*, remain at the default values of 0°, because the helix starts from *Point.2* and its type is cylindrical. The *Taper* angle, for values other than 0°, creates a conical helix. Figure 3.133 shows the helix parameters setting dialog box, the *Revolute.1* surface and the helix drawn around the *Y* axis.

From the same *Wireframe* toolbar, *Plane.1* is inserted in the *Point.2* location and normal to the helix. In the *Plane Definition* dialog box (Figure 3.134), the *Plane type* is *Normal to curve*, and in the *Curve* and *Point* fields, the helix and, respectively, its starting point are selected.

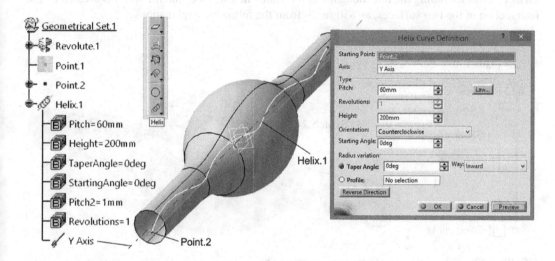

FIGURE 3.133 Features and parameters that define the helix.

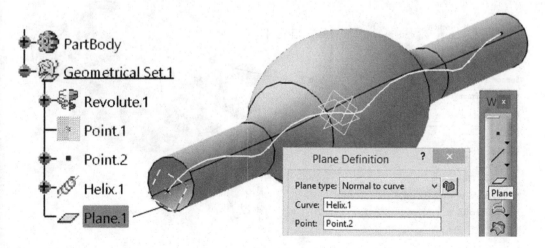

FIGURE 3.134 Inserting a plane normal to helix in the *Point.2*.

Sketch.2 is initiated in this plane. The user draws a vertical line, its lower endpoint coincides with *Point.2*, and the upper endpoint is at a distance of at least 30 mm from it (Figure 3.135). The line is aligned and directed in the positive direction of the *Z* axis.

Sketching the line is also possible outside the sketch. Thus, the *Line* icon in the *Wireframe* toolbar is activated; in the *Line type* list, the *Point-Direction* option is selected; in the *Point* field, it will be the starting point of the helix (*Point.2*); for the direction, right-click in the *Direction* field and select the vertical direction *Z*. As a support, the user can choose the plane at the end of the helix (*Plane.1*) and, then, for the line length, sets the value of 30 mm in the *End* field, according to Figure 3.136.

Depending on the chosen method of drawing the line, a sketch, *Sketch.2*, or a line, *Line.1*, will be inserted in the specification tree. To continue explaining this modelling example, the user keeps the *Sketch.2* feature.

The line represents a normal of the helix at its endpoint. The spiral that forms the ornament part is obtained by the intersection between successions of such normals projected on the *Revolute.1* surface. Thus, extruding the line along the helix results in a surface, and the spiral is created by the intersection of the two surfaces, as will result from the following explanations.

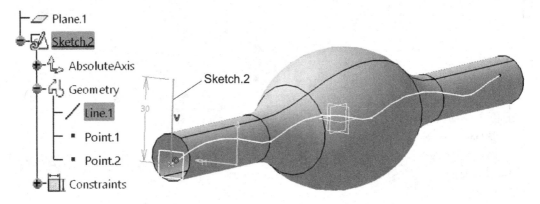

FIGURE 3.135   The second sketch contains a vertical line.

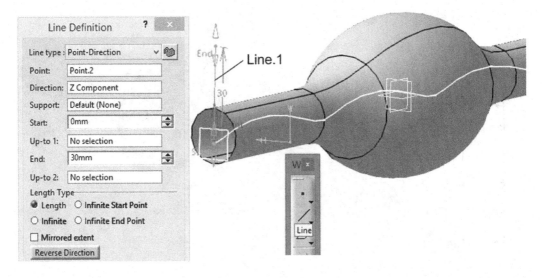

FIGURE 3.136   Another solution to draw the vertical line.

The *Sweep* tool icon in the *Surfaces* toolbar is activated and the *Swept Surface Definition* dialog box opens (Figure 3.137). From the *Profile type* list of icons, the user chooses the first one, *Explicit* and, then from the drop-down list with *Subtype* options, chooses *With pulling direction* so that the surface is oriented along the *Y axis* (selected as *Y Component* by right-clicking in the *Direction* field). The line (*Sketch.2*) is chosen as the profile, and the guide curve is the *Helix.1*, according to the figure.

CATIA v5 provides the user with four options for creating the twisted surface *Sweep.1*, selected in the *Angular sector* area (solution 1 of 4). In fact, the program proposes the creation of the *Sweep.1* surface depending on how its twist is considered: in a trigonometric or clockwise direction, an example (solution 3 of 4) being presented in Figure 3.138. It is very clear how the surface is twisted in the opposite direction to the variant in Figure 3.137. For this model, the variant of solution 3 of 4 will not be kept.

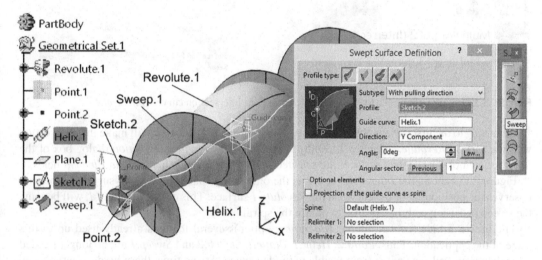

FIGURE 3.137 Creating the *Sweep.1* surface by extruding the line along the helix.

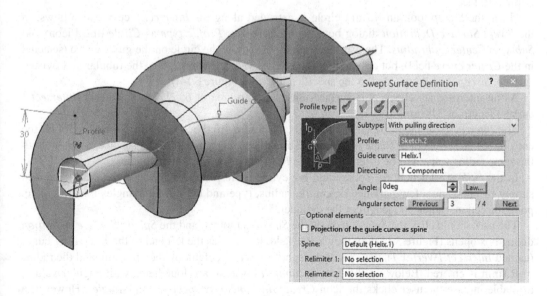

FIGURE 3.138 Another solution of creating the *Sweep.1* surface.

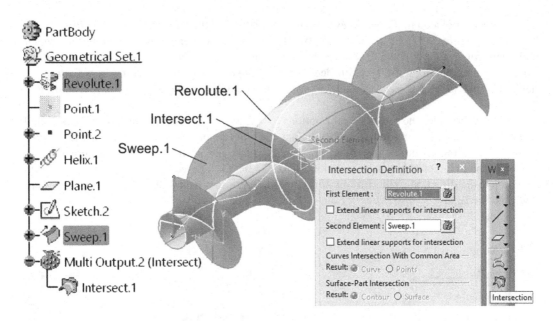

FIGURE 3.139   Intersecting the two surfaces to obtain their intersection curve – *Intersect.1*.

Returning to the first variant obtained (*Sweep.1* in Figure 3.137), the user will make the intersection with the surface *Revolute.1* (Figure 3.130) by opening the *Intersection Definition* dialog box of the *Intersection* tool. The contact surfaces are selected in the *First Element* and *Second Element* fields.

Figure 3.139 shows the *Intersect.1* curve as the only entity shared by the two surfaces. The user observes the positioning of the curve on the *Revolute.1* surface. This curve, *Intersect.1*, will become the sweep extrusion path of a profile to obtain the spiral.

The user can hide (context menu, *Hide/Show* option) several features already used up to this stage of the application. Thus, *Point.2*, *Helix.1*, *Plane.1*, *Sketch.2* and *Sweep.1* are no longer needed to be displayed and can switch from visible to hidden space. At any time, these hidden features can be viewed (Figure 3.140) by clicking the *Swap visible space* icon on the *View* toolbar at the bottom of the interface.

Using the *Sweep* tool, an R3 mm circle is extruded along the *Intersect.1* curve, as follows: in the *Swept Surface Definition* dialog box, the user chooses *Profile type* as *Circle* (third icon) and *Subtype: Center and radius*. The user does not have to draw the circle on the guide curve (selected in the *Center curve* field), but only to enter the radius value, R3 mm. Thus, the tubular and twisted surface, *Sweep.2*, will be added to the specification tree (Figure 3.141).

As an intermediate check, this surface must have an area of 6559.73 mm^2 and the *Intersect.1* curve a length of 348 mm.

The *Intersect.1* surface is open at both ends. According to the 2D drawing in Figure 3.125, the ends have a hemispherical shape of radius R3 mm. Thus, at each end of the surface, a sphere with the same radius value will be inserted, and the *Revolute.1* surface can be hidden.

In the *GSD* workbench, the sphere and the cylinder are primitive surfaces and can be created simply by choosing a few parameters: centre, radius, type and delimitation angles for the sphere, point, direction, radius and length of the cylinder.

The user should click the *Sphere* icon on the *Surfaces* toolbar, and the *Sphere Surface Definition* dialog box opens (Figure 3.142). In the *Center* field, he chooses the left end of the *Intersect.1* curve (named *Intersect.1\Vertex.1*), the axis of the sphere is set by default by the program, and the radius of R3 mm is entered. Below, in the *Sphere Limitations* area, are four fields with angular values, available in case the user clicks the icon *Create the sphere by specifying the angles*. However, a sphere feature is inserted in the model by clicking on the adjacent icon, *Create the whole sphere*.

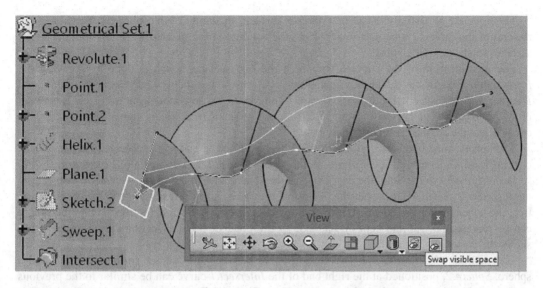

FIGURE 3.140 Swap visible space to display the hidden features: points, curves, planes and surfaces.

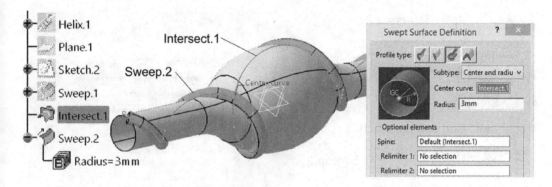

FIGURE 3.141 Creation of the twisted surface *Sweep.2*.

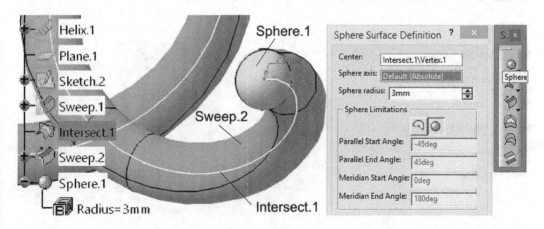

FIGURE 3.142 Creation of a sphere at the left end of surface *Sweep.2*.

The *Sphere.1* is partially inside the *Sweep.2* surface, so an edit of the sphere is needed to fit in the end of the *Sweep.2* tube. Obviously, from the way they were created, it results that these two surfaces intersect after a curve.

The identification of the curve after which the two surfaces touch is done by applying the *Intersection* tool. In the dialog box in Figure 3.143, in the two fields *First Element* and *Second Element*, the user selects the surfaces *Sphere.1* and *Sweep.2*. If the modelling is correct, the result of the intersection is a curve of length 18.85 mm and radius R3 mm and is named *Intersect.2*. This curve is just the edge at the end of the tube.

The sphere must be cut according to this edge to close the end of the *Sweep.2* surface. Thus, clicking the *Split* icon on the *Operations* toolbar opens the *Split Definition* dialog box (Figure 3.144).

In the *Element to cut* field, the user chooses the sphere and, in the *Cutting elements* list, selects the *Intersect.2* curve. After the cutting operation, the *Split.1* surface results, and in the specification tree, the sphere becomes hidden. The dialog box contains the *Other side* button, so that the sphere can be represented, cut and kept on one side or the other of the intersection curve. Figure 3.144 shows the correct *Split.1* surface.

The method of obtaining the intersection curve between the tubular surface *Sweep.2* and a sphere, *Sphere.2*, positioned at the right end of the *Intersect.1* curve can be similar to the previous one: created by the intersection of the two surfaces. The modelling steps present, however, another variant, by applying the *Boundary* tool in the *Operations* toolbar.

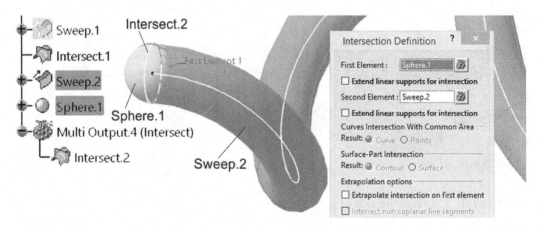

FIGURE 3.143   Intersection between the sphere and the surface *Sweep.2*.

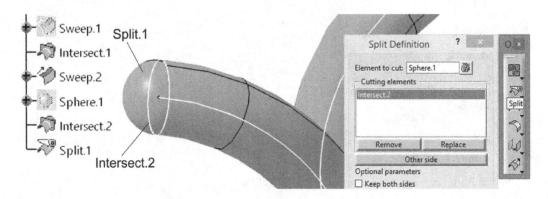

FIGURE 3.144   The sphere feature was cut with the intersection curve.

In the *Boundary Definition* dialog box (Figure 3.145), the user chooses the *Point continuity* option in the *Propagation type* list and then selects the edge at the end of the tube, named *Sweep.2\ Edge.2*. The edge becomes from a 3D element, by extraction, a 2D element, the *Boundary.1* feature. The editing of the sphere with respect to this curve is also done with the help of the *Split* tool, resulting in a closing surface of the tube, named *Split.2*.

By obtaining this surface, the modelling of the spiral elements ends. In the specification tree in Figure 3.146, it is observed that three surfaces are visible and the rest have become hidden. These surfaces, *Sweep.2*, *Split.1* and *Split.2* are joined using the *Join* tool in the *Operation* toolbar to become a single surface, *Join.1*.

In the *Join Definition* dialog box, the user selects the surfaces and then clicks the *Check tangency* and *Check connexity* options in the *Parameters* tab. These options verify if the adjacent surfaces are tangent and that there are no small gaps between them. The checks represent an important stage of the validation of the *Join.1* surface, thus being prepared for the transformation into solid. To comply with our modelling, find out if the resulting surface area is 6672.83 mm².

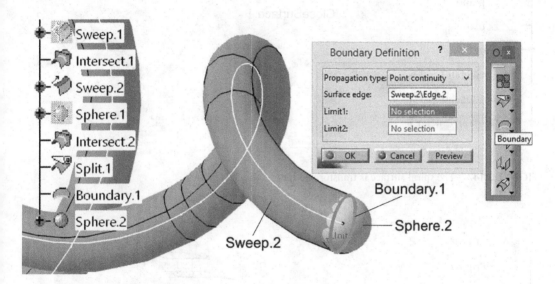

FIGURE 3.145 The intersection curve is extracted from the end of the *Sweep.2* surface.

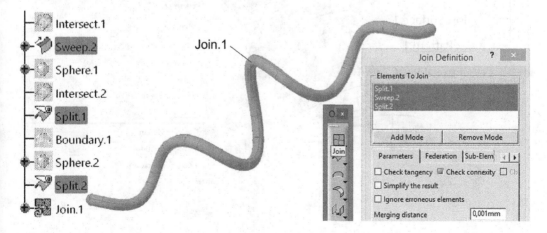

FIGURE 3.146 Joining surfaces to obtain the first spiral 3D element.

The user switches to the *CATIA Part Design* workbench and accesses the *Close Surface* tool in the *Surface-Based Features* toolbar. In the dialog box that opens in Figure 3.147, he selects the *Join.1* surface in the *Object to close* field, and the *CloseSurface.1* feature is added to the specification tree within the *PartBody*. The spiral is complete and has a volume of 9952.66 mm³.

The ornament presented at the beginning of this application consists of five such spirals, and to obtain them, the *Circular Pattern* tool from the *Transformation Features* toolbar is used. First, select the *CloseSurface.1* spiral and then press the circular multiplication icon.

In the *Circular Pattern Definition* dialog box in Figure 3.148, the user can see the solid of the spiral selected in the *Object* field, chooses by right-clicking the *Y Axis* in the *Reference element* field and then sets the multiplication parameters. Thus, the *Complete crown* option is selected in the *Parameters* list and the total number of entities (5) that must result from the multiplication is entered.

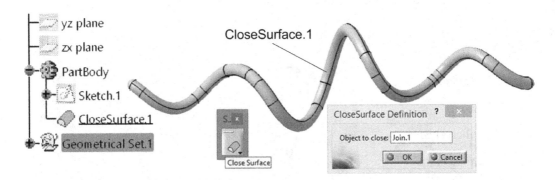

FIGURE 3.147   The spiral surface is transformed into a solid 3D feature.

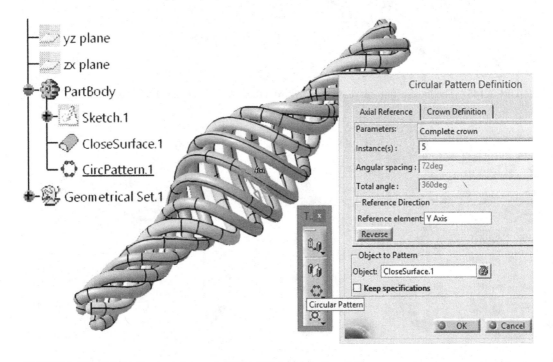

FIGURE 3.148   Multiplication of the spiral 3D solid to obtain the whole ornament.

The *Complete crown* option performs the multiplication and regular arrangement of the selected object in several instances, on the whole angle of 360°. The value of the angles between two successive instances is established by dividing the total angle by the number of instances needed. Thus, the *Angular spacing* and *Total angle* fields are not editable for this option in the *Parameters* list.

The total volume of the ornament formed by the five spirals is 49763.28 mm³, and made of steel, it will have a mass of approximately 0.4 kg.

Modelling solution: https://youtu.be/IexKFnGl1ZI.

3.8 MODELLING OF A FORK-TYPE PART

The modelling process of a fork-type part (2D drawing in Figure 3.149) is presented through the following tutorial. It uses certain tools from the *GSD* and *Part Design* workbenches. Due to the shape of the part, the modelling cannot be done only in *Part Design*; thus, the user needs to apply tools from *GSD*, such as *Intersection*, *Extrude*, *Sweep* and *Join*.

In a sketch created in the *XY Plane* (the *Sketcher* workbench), the profile in Figure 3.150 is drawn and constrained. The positioning towards the *H* and *V* axes is to be noted, as follows: between the short end of the profile and the *H* axis, the dimension of 64.5 mm is set, and between the long end and the *V* axis, it is 18.25 mm (14.25 + 4).

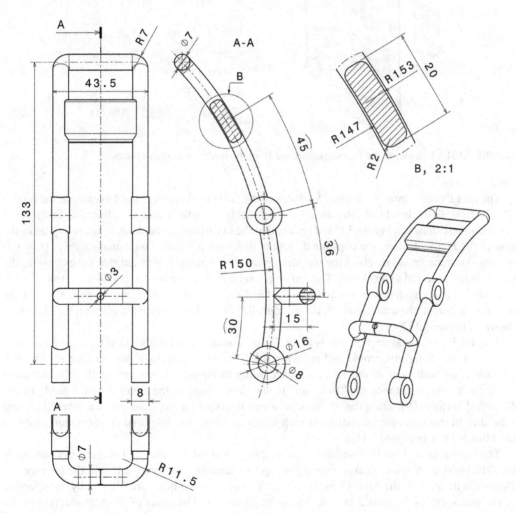

FIGURE 3.149   The 2D drawing of the part to be modelled – fork-type part.

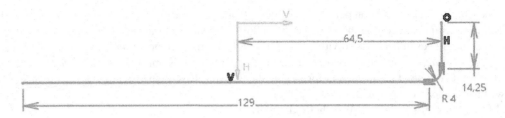

FIGURE 3.150   The first sketch of the part – *Sketch.1*.

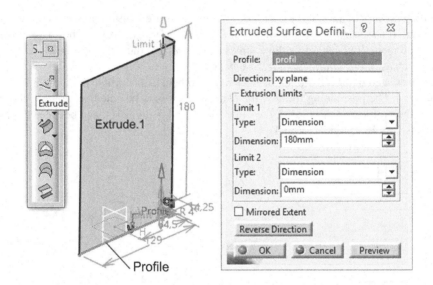

FIGURE 3.151   Launching the *Extrude* command from the *Insert → Surfaces* menu.

The tool *Extrude* from the toolbar *Surfaces* of the *GSD* workbench is used to extrude the profile (*Profile* field) to the height of 180 mm (*Dimension* field) along the plane *XY* (*Direction* field). From the dialog box shown in Figure 3.151, it results that the extrusion can be done, if necessary, as in the case of the *Pad* tool, along the opposite direction, under the *XY Plane*, with another value (*Limit 2*) or with the same value, by checking the *Mirrored extent* option. It is noted that for extrusion, the profile is not required to be closed. The *Extrude.1* surface is added to the specification tree.

A new sketch, *Sketch.2*, is created next in the *YZ Plane*. With the centre at the origin of the coordinate system, the user draws an arc of radius R150 mm, having the endpoints on the *H* axis, as shown in Figure 3.152.

The *YZ Plane* containing the arc is at 18.25 mm distance from the surface *Extrude.1* (see also Figure 3.150). Using this profile and applying the *Extrude* tool again, a new surface is obtained, *Extrude.2*, on both sides of the *YZ Plane*. According to Figure 3.153, it results that the extrusion takes place over a distance of 40 mm (fill in the dimensions in the *Limit 1* and *2* fields in the *Extruded Surface Definition* dialog box), so it can intersect the previous surface, *Extrude.1*, and exceed it. In this case, for the extrusion with the same value, the *Mirrored extent* option could be used (unchecked in Figure 3.153).

The intersection of the two surfaces is created using the *Intersection* tool in the *Wireframe* toolbar. The obtained result is a curve named *Intersect.1*. According to Figure 3.154, in the *Intersection Definition* dialog box, the *First Element* and the *Second Element* fields are filled in by the selection of the two surfaces *Extrude.2* and, respectively, *Extrude.1*. The order of surface selection in the respective fields is not important, the result being the same.

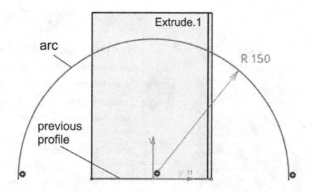

FIGURE 3.152 Drawing the arc of a radius of R150 mm.

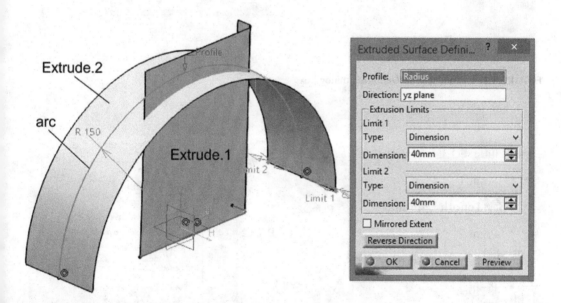

FIGURE 3.153 Extruding the arc of a radius R150 mm.

The sketches *Sketch.1* and *Sketch.2* on which the *Extrude.1* and *Extrude.2* were based can be hidden by right-clicking them in the specification tree and selecting the *Hide/Show* option.

On the curve *Intersect.1*, three points are added, *Point.1*, *Point.2* and *Point.3*, using the *Point* tool (*Wireframe* toolbar) at specified distances from the left endpoint of the curve, as shown in Figure 3.155. When defining the points, the *On curve* option was used (option selected in the *Point type* drop-down list in the *Point Definition* dialog box), *Distance on curve* option was checked, and the values of 24 mm (*Point.1*), 60 mm (*Point.2*) and 105 mm (*Point.3*) were specified, respectively.

These dimensions are measured on the large arc of the curve *Intersect.1*, according to the 2D drawing (Figure 3.149), and are not linear dimensions between the endpoints of arc segments.

It is noted that in the reference point, a red arrow is shown, with the tip pointing in a direction tangent to the arc. By pressing the arrow (or the *Reverse Direction* button, not shown in the figure), the program considers the other endpoint of the curve as a reference point.

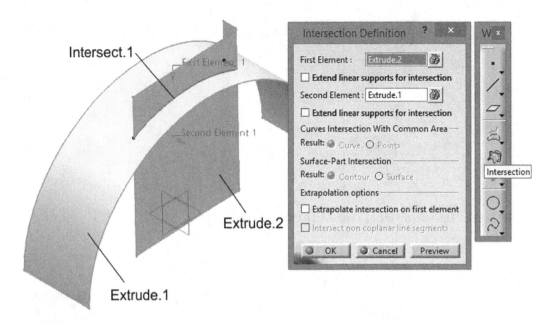

FIGURE 3.154 *Intersection Definition* dialog box.

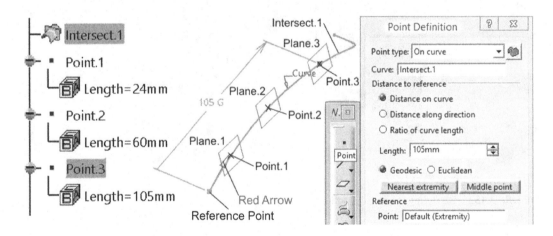

FIGURE 3.155 Creating three points using the *On curve* option of the *Point* tool.

Once the first point has been inserted, if the user considers it as a reference point in creating the second one, the red arrow is positioned on the first point and, while clicking it, causes the direction to change.

For example, considering the *Point.1* as a reference and clicking the arrow, *Point.2* is no longer on its right, but on the left, still keeping the specified distance.

Figure 3.155 also shows three planes, *Plane.1*, *Plane.2* and *Plane.3*, inserted using the *Plane* tool in the same *Wireframe* toolbar. The planes will be positioned in the locations of the three points of the curve *Intersect.1*. In Figure 3.156, the user can learn how to create the plane *Plane.3*, using the *Parallel through point* option in the *Plane type* list. The plane is then obtained parallel to the *YZ Plane* going through the *Point.3*.

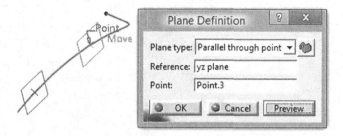

FIGURE 3.156   Launching the *Plane* tool with *Parallel through point* option.

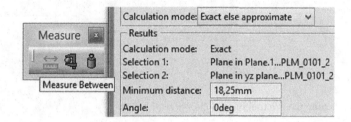

FIGURE 3.157   Launching the *Measure Between* tool.

As a check, there should be 18.25 mm between any of the three planes and the *YZ Plane*, measured using the *Measure Between* ruler in the *Measure* toolbar (Figure 3.157). In the information box with the same name, the distance and the angle of 0° appear in the *Minimum distance* field (the planes are parallel) and the two selections are displayed above.

In the *YZ Plane*, with the centre in the right endpoint of the curve *Intersect.1*, the user draws a circle with a diameter of Ø7 mm (Figure 3.158), contained in the *Sketch.3*.

The next step is to switch to *Part Design* workbench and, using the *Rib* tool, located in the *Sketch → Based Features* toolbar, extrude the circle along the curve to create the solid body in Figure 3.159. In the *Rib Definition* dialog box, in *Profile* field the user selects the *Sketch.3* and in the field *Center curve* selects the *Intersect.1*. The *Keep Angle* option is also chosen from the *Profile control* drop-down list. An arm was obtained, practically, as half of the main body of the fork.

The *Rib* feature is possible due to the sufficiently large fillet radius of R4 mm (according to the 2D drawing; Figure 3.150), transferred to the curve *Intersect.1*. For certain values of this radius (example: R3.5; R3; R2.5 mm or less), the formed solid will self-intersect. Even if the program does not generate an error message, it is impossible to bend the fork arm during the manufacturing step and this can lead to cracking or breaking.

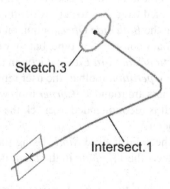

FIGURE 3.158   Creating the *Sketch.3* profile.

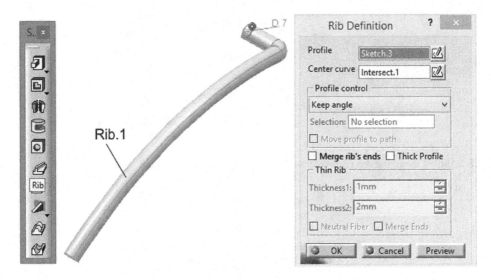

FIGURE 3.159   Launching the *Rib* tool in *Part Design* workbench.

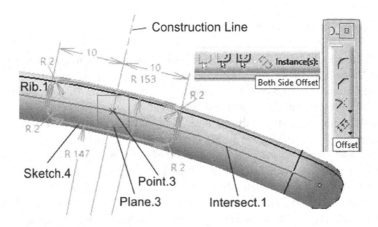

FIGURE 3.160   *Sketch.4* definition dialog box.

In the *Plane.3*, a new *Sketch.4* is created and the profile of the connecting body between the two arms of the fork will be drawn. Thus, with the centres in the origin of the coordinate system, two circles of radii R153 mm and R147 mm are drawn, respectively (Figure 3.160). An auxiliary construction line is created starting from the origin that also passes through *Point.3* and exceeds the circle of radius R153 mm. The line is then copied from side to side at 10 mm distance using the *Offset* tool in the *Operation* toolbar and activating the *Both Side Offset* option on the *Sketch Tools* toolbar.

The two copied lines are of auxiliary construction type, but they can later be converted into standard lines by clicking the *Construction/Standard Element* icon in the same *Sketch Tools* toolbar.

Using the *Quick Trim* tool in the *Operation* toolbar, the user removes unnecessary arcs and line segments outside the profile. Its corners are rounded (*Corner* tool) with a radius of R2 mm, resulting in the final profile in Figure 3.160. It is recommended to check the validity of this sketch using the *Sketch Analysis* tool in the *Tools* menu.

The *Sketch.4* is created inside the solid feature *Rib.1*. Using the *Pad* tool, the user extrudes it with the *Up to plane* option and selects the *YZ Plane* in the *Limit* field of the *Pad Definition* dialog box (Figure 3.161).

The fork is symmetrical with respect to the *YZ Plane*, so that, using the *Mirror* tool in the *Transformation Features* toolbar, the other half of the solid is obtained, according to Figure 3.162.

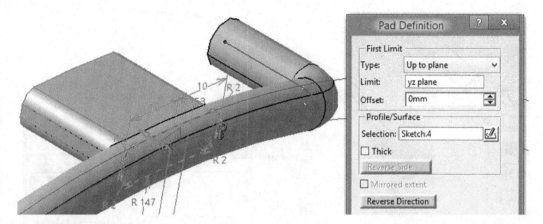

FIGURE 3.161 Applying the *Pad* tool to the *Sketch.4*.

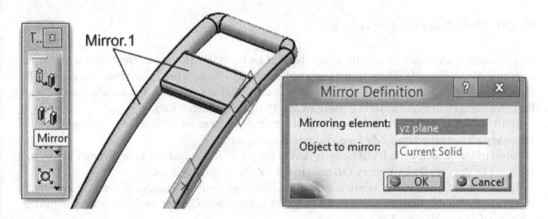

FIGURE 3.162 Applying the *Mirror* tool on the entire solid.

The fork has at the ends of the long arms two identical fastening elements, one for each arm. Thus, in the plane *Plane.1*, a new sketch is initiated that will contain a circle with a diameter of Ø16 mm. The centre of the circle is 6 mm from the left end of the arm *Rib.1*, but also on a circle (of auxiliary construction lines) of radius R150 mm, with the centre at the origin of the coordinate system (Figure 3.163).

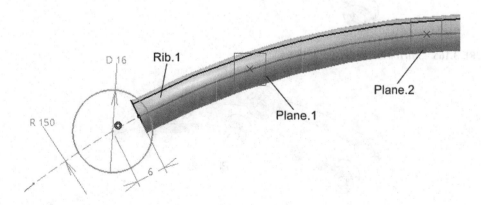

FIGURE 3.163 *Sketch.5* definition dialog box.

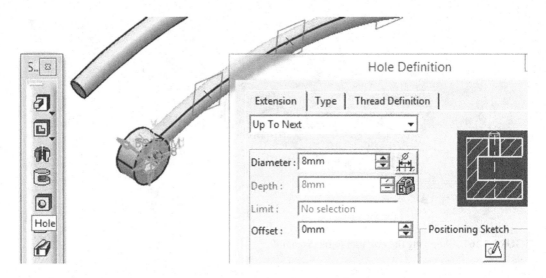

FIGURE 3.164   *Hole.1* definition dialog box.

The circle is extruded over 8 mm using the *Pad* tool and the *Mirrored extent* option (4 mm on each side of the *Plane.1*). Through the obtained cylindrical body, a perforated hole feature (*Up To Next* option), *Hole.1*, with a diameter of Ø8 mm is created (Figure 3.164). The hole is coaxial with the cylindrical element, constraint defined by the *Positioning Sketch* icon.

Both entities (cylinder and hole) that form a clamping element are copied (*Mirror*) having as the mirroring element the same *YZ Plane*, according to Figure 3.165.

Similarly, a second cylindrical body is drawn (in the *Plane.2*) and the hole through its centre, followed by the copy by symmetry (*Mirror*) to the *YZ Plane* (Figure 3.166).

In the *Plane.1*, an auxiliary construction line is drawn, tangent to the curve *Intersect.1* in *Point.1*. Perpendicular to this tangent and with an endpoint in the *Point.1*, a standard line is drawn, having a length of 15 mm (Figure 3.167).

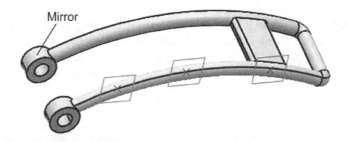

FIGURE 3.165   *Mirror* definition dialog box.

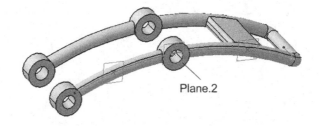

FIGURE 3.166   *Mirror* definition dialog box for the second cylinder and hole.

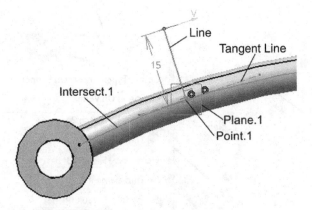

FIGURE 3.167 Creating a line perpendicular to a tangent line.

At the top of the line, a new plane is created, *Plane.4*, perpendicular to it using the *Normal to curve* option in the *Plane Definition* dialog box, according to Figure 3.168.

In a new sketch created in the *Plane.4*, a line is drawn so that one endpoint coincides with the upper end of the line previously drawn and in the *Plane.1*, and the other end coincides with the *YZ Plane* (Figure 3.169). The *YZ Plane* is below the planes *Plane.1* to *Plane.4* due to the initial positioning of the part on a radius of R150mm (Figure 3.152). For checking, the length of this line must be approx. 17.95mm.

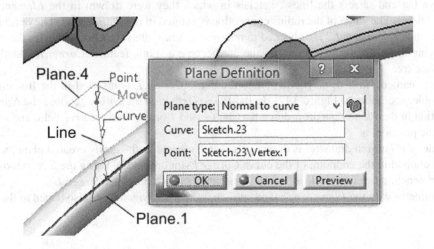

FIGURE 3.168 *Plane.4* definition dialog box.

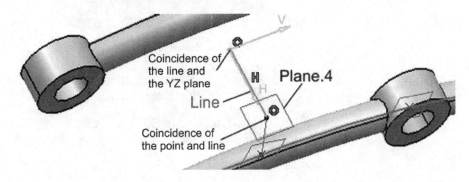

FIGURE 3.169 The line drawn in *Plane.4*.

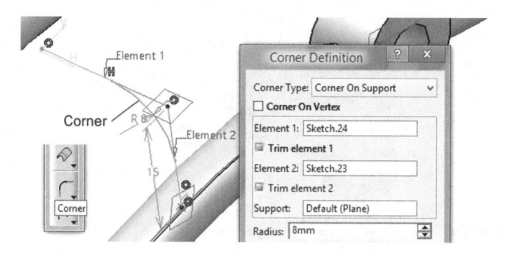

FIGURE 3.170   *Corner Definition* dialog box.

The two lines are perpendicular to each other. It is necessary to create a R8 mm (11.5–3.5 mm) radius connection. Thus, the user uses the *Corner* tool in the *Wireframe* toolbar; in the *Corner Definition* dialog box (Figure 3.170), he chooses the *Corner On Support* option in the *Corner Type* drop-down list and selects the lines (sketches in which they were drawn) in the *Element 1* and *Element 2* fields. The value of the radius connection is entered in the *Radius* field. The options *Trim element 1* and *Trim element 2* are checked to remove the lines intersecting corner after the fillet.

Following the fillet, the two lines and the arc become a single feature, *Corner.1*, added to the specification tree.

A plane, named *Plane.5*, perpendicular to the curve *Corner.1*, is inserted in the free endpoint of the profile, according to Figure 3.171. In the *Plane Definition* dialog box, choose the *Normal to curve* option in the *Plane type* drop-down list, the *Corner.1* profile in the *Curve* field, and the end-point as the positioning.

A circle of Ø7 mm in diameter is drawn in a sketch created in the newly created plane, with the centre positioned in the endpoint of the curve *Corner.1* (Figure 3.172). Using the *Sweep* tool of the *GSD* workbench, move this circle along the curve and obtain the surface *Sweep.1*.

By symmetry with the *Plane.5*, a second surface is created, *Symmetry.1*, also added to the specification tree.

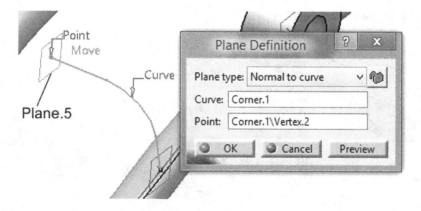

FIGURE 3.171   *Plane.5* perpendicular to *Corner.1*.

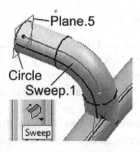

FIGURE 3.172 Launching the *Sweep* tool.

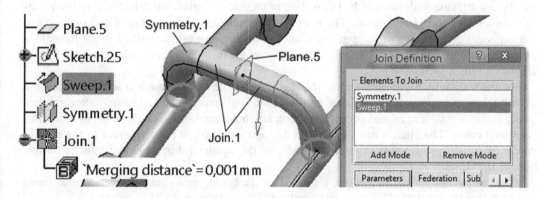

FIGURE 3.173 Applying the *Join* tool to the *Sweep.1* and *Symmetry.1*.

The two surfaces, *Sweep.1* and *Symmetry.1*, are joined into one, using the *Join* tool, according to Figure 3.173. Filling the surface and obtaining a solid is then done in the *Part Design* workbench using the *Close Surface* tool on the *Surface-Based Features* toolbar.

On top of the newly created solid, a perforated hole (Figure 3.174) with a diameter of Ø3 mm should be obtained. Thus, the *Hole* tool is activated and the upper cylindrical surface of the *CloseSurface.1* feature is selected. To position the hole accurately, click the *Positioning Sketch* icon, the *Hole Definition* dialog box disappears, and the centre of the hole (represented by an asterisk) is then constrained to coincide with both the plane *Plane.5* and the end of the *Corner.1* in the same plane.

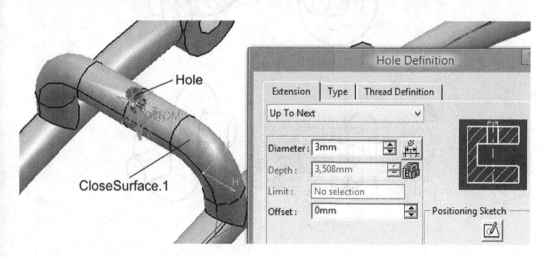

FIGURE 3.174 Creating a hole on the *CloseSurface.1*.

3.9 MODELLING OF A 3D KNOT

The model presented in this tutorial does not necessarily represent the shape of a part, but rather a wireframe modelling problem of a complex 3D geometry and power of surface design capabilities in CATIA. It is observed as the knotted shape of the tube with a diameter of Ø5 mm, and it does not touch itself when it passes through the created features. The user can easily set other dimensions for the geometric elements of the knot.

The drawing of the knot is presented in Figure 3.175 by three orthogonal projections and one isometric view. In the main view, there is an equilateral triangle with sides of 10.1 mm, which represents the directions of the tube's orientation in different planes.

The user accesses the *GSD* workbench and, then in the *Sketch.1* of the *YZ Plane*, draws an equilateral triangle with sides of 10.1 mm. Two of its sides are placed symmetrical to the *V* axis of the sketch's coordinate system. The apex of the triangle is at the origin of the system, and the two sides connected by this apex are spaced at an angle of 60°. Inscribed in the triangle, a circle of auxiliary construction is created tangent to all three sides and centred on the *V* axis (Figure 3.176).

For the rectangle, the *Profile* tool in the toolbar with the same name was applied because it allows the user to draw multiple lines and/or circle segments connected to one after another. These segments, as seen in the specification tree, are, however, individual elements of type: line, point and circle. The circle's role is to mark, through its centre, a point (*Point.4* in the specification tree) that represents the centre of gravity of the equilateral triangle and must be visible outside the sketch.

Thus, an auxiliary construction line was used for the circle, but its centre is transformed into a standard element using the *Construction/Standard Element* icon on the *Sketch Tools* bar.

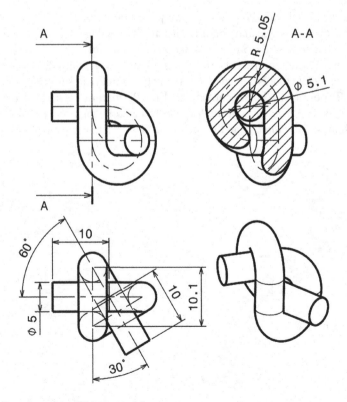

FIGURE 3.175 Drawing of the 3D knot model.

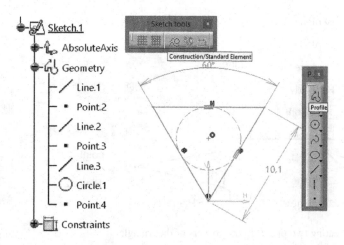

FIGURE 3.176   Equilateral triangle and a circle tangent to its sides to define the centre of gravity.

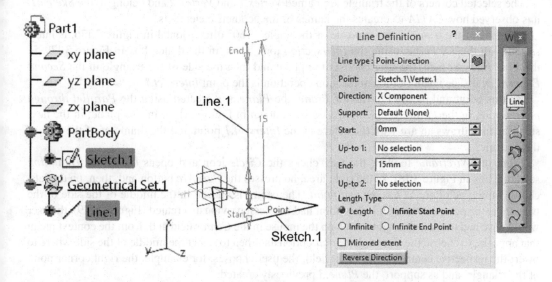

FIGURE 3.177   Drawing a line perpendicular to the triangle's plane from the centre of gravity.

Perpendicular to the *YZ Plane* in which the triangle was drawn, the user draws a line (the *Line* tool in the *Wireframe* toolbar) pointed in the positive direction of the *X* axis. In the *Line Definition* dialog box (Figure 3.177), from the *Line type* list, he chooses the *Point-Direction* option; in the *Point* field, it is the triangle's centre of gravity (defined in the *Sketch.1*). As the direction of the line, the *X Component* is selected by right-clicking in the *Direction* field. The line is not (yet) in a certain plane, so the *Support* field does not need to be completed and it is considered as default by the program. The length of the line is chosen to be 15 mm after entering the value in the *End* field.

With the triangle and the line in the workspace, the user can insert two planes that go through one corner of the triangle and through the *Line.1*, according to Figure 3.178. He accesses the *Plane* icon on the *Wireframe* toolbar and, from the *Plane Definition* dialog box, chooses the *Through point and line* option from the *Plane type* drop-down list and selects a corner of the triangle and then the line in the *Point* and *Line* fields, respectively. Thus, the *Plane.1* and *Plane.2* are inserted in the corners of the triangle and in the specification tree. They have a certain orientation determined by the positions of the elements that define them (points and line).

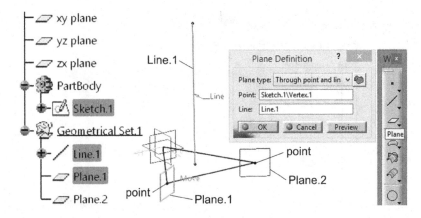

FIGURE 3.178   Creating two planes in the corners of the triangle.

The selected corners of the triangle are named *Vertex.1* and *Vertex.2* and belong to the *Sketch.1*; it is observed how *CATIA v5* creates the names of the geometric elements.

One side of the triangle, the opposite of the angle of 60° dimensioned in Figure 3.176, is intersected with the *ZX Plane*, using the *Intersection* tool. Thus, in the dialog box in Figure 3.179, in the *First Element* field, the user chooses the plane and then the side of the triangle in the *Second Element* field. The result of the intersection operation is the point *Intersect.1*.

By this point and parallel to the *XY Plane*, the *Plane.3* is created using the *Parallel through point* option in the *Plane Definition* dialog box shown in Figure 3.180. In this plane, in the next step, the user draws an arc with the centre at the *Intersect.1* point and the diameter of the side of the triangle.

From the *Wireframe* toolbar, the user clicks the *Circle* icon and opens the *Circle Definition* selection box in Figure 3.181. In the right area, he presses the *Part Arc* button and, then, on the left, chooses the type of arc as *Center and point*. The centre is chosen in the middle of the side of the triangle at the point *Intersect.1*. If this point had not been previously created (Figure 3.179), the user would have had to press the right button on the mouse in the *Center* field and, from the context menu that appears, to choose the *Create Midpoint* option and then to select the middle of the side where to insert the respective centre. In the *Point* field, the user chooses, for example, the right corner point of the triangle, and as support, the *Plane.3* previously created.

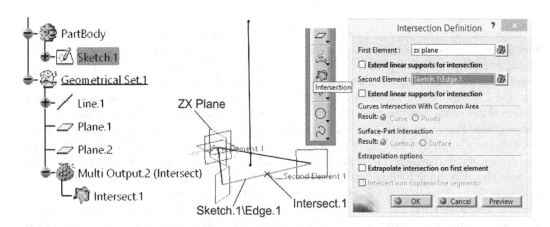

FIGURE 3.179   Intersection between the *ZX Plane* and a side of the triangle.

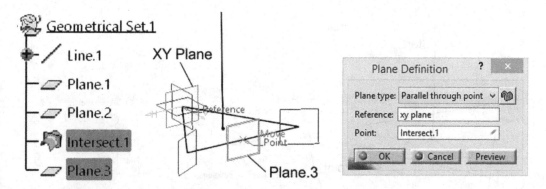

FIGURE 3.180   Plane inserted parallel to the *XY Plane* through the intersection point.

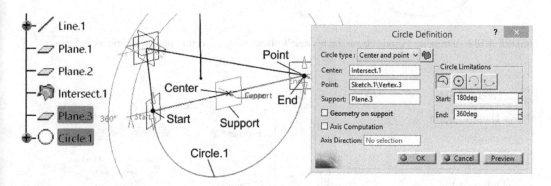

FIGURE 3.181   Drawing a circle under the triangle's plane using a centre and a corner point.

By default, *CATIA* draws an arc above or below the side of the triangle. In order to get the correct representation as it is in Figure 3.181, the *Start* and *End* fields receive the values 180° and 360°, respectively. Regardless of the position of the start and end points of the arc, the difference between the two angular values must be 180°.

At the right end of the arc, another one is drawn, named *Circle.2*, according to the selections in Figure 3.182: the centre of the arc is the lower end of the vertical line, as the *Plane.2* is selected as the support. The angle developed for this arc is set to 90°.

From the point where the arc *Circle.2* touches the vertical *Line.1* (Figure 3.177), a horizontal line is drawn, named *Line.2*, tangent to the arc. In the *Line Definition* dialog box (Figure 3.183), from the *Line type* list, the user chooses the *Tangent to curve* option and selects the arc and then the tangent point in the *Element.2* field. The support for this 10 mm line (value entered in the *End* field) may remain blank, but it is the *Plane.2* (Figure 3.178).

From the left endpoint of the first arc of a circle (Figure 3.181), a vertical line is drawn and its length must be equal to the length of the segment between the point (centre of gravity) at the base of the *Line.1* and the tangent point named *Element.2*.

In the *Line Definition* dialog box, the user chooses the *Point-Direction* option, the point marked by *Start* and the direction along the *X* axis, and in the *End* field, he opens by right-clicking the context menu and chooses the *Measure Between* option, according to Figure 3.184. A selection box with the same name opens and allows the user to select two points (*Measure point 1* and *Measure point 2*). The distance value between them is automatically inserted in the *End* field and determines the length of the line (5.831 mm).

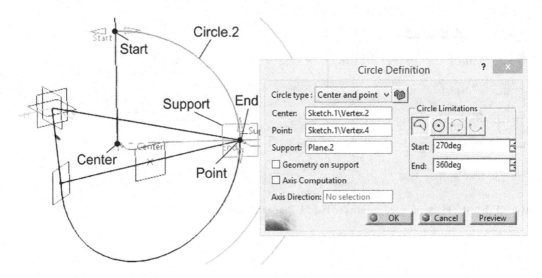

FIGURE 3.182   Drawing a circle above the triangle's plane using the centre of gravity and a corner point.

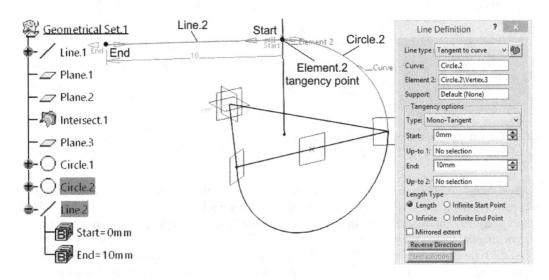

FIGURE 3.183   Drawing a line tangent to the arc *Circle.2*.

Line.3 is added to the specification tree. The top endpoint is projected onto *Line.2* to define the centre of a new arc using the *Projection* tool. In the *Projection Definition* selection box, the user chooses the *Normal* type for projection, the end of the *Line.3* is selected in the *Projected* field, and the *Support* field contains the *Line.2* (Figure 3.185).

The *Nearest solution* option does not influence the projection result in this case, that is *Project.1*, but it is useful if the user should have projected a point on the nearest support.

At this point and perpendicular to *Line.2*, the user inserts a new plane, named *Plane.4*, and then, with the centre at the same point *Project.1*, he draws an arc, *Circle.3*, as shown in Figure 3.186. The arc is placed in the *Plane.4*, and the point selected in the *Point* field is the end of the vertical *Line.3*.

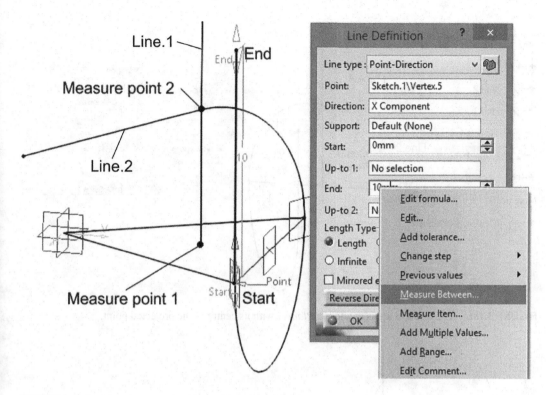

FIGURE 3.184 Drawing a vertical line with its length measured between two points.

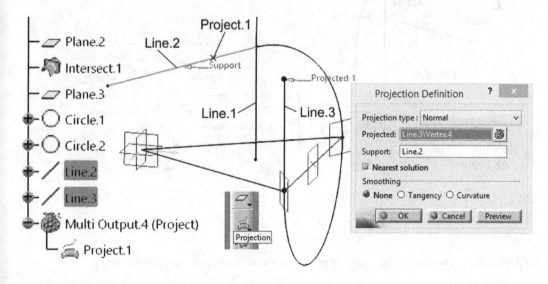

FIGURE 3.185 Projection of a point onto a line.

Similarly, the user draws the last arc, named *Circle.4*, having the centre in the point *Element.2* (as defined in Figure 3.183), the starting point is in the left endpoint of the arc *Circle.3*, and the endpoint is in the endpoint of the vertical *Line.1*'s bottom. The *ZX Plane* is chosen as the support of this arc, according to Figure 3.187.

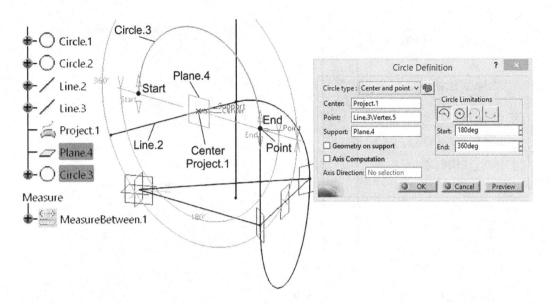

FIGURE 3.186   The *Circle.3* arc is created in *Plane.4* with its centre in the projected point.

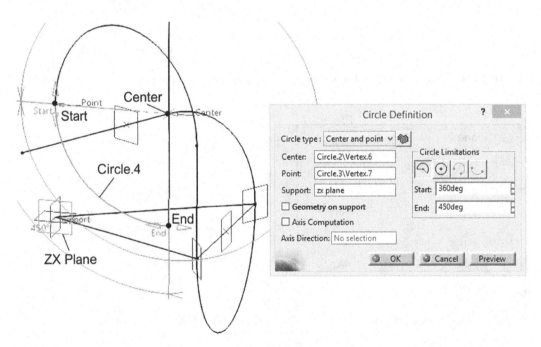

FIGURE 3.187   The *Circle.4* arc is created in *ZX Plane* with its centre in the point *Element.2*.

The last element of the twisted curve of the knot is *Line.4*, drawn from the lower endpoint of the arc *Circle.4*. This point coincides with the centre of gravity of the equilateral triangle defined in Figure 3.176. The line is 10 mm long and tangent to *Circle.4* at this point (noted on drawing as *Element.2*). Figure 3.188 shows how the line was drawn and its options in the *Line Definition* dialog box.

The user has to check the correctness of drawing the entire 3D curve, consisting of connected line and arc segments. If all this has been carefully drawn, *Line.4* passes through the point *Intersect.1* (Figure 3.188). The check is fast using the *Measure Between* tool, and the distance between the two elements should be 0.

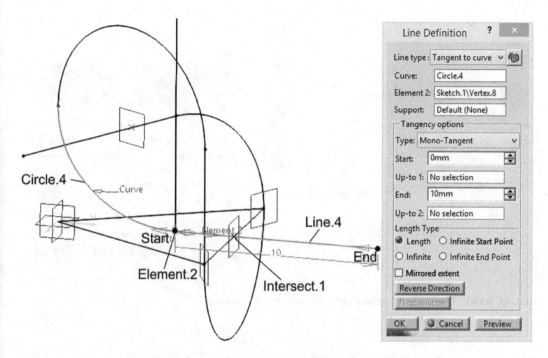

FIGURE 3.188   *Line.4* being drawn tangent to the arc *Circle.4* in the *YZ Plane*.

The specification tree contains many features that helped the user to create the knot 3D curve, but there is no need for them to be represented in this stage. Thus, these elements/features are selected by holding down the *Ctrl* key and the *Hide/Show* option is chosen from the context menu (Figure 3.189). Thus, only the segments that form the curve should remain visible: *Circle.1*, *Circle.2*, *Line.2*, *Line.3*, *Circle.3*, *Circle.4* and *Line.4*.

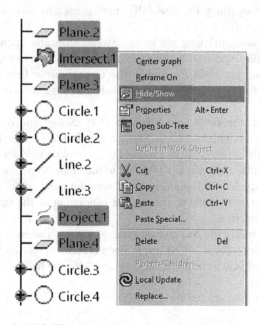

FIGURE 3.189   Accessing the *Hide/Show* option using the context menu.

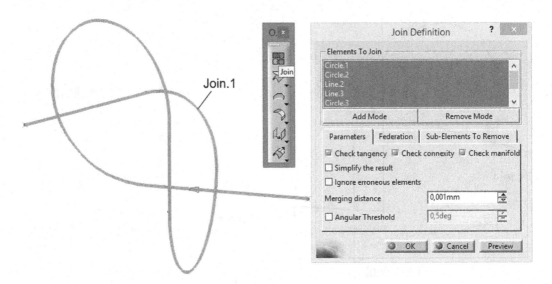

FIGURE 3.190 Line and arc segments joined together in a 3D curve.

The line and arc segments must be grouped into a single entity. From the *Operation* toolbar, the user clicks the *Join* icon and the *Join Definition* selection box opens. In the *Elements To Join* field, he selects all the remaining visible segments, according to Figure 3.190. The user can also try the *Check* options to perform the second control of the knot 3D curve. If this 3D curve turns green after pressing the *Preview* button, it means that all the segments are connected successively, continuously and tangent to each other, so the curve is correct.

The connections and tangents of the curve are very important because on its basis a surface and, subsequently, the solid of the knot are created.

Continuing the hybrid modelling, the *Join.1* 3D curve is used to move a circle along it, and the expected result should be a surface.

From the *Surfaces* toolbar, the user clicks the *Sweep* icon and opens the selection box in Figure 3.191. In the *Profile type* list, he chooses the third icon, *Circle*, then uses the *Center and radius* option, chooses the *Join.1* curve in the *Center curve* field and inserts the radius R2.5 mm for the knot surface. For verification, the surface of the *Sweep.1* must have an area of 1191.93 mm^2.

The user switches from the *GSD* workbench to *Part Design* and opens the *Close Surface Definition* selection box by clicking the *Close Surface* icon in the *Surface-Based Features* toolbar (Figure 3.192). He selects the *Sweep.1* surface in the *Object to close* field and observes how the solid *CloseSurface.1* model results, after which it is added to the model's specification tree.

The program temporarily adds two planes, at the beginning and end of the surface, marked on Figure 3.192. These planes only serve to delimit the generation of the surface and are not part/appear of/in the specification tree.

The solid fills the gap inside the *Sweep.1* surface, as seen at the ends of the knot's tube. The measured volume of the knot solid is 1489.92 mm^3. For a correct representation of the 3D model, all the other elements obtained in wireframe modelling and with the help of surfaces can be hidden, including the default planes.

Modelling solution: https://youtu.be/VrxsD0cZI5A.

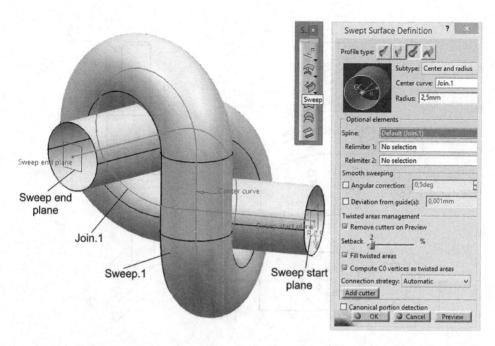

FIGURE 3.191 Surface *Sweep.1* created along the 3D curve *Join.1*.

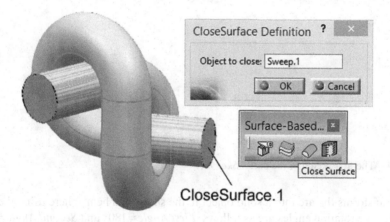

FIGURE 3.192 Closing the surface *Sweep.1* to create the solid model of the knot.

3.10 MODELLING OF AN AXLE SUPPORT

The hybrid 3D modelling process of a shaft-support part (2D drawing in Figure 3.193) is presented through the following step-by-step tutorial. It uses certain tools mainly from the *Part Design* and *GSD* workbenches.

In a sketch in the *XY Plane*, a circle is drawn with the centre at the origin of the coordinate system, and coinciding with the *H* axis, an axis line is inserted (*Axis* tool). The circle is trimmed with this axis so that the lower semicircle is removed (Figure 3.194). The result is an arc with the endpoints on the axis line.

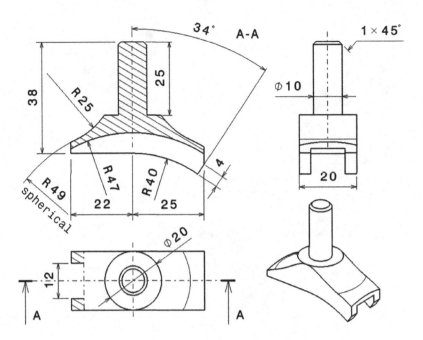

FIGURE 3.193 The 2D drawing of the part to be modelled – shaft-support part.

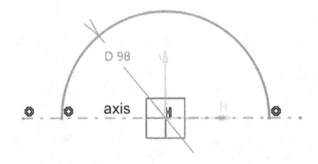

FIGURE 3.194 The first sketch of the part – *Sketch.1*.

The user transforms the arc into a solid object in the shape of a hemisphere using the *Shaft* tool (Figure 3.195). The rotation angles are as follows: *First Angle*=180° and *Second Angle*=0°.

A rectangle of 20 mm wide, symmetrical to the axes of the coordinate system, is drawn on the flat surface of the hemisphere. As can be seen in Figure 3.196, the corners of the rectangle are outside the solid of the hemisphere. This is followed by cutting (with the *Pocket* tool) using the outside of the rectangle (*Reverse Side* option – the horizontal orange arrow is pointing outwards). The user draws two more rectangles on one of the resulting side flat surfaces so that they are positioned at 22 and 25 mm, respectively, from the vertical axis (Figure 3.197). In Figures 3.196 and 3.197, the rectangles are represented by a dashed line. Using a new *Pocket* cut, he removes two more volumes from the hemisphere.

A profile consisting of two lines and a circle with a diameter of Ø80 mm and the centre at the origin of the coordinate system is drawn on the same side surface. As seen in Figure 3.198, the horizontal line is tangent to the circle, and their point of tangency coincides with the vertical axis of the *HV* coordinate system.

The second line was drawn to close the profile. The *Trim* tool was then used to edit it and to remove unnecessary segments.

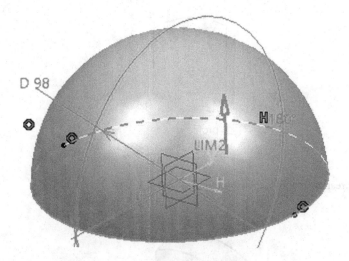

FIGURE 3.195   The first 3D feature of the part – *Shaft.1*.

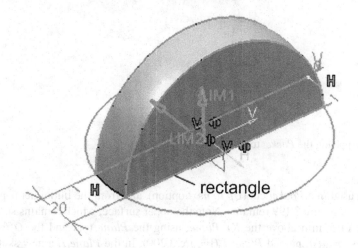

FIGURE 3.196   Applying the first *Pocket* cut.

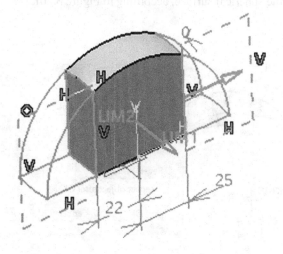

FIGURE 3.197   Applying the second *Pocket* cut.

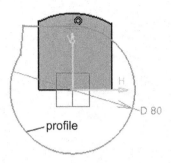

FIGURE 3.198   Creating a more complex profile.

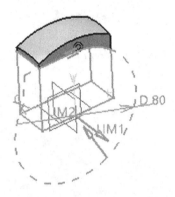

FIGURE 3.199   Applying the *Pocket* tool.

The profile is used in a *Pocket* cut (*Up to next* option), and from the initial hemisphere, only the solid represented in Figure 3.199 remains. Note the upper surface, which remains spherical.

At a distance of 53 mm above the *XY Plane*, using the *Plane* tool and the *Offset from plane* option, a plane is created, named *Plane.1* (Figure 3.200). In the *Plane.1*, a new sketch is created, and with the centre in the sketch's origin, a circle with a diameter of Ø20 mm is drawn, which is extruded *Up to next* to the spherical surface, according to Figure 3.201.

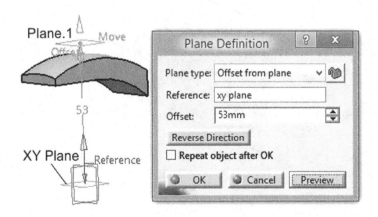

FIGURE 3.200   Creating the *Plane.1*.

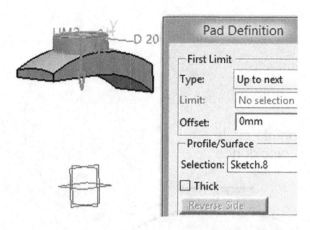

FIGURE 3.201   Applying the *Pad* tool.

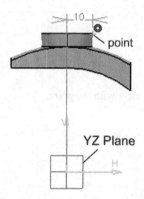

FIGURE 3.202   Creating the sketch and placing the point.

In the *YZ Plane*, a new sketch is created, which contains a point at a distance of 10 mm from the *V* axis of the coordinate system and also coincides with the upper flat surface of the previously obtained cylinder feature (Figure 3.202).

In a new sketch of the same *YZ Plane*, a circle is drawn with a diameter of Ø50 mm (Figure 3.203), which is constrained coincident with the previously created point.

From the origin to the centre of this circle, a line of length 74 mm is drawn (25 + 49 mm, the two radii that define the outer surface of the part, according to the drawing in Figure 3.193).

Figure 3.204 shows both the intersection of the circle and the line, as well as the imposed constraints.

The circle and the line intersect at a point located, according to the constraints, on the spherical surface. From this point, a line is drawn to the point created in the previous step in Figure 3.202.

The two lines are transformed into auxiliary construction elements using the *Construction/ Standard Element* tool in the *Sketch Tools* toolbar. The circle is cut (*Quick Trim* tool) so that in the sketch remains an arc (Figure 3.205) delimited by the two points specified above. Figure 3.206 shows a detail of the line between the points, the chord of the arc.

In the *GSD* workbench, using the *Revolve* tool in the *Surfaces* toolbar and the sketch with the arc, a revolution surface is created, *Revolute.1*, as follows: in the *Profile* field of the dialog box in Figure 3.207, select the profile, and in the *Revolution axis* field, the user right-clicks and chooses the *Z Axis* option.

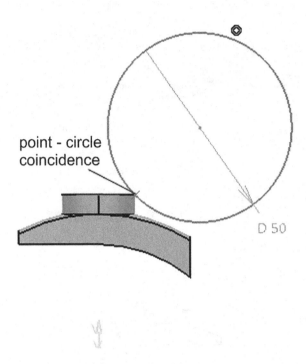

FIGURE 3.203 Creating a circle coincident with the point.

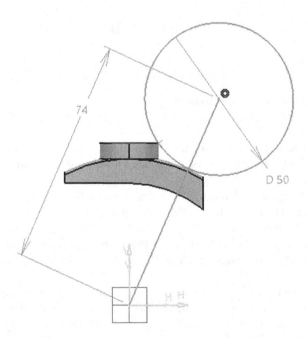

FIGURE 3.204 Drawing a line from the origin to the centre of the circle.

Back in the *Part Design* workbench, the surface is closed and transformed into a solid using the *Close Surface* tool, presented in Figure 3.208. Once the volume delimited by the surface has been obtained, the *Revolute.1* surface can be hidden from the specification tree (right-click and select *Hide/Show* option from the context menu; Figure 3.209).

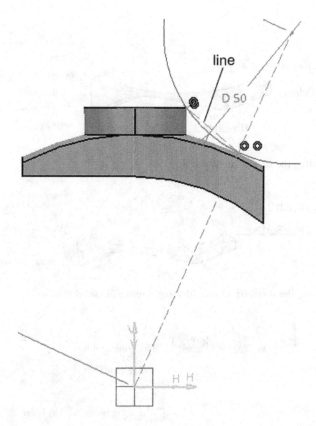

FIGURE 3.205 Creating a construction line.

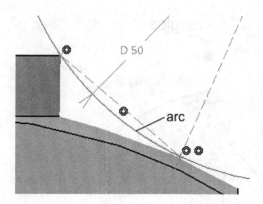

FIGURE 3.206 Details of the line drawn.

In the *Plane.1*, the user draws two rectangles symmetrical about its *Y axis* in a new sketch and applies the *Pocket* tool with the option *Up to next*. The distance between the two rectangles is 20 mm. This way, the volume of the part is cut laterally, according to Figure 3.210.

Also, in a new sketch in the *YZ Plane*, with the centre also at the origin, a circle with a diameter of Ø94 mm is drawn. With this circle, the solid is cut, on both sides of the plane, on a depth (*Depth* field) of 6 mm, using the *Pocket* tool and having the *Mirrored extent* option checked (Figure 3.211).

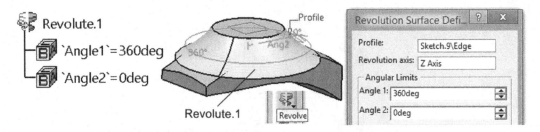

FIGURE 3.207   Creating the revolution surface in *GSD*.

FIGURE 3.208   Closing the surface and transforming it into a 3D solid feature.

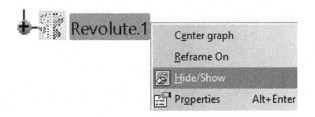

FIGURE 3.209   Hiding the surface using the context menu.

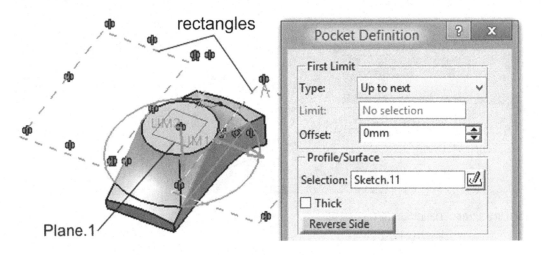

FIGURE 3.210   Cutting the sides of the part.

On the right side of the solid created, on the planar face, a sketch is drawn consisting of two lines and an arc, presented in Figure 3.212. The arc and the vertical line are obtained by projecting the edges of the solid onto the sketch (the *Project 3D Elements* tool in the *Operation* toolbar). The oblique line extends towards the origin of the coordinate system and makes an angle of 34° with respect to the *V axis*.

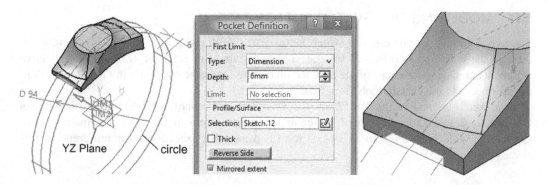

FIGURE 3.211 Applying the *Pocket* tool with *Mirrored extent* option checked.

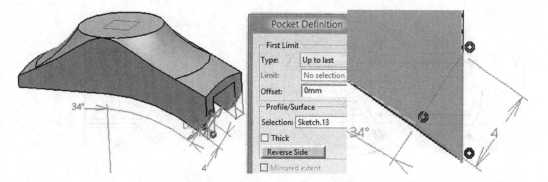

FIGURE 3.212 Creating the sketch for the last step in the modelling process.

The vertical line and the arc are constrained by the projection of the edges, but also by the 4 mm length of the oblique line. This profile is used to cut the corners of the part applying the *Pocket* tool and the *Up to last* option.

On the upper planar surface of the workpiece body, concentric with the circular edge, a circle with a diameter of Ø10 mm is created. The user extrudes this circle (Figure 3.213) to a height of 25 mm (*Pad* tool). Finally, a chamfer of 1×45° is created on the circular edge at the top of the cylinder.

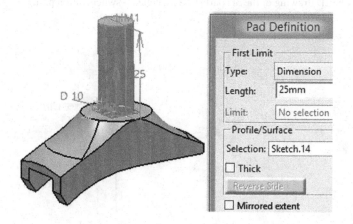

FIGURE 3.213 Creating the final *Pad* of the part.

3.11   MODELLING OF A SWITCH BUTTON PART

The hybrid modelling process of a switch button part (2D drawing in Figure 3.214) is presented through the following step-by-step tutorial. It uses certain tools from the *GSD* and *Part Design* workbenches. Due to the shape of the part, the modelling cannot be done only in *Part Design*; thus, the user needs to use certain tools from *GSD*.

In a sketch created in the *XY Plane* (the *Sketcher* workbench), a circle with a diameter of Ø60 mm is drawn and constrained; its centre is at the origin of the *HV* coordinate system. Coinciding with the *H* axis, an axis (*Axis* tool) is drawn that exceeds the circumference of the circle (Figure 3.215).

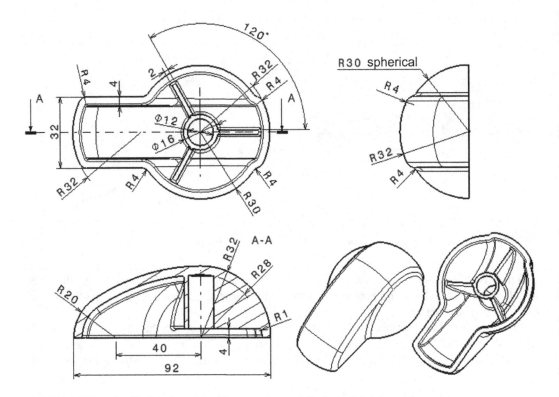

FIGURE 3.214   The 2D drawing of the 3D part to be modelled – switch button part.

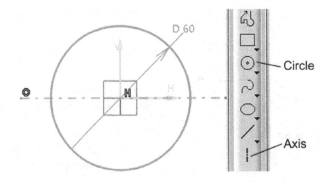

FIGURE 3.215   Creating the first sketch by drawing a circle of diameter Ø60.

Using the *Quick Trim* tool, the user trims the lower half of the circle with the axis line (Figure 3.216). The endpoints of the circle arc remaining on the axis and the coincidence constraints between these endpoints and the axis are observed. The profile of the arc (*Sketch.1*) is rotated around the axis using the *Shaft* tool. In the *Shaft Definition* dialog box (Figure 3.217), angular values can be entered in the *Angle* fields, but their sum should be 180°. To simplify the explanations, enter this value in the *First angle* field.

In the *YZ Plane*, a new sketch is created and a R20 mm radius circle is drawn to the left of the half of the sphere previously obtained using the *Shaft* command. The centre of the circle is positioned 40 mm to the left of the origin of the coordinate system. A second circle with a radius of R32 mm is located with the centre at the origin (Figure 3.218). Thus, the centres of the two circles are placed at a distance of 40 mm from each other.

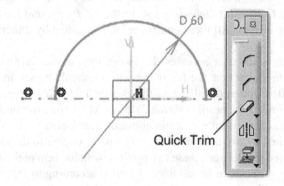

FIGURE 3.216 Trimming the circle with the axis line.

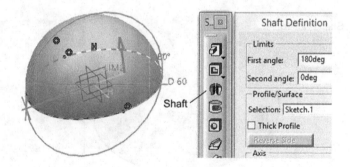

FIGURE 3.217 Applying the *Shaft* command.

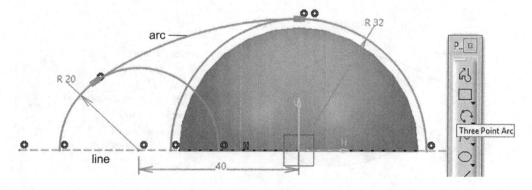

FIGURE 3.218 Creating the second sketch by drawing three arcs.

Using the *Three Point Arc* tool, another arc is created; its endpoints are on the circumference of the left and right circles, respectively. The third point necessary to define the arc can be chosen arbitrarily. The right endpoint of the arc is constrained to be coincident with the *V* axis. Also, two tangent constraints are added between the arc and the two circles of radii R20 and R32 mm, respectively.

Coinciding with the *H* axis, a construction line is drawn that exceeds the circumferences of the two circles. Using the *Quick Trim* tool, the arcs below the line are initially removed, according to Figure 3.218. The arcs that intersect above the line are then trimmed, leaving the profile that forms the *Sketch.2* as in Figure 3.219.

In a new sketch, *Sketch.3* – which is created in the *XY Plane*, an arc of radius R32 mm is drawn, so that it satisfies the three conditions established by constraints: the endpoint of the profile in the sketch *Sketch.2* coincides with the arc, the endpoints of the arc are symmetrical (check the *Symmetry* option in the *Constraint Definition* list) to the *YZ Plane*, and the distance between the two endpoints should be 32 mm. All these constraints are specified by dimensions and symbols in Figure 3.219.

In the *GSD* workbench, the user accesses the *Sweep* icon in the *Surfaces* toolbar, and in the *Swept Surface Definition* dialog box, the following settings should be set: in the *Profile* type area, he chooses the first button option (*Explicit*); in the *Profile* field, selects the arc (*Sketch.3*); and in the *Guide curves* field, selects the profile (*Sketch.2*) along which the extrusion is made. The surface obtained (Figure 3.220) is at a distance of 2 mm above the hemisphere.

The surface is delimited by two arcs: the first is the one belonging to the sketch *Sketch.3* and the second is called *Edge.1* (edge). Between these two profiles, with the help of the *Blend* tool in the same *Surfaces* toolbar, a new surface can be obtained, *Blend.1*, according to Figure 3.221. In the *Blend Definition* dialog box, the *First Curve* and *Second Curve* fields are filled with the two arcs.

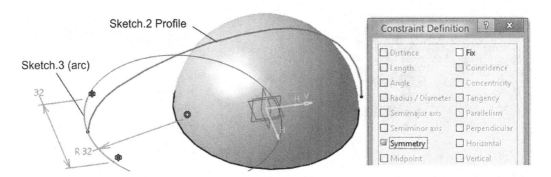

FIGURE 3.219   The final shape of the *Sketch.2*.

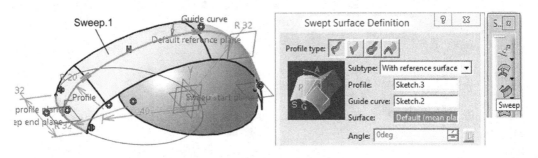

FIGURE 3.220   Creating a swept surface feature.

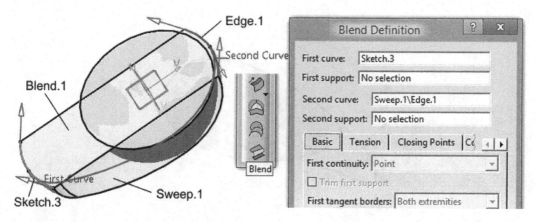

FIGURE 3.221   Creating the blended surface.

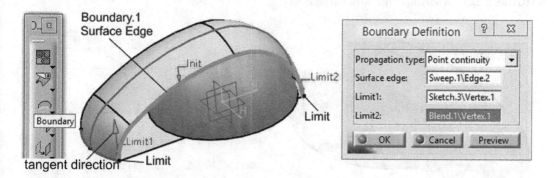

FIGURE 3.222   Extracting the boundary of the surface *Sweep.1*.

There are two surfaces: *Sweep.1* and *Blend.1*. To close the space between them, two side surfaces are used, also of *Blend* type. Such a surface requires two curves in order to be formed.

According to Figure 3.222, the *Boundary* tool in the *Operations* toolbar is used to extract a single side edge of the surface *Sweep.1*. By default, this surface is delimited laterally by three curves (arcs), as they resulted from the *Trim* editing of the circles in Figure 3.218.

In the *Boundary Definition* dialog box, in the *Surface Edge* field, the user should select one of the three arcs; only one selection of the three possible ones is allowed.

Due to the continuity settings (*Point continuity* or *Tangent continuity* in the *Propagation type* list), the other two arcs are automatically selected. In the *Limit* fields, the endpoints of the profile are chosen, as shown in the figure. By clicking the red arrow, the extracted edge can have two variants: the one in Figure 3.222 or the one in Figure 3.223. The user should choose the one in Figure 3.222.

Basically, from a profile consisting of three arcs, only one was obtained, called *Boundary.1*. Using this extracted edge and the corresponding side edge (*Edge.3*) of the previous surface *Blend.1*, a new surface is formed, *Blend.2*. These profiles are selected in the *First curve* and *Second curve* fields of the *Blend Definition* dialog box (Figure 3.224).

Similarly (by going through all the steps in Figures 3.222 and 3.224) or by symmetry (the *Symmetry* tool in the *Operations* toolbar), the second and last closing side surface called *Blend.3* or *Symmetry.1*, as the case may be, is obtained.

Figure 3.225 shows the hemisphere and the obtained surfaces, but also the *Join Definition* dialog box that appeared after clicking the *Join* icon in the *Operations* toolbar.

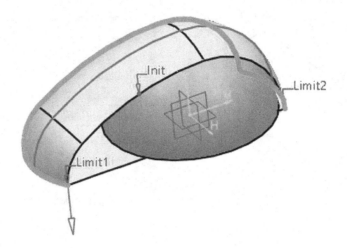

FIGURE 3.223 A variant of the extracted boundary.

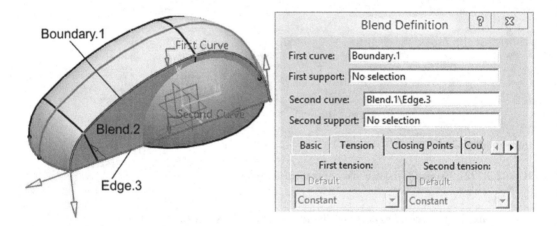

FIGURE 3.224 Creating the *Blend.2* surface feature.

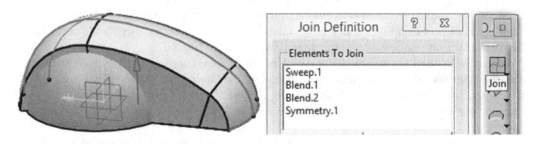

FIGURE 3.225 Joining the surfaces into one single feature.

In the *Elements To Join* field, the user selects the four surfaces to obtain a single surface, *Join.1*. It is closed and empty inside, and to create a solid, the user should switch to the *Part Design* workbench and use the *Close Surface* tool in the *Surface-Based Features* toolbar (Figure 3.226).

In the *Close Surface Definition* selection box, in the *Object to close* field, he selects the surface *Join.1*. Thus, the surface is filled with solid volume, which is added to the specification tree as the feature *CloseSurface.1*.

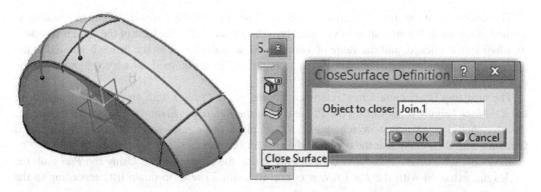

FIGURE 3.226   Applying the *Close Surface* command to obtain the solid.

It is possible and recommended to hide or to remove the visibility of certain features from the specification tree that were used in some of the previous steps. Figure 3.227 shows a part of the context menu for the joined area *Join.1* in which the user can select the *Hide/Show* option. As an effect, the surface (as shown in Figure 3.226) disappears (but is not erased), leaving visible the solid created by the *Close Surface* command. Similarly, the sketches *Sketch.2*, *Sketch.3* and the edge *Boundary.1*, etc. are hidden.

The upper curved edges of the part are connected using the *Edge Fillet* tool in the *Dress-Up Features* toolbar. In the *Edge Fillet Definition* dialog box, the user selects the edges that correspond to extracted boundary (*EdgeFillet.1*) and then the other edges (*EdgeFillet.2*, the result of the intersection of the *Shaft* hemisphere with the solid *CloseSurface.1*) and applies R4 mm radius connections (Figure 3.228).

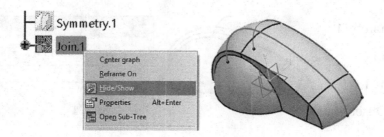

FIGURE 3.227   Hiding the surface *Join.1* and keeping the solid features.

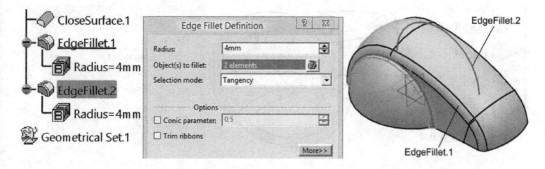

FIGURE 3.228   Applying the *Edge Fillet* command.

The operation of creating the cavity using the *Shell* tool in the *Dress-Up Features* toolbar is applied to the solid obtained up to this stage. Thus, the planar face at the base of the solid is selected, the *Shell* icon is clicked, and the value of 4 mm for the wall thickness on the inside is entered in the *Shell Definition* dialog box. Thus, the cavity in Figure 3.229 is obtained, which corresponds to the external shape of the part.

The next step is to create a new reference feature, plane *Plane.1* with an offset of 4 mm above the *XY Plane* (Figure 3.230) using the *Plane* tool in the *Reference Elements* toolbar. In this plane, the user should create a new sketch, *Sketch.4*, and place two circles with diameters Ø12 mm and Ø16 mm, with their centres at the origin of the current sketch.

Basically, the *Plane.1* and the two circles are inside the cavity *Shell.1*. Using the *Pad* tool, the circles are extruded with the *Up To Next* option from the *Type* drop-down list, according to the representation in Figure 3.231.

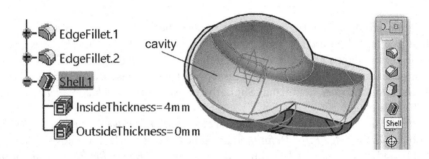

FIGURE 3.229   Creating the cavity using the *Shell* command.

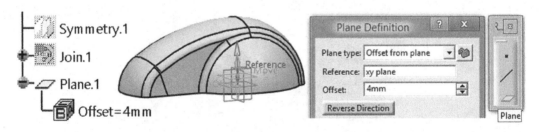

FIGURE 3.230   Creating a new plane with an offset of 4 mm.

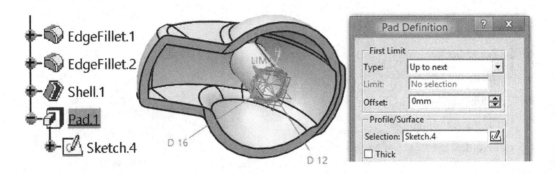

FIGURE 3.231   Applying the *Pad* command to obtain the tube inside the shell.

In the same *Plane.1*, a vertical line is drawn in the *Sketch.5*, coinciding with the *V* axis of the coordinate system. The endpoints of the line also coincide with two circular edges of the solid, as shown in Figure 3.232.

The line is extruded using the *Pad* tool, even if it is an open profile. In the dialog box of Figure 3.233, in the *First Limit* area, in the *Type* list, the user should choose the *Up to next* option, select the *Sketch.5* in the *Profile/Surface Selection* field and check the *Thick* option. As a result, the dialog box expands to the right, making the options in the *Thin Pad* area accessible. The thickness of the extruded solid should be set to 2 mm (*Thickness.1* field), positioned symmetrically to the line by checking the *Neutral Fiber* option. By pressing the *Less* button later, the dialog box can be narrowed.

The figure shows how the line was transformed into a solid element with the role of a pad and follows the curved surface of the cavity.

This pad is multiplied three times around the previously created cylindrical surface based on the two circles with diameters Ø12 mm and Ø16 mm (Figure 3.231). The role of the pads is obvious, to stiffen the cylindrical surface because it has an important functional role of fixing and driving the part.

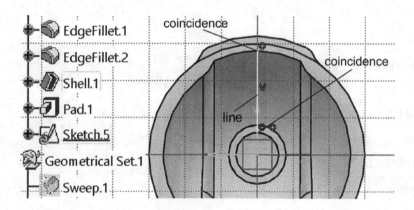

FIGURE 3.232   Creating a new sketch – *Sketch.5*.

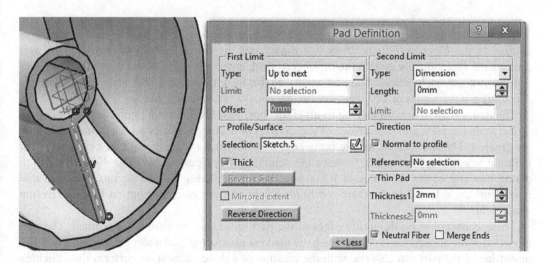

FIGURE 3.233   Creating a pad feature based on an open profile (open sketch).

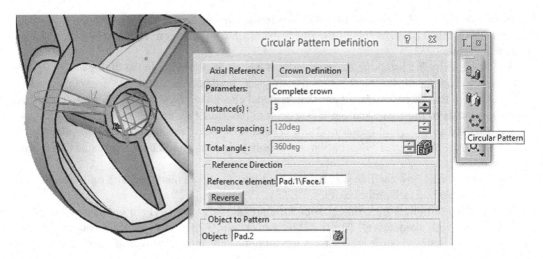

FIGURE 3.234   Applying the *Circular Pattern* command to multiply the *Pad.2*.

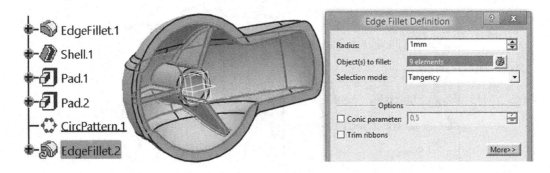

FIGURE 3.235   Applying *Edge Fillet* command on the inside edges of the part.

The pad multiplication is done using the *Circular Pattern* tool in the *Transformation Features* toolbar. In the *Circular Pattern Definition* dialog box, in *the Object to Pattern* field, the user chooses the *Pad.2* feature and, in the *Instance(s)* field, enters the value 3, and then from the list of *Parameters* options, he selects *Complete crown* so that the total angle of 360° is divided equally by 3, resulting in the angles of 120° between two successive (multiplied) pads.

The multiplication options and the result of this operation are shown in Figure 3.234.

Between the pads and the surface of the cavity, connecting surfaces with a radius of R1 mm are created. The user selects all the inner edges on the track (keeping the *Ctrl* key pressed for multiple selection). Figure 3.235 shows the selection. He accesses the *Edge Fillet Definition* dialog box and enters the value of R1 mm in the *Radius* field. The dialog box indicates that nine items are selected in the *Object(s) to fillet* field and the *Tangency* option in the *Selection mode* field.

These connections on the inside edges of the part mean that they are not machined. The part modelled in the application is considered to be made of plastic, injected under pressure in a die. The modelling of the part can also end with the creation of a drive element (to perform the switching function) inside the cylindrical bore.

3.12 MODELLING OF A BALLOON SUPPORT

This tutorial describes the hybrid 3D modelling with surfaces of a balloon support part, represented isometrically and by photographs of a real sample in Figure 3.236. The actual support was used as a source of inspiration, most of the dimensions were taken from this part by measuring with a calliper, and some values were approximated and rounded for a simpler 3D modelling.

The user accesses the *GSD* workbench, from the *Start → Shape* menu. This support requires the creation of a revolution surface; thus, in the *YZ Plane*, the user draws an open profile as in Figure 3.237. The lower end of this profile is placed on the *H* axis. By using the *Revolve* tool, the *Revolute.1* surface is obtained. In the *Revolution Surface Definition* dialog box (Figure 3.238), in the *Profile* field, the user chooses the profile of the *Sketch.1* and, in the *Revolution Axis* field, right-clicks and inserts the *Z Axis* from the context menu. The angle of rotation (*Angle.1*) is equal to 360° for full rotational feature.

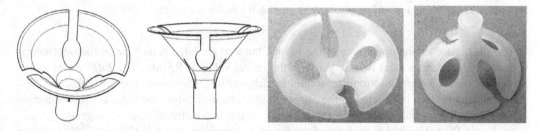

FIGURE 3.236 Isometric views and photographs of a balloon support.

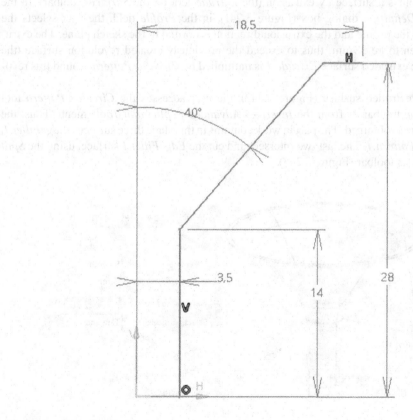

FIGURE 3.237 Open profile drawn in the *YZ Plane*.

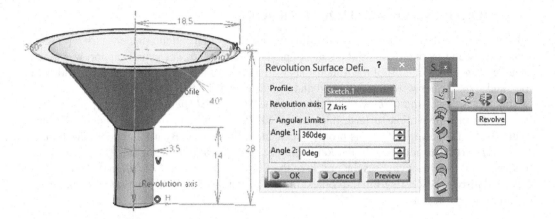

FIGURE 3.238 Revolution surface obtained by rotating the profile around the *Z axis*.

The inner edge on the upper face of the funnel, but also the edge at its base (at the intersection with the cylindrical surface) is rounded with the radius value of R5mm. The *Edge Fillet* tool is accessed from the *Operations* toolbar, and in the *Edge Fillet Definition* dialog box (Figure 3.239), the *Revolute.1* surface is selected in the *Support* field. The user enters the value of the connection radius (in the *Radius* field), and *Object (s) to fillet* contains two elements (the two edges).

In the *XY Plane* (located at the base of the *Revolute.1* surface), a profile (Figure 3.240) is drawn symmetrically with respect to the *V* axis and with the presented dimensions.

This profile becomes a surface by extrusion (the *Extrude* tool on the *Surfaces* toolbar). In the *Extruded Surface Definition* dialog box (Figure 3.241), in the *Profile* field, the user selects the sketch that contains the profile and the extrusion direction is normal to the sketch plane. The extrusion height is chosen to be 38mm, thus to exceed the previously created revolution surface (the funnel surface). The extruded surface *Extrude.1* is multiplied by *Circular Pattern* around this revolution surface.

To multiply the extruded surface (Figure 3.242), the user accesses the *Circular Pattern* tool from the *Replication* toolbar or from the *Insert → Advanced Replication Tools* menu. Thus, the *CircPattern.1* surface is obtained. The part in work contains in this stage three surfaces: *EdgeFillet.1*, *Extrude.1* and *CircPattern.1*. The last two intersect and cut the *EdgeFillet.1* surface, using the *Split* tool on the *Operations* toolbar (Figure 3.243).

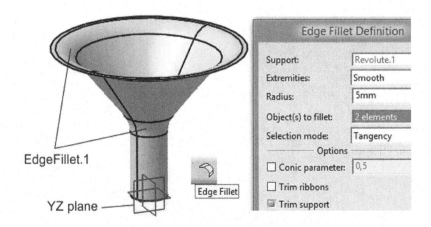

FIGURE 3.239 *Edge Fillet* tool applied to the edge at the base of the funnel.

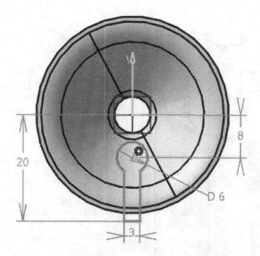

FIGURE 3.240 Profile drawn at the base of the part.

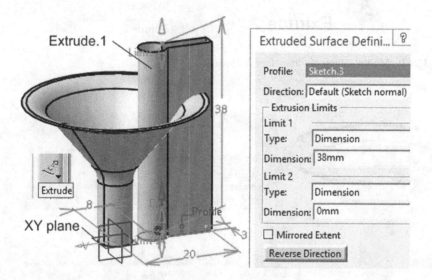

FIGURE 3.241 Profile extruded to exceed the revolution surface.

In the *Split Definition* dialog box, in the *Element to cut* field, the user selects the area to be cut, *EdgeFillet.1*, and in the *Cutting elements* field, he selects the other two surfaces. The result of the *Split* tool should be similar to that in Figure 3.243.

By default, the cutting result is correct and respects the final shape of the balloon holder. However, as an option, by pressing the *Other side* button, the cut-out is completely different, according to Figure 3.244 and cannot be accepted. Thus, the surface to be removed would be the one represented transparently, and the opaque one would be kept.

In the *YZ Plane*, in a new sketch, the *Split.1* surface generators are projected: the top (funnel) and the bottom (cylindrical) using the *Project 3D Silhouette Edges* tool in the *Operations* toolbar. The projections are created on either side, symmetrically, with respect to the *V* axis of the coordinate system. By default, the designed generators are standard elements (continuous and thick line).

The two projections on the left can be removed (no longer needed), and the ones on the right can be transformed into auxiliary construction elements (represented by a dashed and thin line), as in Figure 3.245.

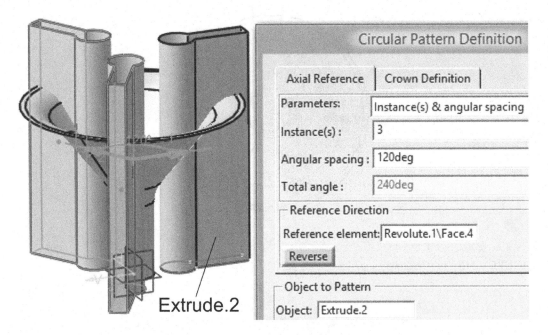

FIGURE 3.242 Multiplication of the surface *Extrude.1*.

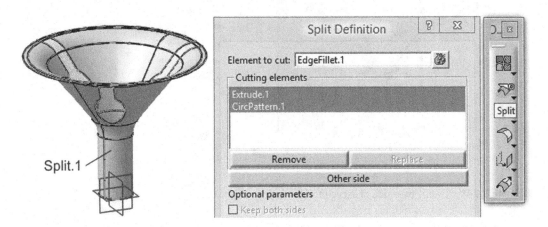

FIGURE 3.243 Surface that resulted after the *Split* tool was applied.

From the *Profile* toolbar, the *Spline* icon is used to draw a curve through three points, as follows: the first point (the lower one) is constrained at a distance of 8 mm from the *H* axis and coincides with the projected generator *Mark.2*, the second point (middle) is 6 mm from the *V* axis, and the third point, which belongs to the generator *Mark.1*, is 11 mm distant from the *V* axis. The spline curve at this point is considered tangent to the generators.

The profile (only the spline curve, not the projected generators) is extruded in the direction of the *Y* axis.

Extrusion takes place towards the inside of the revolution surface (Figure 3.246) over a distance of 6 mm (*Limit 1*), and thus, the *Extrude.2* surface is created. To select the *Y* axis, the user right-clicks in the *Direction* field of the *Extruded Surface Definition* dialog box.

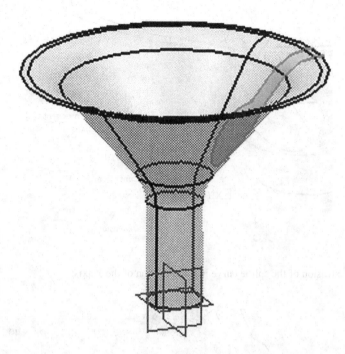

FIGURE 3.244 Surface that resulted after the *Split* tool was applied using the *Other side* button.

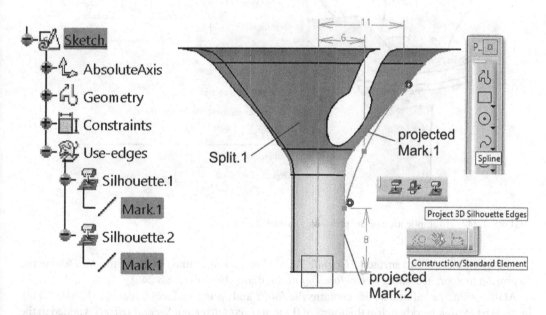

FIGURE 3.245 Projection of *Split.1* surface generators in a sketch.

The extruded surface multiplies circularly around the revolution surface in three identical entities using the *Circular Pattern* tool. The dialog box in Figure 3.247 contains the options for selecting the surface involved in multiplication (*Object to Pattern* field), the type of multiplication (*Instance(s) & angular spacing*), the angle between two successive instances (*Angular spacing*), but also the revolution surface that imposes the axis of rotation (*Reference Direction*). The radial arrangement of the surfaces is observed, but also the fact that they intersect.

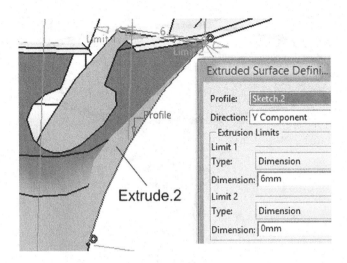

FIGURE 3.246   Extrusion of the spline curve in the direction of the *Y* axis.

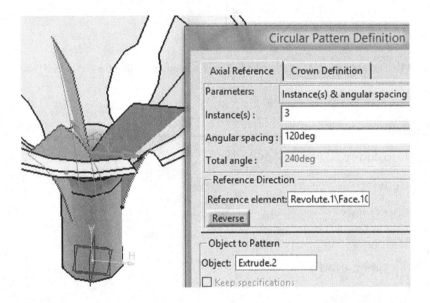

FIGURE 3.247   Multiplication of the previously extruded surface.

The user joins the two surfaces (*Extrude.2* and *CircPattern.2*) into one using the *Join* tool in the *Operation* toolbar, according to the *Join Definition* dialog box (Figure 3.248).

At this point, the balloon holder contains the *Join.1* and *Split.1* surfaces (created in Figure 3.243). In the *Part Design* workbench, a thickness of 0.1 mm (*First Offset* and *Second Offset*) is added to the *Split.1* surface, symmetrical to it. The *Thick Surface* tool from the *Surface-Based Features* toolbar was used (Figure 3.249), and the result is a solid body with a constant thickness of 0.2 mm.

From the *Insert* menu, a new body (*Body.2*) is added to the part, and it appears in the specification tree on the same level as the *PartBody* (this is the default body in model tree). The user observes that the new body is underlined (Figure 3.250), which means that it is the current (active) body. Whatever will be added by the user in the part structure, it will be placed from now on under this body. Thus, the surface *Join.1* is also transformed into a solid by adding 0.2 mm thickness (Figure 3.251), similarly to the surface *Split.1*.

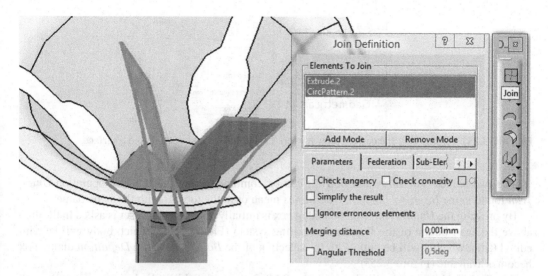

FIGURE 3.248   Joining the extruded and multiplied surfaces.

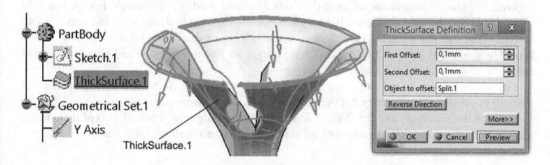

FIGURE 3.249   Adding solid thickness to the resulted surface.

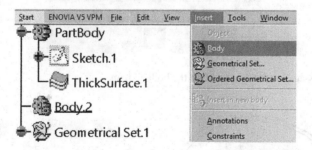

FIGURE 3.250   Inserting a new body in the specification tree of the current 3D model.

The place of the two surfaces is taken by the *ThickSurface.1* and *ThickSurface.2* solids, as seen in the figure. Surfaces can be hidden because they are no longer needed to be visible in the following steps.

Switching between *PartBody* and *Body.2* in terms of editing and/or adding new elements/features (which of the two bodies to become current) is done by right-clicking on the selected body, and from the context menu, the user chooses the option *Define In Work Object*. The underline shifts from one body to another and then descends to the last feature of the specified body. Figure 3.251 shows how the solid was underlined when the *ThickSurface.2* solid was added to the *Body.2*.

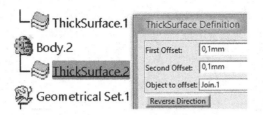

FIGURE 3.251 Specifying the current body to work on and how to add thickness to a surface.

The two bodies (*PartBody* and *Body.2*) can be combined using the *Boolean* operation *Union Trim* found in the *Insert → Boolean Operations* menu or in the toolbar with the same name.

By pressing the *Union Trim* icon, no dialog box is initially displayed; the user is asked in the area above the *Start* button of the *Windows* operating system (Figure 3.252) which body will be trim edited (from which it will be cut). With the selection of the *Body.2*, the *Trim Definition* dialog box becomes available (Figure 3.253).

The *Trim* field is already filled in with the *Body.2*, the *with* field is filled in automatically with *PartBody*, and the user must select the faces that will be removed when the *OK* button is pressed. Figure 3.253 shows these faces, areas of the *Body.2* located inside the *PartBody*. For the last field (*Faces to keep*), the user can select the surfaces outside the *PartBody* (as seen in Figures 3.246 and 3.247) or can leave the field unselected. These selections determine what is cut and what is kept. The colours of the selection are different: purple for *Faces to remove* and blue for *Faces to keep*.

The two bodies are joined into a single one by the *Trim.1* feature (Figure 3.254). It is noted that it is still possible to edit the *Body.2* if the user considers this as necessary.

Between the *PartBody* and the *Body.2*, some intersection edges are created, two of them being marked in Figure 3.254. These edges belong at this point to the *PartBody* solid feature.

FIGURE 3.252 Message for the user to select the body to trim.

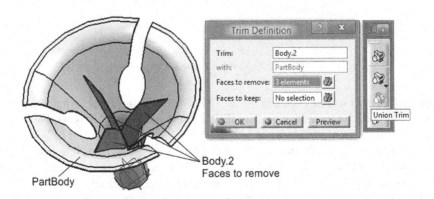

FIGURE 3.253 Applying the *Union Trim* tool to cut and join two solid bodies of the support.

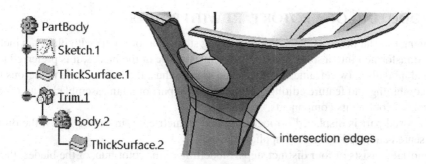

FIGURE 3.254 The result of the *Boolean* operation with two of six intersection edges marked.

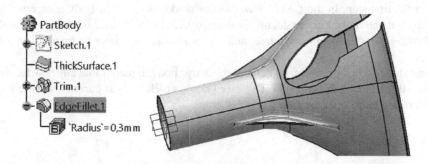

FIGURE 3.255 *Edge Fillet* tool applied to all six intersection edges.

The six intersection edges are rounded (*Edge Fillet* tool) to a radius of R0.3 mm, according to Figure 3.255. These fillets also prove the correctness of the *Boolean* operation.

The solid model of the balloon holder is complete and editable according to its specification tree. Being, however, a part created using surfaces, the user may choose to transform it into a new surface for further editing.

Thus, in the *GSD* workbench, the user applies the *Extract* tool in the *Operations* toolbar or in the *Insert → Operations* menu. In the *Extract Definition* dialog box in Figure 3.256, he chooses the type of surface propagation on the solid as *Tangent continuity*. Other variants can be applied; depending on the selection, certain parameters available are accessible by pressing the *Show parameters >>* button. The *PartBody* solid is selected in the *Element(s) to extract* field, and the specification tree is completed with the *Extract.1* surface.

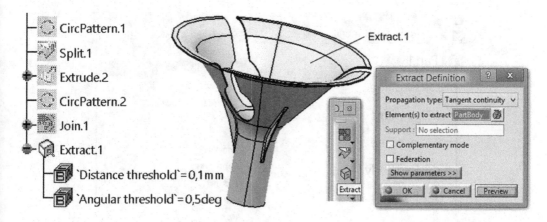

FIGURE 3.256 Extracting surfaces from the solid model.

3.13   MODELLING OF A ROTOR PART WITH BLADES

The following tutorial presents the hybrid modelling steps of a rotor part with blades, which is subsequently transformed into a solid. Due to the complex shape of the blades, it is preferred to implement hybrid modelling (wireframes, surfaces and solids); thus, the user can use numerous tools for creating, combining and feature editing. The rotor can be part of a fan assembled in an installation, with the role of cooling its components.

The proposed part is displayed orthogonally and isometrically in Figure 3.257, the dimensions being presented and used during the application.

This tutorial consists of four distinct steps: modelling of the rotor hub, of the blades, their transformation into solids and, finally, their circular multiplication around the hub. Therefore, both *GSD* and *Part Design* workbenches will be used.

Thus, in the first step, in the *CATIA Part Design* workbench, a new body is inserted from the *Insert → Body* menu in the current document (Figure 3.258). The *Body.2* is positioned on the same level as *PartBody* in the specification tree, and the two bodies will later be involved in a *Boolean* operation.

Once inserted, *Body.2* becomes, by default, the active body; it means that any new feature added to the part will be placed in this body. The user knows which body is current because it is represented underlined in the specification tree.

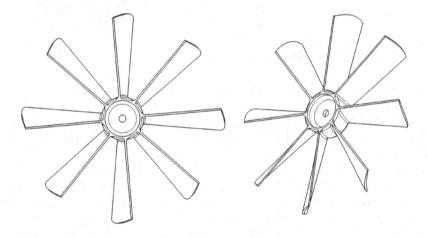

FIGURE 3.257   Views of the rotor part with blades.

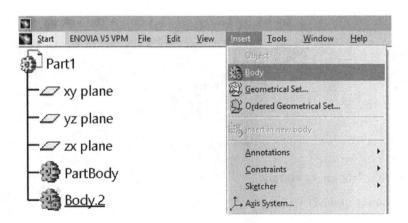

FIGURE 3.258   Inserting a new body in the current part document.

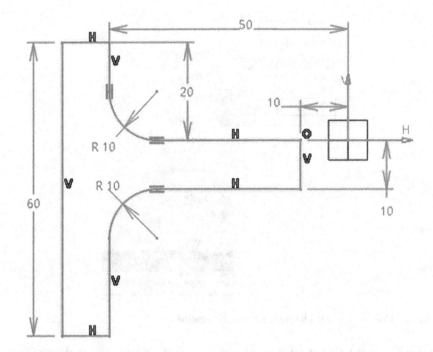

FIGURE 3.259 Profile drawn in *Sketch.1* of the *YZ Plane* – it is the hub profile of the rotor.

In the *Sketch.1* of the *YZ Plane*, the profile from Figure 3.259 is drawn, respecting the dimensional constraints between the elements, but also with respect to the coordinate system.

In the *CATIA Part Design* workbench, by using the *Shaft* tool, a revolution body is obtained using the previously created sketch. In the *Shaft Definition* dialog box (Figure 3.260), the sum of the *Angle* fields is 360°, and the *Profile/Surface Selection* field contains the *Sketch.1*. Figure 3.259 shows that this profile sketch does not contain the axis required for the *Shaft* tool.

To indicate the Z axis as the axis of rotation for the drawn profile, the user right-clicks the *Axis Selection* field. As a result, a context menu opens (Figure 3.261) from which the *Z Axis* option is chosen.

In the second step, from the *Insert* menu (Figure 3.258), another body (*Body.3*) and a geometric set (*Geometrical Set*) are inserted. The name of the geometrical set is changed to *Set.1*. The specification tree is completed with the two features. *Body.3* is the current (active) body, so it is underlined.

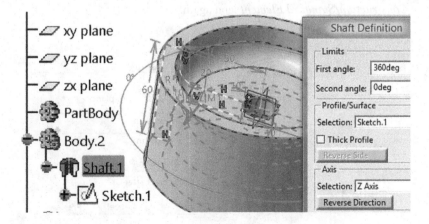

FIGURE 3.260 The solid hub of the rotor is created.

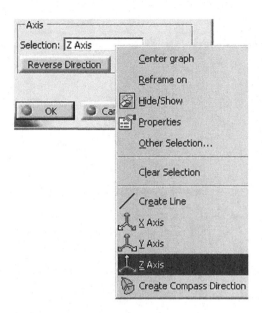

FIGURE 3.261   Context menu to choose the *Z Axis* of rotation.

The user switches to the *CATIA GSD* workbench, selects the *YZ Plane* and creates another plane parallel to it using the *Plane* tool on the *Wireframe* bar. The *Plane.1* will belong to the geometric set named *Set.1*.

In Figure 3.262, the user applied the *Offset from plane* option, and the distance between the two planes (*YZ Plane* and *Plane.1*) is 300 mm. For better highlighting of the planes, the *Shaft.1* feature (the rotor hub) can be hidden.

In this new *Plane.1*, another sketch is created (*Sketch.2*) and the two edges of the upper and lower planar faces are selected.

These edges are projected onto the *Plane.1* and transformed into auxiliary construction lines (numbered 1 and 2 according to Figure 3.263) using the *Project 3D Elements* icon in the *Operation* toolbar and, respectively, *Construction/Standard Element* in the *Sketch Tools* toolbar. Note that the sketch is also included in *Set.1*.

In the same sketch, an arc is inserted (the *Three Point Arc Starting With Limits* tool in the *Profile* toolbar), having the endpoints coinciding (Figure 3.264) with the two auxiliary construction lines obtained by projection. Before drawing the arc, the drawing of standard geometries was resumed by pressing the *Construction/Standard Element* icon again.

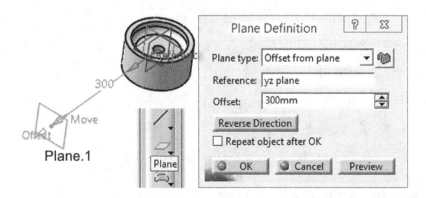

FIGURE 3.262   Creating a new plane parallel to the *YZ Plane*.

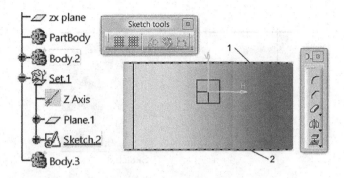

FIGURE 3.263 Planar faces projected as edges in *Sketch.2*.

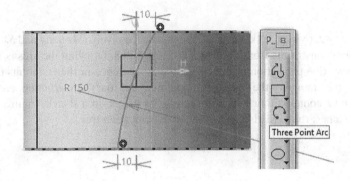

FIGURE 3.264 Drawing an arc between the two projected edges.

The dimensioning and positioning of the arc are established according to the vertical axis *V* of the coordinate system, the endpoints being at distances of 10 mm on both side of the axis, and the radius is assigned the value of R150 mm.

Figure 3.265 shows the *Plane.1* and the arc of the *Sketch.2*. In another sketch (*Sketch.3*) of the same plane another arc is similarly drawn.

The second arc has a radius of R140 mm, its endpoints being 5 mm to the left of the ends of the first arc (Figure 3.266).

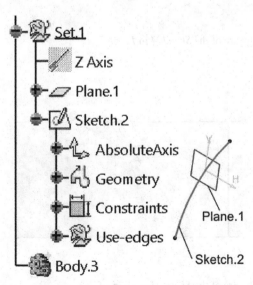

FIGURE 3.265 *Sketch.2* drawn in *Plane.1*.

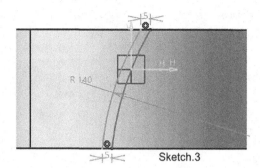

FIGURE 3.266 Drawing a second arc between the two projected edges of *Sketch.3*.

Also, in the *Plane.1*, two more sketches are created successively (*Sketch.4* and *Sketch.5*) respecting the dimensional constraints from Figures 3.267 and 3.268. Drawing the arcs is the same as the one presented above (the projection of the edges, the coincidence of the endpoints of the arcs with these projections, the change of the lines type standard – auxiliary construction, etc.).

Basically, *Plane.1* contains four arcs and each arc is placed in a sketch. Figure 3.269 shows in detail the four sketches; they are also placed in the specification tree.

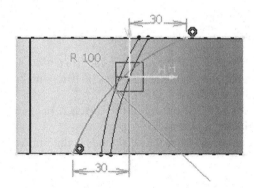

FIGURE 3.267 Drawing the arc of the *Sketch.4* in *Plane.1*.

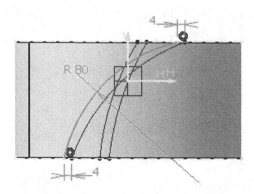

FIGURE 3.268 Drawing the arc of the *Sketch.5* in *Plane.1*.

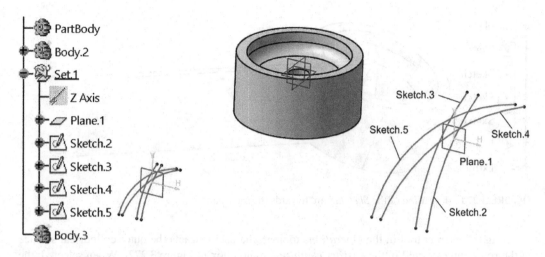

FIGURE 3.269 Representation in detail of the four arcs in the *Plane.1*.

The arc of the *Sketch.2* is extruded (the *Extrude* tool on the *Surfaces* bar) to obtain a first surface. In the *Extruded Surface Definition* dialog box (Figure 3.270), the user selects the sketch in the *Profile* field and the direction (*Direction* field) is along the *X* axis (chosen from the context menu of the field by an operation similar to the one in Figure 3.261). Below, in the *Type* field, the user chooses the *Dimension* option, and in the field with the same name, he enters the value of 300 mm and presses the *Preview* button to see the result. Extrusion must take place towards the rotor hub, and the *Reverse Direction* button is used, which changes the direction of the extrusion.

Similarly, the arc of the *Sketch.3* is extruded on the same distance and direction, and the result will be a second surface. The surfaces thus obtained are named *Extrude.1* and *Extrude.2*, according to Figure 3.271.

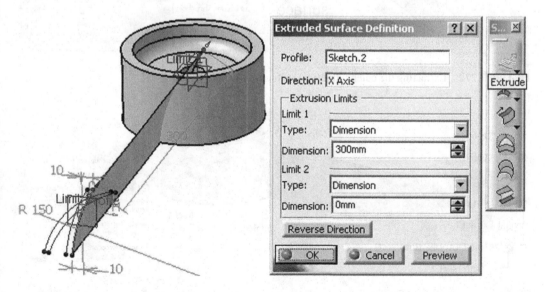

FIGURE 3.270 Extrusion of the *Sketch.2* arc towards the rotor hub.

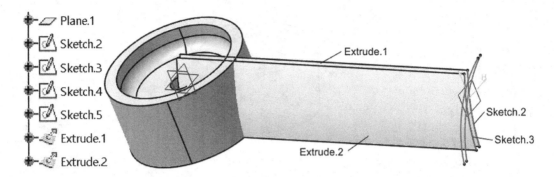

FIGURE 3.271 Extrusion of the *Sketch.3* arc towards the rotor hub.

Using the *Extract* tool in the *Operations* toolbar, the user extracts the outer cylindrical surface of the rotor hub (selected in the *Extract Definition* dialog box in Figure 3.272). When selected, the surface turns green, and then, after pressing the *OK* button, it will turn light yellow (usual colour of surfaces), suggesting to the user that he obtained a surface. The presence of the new feature of surface type, the *Extract.1*, is also observed in the specification tree.

An intersection operation is applied between this surface and the one obtained by extruding the arc of the *Sketch.2*, using the *Intersection* tool in the *Wireframe* toolbar. In the *Intersection Definition* dialog box in Figure 3.273, the *First Element* and *Second Element* fields contain the surfaces *Extrude.1* and *Extract.1*, respectively.

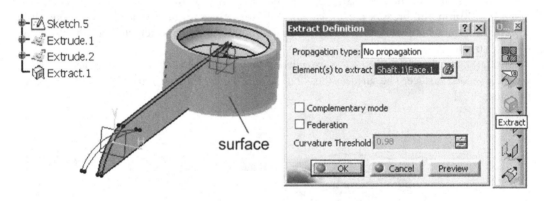

FIGURE 3.272 Extracting the outer cylindrical surface of the rotor hub.

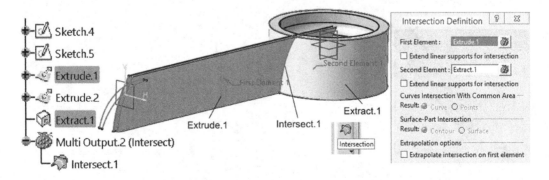

FIGURE 3.273 Intersection of the surfaces *Extrude.1* and *Extract.1*.

The result of the intersection of the two surfaces is also a curve, named *Intersect.1*. Similarly, the second intersection between the cylindrical surface *Extract.1* and the surface *Extrude.2* (obtained from *Sketch.3*) is applied. Following the two intersections, the specification tree contains two new features, *Intersect.1* and *Intersect.2* (Figure 3.274). The two intersection curves are located on the surface *Extract.1* of the rotor hub.

In order to simplify the following representations and explanations, the *Extrude.1* and *Extrude.2* surfaces are hidden by the user, as well as the *Sketch.2* and *Sketch.3* by using the *Hide/Show* option from the context menu of the respective features (Figure 3.275).

Using the *Blend* tool on the *Surfaces* bar, the user creates a surface between the *Intersect.1* and *Sketch.4* curves and, respectively, another surface between *Intersect.2* and *Sketch.5*. Thus, in the *Blend Definition* dialog box, he selects the *Intersect.1* feature in the *First curve* field, and *Sketch.4* in the *Second curve* field. In the representation of Figure 3.276, it is checked if the arrowheads point in the same direction.

The user proceeds in a similar manner for the second surface (created from *Intersect.2* and *Sketch.5*). The result of the two previous operations is the completion of the specification tree with two new features, *Blend.1* and *Blend.2*, presented in Figure 3.277.

The *Intersect.1* and *Intersect.2* curves have their endpoints on the circular edges of the cylindrical surface *Extract.1*. Using the *Boundary* tool in the *Operations* toolbar, the user extracts two segments of these edges, located between the endpoints of the curves.

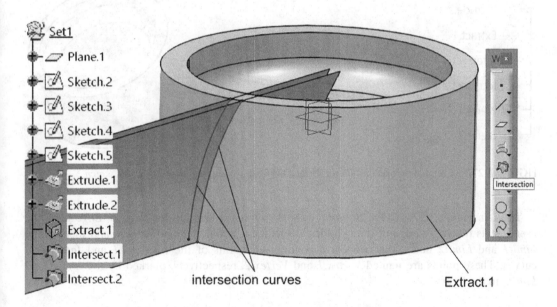

FIGURE 3.274 Intersection curves placed on the *Extract.1* surface of the rotor hub.

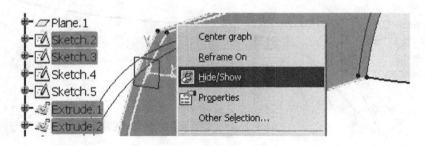

FIGURE 3.275 Hiding sketches and surfaces using the *Hide/Show* option.

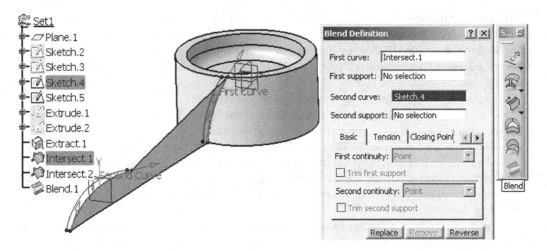

FIGURE 3.276   Creation of a first *Blend* surface between two curves *Intersect.1* and *Sketch.4*.

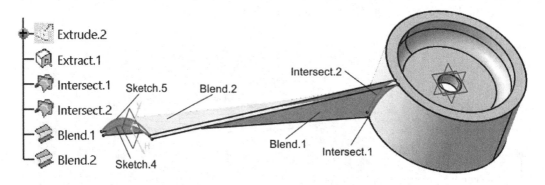

FIGURE 3.277   Creation of a second *Blend* surface between two curves *Intersect.2* and *Sketch.5*.

Thus, in Figure 3.278, in the *Boundary Definition* dialog box, in the *Surface Edge* field, the user selects the upper edge of the cylindrical surface *Extract.1*, named *Edge.1*, and then, in the *Limit1* and *Limit2* fields, the two corresponding endpoints of the *Intersect.1* and *Intersect.2* curves. These points are named *Vertex.1* and *Vertex.2*, respectively, marked by the *Limit1* and *Limit2* in the figure.

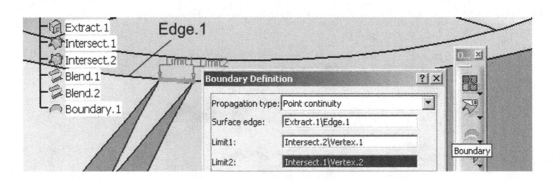

FIGURE 3.278   Small arc segment extracted between two points using the *Boundary* tool.

During the selection of points, an arrow appears, pointed from the *Vertex.1* to *Vertex.2*. This arrow indicates the direction in which a curve (*Boundary.1*) is created, superimposed on the circular edge marked as *Edge.1*. The selection of points and the creation of the curve are shown in the figure and in the specification tree.

Similarly, the *Boundary.2* curve is created, using the lower circular edge of the cylindrical surface *Extract.1*, named *Edge.2* (Figure 3.279).

Also in this case, it is important to check the direction of the arrow from point *Vertex.4* (marked on the figure with *Limit1*) to point *Vertex.5* (*Limit2*).

The cylindrical surface *Extract.1* can be hidden (*Hide/Show* option from its context menu) because it no longer intervenes during the application. Also, at the moment, the solid represented by the rotor hub can be hidden, but also the two curves *Intersect.1* and *Intersect.2*.

Using the *Fill* tool in the *Surfaces* toolbar, a new surface is created at the end of the blades near the rotor hub. Thus, in the *Fill Surface Definition* dialog box, in the *Boundary* field, the *Curves* tab, the user selects the curves that delimit a contour in a certain order, so that each curve has the same endpoint with the precedent and with the next one.

Starting from the upper side, with *Boundary.1* (marked *Curve1* on the 3D model), the user selects, by clicking, the following: the *Edge.3* (*Curve2*) belonging to the *Blend.1* surface, the *Boundary.2* (*Curve3*) and the *Edge.4* (*Curve4*) of the surface *Blend.2* (Figure 3.280).

Before validation, the user must receive the system confirmation of *Closed Contour* (in the figure it is written in white on a blue background).

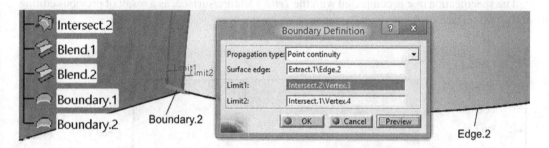

FIGURE 3.279   Small arc segment extracted between two points using the *Boundary* tool.

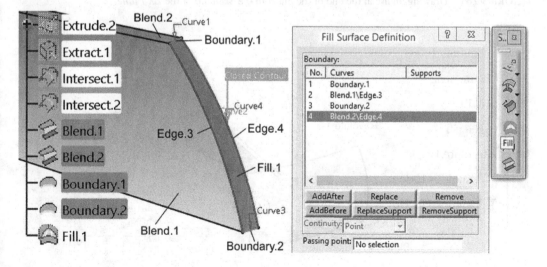

FIGURE 3.280   Selection of the four edges to fill the space between them.

After pressing the *OK* button, a new surface is created between the four edges, named *Fill.1*, which is also added to the specification tree.

At the other end of the blade, where the *Sketch.4* and *Sketch.5* are located, a R50 mm radius arc will be drawn in the new *Sketch.6* of the *ZX Plane*. The endpoints of the arc are considered to coincide with the edges of the surfaces *Blend.1* or *Blend.2*, represented according to Figure 3.281.

The sketch is then used in an operation to create a circular surface using the *Revolve* tool in the *Surfaces* toolbar. In the *Revolution Surface Definition* dialog box in Figure 3.282, in the *Profile* field, the user selects the arc from the *Sketch.6* and chooses the Z axis by right-clicking in the *Revolution Axis* field. In the *Angle.1* and *Angle.2* fields, the value of 10° is entered.

The blade representation and specification tree will contain a new surface, named *Revolute.1*, at an angle of 20°. The angular values may be different, but the user must keep in mind that the *Revolute.1* surface should exceed the *Blend.1* and *Blend.2* surfaces.

Revolute.1 is involved, along with the *Blend.1* surface, in a *Trim* edit using the tool of the same name and the icon placed in the *Operations* toolbar.

In the *Trim Definition* dialog box in Figure 3.283, in the *Trimmed elements* field, the user selects the two surfaces and then clicks the *Other side/next element/previous element* buttons to determine the areas of the surfaces to be removed. The areas of each surface are deleted to the limit/contact with the others.

Thus, by this operation, an area of the surface *Revolute.1* located to the right of the surface *Blend.1* is removed, as shown in Figure 3.283. Also, the area beyond the *Revolute.1* surface is removed from the *Blend.1* surface.

The specification tree is completed with the *Trim.1* feature/surface, as a result of previous editing of the two surfaces.

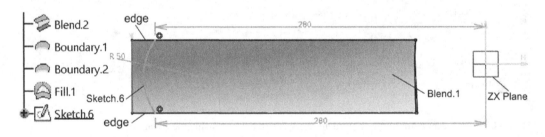

FIGURE 3.281   Drawing an arc at the end of the blade in the *Sketch.6* of the *ZX Plane*.

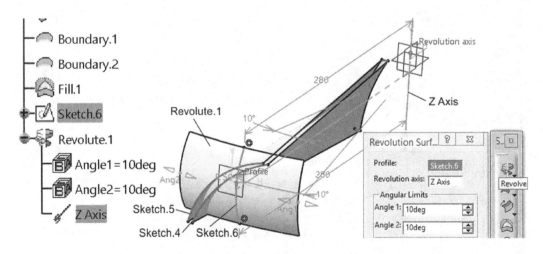

FIGURE 3.282   Creation of a circular surface using the *Sketch.6* and the *Revolve* tool.

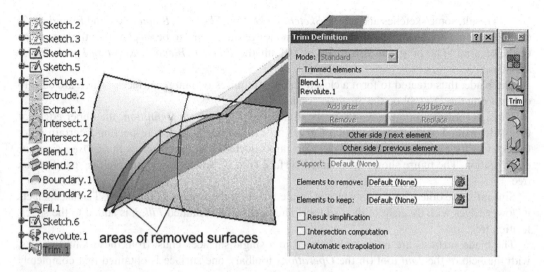

FIGURE 3.283   Trimming of the surfaces *Revolute.1* and *Blend.1*.

Similarly, the user performs an edit between the *Blend.2* and *Trim.1* surfaces to remove the area on the left of the *Blend.2* surface and to keep the one between *Blend.1* and *Blend.2*. In the dialog box shown in Figure 3.284, in the *Trimmed elements* field, the user selects *Blend.2* and *Trim.1* and then presses the *Other side/next element/previous element* and *Preview* buttons until the correct result is obtained.

In this second operation, the *Revolute.1* surface no longer intervenes because it has already been previously edited, and its feature in the specification tree was automatically hidden, the *Trim.1* being the remaining surface from the initial one. The result of this second trimming is the surface *Trim.2*, added to the specification tree.

In the two previous figures, it is observed that the surface areas that are removed become transparent when the *Preview* button is pressed. Thus, the program visually supports the user in the correct choice of the areas that are kept and, respectively, those that are removed.

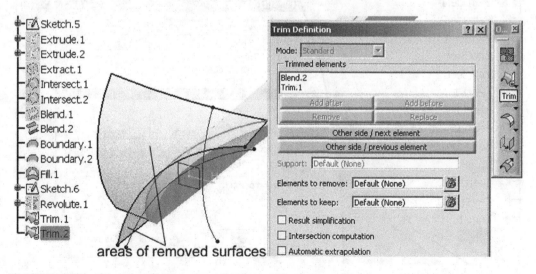

FIGURE 3.284   Trimming of the surfaces *Trim.1* and *Blend.2*.

As a result, some sketches and edges (*Sketch.4*, *Sketch.5*, *Sketch.6*, *Boundary.1* and *Boundary.2*) can be hidden with the context menu; they are no longer necessary to be displayed in the next steps of the tutorial. The program also hides by default the *Blend.1*, *Blend.2*, *Revolute.1* and *Trim.1* surfaces.

The blade, thus created to form a closed volume, requires two more surfaces, which are easy to obtain using the *Fill* tool on the *Surfaces* toolbar.

Thus, according to Figure 3.285, in the dialog box *Fill Surface Definition*, the upper edges of the blade (named on the representation *Curve1*,..., *Curve4*) are selected, respecting a certain order (clockwise, for example). Once all four edges are selected, the user receives the *Closed Contour* confirmation. The figure shows the resulted surface (*Fill.2*), added as a feature in the specification tree.

Similarly, the other (bottom) surface is created to close the volume of the blade. The user selects its lower edges with the help of the *Fill* tool. The new surface, named *Fill.3*, is added to the specification tree.

The blade surfaces are ready to be joined in a single surface. Thus, from these four surfaces, with the help of the *Join* tool (in the *Operations* toolbar), one surface is obtained that completely delimits the blade.

In the *Join Definition* dialog box in Figure 3.286, in the *Elements To Join* field, the surface-type features are added by selecting in the specification tree: *Fill.1*, *Trim.2*, *Fill.2* and *Fill.3*. Their identification is simplified by the fact that all the other elements of the blade, used in the previous steps, were hidden manually by the user or automatically by the program. The result of this join is a single continuous surface, *Join.1*, which delimits a future solid volume of the blade.

The stage of working with surfaces is considered completed, followed by the transformation of the blade into a solid 3D body. To do this, the user accesses the *CATIA Part Design* workbench and right-clicks on the *Body.3* in the specification tree, and then, from the context menu that appears, he chooses the *Define In Work Object* option. It is noticed that the *Body.3* feature becomes underlined, being the current body in which the user can continue modelling the complex part. From the *Surface-Based Features* toolbar, he chooses the *Close Surface* icon and, then, in the *Close Surface Definition* dialog box (Figure 3.287), in the *Object to close* field, selects the *Join.1* feature from the specification tree.

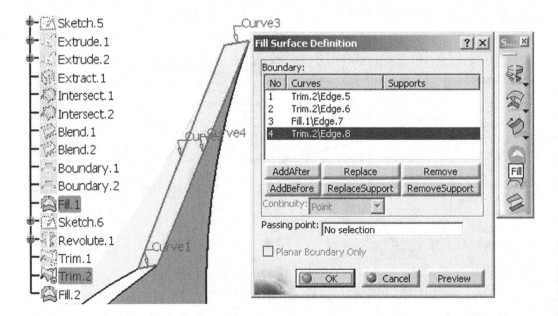

FIGURE 3.285   Creation of a surface by filling a certain area delimited by four curves.

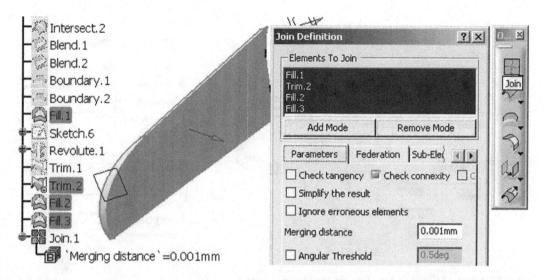

FIGURE 3.286 Joining the four component surfaces of the blade.

After filling the surface, a solid body resulted, being delimited and located inside the surface. This solid, named *CloseSurface.1* and contained in the *Body.3*, is one of the eight blades of the rotor. The geometric set *Set.1*, which stores the created surfaces, can be hidden using its context menu.

Also, the rotor hub (*Shaft.1*), located in the *Body.2*, must become visible (Figure 3.288). In the last stage, a circular multiplication of the blade is applied around the hub axis.

The *Circular Pattern Definition* dialog box shown in Figure 3.289 opens after clicking the *Circular Pattern* icon on the *Transformation Features* toolbar.

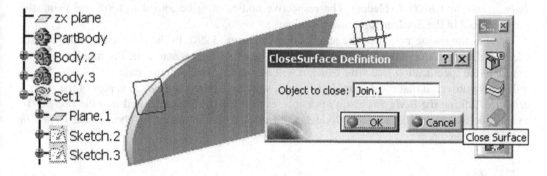

FIGURE 3.287 Transformation of the *Join.1* surface into a solid.

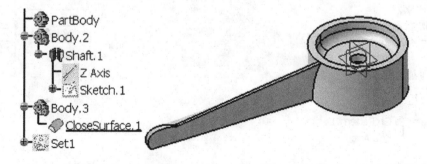

FIGURE 3.288 The first solid model of a blade represented near the rotor hub.

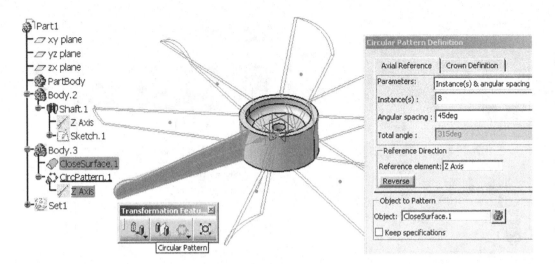

FIGURE 3.289   Multiplication of the first model of a blade.

In the *Parameters* field, the user chooses the option *Instance(s) & angular spacing*; there will be eight blades around the rotor hub, arranged equidistantly (*Instance(s) field*), at angles of 45° (*Angular spacing*).

To set the axis of rotation, the user right-clicks in the *Reference element* field and chooses *Z Axis* from the context menu. The object involved in multiplication (*Object to Pattern*) is the solid *CloseSurface.1* from the specification tree.

Figure 3.289 shows a preview of the circular multiplication; the *Body.3* will contain in its tree structure the *CircPattern.1* feature.

The result of the modelling steps is currently represented by two bodies with solid features: *Body.2* (hub) and *Body.3* (blades). The respective bodies must be joined to form one using the *Assemble* tool in the *Boolean Operations* toolbar.

Thus, in the dialog box with the same name in Figure 3.290, in the *Assemble* field, the user selects the *Body.3*, and in the field *To:* the *Body.2*. The *After:* field is automatically filled in with the feature in the specification tree after which it will be added (*Shaft.1* in this case).

It is also observed that the *Assemble.1* feature, which contains the *Body.3*, is part of the *Body.2* structure. Editing the *Body.3* is always possible, even if it has been integrated into the *Body.2*. To do this, the user right-clicks on the *Body.3*, and from the context menu, he chooses the *Define In Work Object* option.

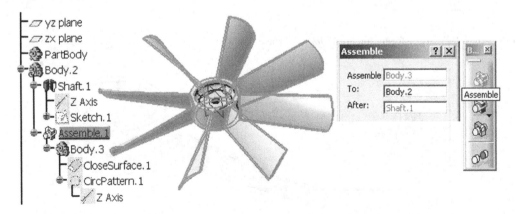

FIGURE 3.290   Result of the blade multiplication around the hub and joining of all features.

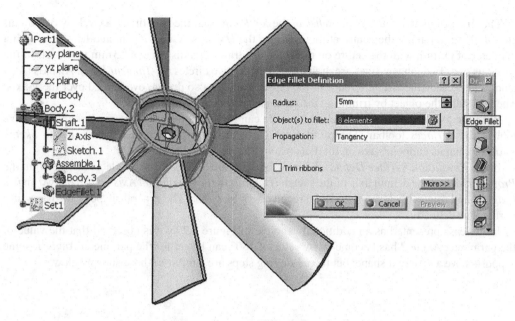

FIGURE 3.291 The *Edge Fillet* tool applied on the edges between the blades and the rotor.

The last operation applied to this rotor with blades is to round some edges in the area where the blades touch the hub, using the *Edge Fillet* tool (found on the *Dress-Up Features* toolbar). In the dialog box in Figure 3.291, the user enters the value of R5 mm in the *Radius* field, and in the *Object(s) to fillet* field, he selects the edges at the intersection between the blades and the hub.

3.14 MODELLING OF A HANDLE KNOB

The hybrid modelling process of a screwed handle knob part (2D drawing in Figure 3.292) is presented through the following step-by-step tutorial. It uses certain tools from the *GSD* and *Part Design* workbenches. The part is used in the fastening-tightening systems of the handles for handling electric and thermal lawn mowers.

The hexagonal head of the screw is fixed in the housing of the holder, which, by manual actuation, tightens the screw in the nut of the handle. The part is also represented isometrically by the sequence of pictures in Figure 3.292, and the dimensions were measured and rounded, when needed.

The part modelling begins in the *GSD* workbench and will continue in *Part Design* to add solid features (thickness, holes, edge fillets, etc.).

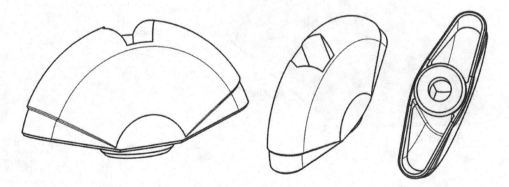

FIGURE 3.292 Isometric representations of the 3D part to be modelled – handle knob.

The first step is to create the *Sketch.1* in the *XY Plane*, and the user draws a circle with a diameter of Ø16 mm having the centre at the origin of the *HV* axis system. Then, another circle with a diameter of Ø8 mm with the centre on the *H* axis is drawn at a distance of 23 mm to the left of the previous one. Two tangent lines are drawn between the two circles (*Bi-Tangent Line*). An axis (*Axis* icon) that exceeds the circumference of the bigger circle is also drawn, overlapped over the *V* axis. The user edits the object by trimming the elements as shown in Figure 3.293.

With the axis drawn in the sketch, the profile can be rotated around the axis using the *Revolve* tool in the *Surfaces* toolbar. The profile in Figure 3.293 is open (*Opened*, according to *Sketch Analysis*), but its endpoints are on the *V* axis.

In the *Revolution Surface Definition* dialog box (Figure 3.294), the user selects the sketch in the *Profile* field, and the default axis of the sketch (*Default*) in the *Revolution Axis* field. The two angles (*Angle 1* and *Angle 2* fields, respectively) receive 165° and −15°. The general shape of the basic support is thus defined.

In the case presented as an additional example in Figure 3.295, it is observed that the value of the parameter *Angle 2* has become 0°. A value of 180° can be set for the parameter *Angle 1*, some supports have a different shape, but the modelling steps are similar to those shown below.

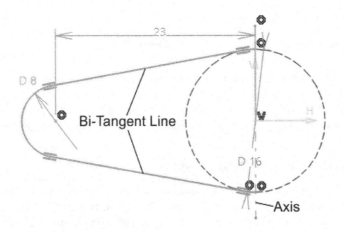

FIGURE 3.293   Creating the first sketch – *Sketch.1*.

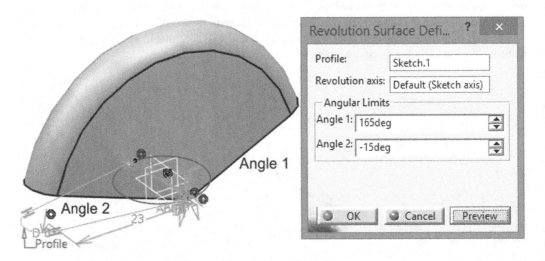

FIGURE 3.294   Creating a revolution surface using the *Revolve* command.

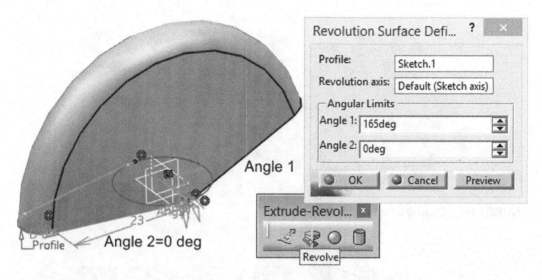

FIGURE 3.295   Creating a revolution surface using different values for the parameters.

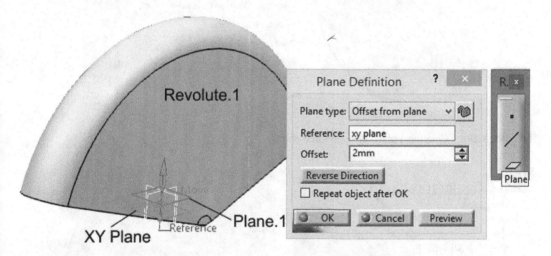

FIGURE 3.296   Creating the *Plane.1* with an offset of 2 mm above the *XY Plane*.

Returning to the angular values in Figure 3.294, the modelling of the part continues with the creation of *Plane.1* at a distance of 2 mm above the *XY Plane* (Figure 3.296).

The plane is used to cut surface *Revolute.1* using the *Split* tool. The area above the plane is then kept (Figure 3.297).

The selection of the surface to be removed and the one to be kept is done using the *Other side* button, the removed area going transparent.

After the cutting, some edges at the base of the surface result. The three points (marked *Point 1* to *Point 3*, Figure 3.298) are used as inputs for creating a new plane, *Plane.2*, by launching the *Plane* tool with the option *Through three points*. Points were not previously created by the user, but selected successively, as edge intersections, in the *Point* fields.

In the *Sketch.2* created in the *Plane.2*, a circle of diameter Ø9 mm is drawn, concentric with the circular edge of diameter Ø8 mm of the surface *Split.1* (edge on the right, Figure 3.299).

Another circle is drawn tangent to the circular edge resulted from the cut (details in Figure 3.300), even if the two objects are not in the same plane. The edge is in the *Plane.1*, and the circle in the *Plane.2*. For checking, this second circle tangent to the edge has a radius of about R7.738 mm.

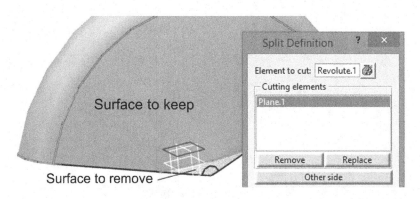

FIGURE 3.297   Creating a cut by using the *Split* tool.

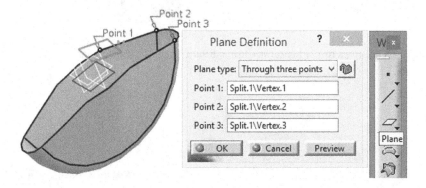

FIGURE 3.298   Creating a plane through three points.

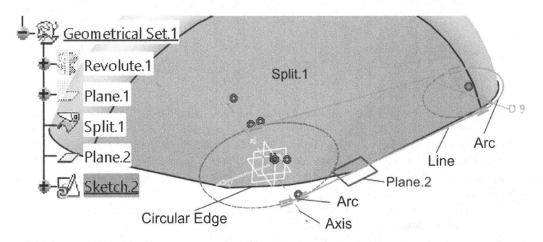

FIGURE 3.299   Creating the *Sketch.2*.

The tangent constraint (inward) between the circle and the edge was chosen because a coincidence constraint is not possible.

Two lines (*Bi-Tangent* Lines) are drawn between the circles, but also an axis (*Axis*) that goes through the centre of the large circle (Figure 3.299). After trimming some arcs (represented with a dashed line in the figure), an open profile results, with the endpoints on the axis. The profile of the sketch becomes a surface by rotating it around the axis with the *Revolute* tool (Figure 3.301).

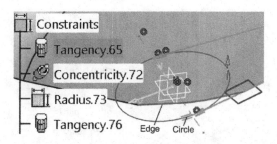

FIGURE 3.300 Creating the tangency between the objects in *Sketch.2*.

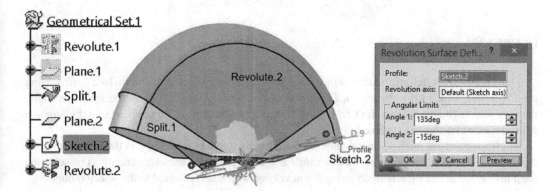

FIGURE 3.301 Creating the revolution surface *Revolute.2*.

In the *Revolution Surface Definition* dialog box, the sketch is selected in the *Profile* field, its axis is filled in automatically in the *Revolution Axis* field, and the two angles (*Angle* fields) receive the values of 135° and −15°, respectively. It is observed that the surface *Revolute.2* does not start from the plane of the sketch (*Plane.2*), but above it, at an angle of 15°.

Between the two surfaces *Split.1* and *Revolute.2* remains a space (Figure 3.302) due to the different values of diameters (Ø8 mm, Figure 3.293; respectively, Ø9 mm, Figure 3.299). The space must be closed with a surface using the *Fill* tool, but it requires edges to be selected to create a boundary.

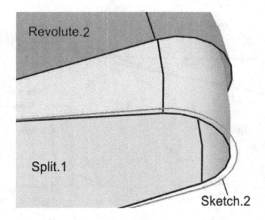

FIGURE 3.302 Showing the *Sketch.2*.

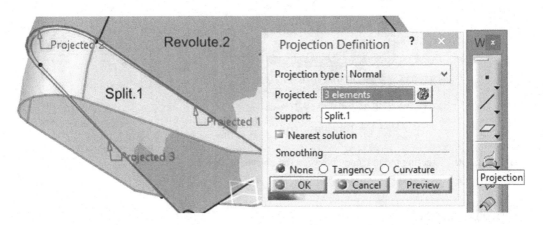

FIGURE 3.303   Projecting the *Revolute.2* edges on the *Split.1* surface.

Thus, three edges delimiting the surface *Revolute.2* will be projected on the surface *Split.1* using the *Projection* tool. In the *Projection Definition* dialog box (Figure 3.303), in the *Projected* field, the user selects (holding down the *Ctrl* key) the three edges marked in Figure 3.303. In the *Support* field, the user chooses the surface *Split.1*.

By clicking the *Fill* icon, a selection box is opened as shown in Figure 3.304; in the *Boundary* area, the user selects the edges of the surface *Revolute.2* and the projected ones sequentially. The selection must take place in one direction, choosing the next edge which is connected to the selected one.

For example, according to Figure 3.304, edges 1, 6 and 5 belong to the surface *Revolute.2*, whereas edges 4, 3 and 2 are the projected ones and belong to the surface *Split.1*. Thus, a new feature, *Fill.1,* is added to the specification tree (Figure 3.305).

By using the *Symmetry* tool in the *Operations* toolbar, a surface, identical to *Fill.1*, is obtained, named *Symmetry.1*, and positioned between the surfaces *Revolute.2* and *Split.1*, at the other end of the part (Figure 3.306). The reference feature for the mirroring is the *YZ Plane*.

The three surfaces, *Revolute.2*, *Fill.1* and *Symmetry.1*, are joined into a single one using the *Join* tool (Figure 3.307). This operation results in the surface *Join.1*, also added to the specification tree.

According to Figure 3.308, two surfaces are available: *Join.1* and *Split.1* (previously obtained; Figure 3.297). A trimming operation is applied between them, and the tips at the base of the part are removed. The surface *Trim.1* is added to the specification tree, and the user can hide (*Hide*) the other surfaces. The surfaces *Fill.1* and *Symmetry.1* closed the space between *Revolute.2* and *Split.1*, thus making the trimming operation possible.

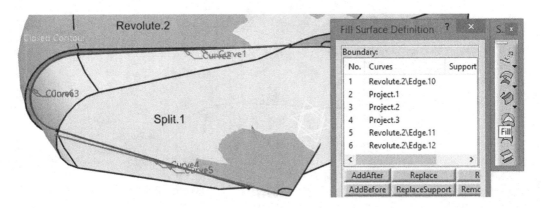

FIGURE 3.304   Creating a filled surface by using the *Fill* command.

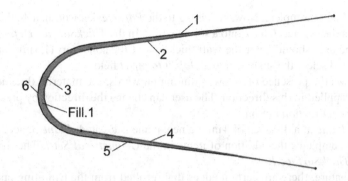

FIGURE 3.305 Explanation on how to select the edges for creating a boundary used by the *Fill* tool.

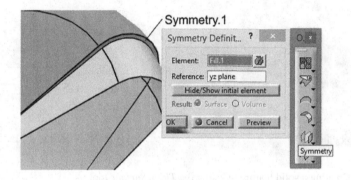

FIGURE 3.306 Mirroring a surface using the *Symmetry* command.

FIGURE 3.307 Creating a joined surface by using the *Join* command.

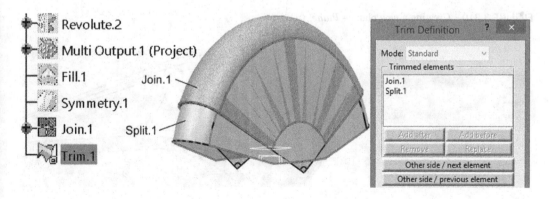

FIGURE 3.308 Applying the *Trim* command.

Surface modelling is complete. Now, switching to the *Part Design* workbench, the *Thick Surface* tool is used to turn the surface *Trim.1* into a solid feature. In the *ThickSurface Definition* dialog box (Figure 3.309), the user should enter the wall thickness of the solid part (1.3 mm outwards) in the *First Offset* field and select the surface in the *Object to offset* field.

The figure shows the presence of arrows pointing outwards, the material thickness of the *First Offset* field being applied in this direction. The user can change the direction by pressing the arrowhead or the *Reverse Direction* button.

Above the *XY Plane* at a distance of 4 mm, a new plane is created, *Plane.3*, according to Figure 3.310. This plane completes the addition of items in the *Geometrical Set.1*. The figure also shows the solid feature *ThickSurface.1*.

On this solid feature, there are certain edges that resulted from the trimming operation (Figure 3.308). These edges are connected using the *Edge Fillet* tool with the value of 0.3 mm (Figure 3.311).

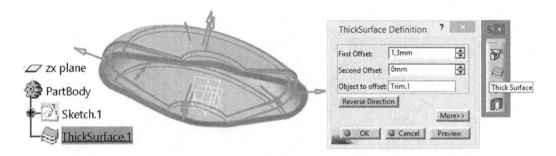

FIGURE 3.309 Creating a solid feature by using the *Thick Surface* tool.

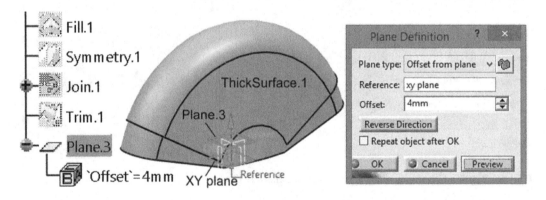

FIGURE 3.310 Creating a new plane – *Plane.3*.

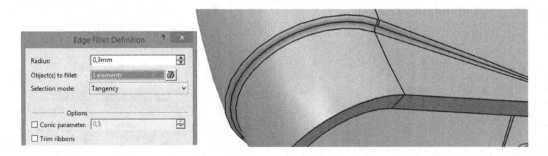

FIGURE 3.311 Applying the *Edge Fillet* tool.

According to the figure, these edges resulted from the initial delimitation of the surfaces *Fill.1* and *Symmetry.1* (Figures 3.305 and 3.306).

In the *XY Plane*, with the centre at the origin, a circle with a diameter of Ø16.4 mm (*Sketch.3*) is drawn, which is then extruded according to Figure 3.312. In the *Pad Definition* dialog box, in the *Selection* field, this sketch is selected and the extrusion type is *Up To Next*.

Basically, by this type of extrusion, the inner surface of the part is chosen as the limit, and the feature *Pad.1* takes its shape in its upper part.

In the *Plane.3*, a hexagon with the side dimension of 11 mm is drawn (*Sketch.4*, Figure 3.313), with the centre at the origin of the coordinate system (on the axis of the cylinder *Pad.1*). A perforated cut, *Pocket.1,* is made based on this hexagon, according to Figure 3.314. The screw head will enter this hexagonal cut and will only be allowed to rotate together with the handle knob.

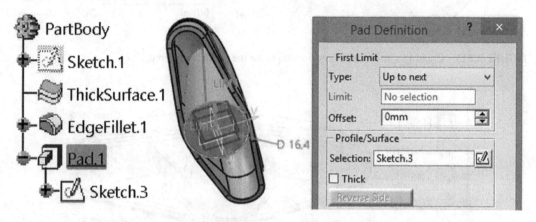

FIGURE 3.312 Creating a pad using the *Pad* tool with the option *Up to next*.

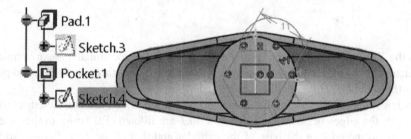

FIGURE 3.313 Creating a hexagon profile – *Sketch.4*.

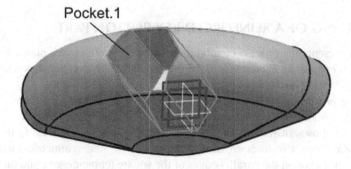

FIGURE 3.314 Creating a cut using the *Pocket* tool.

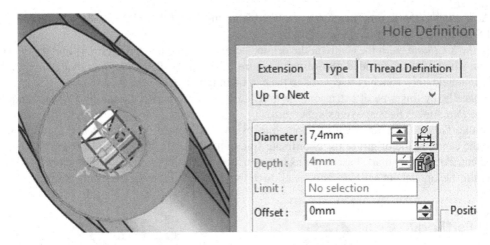

FIGURE 3.315   Creating a go-through hole – no input for the depth is required.

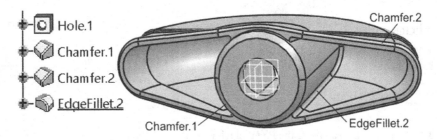

FIGURE 3.316   Applying the *Chamfer* and *Edge Fillet* tools as the final step of the modelling process.

Also, in the *XY Plane* and with the centre at the origin of the coordinate system, a hole (*Hole.1*) with a diameter of Ø7.4 mm is created (Figure 3.315). The hole goes through all the solid part. Being a simple go-through hole, *Hole.1* will not be threaded.

The part has rounded corners on the inside and two chamfers on the outside edges. According to Figure 3.316, the edges between the cylinder *Pad.1* are filleted (R0.5 mm) to the inner surface of the support and the edge at the base of the cylinder and that of the *ThickSurface* solid feature are chamfered (1×45° for *Chamfer.1* and 0.5×45° for *Chamfer.2*). Finally, the modelled part has a volume of 5.129 cm³.

3.15   MODELLING OF A REINFORCED KEY BUTTON PART

This application presents the process of solid and surface modelling of a plastic cover, reinforced inside. Figure 3.317 shows its 2D drawing. *Part Design* and *GSD* workbenches will be used for 3D hybrid modelling, while presenting certain modelling tools and techniques.

The modelling procedure starts in the *GSD* workbench by drawing a square of 30×30 mm in the *XY Plane* (*Sketch.1*), symmetrical to the *H* and *V* axes, according to Figure 3.318. Its corners are rounded with radiuses of R5 mm and R6 mm, and they are symmetrical to the *H* axis. In the *Sketch.2* of the *ZX Plane*, two lines are drawn inclined at 10° and symmetrical to the *V* axis. The lower ends of the lines are on the parallel edges of the square (coincidence), and the upper ones at a distance of 25 mm from the *H* axis (Figure 3.319).

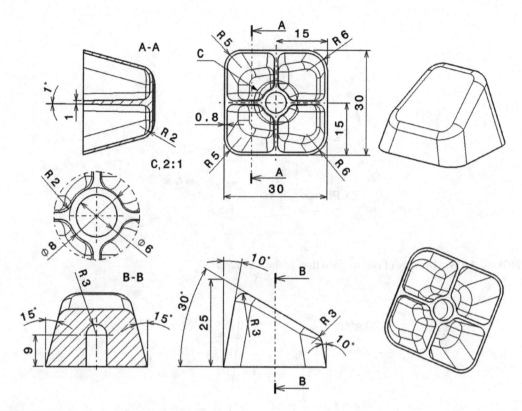

FIGURE 3.317   Drawing of the button part.

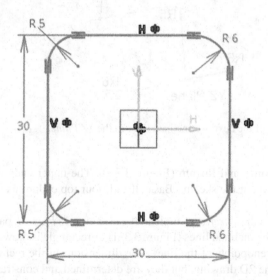

FIGURE 3.318   Rectangle sketch with rounded corners.

For the correct drawing of the lines, it is recommended first to measure the corners arcs of the base using the *Measure* tool. It can be seen in Figure 3.319 that one line has the lower endpoint between the R5 mm radius corners and the other line between the R6 mm radius corners, according to the 2D drawing. Similarly, in the *YZ Plane*, two other inclined lines are drawn, by 15° with respect to the *V* axis of *Sketch.3*. The lower endpoints of the lines are on the square edges, placed

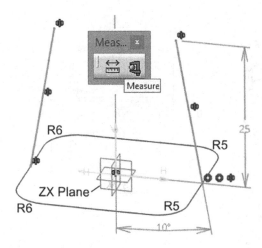

FIGURE 3.319   Drawing of two inclined lines in the *ZX Plane*.

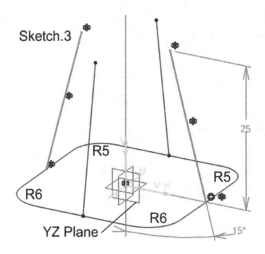

FIGURE 3.320   Drawing of the next two inclined lines in the *YZ Plane*.

between the radii of R5 mm and R6 mm (Figure 3.320). The upper endpoints are also 25 mm up from the *XY Plane* (*H* axis of the sketch). Basically, all four top endpoints of the inclined lines are in the same plane.

Using the *Plane* tool in the *Wireframe* toolbar, the user creates *Plane.1*, parallel to the *XY* through one of the endpoints of the inclined lines (Figure 3.321). A rectangle is drawn in this plane so that its edges coincide with the endpoints of the lines. The dimensions of the rectangle do not result from measures presented in the 2D drawing, but they are determined and constrained by the positions of the top ends of the previously created lines.

The corners of the rectangle that is located in the *Sketch.4* (Figure 3.322) correspond to those at the base: if the corner of the rectangle in the *XY Plane* has a R6 mm radius, then the corresponding corner of the top rectangle must have the same value (Figure 3.322). The correspondence of the values for these radii results from the 2D drawing.

Using the *Multi-sections Surface* tool, the user opens the selection box presented in Figure 3.323. On the *Section* column, the profiles from *Sketch.1* and *Sketch.4* should be selected. On the *Closing Point* column, the presence of two points: *Closing Point1* and *Closing Point2* can be noticed.

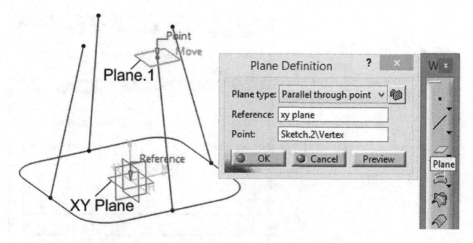

FIGURE 3.321 *Plane.1* parallel to the *XY Plane* at the top end of the lines.

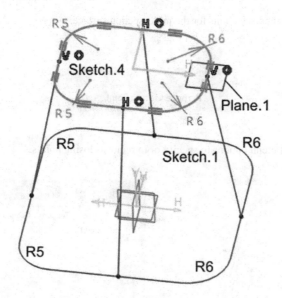

FIGURE 3.322 Rectangle with rounded corners in the *Plane.1*.

In this case, the points are correctly aligned by default, but if another one is needed, right-click on it in the *Closing Point* column and the context menu in Figure 3.324 appears to choose the *Replace Closing Point* option. The surface also requires the establishment of the other paired points in the *Coupling* tab. The user should select corresponding points found in the sketches *Sketch.1* and *Sketch.4* (Figure 3.325).

Figure 3.325 shows how seven coupling correspondences between points of the two sketches were established. The user should note that no correspondence is established or needed between the two *Closing Points*. The representation of the points, in figure, was highlighted graphically using larger black dots, and the correspondence between them is marked by the program with a thin green line. Along with these lines, the names *Coupling1*, *Coupling2*, etc. will appear.

When pressing the *Preview* button, a pre-visualization of the surface between the two sketches *Sketch.1* and *Sketch.4* will appear. Figure 3.326 shows how the surface was generated in the correct way. Because they are no longer needed, *Sketch.1* to *Sketch.4* and *Plane.1* can be hidden (*Hide/ Show* option in the context menu of each feature).

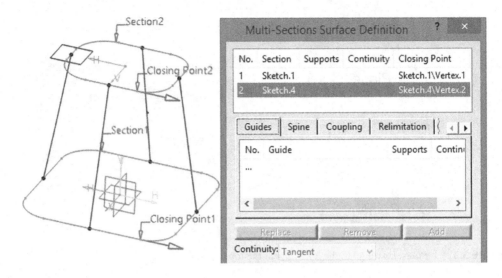

FIGURE 3.323   Settings and selections for the multi-section surface.

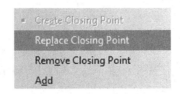

FIGURE 3.324   Context menu options to specify or to remove closing points.

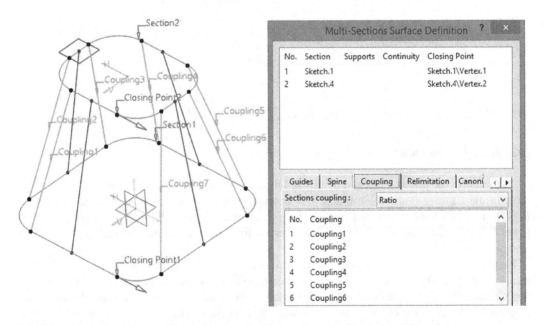

FIGURE 3.325   Choosing coupling points in pair of points from the *Sketch.1* and *Sketch.4*.

In the new *Sketch.5* of the *ZX Plane*, a vertical 25 mm long construction line is drawn using the *Construction/Standard Element* tool from the *Sketch Tools* toolbar. Figure 3.327 shows this construction line as a dashed line and how it is distanced 15 mm from the *V* axis of the sketch coordinate system.

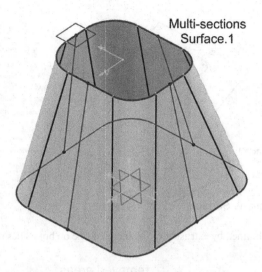

FIGURE 3.326 The *Multi-sections Surface.1* is created and displayed between the two sketches.

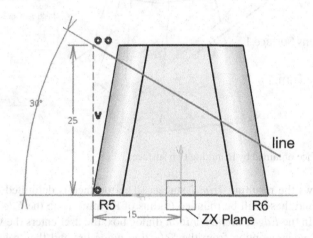

FIGURE 3.327 Drawing of an inclined line in the *ZX Plane*.

From the upper end of the previously created construction line, the user begins to draw a second line of standard type (represented as a continuous line), inclined at 30° from the *H* axis, and it is located between the connected faces with radii of R5 and R6 mm, respectively. The line exceeds *Multi-sections Surface.1*. Using the *Extrude* icon in the *Surfaces* toolbar, the inclined line of *Sketch.5* is extruded on both sides of the *ZX Plane*, for a distance of 15 mm.

Figure 3.328 shows how the flat surface *Extrude.1* intersects and exceeds *Multi-sections Surface.1*. In the *Extruded Surface Definition* dialog box, in the *Profile* field, *Sketch.5* should be selected. Also, in the *Direction* field, by right-clicking from the list, the *Y* axis (*Y Component*) is chosen and the *Mirrored extent* option is checked. Above, in the *Dimension* field, a value of 15 mm is entered.

Between the two surfaces *Extrude.1* and *Multi-sections Surface.1*, the *Trim* editing operation is applied (Figure 3.329). Thus, in the *Trim Definition* dialog box, the user selects the surfaces in the *Trimmed elements* field, and then he presses the *Other side/next element/previous element* buttons until the areas of the surface marked in the figure are removed. Their appearance is transparent in a preview, and the kept areas (and that are forming the *Trim.1* surface) are opaque.

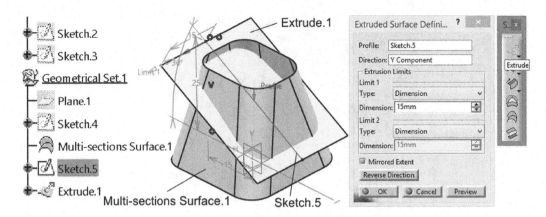

FIGURE 3.328   Surface obtained by extrusion of the inclined line on both sides of the *ZX Plane*.

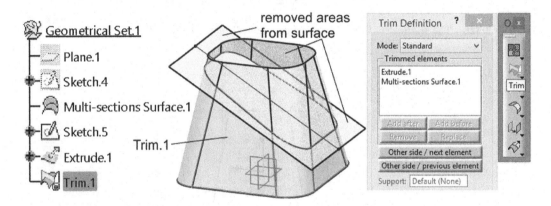

FIGURE 3.329   Surface obtained by trimming two surfaces.

Figure 3.330 shows the resulting *Trim.1* surface. The top face is delimited by certain straight and circular edges, and these will be rounded (radius of R3 mm) using the *Edge Fillet* icon on the *Operations* toolbar. In the *Edge Fillet Definition* dialog box, the user enters the value in the *Radius* field, chooses the *Tangency* option from the *Selection mode* list and then selects the face edges (multiple selection with the *Ctrl* key pressed).

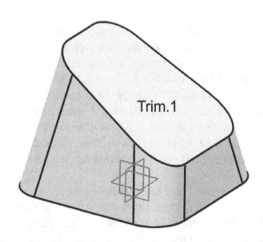

FIGURE 3.330   *Trim.1* surface after the removal of certain areas.

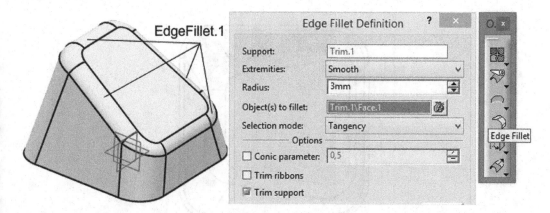

FIGURE 3.331   *Edge Fillet* tool applied on the edges of the top surface.

Also, instead of this selection of all edges, the flat face can be directly selected by clicking on it (which, of course, implies the default selection of edges; Figure 3.331) by choosing only a single edge. The *Tangency* option has the role of automatically filleting all edges that are tangent consecutively. The result is a single surface, *EdgeFillet.1*, placed in the specification tree.

In this stage, the user switches to the *Part Design* workbench and uses the *Thick Surface* tool on the *Surface-Based Features* toolbar to transform the *EdgeFillet.1* surface into a solid 3D object. In the *ThickSurface Definition* dialog box, in the *First Offset* field, he enters the part wall thickness of 0.8 mm, according to the 2D drawing (Figure 3.317).

By pressing the *Preview* button, the button is displayed with several arrows pointing inwards or outwards the part. Since the previously modelled surfaces are considered for the outer dimensions of the part, these arrows must be pointed inwards. Changing the direction is done by clicking on the arrows' tip or by pressing the *Reverse Direction* button (Figure 3.332). The *EdgeFillet.1* surface can now be hidden in the specification tree using the *Hide/Show* option in its context menu.

From the *Insert* menu, a new body, named *Body.2*, is added to the specification tree. Once inserted, it is observed that the new body is underlined, being considered therefore as the current body. At the base of the solid, in the *XY Plane*, a profile is drawn in the *Sketch.6*, consisting of two rectangles 1 mm wide and 34 mm long and a circle having the diameter of Ø8 mm, with the centre at the origin of the coordinate system. These two rectangles have their edges equidistant from this origin and perpendicular to each other. Certain line segments and arcs are removed by trimming to obtain a closed contour (Figure 3.333), according to the 2D drawing.

This profile of the *Sketch.6* is extruded on a distance of 23 mm, resulting in the *Pad.1* feature in the specification tree. Figure 3.334 shows that *Pad.1* is contained in *Body.2* and intersects *PartBody*, and it is represented in a different colour. The extrusion direction is on the Z+ axis.

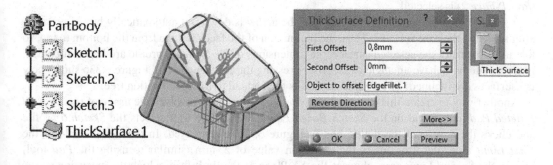

FIGURE 3.332   The surface feature *EdgeFillet.1* is transformed into a solid part.

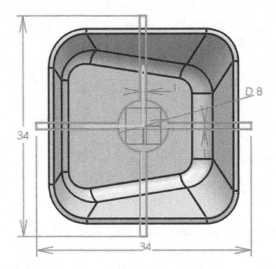

FIGURE 3.333 Drawing of two rectangles and a circle in the *XY Plane.*

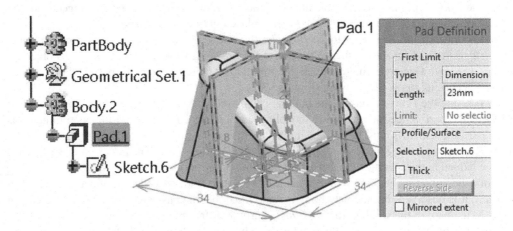

FIGURE 3.334 Profile of *Sketch.6* is extruded to obtain the *Pad.1* feature in *Body.2.*

According to the section *A-A* of the 2D drawing, an inclination of 1° of the walls of *Pad.1* is observed. Thus, the *Draft* tool in the *Dress-Up Features* toolbar is used. In the *Draft Definition* dialog box, the user enters the inclination angle in the *Angle* field and then selects the twelve side faces (flat and cylindrical) of the *Pad.1* feature. Figure 3.335 shows three of these faces, as an example. As a neutral element, against which the inclination is calculated, the flat surface at the base of *Pad.1* (*Pad.1\Face.2*) is selected.

It should be observed that the direction of the arrow is down (set automatically by the program after selecting the base surface). Thus, the inclination of the faces is done from the bottom to top, so that in the plane of the selected surface, the dimensions of the *Sketch.6* profile are kept. Of course, two opposite faces have an angle of 2°. After pressing the *Preview* button, Figure 3.336 shows how the surfaces are inclined. The *Draft.1* feature is then added to the specification tree.

Another way to create this extrusion and drafting in one step involves the usage of the *Drafted Filleted Pad* tool, found on the *Sketch-Based Features* bar. The user selects the *Sketch.6* profile and clicks the icon, and the dialog box in Figure 3.337 is displayed. In the *Length* field in the *First Limit* area, the user enters the extrusion value of 23 mm (similar to using the *Pad* tool) and, in the *Second Limit* area, chooses the *XY* Plane as the limit from which the extrusion starts.

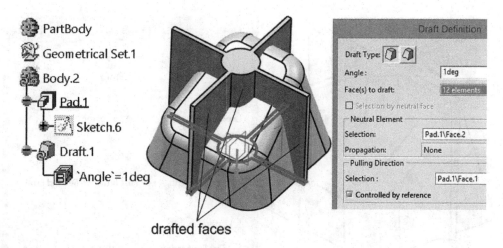

drafted faces

FIGURE 3.335   Selection of faces to be drafted.

The *Angle* and *Second limit* options in the *Draft* area are checked, and then a value of −1° is entered to set the inclination.

The angle value is negative because only through this parameter the inclination can be controlled. Pressing the arrow tip (pointing downwards) causes the extrusion and inclination to be created below the *XY Plane* that contains the profile. The *Radius* options in the *Fillets* area must also be unchecked because the features contained in the *Body.2* do not have fillets.

The specification tree is updated in both cases with the features *Pad.1* and *Draft.1*.

A *Union Trim* operation is performed between *Body.2* and *PartBody* to combine the two bodies. Thus, in the *Trim Definition* selection box (Figure 3.338), *Body.2* is selected in the *Trim* field and the *With* field is automatically filled in with *PartBody* (non-editable field). The *Faces to keep* field contains five side end surfaces and the upper face of the *Body.2* (three of these surfaces are marked in the figure). The *Faces to keep* field contains outer surfaces of the *PartBody* (marked also on the figure).

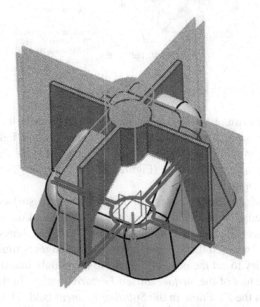

FIGURE 3.336   Direction of drafting and a preview on how the faces will be inclined.

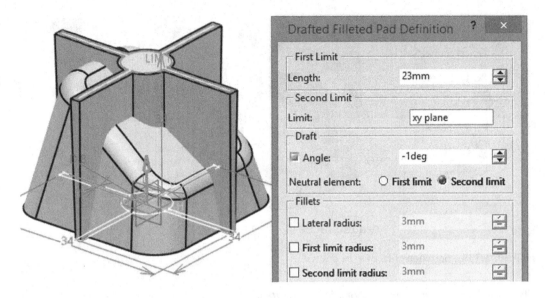

FIGURE 3.337 Creating a drafted pad.

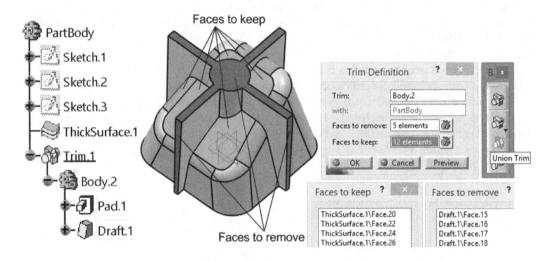

FIGURE 3.338 Selection of faces to keep and faces to remove for the *Union Trim* operation.

When selecting surfaces (which will be removed and/or kept), they will change colour; for example, the removed surfaces become purple and the retained ones become light blue.

Following the union trim of the two bodies, *Body.2* is included in the *PartBody* structure under the *Trim.1* feature in the specification tree (Figure 3.338). Only the geometry placed inside the button part remains from *Body.2*.

Due to the way in which the *Thickness* of the *EdgeFillet.1* surface was added (inwards, Figure 3.332), the lower edge of the solid exceeds the *XY Plane* and the inner faces (obtained by combining with *Body.2*), as it is presented in Figure 3.339. The figure shows, for instance, two areas (of the four existing ones) in which the difference between the edges is illustrated.

It is, therefore, necessary to cut the edge of the part to be brought into the *XY Plane*. The fastest solution is to use the *Split* tool of the *Surface-Based Features* toolbar. In the *Split Definition* selection box, the user chooses the *XY Plane* in the *Splitting Element* field. The arrow that defines what to be kept must be positioned with the tip inwards. Following the elimination of the volume beyond the *XY Plane*, a continuous surface results, a fragment of it is shown in Figure 3.340.

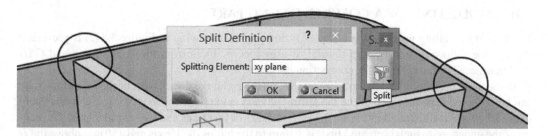

FIGURE 3.339 Problems at the base of the button – the lower edge of the solid exceeds the *XY Plane*.

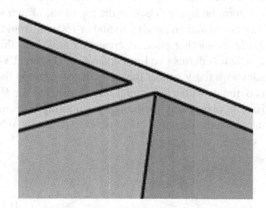

FIGURE 3.340 After splitting, the base of the part is in the *XY Plane*.

The *Split.1* feature is added to the specification tree (Figure 3.341). Using the *Split* tool is very easy and fast, with a just few selections. Another option would be for the user to create a sketch in the *XY Plane* (a rectangle) and then to use it in a *Pocket* cut of the respective volume. The interior of the part should be filleted, according to the 2D drawing in Figure 3.317.

Figure 3.341 shows on the left image certain edges to be selected, and on the right image the result of applying the *Edge Fillet* tool with a radius of R2 mm.

To check the final volume of the reinforced cover part, the correct value is 3.241 cm^3 (using the *Measure Item* tool).

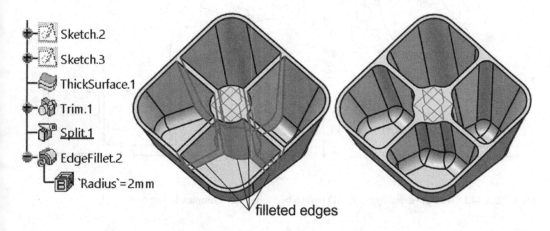

FIGURE 3.341 Interior of the button part before and after filleting the inner edges.

3.16   MODELLING OF A COMPLEX FITTING PART

The hybrid modelling process of a complex fitting part (2D drawing in Figure 3.342) is presented through the following step-by-step tutorial. It uses specific tools from the *Part Design* and *GSD* workbenches. The part is used for connection purposes.

In the main projection and in the adjacent one, on the right, the dimensions of the axis of the tube can be observed. The first step of the modelling process is to create the *Sketch.1* (*GSD* workbench) in the *YZ Plane* as illustrated in Figure 3.343. It consists of two lines and two circular arcs. The centre of the upper arc (which is created first) is 10 mm to the left of the *V* axis and 150 mm above the *H* axis. The centre of the lower arc is 50 mm above the *H* axis, and the arc is tangent to the vertical line of 50 mm length (coinciding with the *V* axis). A second line is drawn between the two arcs, tangent at both ends. Instead of the arcs, the user is advised, as an alternative, to draw circles that will later be trimmed to the lines shown in the figure. Thus, at the top of the sketch was drawn a horizontal line with the role of auxiliary construction, parallel to and at 150 mm from the *H* axis.

At 50 mm from the *ZX Plane*, another plane is created, *Plane.1*. In this new plane, a second sketch is created, *Sketch.2*, which contains two lines and an arc (Figure 3.344).

The vertical line coincides with the *V* axis of the coordinate system in the *Plane.1*, and it is tangent to an arc, which then continues with a horizontal line (parallel to the *H* axis).

Both sketches are in different planes and obviously have nothing in common. They were drawn according to the representations in the drawing, but in order to create the tube, the user must obtain a single path from the two curves.

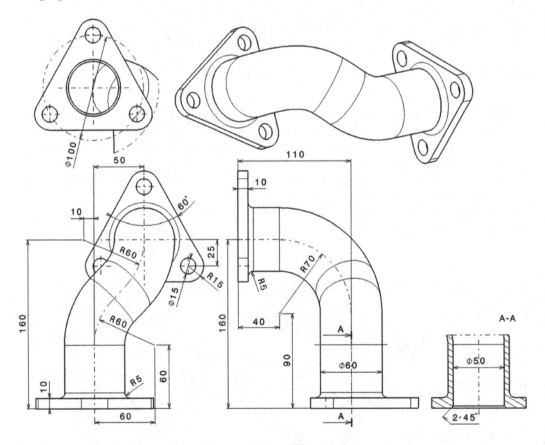

FIGURE 3.342   The 2D drawing of the 3D part to be modelled – complex fitting part.

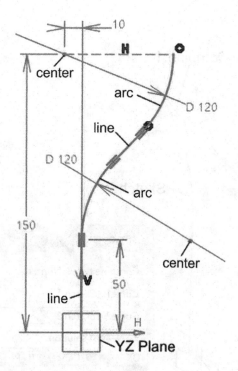

FIGURE 3.343   Creating the first sketch – *Sketch.1*.

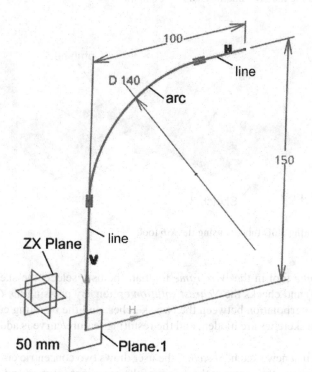

FIGURE 3.344   Creating the second sketch, *Sketch.2*, in the *Plane.1*.

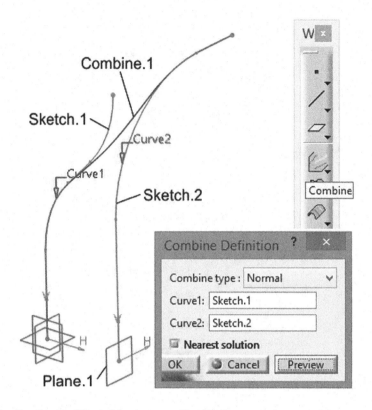

FIGURE 3.345 Creating the 3D combined curve – *Combine.1*.

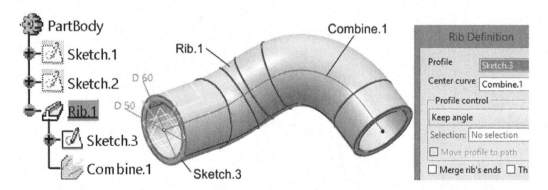

FIGURE 3.346 Creating a 3D tube by using the *Rib* tool.

Using the *Combine* tool in the *Wireframe* toolbar, the user selects the sketches in the *Curve* fields (Figure 3.345) and checks the *Nearest solution* option. By pressing the *OK* button, the program performs an interpolation between the two sketches, and the resulting curve, *Combine.1*, is tangent to both. The sketches are hidden, and the resulting feature/curve is added to the specification tree.

In the *XY Plane*, in a new sketch, *Sketch.3*, the user draws two concentric circles, with diameters Ø50 and Ø60 mm, having the centre at the origin of the coordinate system and coinciding with the lower end of the curve *Combine.1*. In the *Part Design* workbench, the two circles are extruded along the curve, using the *Rib* tool (Figure 3.346).

Following this extrusion, the tube is obtained, hollow inside, because two concentric circles were used in the profile sketch; the volume obtained from the rib extrusion of the small circle is automatically extracted from the resulting volume similarly involving the large circle.

The part contains two triangular flanges at the ends of the tube. Thus, at the top end, in a new sketch, *Sketch.4*, the profile from Figure 3.347 is drawn. The lower side of the triangle is parallel to the *XY Plane*. The corners of the triangle are rounded (*Corner* tool of the *Sketcher* workbench) with a radius of R15 mm. By extruding the triangle using the *Pad* tool, the body of the flange is obtained, which is then connected (R5 mm) with it. The features *Pad.1* and *EdgeFillet.1* (Figure 3.348) were added to the specification tree.

Through the flange, in order to ensure the opening of the tube, a hole is created, *Hole.1* (Figure 3.349), with a diameter of Ø50 mm. The hole uses the *Up To Next* option and has its centre on the flat face of the flange. There are several ways to position this centre, but the simplest and most used is to make it concentric with the circular edge on the other side, the back of the flange, obtained by connecting *EdgeFillet.1* (Figure 3.350). The figure also shows the context menu for setting the geometric constraint.

Another option would be the coincidence of the centre with the axis of the cylindrical surface of the tube. The axis is selectable in the positioning sketch only after the constraint tool is activated.

The edge of the hole is chamfered 2×45° (*Chamfer.1*), and then the *Hole.2* is created (Figure 3.351) with a diameter of Ø15 mm, throughout the whole flange, with the centre in one of the centres of the circular edges of the flange. In a new *Sketch.7,* on the flat face of the flange, two points are positioned in the centres of the other two circular corners. Using the *User Pattern* tool, the user multiplies the *Hole.2* in the two new positions (Figure 3.352).

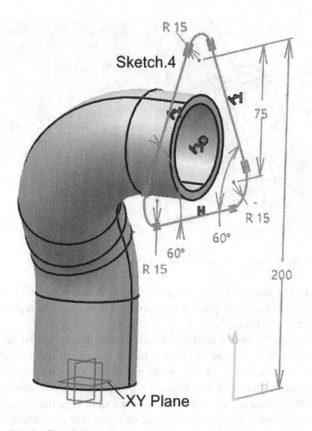

FIGURE 3.347   Creating the *Sketch.4*.

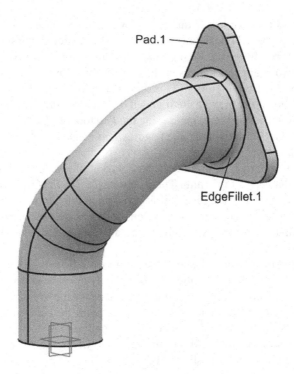

FIGURE 3.348   Creating a pad and applying the *Edge Fillet* tool.

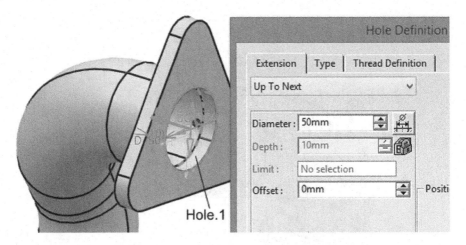

FIGURE 3.349   Creating the *Hole.1* on a planar face of the flange.

At the other end of the tube, the part has an identical flange. Of course, following the steps above, it can be modelled again, but a quick way to obtain it is to copy and reposition the first flange, according to the 2D drawing (Figure 3.342).

Therefore, the following method is applied: the user holds down the *Ctrl* key, clicks on the feature *Pad.1* in the specification tree and moves the mouse cursor on the flat surface at the second end of the tube. The representation of the 3D model must be at a zoom factor large enough that the selection of the flat surface between the two circular edges (a relatively narrow area according to the size of the tube) with the mouse is easy. When the cursor touches the flat surface, its representation contains an arrow (cursor) and a plus sign (+) with the meaning that the feature *Pad.1* will be copied to the new location (on the flat surface), resulting in the feature *Pad.2* and the *Sketch.8*. An error message is displayed on the screen because the *Sketch.8* requires editing (Figure 3.353).

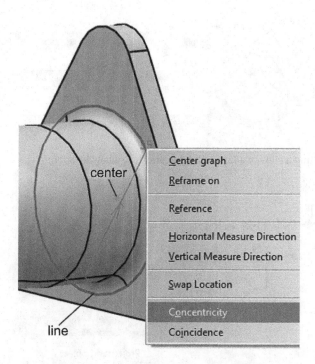

FIGURE 3.350 Positioning the *Hole.1* after applying the *Edge Fillet* tool.

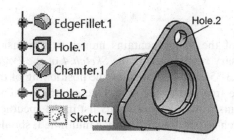

FIGURE 3.351 Creating the *Hole.2* on a planar face of the flange.

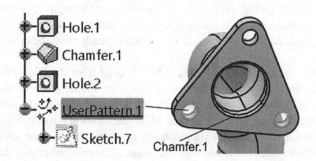

FIGURE 3.352 Applying the *User Pattern* tool to multiply the *Hole.2*.

Thus, in the *Update Diagnosis* dialog box, the user clicks the *Edit* button and opens the sketch. Note that a certain feature, *Projection.1*, is the error generator. This feature is removed from the specification tree (with *Delete*), which is in the *Use-edges* geometrical set (Figure 3.354).

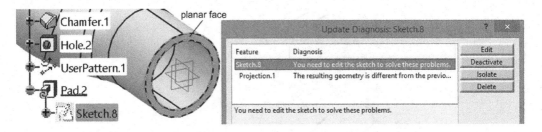

FIGURE 3.353   Creating a copy of a feature by dragging it to a new location.

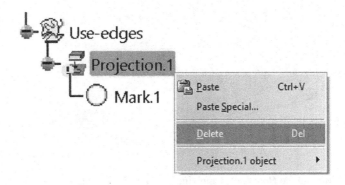

FIGURE 3.354   Deletion of the *Projection.1*, the error generator, from its context menu.

The triangular profile of the flange contains numerous dimensional and geometric constraints, which were necessary when drawing the *Sketch.4* of the *Pad.1* feature (Figure 3.347). These are shown in Figure 3.355 and do not allow the positioning of the sketch (and, implicitly, of the feature *Pad.2*) at the free end of the tube. One solution would be to remove the *Angle*, *Radius* and *Offset* constraints from the *Constraints* list in the specification tree. The user must also be careful that the sketch he has drawn (and then multiplied) should not contain coincidence constraints.

Figure 3.356 shows the sketch without the dimensional constraints, but with an auxiliary construction circle added (*Construction Element* option in the *Sketch tools* toolbar). The circle is drawn with the *Three Point Circle* tool of the *Profile* toolbar and goes through the centres of the three arcs (the fillets of the triangle corners). Its construction is needed because in the next step, the profile rotates around the centre, and then it is positioned concentrically with the circular edge at the end of the tube.

Thus, in the sketch, the *Rotate* tool (Figure 3.357) is used from the *Operation* toolbar to rotate the profile. In the *Rotation Definition* dialog box, the user should uncheck the *Duplicate mode* option (the *Instances* field turns out grey), select the initial profile through the selection window and choose the centre of the circle as the point around which the rotation is made. Then, the user chooses a reference line (*reference line for angle*) horizontally to the right and enters a value of 300° below in the *Value* field. All selections are made with the mouse. The original profile disappears, and only the rotated one remains.

The profile must be positioned at the free end of the tube, according to Figure 3.358. A concentric constraint is established between the auxiliary construction circle and the circular edge at the free end of the tube.

The profile is extruded 10 mm to create the second flange of the part, the direction being to its outside. The feature *Pad.2* is added to the specification tree (Figure 3.359).

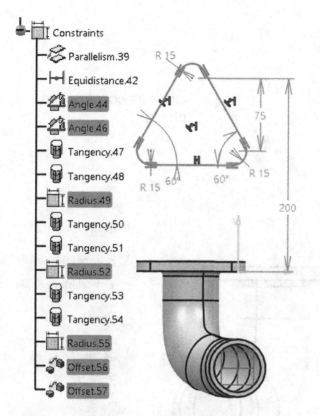

FIGURE 3.355 Editing the *Pad.2* sketch by removing some constraints.

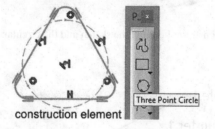

FIGURE 3.356 Creating a circle as a construction element.

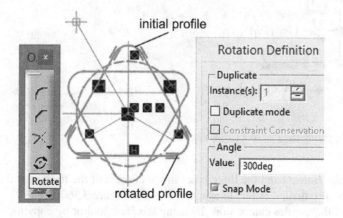

FIGURE 3.357 Applying a rotation to the flange profile.

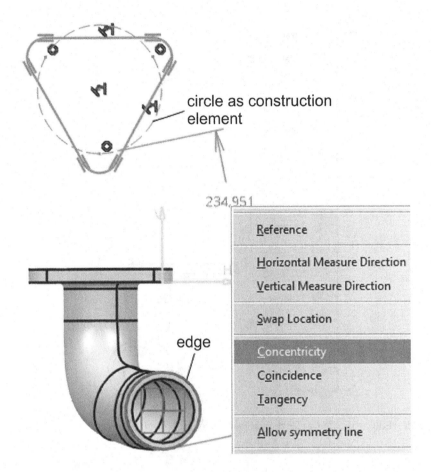

FIGURE 3.358   Creating the constraint between the sketch and the circular edge.

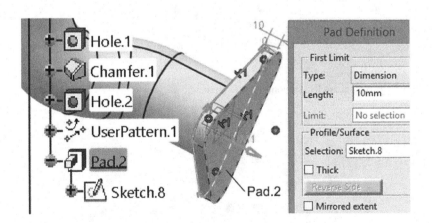

FIGURE 3.359   Applying the *Pad* tool to the sketch.

The central hole (*Hole.3*) and the three holes in the corners of the triangular flange (*Hole.4* and *CircPattern.1*) are also drilled in this flange, according to Figure 3.360.

The creation of these holes can be done by using the *Hole* tool or by copying and repositioning the previous ones (*Hole.1* and *Hole.2*).

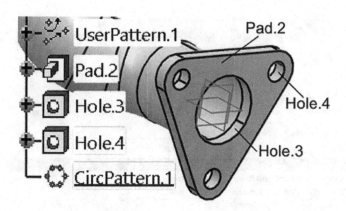

FIGURE 3.360 Drilling the second flange.

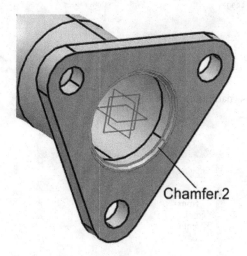

FIGURE 3.361 Applying the *Chamfer* tool on a circular edge.

The *Hole.3* has also a 2×45° chamfered edge. The *Chamfer* tool is used to add a chamfer, but this can also result from copying the *Chamfer.1* feature to that edge. Thus, the *Chamfer.2* feature is obtained in the specification tree (Figure 3.361), but the chamfer values, even if they are initially equal (by copying), do not remain the same if, for example, the *Chamfer.1*'s value changes.

To preserve the same value of the chamfers, the user can create a parameter (*user parameter*), with a single value or multiple values, and then apply this parameter to other chamfer features.

First, it is necessary to select certain options in the menu *Tools → Options → General → Parameters and Measure → Knowledge* tab. The user checks the options *With value* and *With formula* and, then in the menu *Tools → Options → Infrastructure → Part Infrastructure → Display* tab, checks the options *Constraints*, *Parameters* and *Relations*. Thus, the parameters and relations that will be entered are visible in the specification tree.

Clicking on the *Formula f(x)* icon in the *Knowledge* toolbar opens the *Formulas* dialog box. There is a list of parameters (Figure 3.362) that can be sorted by several criteria using the *Filter Type* field. A new *Length* parameter will be created with *Multiple Values*. When pressing the *New Parameter of Type* button, the *Value List* dialog box opens (Figure 3.363), in which the user will enter values for the bevel chamfer. After each value entered, the user presses *Enter*, and it moves from the top field to the list below. The final list will contain four values (1.5, 2, 2.5 and 2.8 mm).

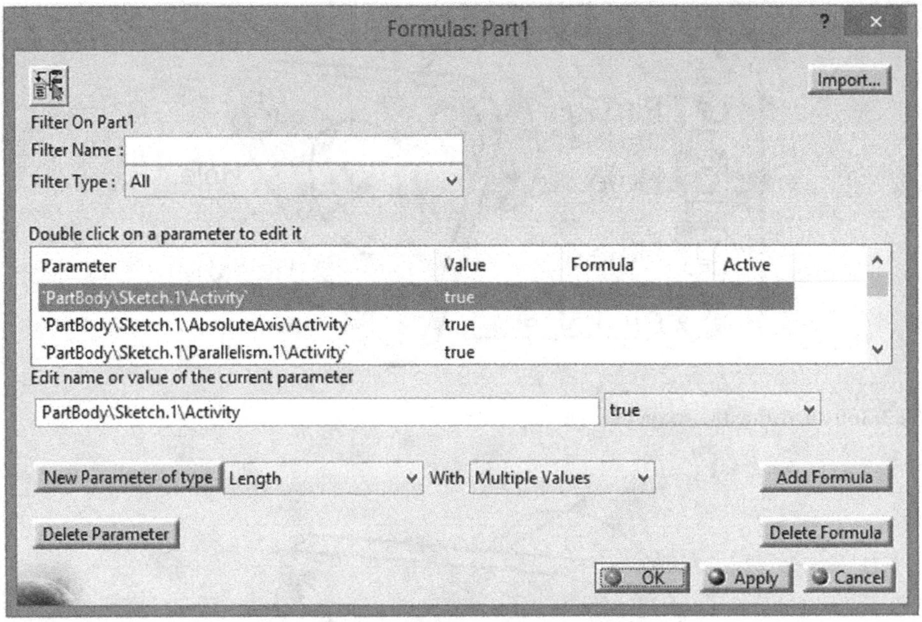

FIGURE 3.362   Adding a new parameter of the *Length* type.

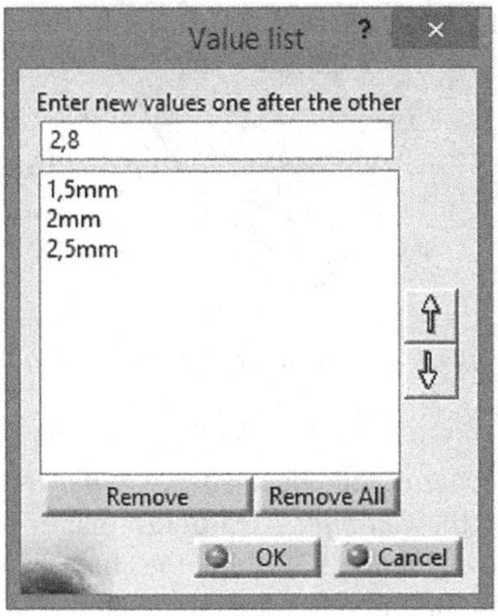

FIGURE 3.363   Populating the list of values for the new parameter.

Each entry in the list contains the value and the unit of measurement (mm). The unit of measurement is added by default by the program (or by the user if he wishes, but is not required) and corresponds to the type of parameter chosen (length), but also conforms to the choice made by the user in the menu *Tools → Options → General → Parameters and Measure → Units* tab (Figure 3.364).

The *Length.1* parameter becomes available in the specification tree (Figure 3.365).

If needed, the name of this parameter can be changed from its context menu. Thus, by right-clicking on the name, the menu opens, and from the list of options, the user chooses *Properties*. The name can be changed in the *Local name* field (Figure 3.366), and a comment related to the parameter can be entered in the *Comment* field also. Below, the *Constant* option locks the parameter to the current value, and *Hidden* hides it from the specification tree.

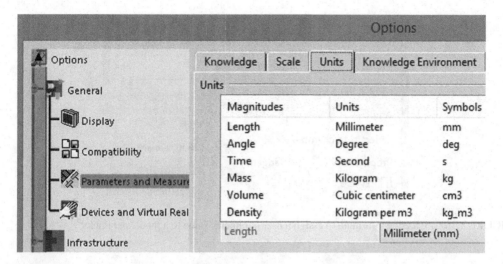

FIGURE 3.364   The settings for the units in the *Tools* menu.

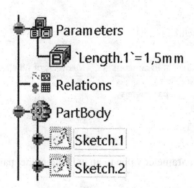

FIGURE 3.365   Displaying the parameter *Length.1* in the specification tree.

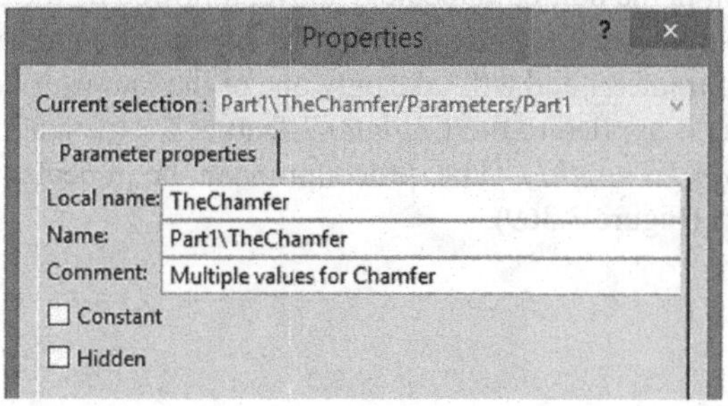

FIGURE 3.366   Editing a parameter name.

This parameter (called *TheChamfer*, which can be found also in the specification tree) must be applied to both chamfer features. Thus, a double-click on *Chamfer.1* will open the dialog box in Figure 3.367, where the user can observe the value of 3 mm in the *Length.1* field. In this field, the right mouse button is pressed and the context menu becomes available (Figure 3.367), from which the *Edit formula* option should be clicked.

Chamfer Definition ? ✕

Mode:	Length1/Angle ⌄
Length 1:	3mm ▲
	Edit formula...
Angle:	45deg
	Edit...
Object(s) to chamfer:	CircPattern.1\
	Add tolerance...
Propagation:	Tangency
	Change step
☐ Reverse	
	Measure Between...

FIGURE 3.367   Creating a formula to assign a user parameter value to a model parameter.

Formula Editor : `PartBody\Chamfer.1\ChamferRibbon.1\Length1`   ? ✕

PartBody\Chamfer.1\ChamferRibbon.1\Length1 =

`TheChamfer`

Dictionary	Members of Parameters	Members of All
Parameters	All	`TheChamfer`
Part Measures	Renamed parameters	
Circle Constructors	Length	
Design Table		
Direction Constructors		

FIGURE 3.368   Assigning a user parameter value to an existing model parameter.

According to Figure 3.368, a parameter selection box becomes available (based on criteria/filters: *Dictionary*, *Members of Parameters* and *Members of All*), the parameter that defines the chamfer value is already present in the non-editable field above, followed by the equal sign (=), and in the lower editable field, the parameter *TheChamfer* from the specification tree is selected.

TheChamfer is a user parameter and will take only one of the four user-defined values, according to Figure 3.363. The value is applied to the *Chamfer.1* feature for the length parameter (*PartBody\Chamfer.1\ChamferRibbon.1\Length1*). The procedure can be repeated to create an equality between the two chamfers (Figure 3.369).

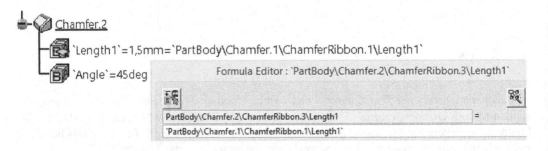

Chamfer.2

`Length1`=1,5mm=`PartBody\Chamfer.1\ChamferRibbon.1\Length1`

`Angle`=45deg   Formula Editor : `PartBody\Chamfer.2\ChamferRibbon.3\Length1`

PartBody\Chamfer.2\ChamferRibbon.3\Length1 =

`PartBody\Chamfer.1\ChamferRibbon.1\Length1`

FIGURE 3.369   Creating a formula for both chamfer features.

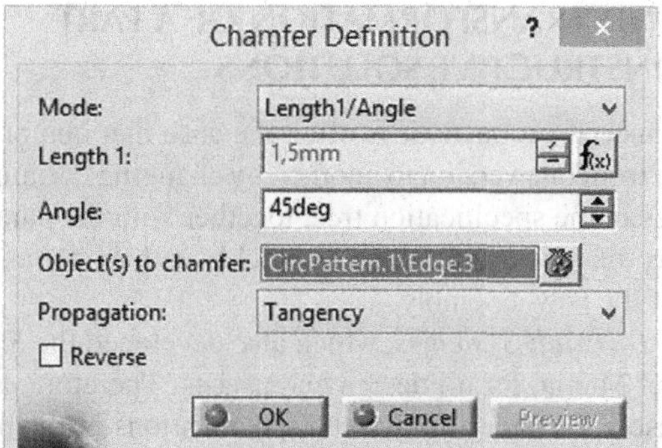

Chamfer Definition	?	×

Mode: Length1/Angle ˅

Length 1:   1,5mm $f_{(x)}$

Angle: 45deg

Object(s) to chamfer: CircPattern.1\Edge.3

Propagation: Tangency ˅

☐ Reverse

⬤ OK   ⬤ Cancel   Preview

FIGURE 3.370 Modifying an already established formula.

Of course, *Chamfer.2* takes the value of the parameter *TheChamfer*. If the user tries to change directly (by double-clicking) the value of the chamfer for either of the two features *Chamfer.1* or *Chamfer.2*, he will notice that it is not editable (*Length.1* field in Figure 3.370), the only possibility being through the formula (the *f(x)* icon).

The previously established formulas appear in the list of the *Formulas* dialog box (Figure 3.371) displayed by clicking the *Formula f(x)* icon in the *Knowledge* toolbar, but also in the specification tree → *Relations* (Figure 3.372).

Modelling solution: https://youtu.be/LyILHjZYbEA.

Double click on a parameter to edit it

Parameter	Value	Formula	Active
`PartBody\Hole.4\Tap depth`	10mm		
`PartBody\Hole.4\Tap diameter`	55mm		
`PartBody\Hole.4\Pitch`	1mm		
`PartBody\CircPattern.1\CircleSpacing`	20mm		
`PartBody\Chamfer.2\ChamferRibbon.3\Length1`	1,5mm	= `PartBody\Chamfer.1\ChamferRibbon.1...`	yes
`Geometrical Set.1\Plane.1\Offset`	50mm		
`TheChamfer`	1,5mm		

FIGURE 3.371 Displaying the parameters list and formulas.

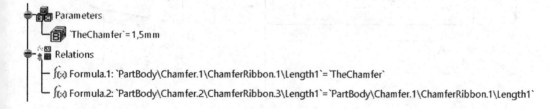

◆ Parameters
  `TheChamfer`=1,5mm
◆ Relations
  $f_{(x)}$ Formula.1: `PartBody\Chamfer.1\ChamferRibbon.1\Length1`=`TheChamfer`
  $f_{(x)}$ Formula.2: `PartBody\Chamfer.2\ChamferRibbon.3\Length1`=`PartBody\Chamfer.1\ChamferRibbon.1\Length1`

FIGURE 3.372 Displaying the *Relations* feature in the specification tree.

3.17 MODELLING AND TRANSFORMATION OF A PART INTO TWO CONSTRUCTIVE SOLUTIONS

In all design stages of industrial products, it is often the case that one or more components of an assembly change or evolve from one version to another, by changing certain dimensions, and also by appearance of new features in the specification tree, together with the parameters that define them. Thus, a 3D model of a part must be transformed as quickly and simply as possible to meet various functional requirements of the new assembly.

Every year since 2000, *Dassault Systèmes*, which also developed the *SolidWorks* solution, organizes a competition, Model Mania, for all design enthusiasts. Therefore, a 2D drawing of a part is proposed, which is in phase I. Considering that the specifications of the project change, a new 2D drawing of the same part is later proposed, which, however, contains some modifications in geometry and dimensions. This is the second phase of the part, and the community is looking for solutions to transform the 3D model of the part of phase I and obtain the model proposed by phase II with a minimum of modifications and features added/changed in the specification tree. Every year, the parts of both phases are very interesting, and the challenges of 3D modelling are high. More details are available at: *https://edu.3ds.com/en/challenges/model-mania*.

The application shows the phase I modelling of the part in Figure 3.373. The 2D drawing for the 2019 competition does not reveal a difficult 3D model, but the user must adopt a simple working strategy, with the lowest possible number of features in the specification tree, in an order that allows further modifications, increasing the robustness of the model.

Phase I modelling is not a problem for an experienced user, but the main question is how it will turn into phase II (Figure 3.383) without deleting the existing features from the specification tree. It is only allowed to add new features to obtain the new 3D model and, of course, to edit the existing ones. This restriction is due to the fact that for the part model of phase I, finite element analyses,

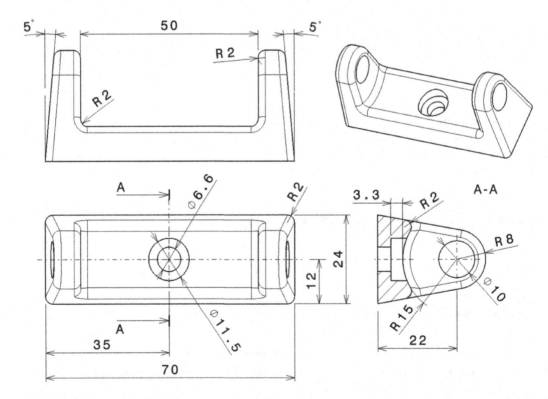

FIGURE 3.373   The 2D drawing of the part in phase I.

processing simulations, etc. can be created, which refer to certain features in the specification tree. Their disappearance or faulty editing leads to the need to start those simulations from the beginning.

Let's suppose that at the time of modelling the part according to the 2D drawing of phase I, the modification requirement for obtaining phase II is not known. Therefore, it is recommended that the user should apply a careful and clear modelling. The *Part Design* workbench will be used for both phases.

In the *ZX Plane*, in the *Sketch.1*, the profile from Figure 3.374 is drawn. The arc is obtained by trim editing of a circle with a diameter of Ø16 mm with respect to two oblique lines, symmetrical relative to the *H* axis of the coordinate system. The centre of the circle is at the origin of this system. Oblique lines start at the ends of a vertical line on the left and are tangent to the circle. The vertical line has a length of 24 mm, and a distance of 22 mm is established between the centre of the circle (implicitly, the *V* axis) and the line. The figure shows all the constraints symbols applied to the profile – it is closed and can be checked from the *Tools → Sketch Analysis* menu.

Using the *Pad* tool, with the *Mirrored extent* option checked, the profile is extruded on both sides of the *ZX Plane* with 35 mm each (Figure 3.375). The option can only be used if the user has previously selected the *Dimension* option in the *Type* options drop-down list.

The solid body is pierced (using the *Up To Next* option) by *Hole.1* with a diameter of Ø10 mm, concentric (Figure 3.376) with the circular edge of the planar face at one end of the solid *Pad.1*. The hole is simple and not threaded.

In the same *ZX Plane*, in the *Sketch.3*, the user draws a circle with a diameter of Ø30 mm, placing its centre at the origin of the coordinate system, as it results from the 2D drawing of the part.

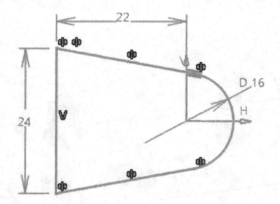

FIGURE 3.374   Profile of the *Sketch.1*.

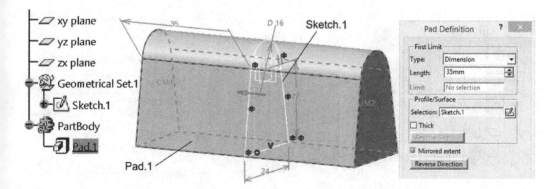

FIGURE 3.375   Extrusion of the *Sketch.1* profile on both sides of the *ZX Plane*.

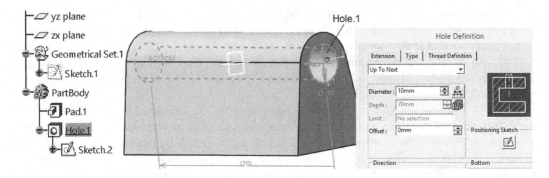

FIGURE 3.376 Adding the *Hole.1* through the whole *Pad.1* solid.

The circle is then involved in a 25 mm depth *Pocket.1* cut on both sides of the plane, using the *Mirrored extent* option checked (Figure 3.377).

This symmetry of the cut is similar to the extrusion required to create the solid *Pad.1*.

Below the *YZ Plane* at 15 mm (radius R15 mm of the circle in the *Sketch.3*), a new *Plane.1* is created using the *Plane* tool in the *Reference Elements* toolbar. In the *Plane Definition* dialog box (Figure 3.378), the user chooses the *Offset from plane* option from the *Plane type* drop-down list and then the reference plane *YZ* and enters the value in the *Offset* field. The plane is necessary because it represents the support of the counterbored *Hole.2* with the diameters Ø6.5 mm and Ø11.5 mm, respectively.

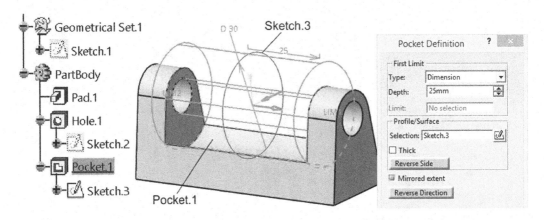

FIGURE 3.377 Adding the *Pocket.1* cut on the *Pad.1* solid.

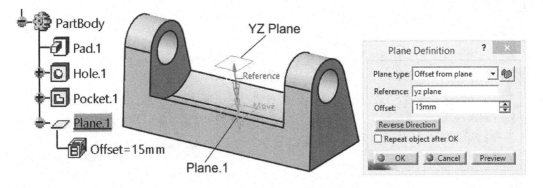

FIGURE 3.378 Adding the *Plane.1* below the *YZ Plane*.

The depth of the reamed surface is 3.3 mm. The centre of the hole, in *Sketch.4*, is 35 mm from one end of the part and 12 mm from the side edge at its base (Figure 3.379). Also, depending on the modelling strategy, instead of the linear distance constraint of 12 mm, a geometric coincidence constraint can be used between the centre of the *Hole.2* and the axis of the semi-cylindrical surface *Pocket.1*. The advantage is that, regardless of the width of the part's base (other than 24 mm), the centre of the hole will remain positioned in the middle.

Figure 3.380 shows the parameters of the large diameter of the counterbored *Hole.2* and its position in the part representation.

At the ends, the part has two inclined surfaces of 5° each, which can be easily achieved using the *Draft* tool. In the *Draft Definition* dialog box (Figure 3.381), the user enters the draft value in the *Angle* field; in the *Face(s) to draft* field, selects the two flat end faces; and then, as a reference element (the *Neutral Element* field → *Selection*), selects the flat surface at the base of the part.

Then, the user must check that the *Pulling Direction* arrow is vertical and pointing upwards; otherwise, he must change its direction by pressing the arrowhead. If it remains oriented towards the base, the two planes will be flared, leading to a wrong inclination of the ends of the part.

The edges to which the inclination is made (of reference) are the two coloured with purple at the base of the part. These edges belong to the flat end faces, one of them being marked in the figure.

The figure shows the two blue inclination planes, marked by the feature *Draft.1* and the angle of inclination of 5° of these planes and each of the surfaces at the ends of the part.

According to the 2D drawing of the part, it has numerous R2 mm rounded edges, practically all the edges of the part, except those belonging to the holes and to the flat surface at its base. To make the specification tree as simple as possible and because all fillet radii have the same value, it is recommended to use the *Edge Fillet* tool only once.

The user should select the edges of the part in the *Object(s) to fillet* field and apply the *Tangency* option so that if he chooses an edge that continues with others in a tangent manner, the fillet operation includes them without the need to explicitly select all edges. There is a (relatively) large number of rounded edges, out of which the user has selected only eight.

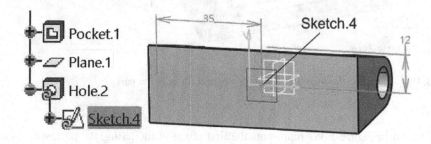

FIGURE 3.379   Placing the centre of the *Hole.2* in the *Plane.1*.

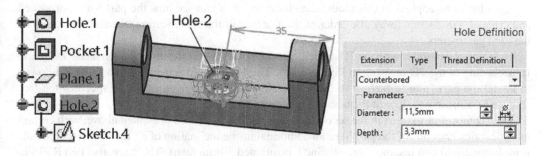

FIGURE 3.380   Parameters and position of the *Hole.2* on the part.

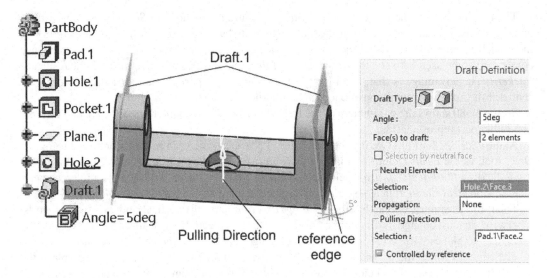

FIGURE 3.381   Adding the draft angle on the two faces of the part.

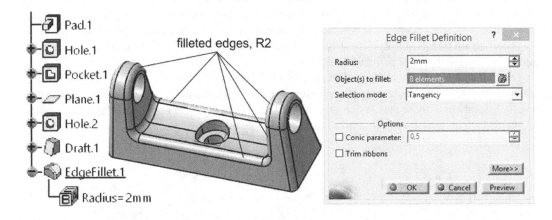

FIGURE 3.382   Applying the *Edge Fillet* tool on selected edges of the part.

The 3D solid in Figure 3.382 represents the first phase of modelling the part, which ends with the determination of its volume and area. Thus, with the help of the *Measure Item* tool, the user clicks on the *PartBody* feature in the specification tree and the results are as follows: volume: 16,812.6 mm³ and area: 6,759.19 mm². The part is saved on hard drive according to phase I.

From the steps applied in this modelling phase, the user can see how the part was approached and created – in a simple way, the order of the features in the specification tree and the values of the used parameters.

The tutorial continues with the 3D modelling of the part, by transforming it from phase I to phase II, according to the 2D drawing in Figure 3.383. The user should remember that it is not allowed to model the part from scratch or to delete the already existing features in the specification tree (Figure 3.382), but only to edit them and to add new ones.

The 3D model of the part in the second phase is similar to the one presented above, with certain modifications: the ends of the part have an additional inside inclination of 8°, the counterbored hole is multiplied, and two instances are obtained, positioned 30 mm apart. There are also two R30 mm radius fillets of the part's end surfaces.

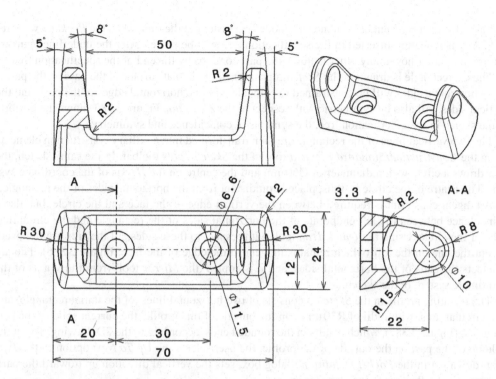

FIGURE 3.383 The 2D drawing of the part in phase II.

The 3D model can be edited by features in its specification tree, especially by inserting new features between the existing ones.

Thus, the user right-clicks on the *Pocket.1* feature and, from the context menu, chooses the *Define In Work Object* option.

Pocket.1 is then represented underlined, becomes current, and the part matches Figure 3.384. It is noted that the features placed below *Pocket.1* in the specification tree are not (yet) taken into account for the 3D model.

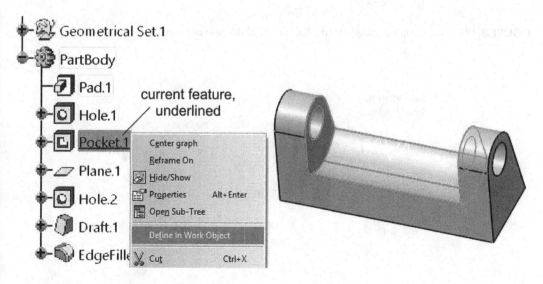

FIGURE 3.384 Defining the feature to work on in the specification tree.

Basically, the user went back in the part modelling history at the time when the *Pocket.1* was created. Any new feature inserted in the specification tree will be placed after the underlined/current feature, no matter how many other features the part contains by the end of the specification tree.

Thus, a rectangle is drawn in the *Sketch.5* positioned on the flat surface at the base of the part.

Its horizontal sides will be constrained to coincide with the horizontal edges of the part, and the vertical sides will also be symmetric with respect to the *ZX Plane*. Figure 3.385 shows the position of the sketch in the specification tree, the symbols of coincidence and symmetry.

The two vertical lines of the rectangle are then transformed into auxiliary construction elements using the *Construction/Standard Element* tool in the *Sketch Tools* toolbar. In the same sketch, the user draws a circle with a diameter of Ø60 mm and the centre on the *H* axis of the coordinate system. The centre of the circle is, thus, placed equidistant from the horizontal sides of the rectangle.

A tangent constraint is required between the vertical edge of the face and the circle, but also a coincidence between the left endpoints of the horizontal sides of the rectangle and the circumference of the circle. Using the *Quick Trim* tool, with respect to these sides, all the arcs are removed, except the one on the left (which remains tangent to the edge of the face; Figure 3.386). The arc connects the ends of the horizontal sides, and it is copied (the *Mirror* tool) from the *V* axis of the coordinate system of the sketch.

The resulting profile in the *Sketch.5* consists of two horizontal lines (of the initial rectangle) and two circular arcs, with radii of R30 mm. On the outside of this profile, the current solid of the part is cut out (Figure 3.387), which results in the rounded ends, according to the 2D drawing. To cut the volume of the part on the outside of the profile, the user selects the *Up To Next* option in the *Type* drop-down list in the *Pocket Definition* dialog box, sets the vertical direction up (toward the part) and presses the *Reverse Side* button (or arrow).

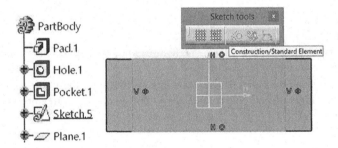

FIGURE 3.385 Sketching a rectangle on the flat surface at the base of the part.

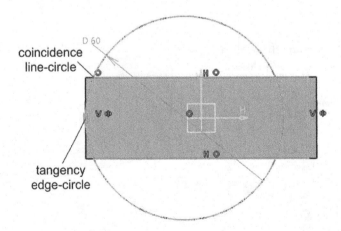

FIGURE 3.386 Sketching a circle tangent to the edge of the flat face.

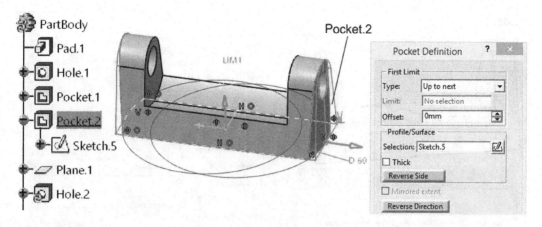

FIGURE 3.387 Cutting the part with the outside of the *Sketch.5* profile.

It is noted that cutting the part takes place only at its ends, the *Pocket.2* feature is underlined (it became current), and it is added to the specification tree after the *Pocket.1* feature and before the *Plane.1* and *Hole.2* features.

After the confirmation of the *Pocket.2*, the information/error message in Figure 3.388 is displayed because the *Draft.1* feature cannot be updated. The two flat faces at the ends of the part, *Face.1* and *Face.2*, obtained according to Figure 3.381, no longer exist (*Edge or face not found*).

The user should click the *Close* button of the *Update Diagnosis: Pocket.2* box and, from the context menu of the *EdgeFillet.1* feature, chooses the option *Deactivate* for the moment. Even if the features in the specification tree, placed after the current feature (*Pocket.2*), are not taken into account for displaying the 3D model of the part, the program constantly checks their validity according to the changes made and the new inserted features.

The place of the flat surfaces was taken by two rounded surfaces R30 mm, and these must be explicitly selected in the *Draft.1* feature. The user double-clicks it in the specification tree, and the *Draft Definition* dialog box and an error message open (Figure 3.389). The flat surfaces *Face.1* and *Face.2* to which these error messages refer are observed, and by pressing the *OK* button, they disappear from the *Face(s) to draft* field (previously selected in Figure 3.381).

Instead of flat surfaces, the user selects the rounded surfaces at the ends of the part to achieve an inclination of 5° (the value is kept in the *Angle* field). The flat surface at the base of the part also remains selected.

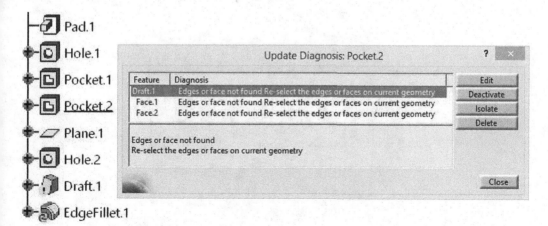

FIGURE 3.388 Error messages displayed when the user tries to update the part.

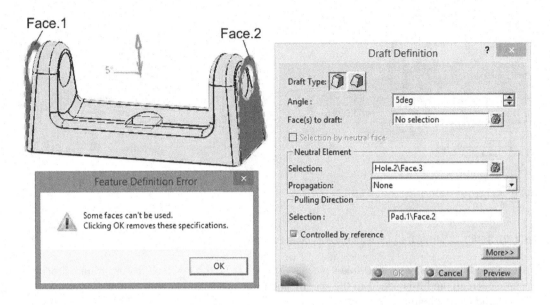

FIGURE 3.389 Removing the unused flat surfaces at the both ends of the part.

The orange arrow should point vertically upwards, and when the *Preview* button is pressed, two transparent, blue truncated cone surfaces are displayed (Figure 3.390).

From the context menu of the *Draft.1* feature, it is highlighted (becomes current – *Define In Work Object* option) and it is followed by the *Point.1* in the specification tree, inserted using the *Point* tool in the *Reference Elements* toolbar. In the *Point Definition* selection box in Figure 3.391, in the *Point type* list, the user chooses the *Circle/Sphere/Ellipse center* option and then selects the circular edge marked on the representation in the figure.

The point is added in/as the centre of the circular edge and in the specification tree, becoming the current feature, being the last created within the part. Although the specification tree ends (at this stage) with *EdgeFillet.1* (from phase I), the point is not added after it, but after the current feature (*Draft.1*).

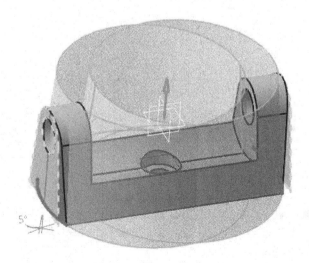

FIGURE 3.390 Representation of two transparent, blue truncated cone surfaces.

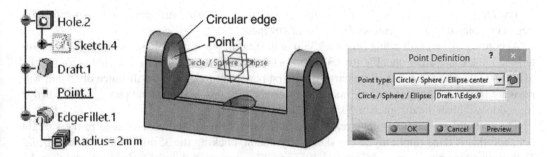

FIGURE 3.391 Adding a new point as the centre of a circular edge.

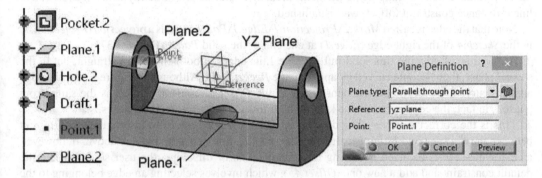

FIGURE 3.392 Placing a new plane in the *Point.1* and parallel to *YZ Plane*.

In the *Plane Definition* dialog box (Figure 3.392), the user applies the *Parallel through point* option from the *Plane type* drop-down list, and then he chooses the *YZ Plane* as a reference and the *Point.1* for positioning the plane. Thus, the *Plane.2* is added to the specification tree after the *Point.1* and on the part in the same place with it (the plane symbol/frame contains the point symbol), as shown in the figure.

Plane.2 is required to draft the inner flat faces of the part's ends. In the *Draft Definition* dialog box (Figure 3.393), the user selects one of these two faces in the *Face(s) to draft* field, sets the angle of 8° in the *Angle* field and chooses the *Plane.2* as the neutral/reference element.

The arrowhead should point vertically upwards, and by pressing the *Preview* button, a blue plane is displayed, which previews the drafted surface to be obtained. It is noticed that above the plane, a volume is cut from the end of the part and, below the plane, a volume is added, because the plane intersects the surface.

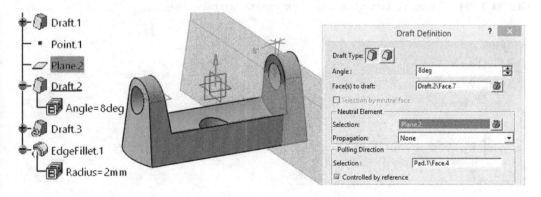

FIGURE 3.393 Drafting two inner flat faces of the part's ends.

The *Draft* tool is used twice, similarly, for each of the two drafted surfaces, and the specification tree is completed with the features *Draft.2* and *Draft.3*.

According to the 2D drawing (Figure 3.383) of the part in the second phase, it contains two counterbored holes instead of one. Figure 3.379 shows the positioning of the *Hole.2* (double-click on the *Sketch.4*) relative to one end of the part. In the first phase of modelling, the distance of 35 mm was measured between the centre of the hole and a straight edge, but in the second phase, a curved edge is used to establish the distance of 20 mm (Figure 3.394).

The user did not explicitly intervene in the selection of the curved edge instead of the straight one, and this is done implicitly in the sketch. By double-clicking the 35 mm value, the *Constraint Definition* dialog box is opened (Figure 3.395) and 20 mm is entered in the *Value* field, and then, by pressing the *More >>* button, the box is extended to the right. The *Name* field contains the name of the constraint, and in the *Supporting Elements* area, there are sketch entities by which the respective linear distance constraint (offset) was established.

Note that the edge is called *Mark.2\Projection.1\Edge.1\Pad.1* and it is a projection (*Projection.1*) in the *Sketch.4* of the right edge (*Edge.1*) at one end of the solid *Pad.1* (Figure 3.375).

At the time of creating this solid in the phase I, its edges at both ends were straight, but in the second phase, they become curved by applying the *Pocket.2* cut. Although in Figure 3.394, the edge appears curved, the geometric element to which the *Offset.17* constraint refers is the same from phase I; in fact, the edge name is also kept at the transition between the two phases.

This is the correct manner of modelling, in which the user carefully inserts the *Pocket.2* cut between the *Pocket.1* and *Plane.1* features in the specification tree, as explained above. If, however, an error message is received regarding the identification of this edge, the user should delete the default constraint and add a new one (*Offset.45*), which involves selecting an edge belonging to the *Pocket.2* feature (Figure 3.396).

The names of the constraints and edges shown in figures above may differ from those of the user, obtained during the modelling steps.

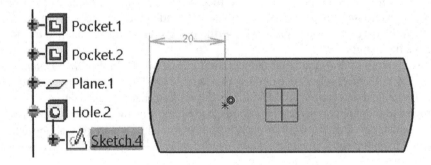

FIGURE 3.394   Placing a centre of a hole at 20 mm from the left curved edge.

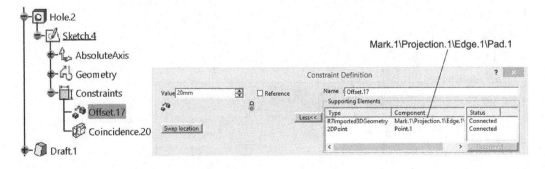

FIGURE 3.395   Establishing the offset value between a point and an edge.

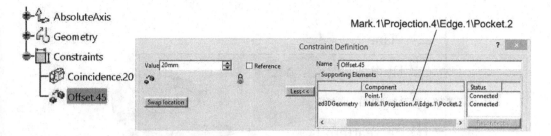

FIGURE 3.396 Replacing an offset constraint by selecting another edge.

The *Hole.2* was, therefore, moved to one end of the part, according to Figure 3.397. Of course, as a feature that was worked on by the user, the *Sketch.4* was underlined (Figure 3.395). Once the editing is complete, the *Draft.3* feature is underlined again, and the *RectPattern.1* multiplication feature is added after it and underlined when the dialog box in Figure 3.397 is displayed. In the *Object to Pattern* field, the user selects *Hole.2*, chooses the *YZ Plane* (or a flat side or the base surface) as the reference element and creates another instance at a distance of 30 mm from *Hole.2*.

The last stage of modelling the part in the two phases consists in activating the *EdgeFillet.1* feature (Figure 3.398), followed by its underlining (it becomes the current working feature; Figure 3.399).

An error message may appear after activation that the rounded outer edges at the ends are not found. The user should click the *Edit* button in the message box and select the respective edges, and the fillet feature will include them.

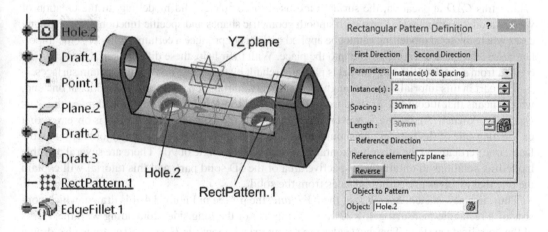

FIGURE 3.397 Multiplication of the *Hole.2* feature.

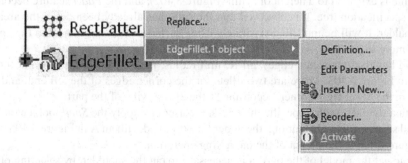

FIGURE 3.398 Activation of the *EdgeFillet.1* feature.

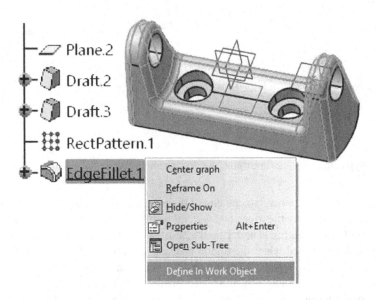

FIGURE 3.399 The *EdgeFillet.1* feature becomes the current work object in the specification tree.

3.18 EDITING AND RECONSTRUCTION OF SOLIDS USING SURFACES – TWISTED AREA

Within this *CAD* application, the surfaces are used to complete solid modelling, in the creation of various modern, attractive features with smooth geometric shapes and specific functionalities. In many cases where the solid modelling cannot be applied or does not produce a certain quality of some important areas, the surface features come into the place. With their help, these difficult surface patches are created from scratch or the solid model is partially rebuilt by extracting and editing certain surfaces.

The part in this tutorial has a rectangular shape, certain rounded edges and a shoulder at one end, which means that it could be created using the standard tools and modelling procedure in the *Part Design* workbench. Analysing, however, carefully the 2D drawing (Figure 3.400), it is observed that on the long side of the bottom, the shoulder has a twisted area. It is very tight and joins two narrow faces, one vertical and the other horizontal, by winding at an angle of 90°. There are several possible modelling solutions to obtain the respective area of the 3D solid part, and this tutorial will explain the one that involves extracting surfaces from the solid.

Thus, in a new sketch, *Sketch.1*, of the *XY Plane*, the profile in Figure 3.401 is drawn, which consists of a rectangle measuring 46×22 mm, having one of the long sides coinciding with the *V* axis of the coordinate system. The short sides are symmetrical about the *H* axis. A square is also drawn with its sides of 10 mm and two of them symmetrical about the *H* axis.

The profile is extruded to a height of 5 mm (Figure 3.402), and the *Pad.1* feature becomes available in the specification tree. It is observed that the solid has already been cut in the area with the twisted shoulder. It will be modelled separately using surfaces at some moment in the application, after a few more steps.

The solid has five connected edges, all R3 mm radius, made with the *Edge Fillet* tool, as it is illustrated in Figure 3.403. There are two fillets for the corner edges of the left end, and the other three meet in the lower right corner, according to the 2D drawing of the part.

To obtain a shoulder with a specific thickness, the user can apply the *Shell* tool, enters the value of 2 mm and selects, for the beginning, the upper face (marked with an A in Figure 3.405). Thus, the solid in Figure 3.405 is the output of the modelling operation.

In order to get the model of the part, it is necessary to cut the shoulder, by selecting other faces, according to the representation in Figure 3.405.

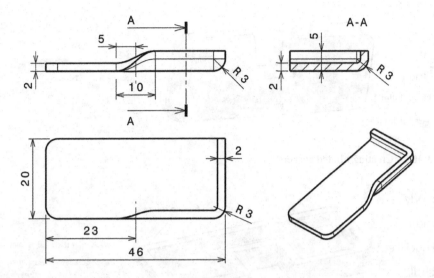

FIGURE 3.400   The 2D drawing of the 3D part to be modelled.

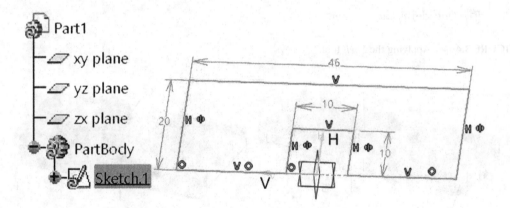

FIGURE 3.401   Creating the first sketch of the part – *Sketch.1*.

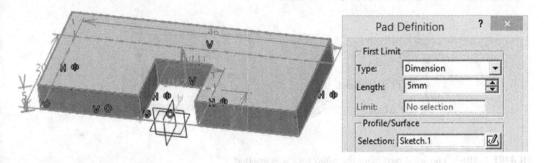

FIGURE 3.402   Creating *Pad.1*, the first solid feature of the part.

In the *Faces to remove* field of the *Shell Definition* dialog box, surface *A* was initially selected, with the effect in Figure 3.404. The other surfaces *B, C* and *G* are further selected, and then when any other *D, E, F, H* or *I* is included in the field (a tangent property is used), the complete selection *A–I* results. By confirming the thickness of the wall and the surfaces to be removed, the solid in Figure 3.406 results.

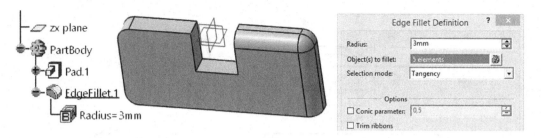

FIGURE 3.403   Creating filleted corners.

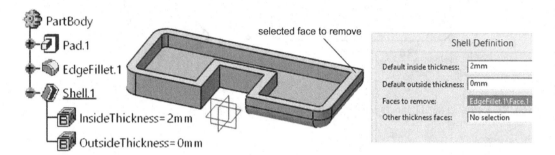

FIGURE 3.404   Applying the *Shell* tool.

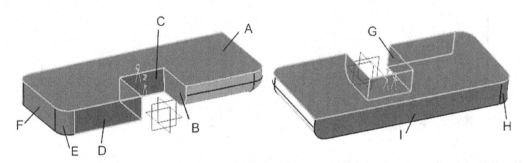

FIGURE 3.405   Marking the faces of the solid model.

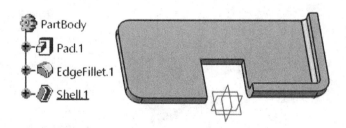

FIGURE 3.406   The solid part after the *Shell* tool was applied.

The solid intentionally lacks the area that will contain the twisted surface. To close the cut, a working method is applied that involves extracting certain surfaces of the solid, so the *GSD* workbench is needed to be activated.

From the *Insert → Operations* menu, the user chooses the *Extract* tool and, then in the *Extract Definition* selection box, in the *Propagation type* drop-down list, chooses the *Tangent continuity* option for an easier and faster selection of surfaces. The first extracted surface is the one marked

Ex1 (Shell\Face.6, but this may differ in the user's application). After this first selection is done, by default, the *Element(s) to extract* field no longer allows other surfaces to be selected, but the icon to its right is clicked and the selections can continue in the adjacent selection box (Figure 3.407).

Thus, the other surfaces marked *Ex2*, *Ex3* and *Ex4* are selected; in fact, most of the surfaces of the solid part are selected, except for those that define the cut. For example, the selection of the *Ex3* surface, which is a flat face, also leads to the selection of rounded and flat surfaces up to *Ex1*. The selection of *Ex2* also determines the selection of the flat surface at the base of the part, and by choosing *Ex4*, the side surfaces of the part are selected (marked with a thick red line in the figure). There are also cases where the extraction needs to be more restricted, and this is possible by choosing the *No propagation* option from the *Propagation type* drop-down list.

The selection of all these surfaces is green, and by confirming their extraction from the solid model, the features *Extract.1,..., Extract.4* are obtained in the specification tree. The colour of the features (and of the extracted surfaces) is, of course, yellow. It is observed that each feature represents, in fact, a group of surfaces, of the type: flat, cylindrical and spherical.

Surfaces are bordered by edges, which must also be extracted so that they can be used at a later stage in the creation of other surfaces for an appropriate solid generation.

Thus, the *Boundary* tool from the *Operations* toolbar or from the *Insert → Operations* menu is used. In the *Boundary Definition* box (Figure 3.408), it is chosen that the selection of the edge does not propagate to the others with which it is connected (*No propagation* option), and then the user clicks on the leftmost edge (*Edge.1*) of the surface *Extract.1*. As a result, the feature *Boundary.1* is added to the specification tree and appears as a line (white and thin) in the graphic area. For a better view of the extracted surfaces and edges, the solid is hidden (right-click on *PartBody* and, from the context menu, click on the *Hide/Show* option).

Figure 3.408 also shows the previously extracted surfaces, and it is observed that the cut area is not in any way closed by any surface.

Next, the user extracts all the edges surrounding this cut area using the *No propagation* option for single edges (lines only) and *Tangent continuity* for those containing lines and arcs. There are eight such extracted edges on the contour of the cut area, according to Figure 3.409, added to the specification tree.

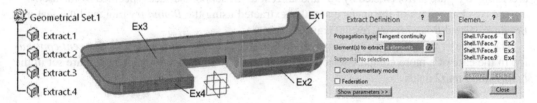

FIGURE 3.407 Applying the *Extract* tool.

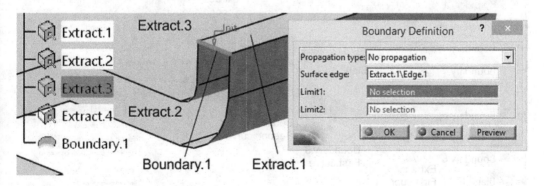

FIGURE 3.408 Definition of a boundary.

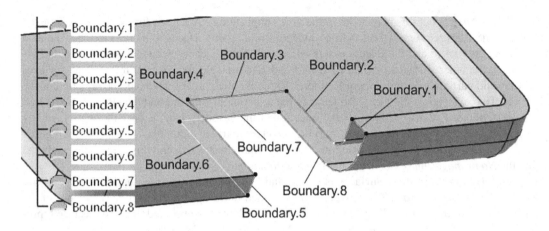

FIGURE 3.409   Displaying the *Extract* result.

The *Boundary.2* and *Boundary.8* edges consist of two lines and a circular arc; all other edges are drawn as lines. Each extracted edge has two black dots at the ends, meaning that the elements are individual and do not form a connected contour.

The first surface that encloses the cut area is created using the *Blend* tool. In the *Blend Definition* dialog box (Figure 3.410), in the fields *First Curve* and *Second Curve*, the user should select the extracted edges *Boundary.5* and, respectively, *Boundary.1*. The selection can be stopped here as well, but the surface obtained *Blend.1* would not be tangent (as it is correctly presented in the 2D drawing) to the extracted surfaces *Extract.4* and, respectively, *Extract.1*. Thus, they must be selected in the *First Support* and *Second Support* fields. In the *Basic* tab, for a good continuity of the cut surfaces, in drop-down lists the user should choose the *Tangency* option.

At the end of the selections, two red arrows appear on each edge of the surface *Blend.1* and they control the way the surface is formed. Shorter arrows should point to the newly created surface *Blend.1*, and both long arrows should point in the same direction, inward or outward. The surface *Blend.1* is, according to Figure 3.410, twisted by 90° and tangent to the support surfaces specified by the user.

The two long edges of the surface are also extracted using the *Boundary* tool, and the features *Boundary.9* and *Boundary.10* result (Figure 3.411).

The choice of the selected edge propagation option can be *No propagation* or *Tangent continuity*. In the case of a more complex selection, the endpoints of an edge to be extracted are set in the *Limit1* and *Limit2* fields.

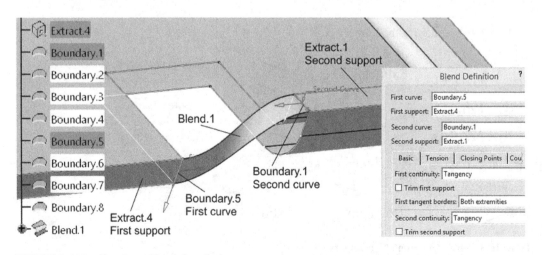

FIGURE 3.410   Creating a blended surface.

The cut area will be closed with two patches, as follows: on the upper surface (*Extract.3*) of the part are the extracted edges *Boundary.2*, *Boundary.3*, *Boundary.4* and *Boundary.9*, and on the bottom surface (*Extract.2*, Figure 3.408) edges *Boundary.6*, *Boundary.7*, *Boundary.8* and *Boundary.10*. All these edges represent the contours of the two patches.

For the top edges to form a surface, the *Multi-Sections Surface* tool in the *Surfaces* toolbar is used.

The selection box in Figure 3.412 contains two main fields: the first (*Section*) contains the sections through which the surface passes, and the second (*Guide*) contains the guidance curves between these sections. These are, in fact, individual curves, which do not intersect, conveniently positioned in certain planes of the workspace, and the contour of the created surface will have their exact shape as it passes through those planes. From one section to another, the program extrapolates the surface to keep it continuous and tangent to other user-selected surfaces.

To create the closing surface in the figure, the user selects the *Boundary.4* and *Boundary.2* edges in the *Section* column, and then the *Extract.3* surface for tangency/continuity.

This means that the program must initiate a *Multi-sections Surface.1* starting from the simple edge *Boundary.4*, tangent to the surface *Extract.3*, to the more complex edge *Boundary.4* and remaining tangent to the same flat surface *Extract.3*. The two edges/sections are very different, and the task of the program is not simple at all, but it is possible by selecting the two guide curves *Boundary.3* and *Boundary.9*. For the first guiding curve, the tangent direction is specified as *Extract.3*.

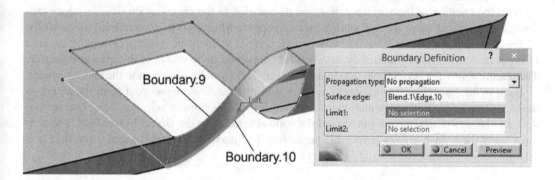

FIGURE 3.411 Creating two more boundaries from the blended surface.

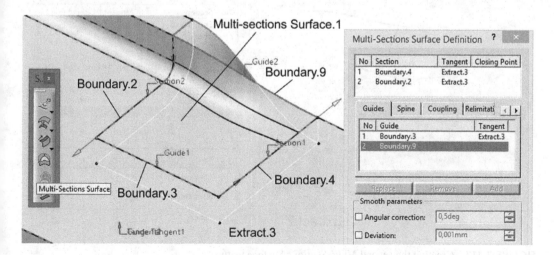

FIGURE 3.412 Creating the first *Multi-Sections Surface* feature.

The guiding curves and sections have three main roles, namely: to make it possible to create the surface based on the settings, but with a minimum of user intervention; to keep it within the selections of the respective fields; and to stretch/extrapolate it correctly to observe the imposed tangent conditions and the closing points (*Closing Point* – option/selection not used in this application).

Similarly, the second surface is created, *Multi-sections Surface.2*, using as sections the *Boundary.6* and *Boundary.8* edges, tangent to the surface *Extract.2*, and the *Boundary.7* and *Boundary.10* as guide curves (Figure 3.413).

From both figures, it is observed the correct way in which the existing connections on the solid part (and, therefore, on the extracted surfaces) continue smoothly, naturally, in the multi-section surfaces obtained.

Excepting the three surfaces (*Blend.1*, *Multi-sections Surface.1* and *Multi-sections Surface.2*), created to close the cut-out area of the solid part, all other features in the specification tree are hidden (*Hide/Show* option in the context menu).

To group/join the remaining visible surfaces (Figure 3.414) into one, the user can apply the *Join* tool in the *Operations* toolbar.

The *Elements To Join* field contains the selected areas, and the user can select the *Check connexity* option below. Thus, it is checked if the point where one edge ends coincides with the point where another edge begins, if two adjacent surfaces have a common edge and there are no spaces between them. *Simplify the result* can also be checked to allow the program to automatically reduce the number of faces and edges needed to create the surfaces.

With the creation of the surface *Join.1* and its addition to the specification tree, its edges turn green, meaning that the process of joining the component surfaces has been completed successfully.

Based on this surface, a 3D solid body will be created and will be integrated into the cut area of the part. The previously hidden *PartBody* solid is restored to be visible in the specification tree. The feature *Join.1* fits perfectly in the cut area (Figure 3.415), and it is tangent to the part's surfaces.

Back in the *Part Design* workbench, the user accesses the *Sew Surface* tool from the *Insert* → *Surface-Based Features* menu and opens the *Sew Surface Definition* selection box. The surface *Join.1* to be 'sewn' into the cut-out area of the solid is selected in the *Object to sew* field. The user must press the orange arrows perpendicular to *Join.1* to make them point inwards, and then it is recommended to check the *Simplify geometry* option.

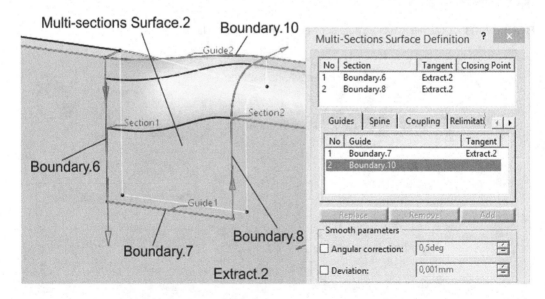

FIGURE 3.413 Creating the second *Multi-Sections Surface* feature.

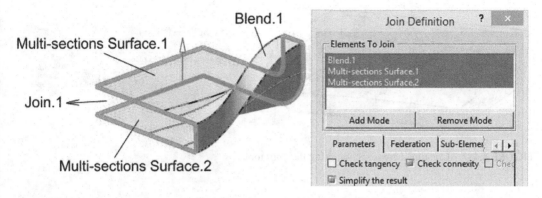

FIGURE 3.414 Creating a joined surface.

The feature *SewSurface.1* is added to the specification tree, but to make it fully visible, the surface *Join.1* is hidden.

Figure 3.416 shows the final model of the solid part in two different representations. The edges that created the component surfaces for *Join.1* are visible on the left image, and the view mode is *Shading with Edges* (in the *View modes* toolbar).

The solid is created correctly, without discontinuities; it is a single entity (in Figure 3.417, a cross section was made through the feature *SewSurface.1*), but the edges are visible and can create misunderstandings for less experienced users.

In such cases, another simplified viewing mode may be applied, *Shading with Edges without Smooth Edges* (Figure 3.416, on the right image). It removes (visually) the less important edges of the part (such as the transition from one type of surface to another: from one flat to another cylindrical, etc.), keeping only those delimiting the solid (only outer edges).

Modelling solution: https://youtu.be/EcE1R7gEZGI.

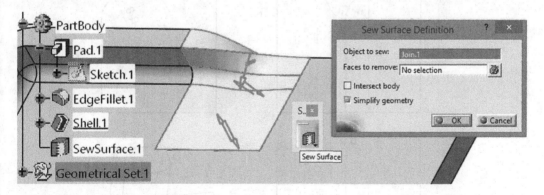

FIGURE 3.415 Creating a 'sewn' surface.

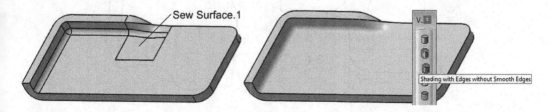

FIGURE 3.416 Displaying *Shading with Edges* vs *Shading with Edges without Smooth Edges* results.

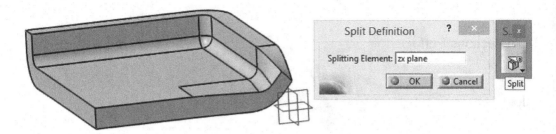

FIGURE 3.417 Creating a cross section using the *Split* tool.

3.19 EDITING AND RECONSTRUCTION OF SOLIDS USING SURFACES – CONNECTED SURFACES

The part in this *CAD* application can be considered as a part of more complex part structures or assemblies (punch of die tool, for example). Its 3D model contains some flat surfaces resulting from the extrusion of rectangular contours, straight and filleted edges.

After the analysis of the 2D drawing (Figure 3.418), it is observed that the geometry of the part is not very complicated, but the area where the three connections (edge fillets of radii R8 and R10) meet can generate some problems (especially due to obvious discontinuities) during the stages of

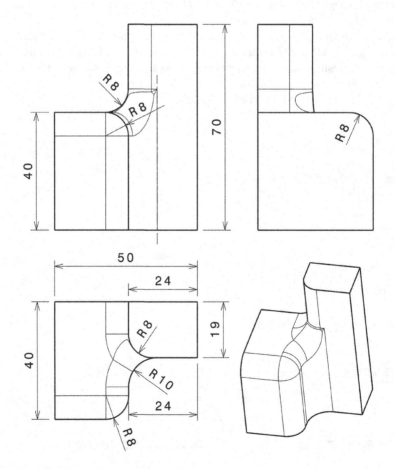

FIGURE 3.418 The 2D drawing of the part to be edited.

creating the semi-finished product by casting or processing the filleted surfaces with a milling cut-
ter having a spherical end. Figure 3.419 shows two details (a and b) of these surfaces, as they were
created by default by the program.

The solid part that resulted from the 3D modelling (*Part Design*) cannot be accepted by the production department with such discontinuous surfaces and must be edited/corrected, with minimal
deviations from the general shape of the part presented in the 2D drawing.

To obtain the solid model, the user starts by drawing a profile (Figure 3.420) in the *Sketch.1*
located in the *XY Plane*.

The first fillet radius of value R10 appears in the sketch. The profile may have one of the corners
(top left, for example) at the origin of the coordinate system of the sketch. By extruding the *Pad* with
a length of 40 mm, the solid from Figure 3.421 is obtained.

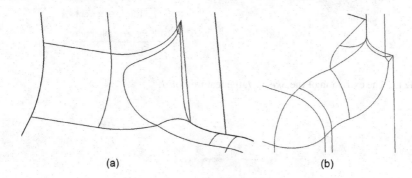

(a)   (b)

FIGURE 3.419   Details of the 3D part that should be reconstructed.

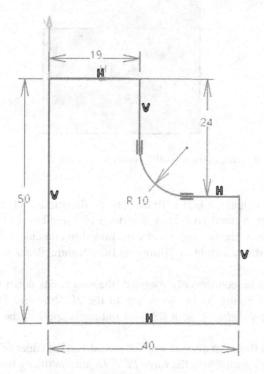

FIGURE 3.420   The first sketch in the modelling process – *Sketch.1*.

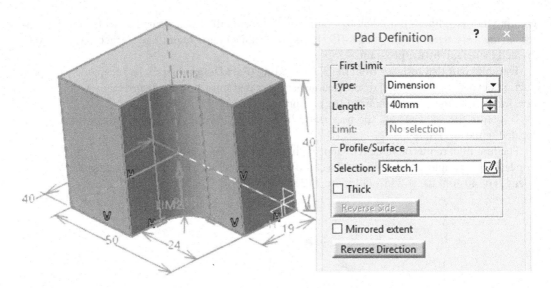

FIGURE 3.421   The extrusion of the *Sketch.1*, the feature *Pad.1*.

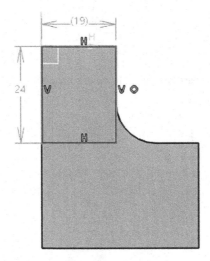

FIGURE 3.422   The second sketch in the modelling process – *Sketch.2*.

A rectangular profile (Figure 3.422) is drawn on the flat surface of the feature *Pad.1* in the *Sketch.2*. The 19 mm dimensional constraint is written in parentheses as it is a reference value, and the vertical right side of the rectangle was constrained to coincide with the edge of the solid. The rectangle is extruded to a height of 30 mm, and the feature *Pad.2* is obtained, according to Figure 3.423.

To select the edges to be continuously rounded, the user holds down the *Ctrl* key and clicks on each edge marked in Figure 3.424, according to the 2D drawing. There are seven of such edges to which the *Edge Fillet* tool, with R8 fillet radius, is applied, the result being the one in Figure 3.425.

It can be observed that there is a discontinuity of R8 radii on the edges of the part, but this cannot be resolved by the settings available in the *Edge Fillet Definition* dialog box. Tangency was chosen as the way to propagate the connections of the edges.

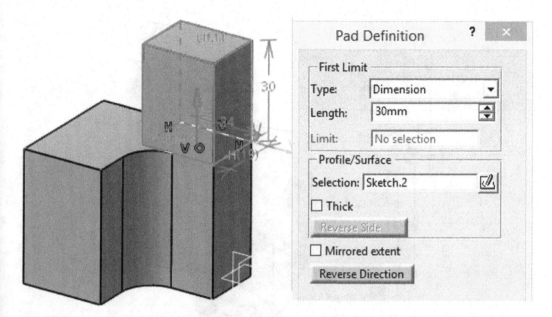

FIGURE 3.423   The extrusion of the *Sketch.2*, the feature *Pad.2*.

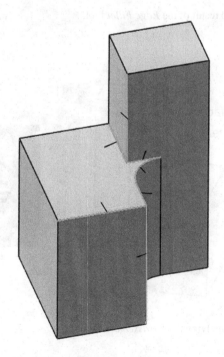

FIGURE 3.424   Selecting the edges to be continuously rounded.

The user switches to the *GSD* workbench to disassembly the solid into individual surfaces. The user clicks the *Disassemble* icon in the *Operations* toolbar and selects the last solid feature in the specification tree (*EdgeFillet.1*), according to Figure 3.426. In the pop-up box, it is seen that the input entity (*input elements*) is one (the solid part), and then the option *All Cells* is selected in order to extract the 22 surfaces that cover the solid. The figure also contains some areas highlighted with their names established by the program.

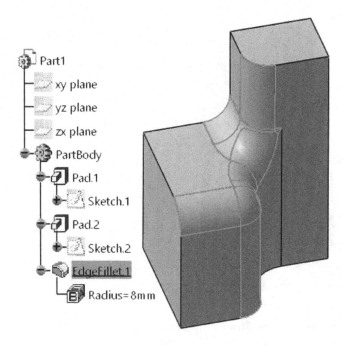

FIGURE 3.425   Displaying the result of the *Edge Fillet* tool.

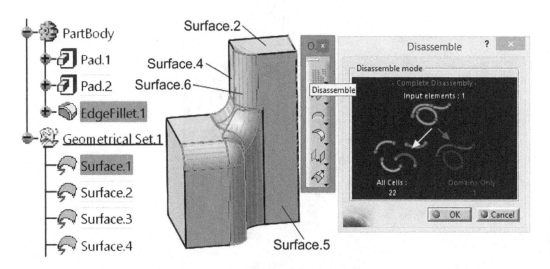

FIGURE 3.426   The result of applying the *Disassemble* tool.

By confirming the selection, the specification tree is updated (in a geometric set) with the features *Surface.1,..., Surface.22*, some of which will be removed or furtherly edited.

When extracted, the surfaces are no longer linked to the solid, as they have lost the parametric connection. Any change in the solid geometry will not be propagated to individual surfaces.

The solid body can be hidden by right-clicking the *PartBody* feature in the specification tree and, then from the context menu, choosing the *Hide/Show* option (Figure 3.427). Only surfaces are kept visible for later selection.

The user will first remove certain surfaces and then edit others, according to Figure 3.428. The representation and numbering of surfaces may differ from one application to another, but the surfaces in the central area of the connection will be deleted and the adjacent ones will be edited and cannot be preserved in the form in which they were extracted from the solid.

After deleting the surfaces *Surface.8*, *Surface.13*, *Surface.9* and *Surface.12*, the gap left by them in the structure of extracted surfaces can be observed (Figure 3.429). This gap will be filled later by a new user-created surface. Two simple auxiliary constructions are required for editing *Surface.10* and *Surface.18*.

Next, the point marked in Figure 3.429, located at the intersection of three edges, must be identified. It will not be found as a *Point* feature in the specification tree, but it can be selected directly with the mouse on the 3D model. The point is then projected (*Projection* tool) on a close vertical edge, located to its right. The selection box in Figure 3.429 shows the point name, as taken from the 3D model (*Surface.11\Vertex*), the edge name (*Surface.10\Edge.1*) and the projection itself.

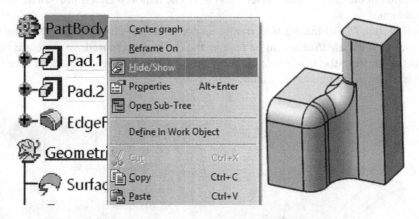

FIGURE 3.427 Hiding the solid body and keeping only the extracted surfaces.

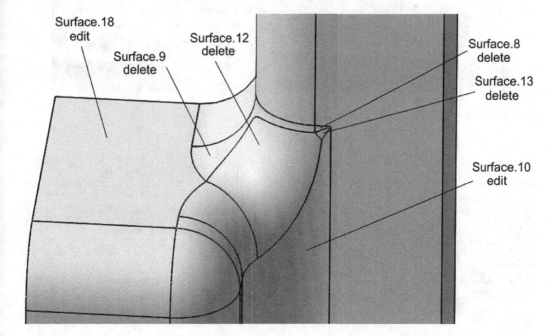

FIGURE 3.428 The extracted surfaces and the actions to be performed on them.

The projection type (*Normal*) and the *Nearest solution* option are checked. The projection of the point on the selected edge can be only done in this way, without selecting a certain direction, because a line (which will be drawn in the next step) that joins the two points is not along an implicit direction of any of the axis *X*, *Y* or *Z*. Only after drawing the line, it can be considered and selected as a possible direction. Basically, the point marked in the figure is projected perpendicular to the nearest position on the edge.

A line is drawn between the two points using the *Point-Point* option (Figure 3.430). In that case, the line is not on *Surface.10* (Figure 3.428) and therefore cannot be used for its editing. The line is then projected on this surface, according to Figure 3.431.

The specification tree is updated with the features *Project.1* (point on the right edge), *Line.1* and *Project.2*. The line and point can be hidden (*Hide/Show* option in the context menu). The *Surface.10* is divided into two areas using the *Split* tool: one area is kept (the lower one, Figure 3.432) and the other one is removed (the upper one). The appropriate area selection is possible by pressing the *Other side* button in the *Split Definition* selection box. The removed area is previewed by transparent representation.

The *Multi Output (Split)* feature is currently visible in the specification tree because the user has not yet selected the area/surface to be kept or the one to be removed, so there are two selection options (by pressing the *Other side* button). Once the confirmation of the kept area has been received, it is added to the tree as the feature *Split.1*.

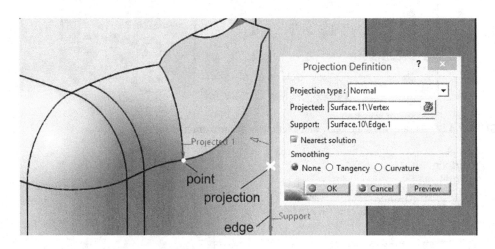

FIGURE 3.429   The projection of a point on an edge.

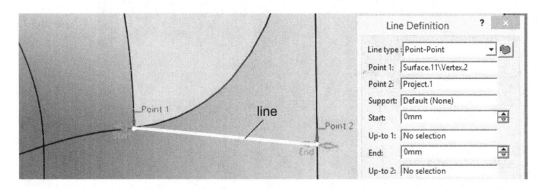

FIGURE 3.430   Drawing a line between a point and its projection.

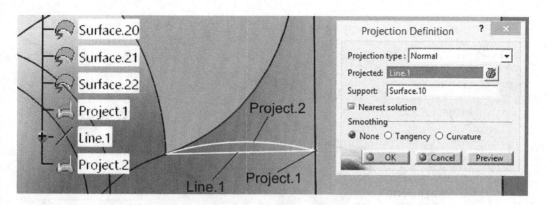

FIGURE 3.431 Projecting a line on a surface.

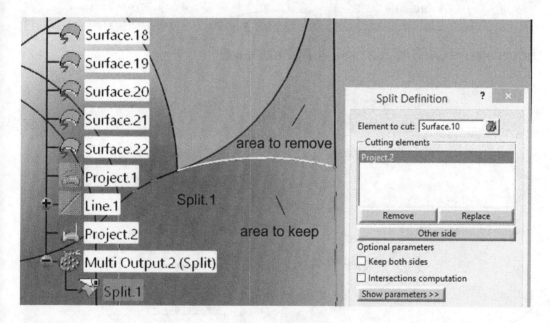

FIGURE 3.432 Applying the *Split* tool on the *Surface.10*.

However, there are also cases in which the aim is to keep both resulted surfaces using the *Split* tool, by checking the *Keep both sides* option. Thus, two features appear in the specification tree, *Split.1* and *Split.2*.

Surface.18 (Figure 3.428) is similarly edited using the *Split* tool, and implementing as *Cutting elements* the *Line.2* that joins the ends of curved edges (Figure 3.433).

The line is drawn using the same *Point-Point* option as before. In the *Support* field, the user can also keep the default selection (*Default None*) or choose the *Surface.18*.

Using the *Split* tool (*Split Definition* box, Figure 3.434) and selecting *Line.2* (*Cutting elements*) as the separating element, the user can edit *Surface.18* (*Element to cut*) to result in the rectangular surface *Split.2*.

Surface.11 must be also removed. It is not possible to simply delete it from the specification tree or with the *Delete* option in the context menu because at least two descending elements (children) are connected to this surface: *Line.1* and *Project.1*.

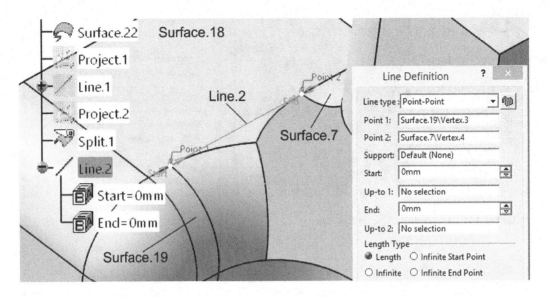

FIGURE 3.433   Drawing the *Line.2* using the *Point-Point* option.

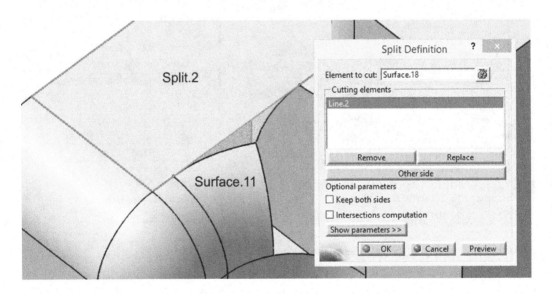

FIGURE 3.434   Editing the *Surface.18* by using the *Split* tool.

Figures 3.429 and 3.430 contain in the selection fields this *Surface.11*. Depending on the way the user works, other elements of the specification tree (e.g. *Line.2*) may also be surface dependent ('parent-child' relation).

In an attempt to delete the surface, the user receives a warning message regarding the geometric features connected to the *Surface.11*. The *Delete* dialog box (Figure 3.435) initially opens in a small format and proposes the standard option, to remove them as well (*Delete all children*). However, this is impossible because deleting the projected point and line has an impact on the features below in the specification tree: *Project.2* and *Split.1* will be also removed. On the other hand, unchecking the option and, however, removing the surface affects (error messages, Figure 3.436) the same features.

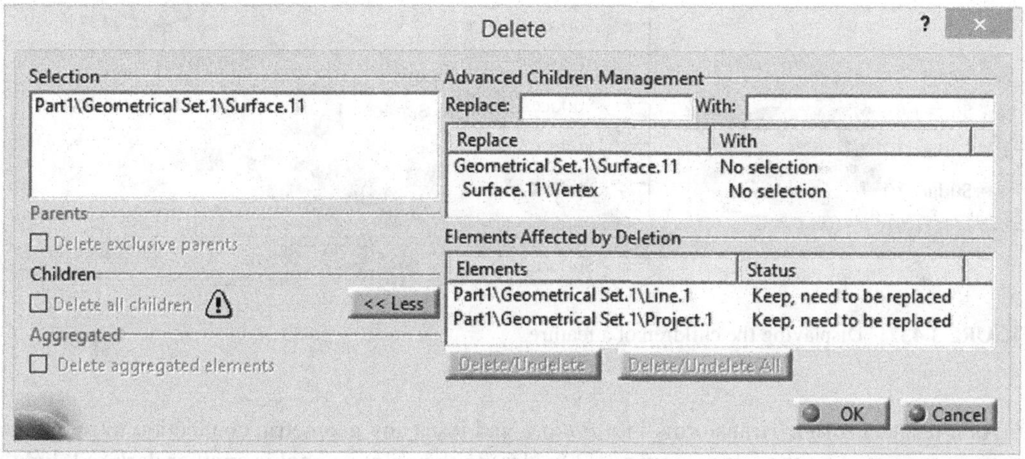

FIGURE 3.435 Warning on deletion of the *Surface.11*.

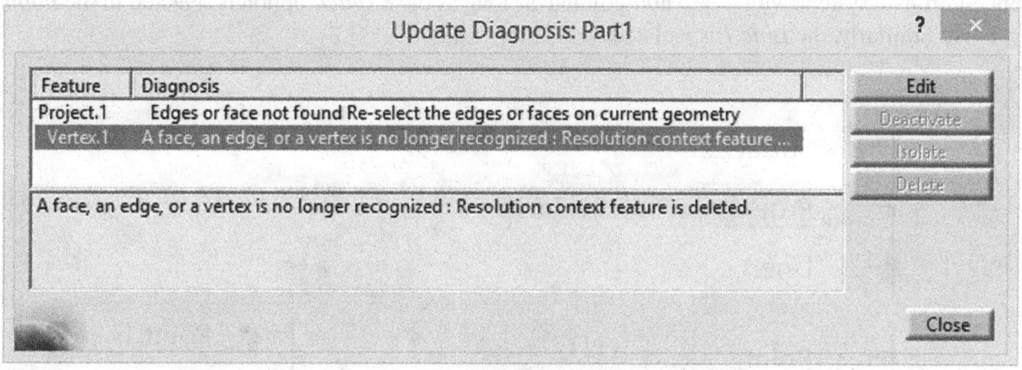

FIGURE 3.436 Warning on dependent features when deleting geometric features.

The error messages show that certain edges, points or faces, if they disappear, affect the features which depend on them. The user has several options for handling error messages by editing, disabling, deleting or isolating items, and he should be aware that closing (*Close* button) the message boxes does not solve the problem.

To find out the descendants that cause these errors, the user clicks the *More >>* button, the *Delete all children* option is unchecked by default, and the *Delete* box expands, according to Figure 3.437. The available options are complex and involve/allow the replacement of items listed in the *Advanced Children Management* area with others selected by the user (*Replace* and *With* columns). Below, in the *Elements Affected by Deletion* area, the features affected by removing the *Surface.11* are listed. These features are also displayed by right-clicking directly on the surface (or in the context menu) and choosing the *Parents/Children* option (Figure 3.437). An information box opens, in which it is observed that the two features *Line.1* and *Project.1* are linked to the surface.

However, the surface must be removed, regardless of the features that depend on it. The simplest solution is to transform them into simple geometric entities, without having connections to the surface and/or other features. Thus, the *xyz object → Isolate* option is chosen from the context menu of each feature (Figure 3.438).

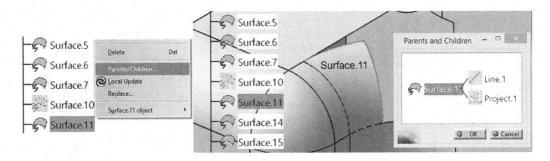

FIGURE 3.437   Displaying the children of a feature.

As a result, *Project.1* transforms into *Point.1* and loses any geometric connection to the point that was projected on the edge (Figure 3.429). Next to the point symbol is another dark red, lightning-shaped symbol. Its significance is that the geometric feature has become a simple, independent, isolated geometric entity. Double-clicking on *Point.1* in the specification tree opens the *Point Definition* box (Figure 3.439), and the user can observe the point's absolute coordinates in the current coordinate system. Values are non-editable as long as the *Explicit* option is selected in the *Point type* list. Similarly, the *Line.1* is isolated.

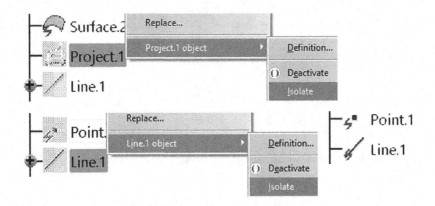

FIGURE 3.438   Isolating features/objects.

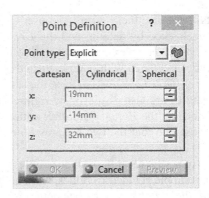

FIGURE 3.439   Displaying the *Point Definition* dialog box.

Following these two transformations and when the user removes the *Surface.11*, the 3D surface model of the part shows a complex cut (Figure 3.440). It must be filled so that the model can be transformed into a solid object.

Activating the *Fill* tool opens the *Fill Surface Definition* box (Figure 3.441), in which the edges marked from 1 to 8 are selected (*Curves* column), and for each edge, a support surface (*Supports* column) is also selected. The edge belongs to it and causes the obtained surface, *Fill.1*, to be tangent to the supporting surface.

Selecting the edge-support pair is simple: the user first clicks on the edge and then on the support, in logical order, to close the contour. The edges are chosen successively, for example, in reverse trigonometric direction. This creates edge-support pairs with very fast checking: the edge must belong to the support surface (example: *Split.2\Edge.1* and *Split.2*, *Surface.7\Edge.2* and *Surface.7*, *Surface.6\Edge.3* and *Surface.6,* and *Surface.5\Edge.4* and *Surface.5*). The resulting surface, *Fill.1*, correctly and completely covers the area where the part is reconstructed.

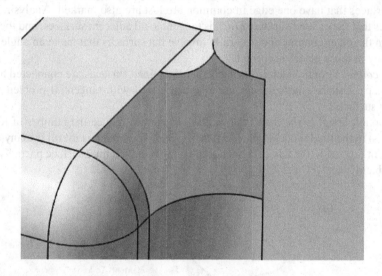

FIGURE 3.440   Displaying the gap to be filled.

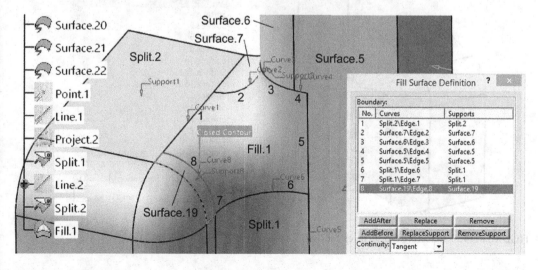

FIGURE 3.441   Selecting the chain of edges that form the boundary for the *Fill* tool.

The 3D model of the part now contains many surfaces: some have remained intact since the extraction from the solid, others have been removed by the user, some were edited according to certain geometric features, and the last surface, *Fill.1*, has filled the remaining gap. The specification tree contains all these individual surfaces.

In order to recreate the solid, the user must merge all these surfaces into one unique patch using the *Join* tool. In the dialog box in Figure 3.442, the surfaces are selected using a selection window with the mouse. This method is recommended when the number of surfaces is relatively large. If the first entity selected in the *Join Definition* box is a surface, then only this entity type is still allowed to be selected (not for, e.g., combining with lines).

The box contains several important options that can be checked when joining surfaces. Thus, when *Check tangency* is checked, the joined features/entities are verified to be continuously tangent to each other.

In this case, according to Figure 3.443, when the user clicks the *Preview* button in the *Join Definition* box, an error message is displayed. Consequently, the edges where the tangential property of two surfaces that have one edge in common are lost are also marked. Analysing the image, the user notices that the created surface, *Fill.1*, is tangent to all adjacent surfaces, and the errors refer to the edges on the outer contour and generally involve flat surfaces that make an angle of 90°. The lack of tangency in those places is correct.

The *Check connexity* option determines whether the joined surfaces are connected to each other and have no gaps. Usually, these spaces are also associated with tangential problems (partial or total) between surfaces.

Also, the option *Simplify the result* allows the program to reduce the number of features (surfaces and edges) in the final model (*Join.1*) when possible. The option is useful if many surfaces are joined, some not tangent to the adjacent ones, and simplifying the final surface partially sometimes solves the problem.

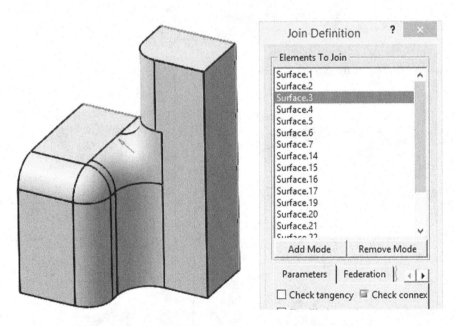

FIGURE 3.442   Selecting the surfaces for the *Join* tool.

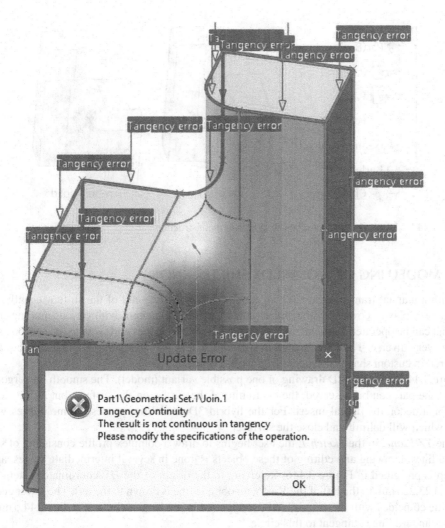

FIGURE 3.443 Displaying an error when checking the tangency of the joined entities.

The user can also define a minimum distance, considered as a tolerance, up to which two joined surfaces are considered to have become one. The default value, but editable in the field *Merging distance*, is 0.001 mm and corresponds to the program settings, under the menu *Tools → Options → Shape → Generative Shape Design → General* tab. Users should follow this tip: it is not recommended to join two surfaces that have a distance exactly equal to the value set in that field.

Finally, after the tangency and connection checks, the *Join.1* surface is added to the specification tree and, in this case, it has an area of 12,169.2 mm² (the *Measure Item* tool was used). To turn it into a solid, the *Close Surface* tool in the *Part Design* workbench is applied to the surface and the *Join.1* feature can be hidden.

The obtained solid, *CloseSurface.1*, is moved by the user to a new body, *Body.2*, and has a volume of 73,113.124 mm³ compared to 73,134.011 mm³, the measured volume of the original part. The difference is small, only 20.88 mm³, and the pre-edit area of 12,169,884 mm² is larger than the area of the reconstructed part, but with an insignificant value of 0.68 mm².

Figure 3.444 compares the two solid bodies/parts.

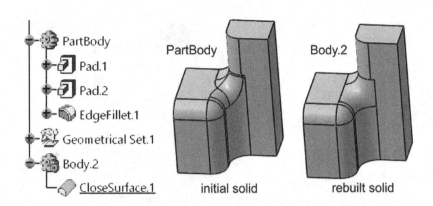

FIGURE 3.444   Displaying the solid before and after reconstruction.

3.20 MODELLING OF A GEARBOX SHIFTER KNOB

Cars with a manual transmission have a gearbox shifter. At one end of the shifter, usually, it is a knob made of plastic, covered with leather, with some metal elements, with various moving components that can be operated by the driver to change gears. The shapes and sizes of the gearbox shifter knob are very diverse. They are created in standard shapes for the automotive industry, but also to be ordered in custom shapes by specialized workshops.

Figure 3.445 shows the 2D drawing of one possible variant (model). The smooth and ergonomic shape of the part can be observed, the continuity of the curves on the surface, but also a slot provided for a decorative metal insert. For the hybrid 3D modelling process, some curves will be drawn, which will delimit and close the surface.

In the *YZ Plane*, in the *Sketch.1*, the user begins to draw a complex profile consisting of several arcs and lines. Drawing and editing of these objects is done in several intermediate stages, and the first step is presented in Figure 3.446. Referring to the origin of the *HV* coordinate system of the sketch, at 23.225 mm to the left of the *V* axis, an oblique line is drawn to the left. The lower endpoint of the line coincides with the *H* axis, and the upper endpoint is on a circle of radius R44.5 mm; the straight standard line is tangent to this circle.

Using the *Construction/Standard Element* icon in the *Sketch Tools* toolbar, the user draws an auxiliary construction line inclined at 16.5° from the *H* axis, according to Figure 3.446. The centres of two circles (standard line elements) of R24.5 mm (right) and R36 mm (left) are positioned on this line. The centre of the circle R24.5 mm is constrained to 72.4 mm from the *V* axis, and the circle R36 mm becomes tangent to the circle R44.5 mm.

The sketch is continued in the second stage by drawing a circle of radius R17 mm inside the circle R36 mm and tangent to it (Figure 3.447). A second auxiliary construction line is drawn horizontally and at a distance of 111.25 mm from the *H* axis. The user also draws and constrains a circle of radius R139 mm tangent to this horizontal line, to the circle R24.5 mm and to the circle R17 mm.

To add the constraints, the circles drawn in the first stage can move freely with the mouse and they are brought into positions similar to those in Figure 3.447.

In the third stage of drawing of the *Sketch.1*, the user adds a circle of radius R24 mm with the centre on the *H* axis and with the circumference at 23.225 mm from the *V* axis. Another circle of radius R88 mm is drawn tangent to the previous circle and to the circle of radius R24.5 mm, according to Figure 3.448. The auxiliary construction line in Figure 3.446 is constrained as follows: the endpoints of the line become coincident with the circles of radii R24.5 mm and, respectively, R17 mm, and the length of the line is imposed at the value of 149.3 mm. Also, the oblique line tangent to the circle of radius R44.5 mm makes with the *V* axis an angle of 5.3°. In the figure, the points of tangency between the geometric elements of the sketch were marked with a *T*.

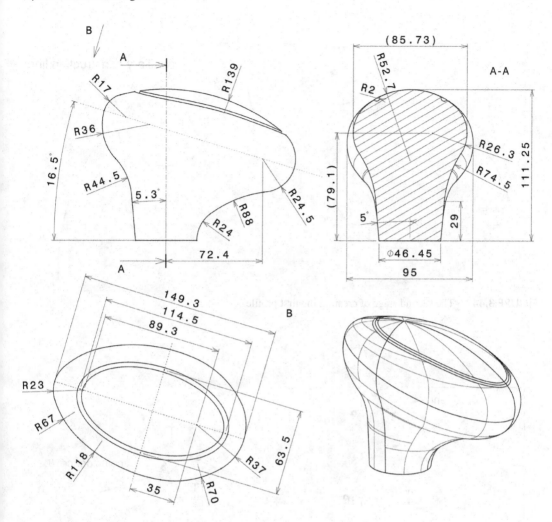

FIGURE 3.445 The 2D drawing of the part to be modelled.

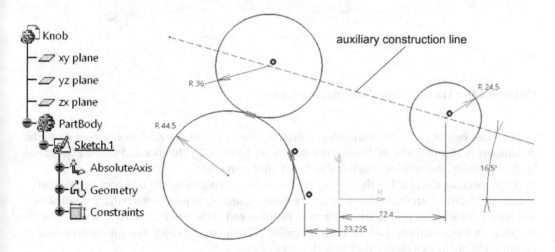

FIGURE 3.446 The first stage of creating the first profile.

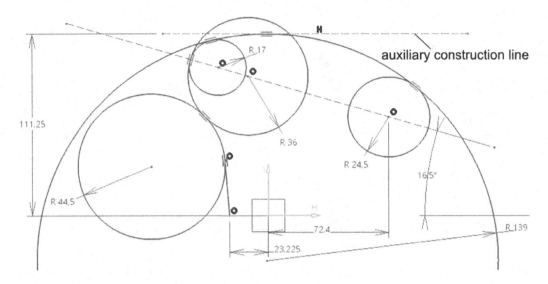

FIGURE 3.447   The second stage of creating the first profile.

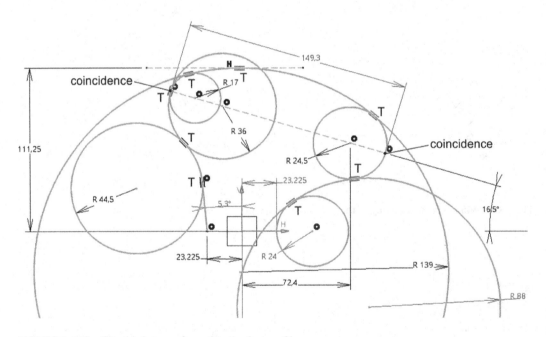

FIGURE 3.448   The third stage of creating the first profile.

The three figures show the main steps of drawing the sketch in the *YZ Plane*, according to the dimensions in Figure 3.445. If the drawing steps were followed by the user and all the constraints applied correctly, the profiles drawn in the sketch must turn green.

After checking the profiles, the user starts the profile editing using the *trim* tool to get a curve consisting of several arcs and line segments. Editing, however, requires an auxiliary construction line coinciding with the *H* axis, which starts at the lower end of the line inclined 5.3° from the vertical, then intersects and exceeds the circle of radius R24 mm to the right. The line is delimited in Figure 3.449 by the two points marked with *start* and *end*.

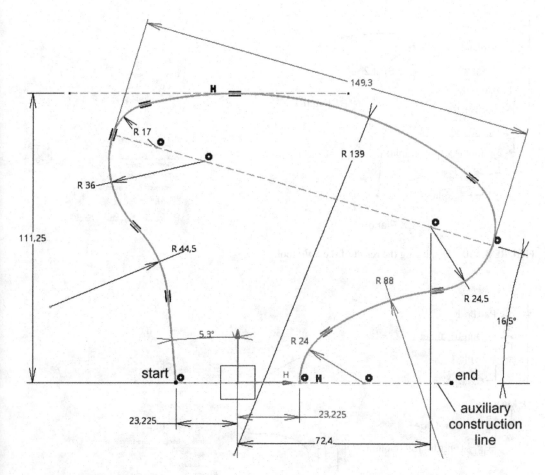

FIGURE 3.449 Displaying the first profile of the modelling process.

The *Quick Trim* tool in the *Operation* toolbar is used to remove certain geometric elements and keep only those in the figure. By double-clicking its icon, the trim function remains active and the user can select the segments to be deleted successively and quickly.

Also, if the user has applied the *Quick Trim* tool correctly, the final profile is as it is shown in Figure 3.449. This is an open profile at the two lower endpoints coinciding with the *H* axis; it keeps its geometric and dimensional constraints, but also the auxiliary construction lines. A quick check is possible by accessing the *Sketch Analysis* option in the *Tools* menu, and the information message must be *Default Profile: Opened.*

In addition to the *Sketch.1*, its profile is split into two separate profiles using the *Split* tool in the *Operations* toolbar. In the *Split Definition* dialog box in Figure 3.450, in the *Element to cut* field, the user selects the *Sketch.1*, and in the *Cutting elements* field, he selects the *ZX Plane* that is perpendicular to the plane in which the profile was drawn. The user also ticks the *Keep both sides* option for the program to create two curves, *Split.1* and *Split.2*, marked in the figure and added to the specification tree. The length of the two curves can be measured as follows: *Split.1* is 221.12 mm and *Split.2* is 154.65 mm, an example being shown in Figure 3.451. To choose the measurement curve, the *Measure Item* tool is used; the user will click on its name in the specification tree and not on the 3D model. In this second case, only the arc or line segment that was selected with the mouse will be measured.

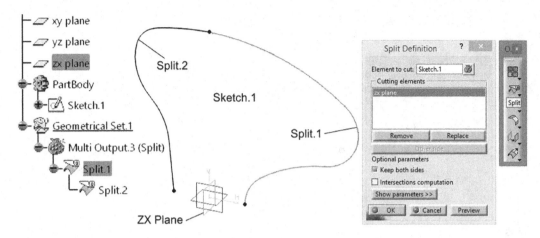

FIGURE 3.450   Displaying the result of the *Split* tool.

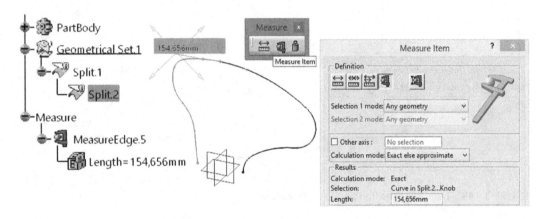

FIGURE 3.451   Measuring the *Split* features.

The next step consists in selecting the *ZX Plane* and creating a new sketch, *Sketch.2*. In Figure 3.452, the view of the sketch is from the negative direction of the *Y* axis; thus, it can be observed the directions of the axes *H* and *V*. To change the view so the user looks perpendicular on the *Sketch.2*, he should click *Normal view* button in the *View* toolbar.

From the origin of the sketch's coordinate system and coinciding with the *V* axis, an auxiliary construction line is drawn that exceeds the profile of the *Sketch.1*. To the right of the *V* axis and with one of the ends coinciding with the *H* axis, an oblique line is drawn, inclined by 5° with respect to the vertical axis. This point is 23.225 mm from the *V* axis. The upper end of the line is positioned 29 mm above the *H* axis and belongs to a circle of radius R74.5 mm, so that the oblique line is tangent to it.

With the centre on the vertical line of auxiliary construction, the user draws a circle of radius R52.7 mm. Its circumference is constrained coincident with the point of intersection of the curves *Split.1* and *Split.2*.

A third circle is drawn between the two circles, with a radius of R26.3 mm and tangent to them, according to Figure 3.452. It does not reach the centre of the circle with a radius of R52.7 mm. The tangency points were marked with *T* as in Figure 3.448. For Figure 3.452, the user is recommended to have an isometric view for a better understanding of the circles' positions.

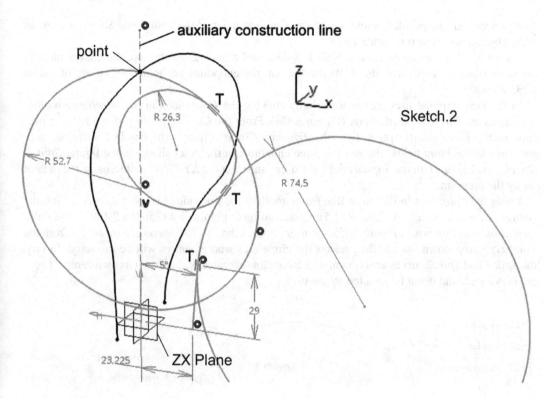

FIGURE 3.452 Creating the *Sketch.2*.

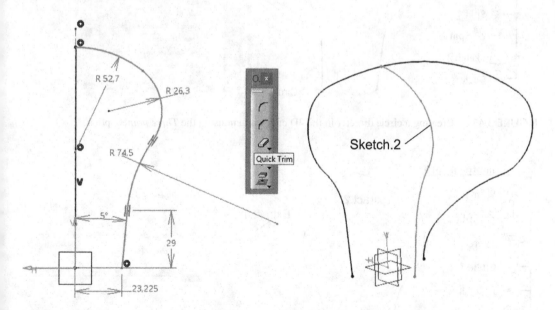

FIGURE 3.453 Displaying the final shape of the *Sketch.2* and the isometric view.

Using the *Quick Trim* tool, many arcs are removed from the circles drawn in Figure 3.452 and only one contour is retained, to the right of the *V* axis, according to Figure 3.453. The line and arc segments that make up this profile are successively tangent to a very good continuity of gearbox surface. Furthermore, the profile of the *Sketch.2* is open and corresponds to the *A-A* section in Figure 3.445.

At its upper end, the profile coincides with the common point of the *Split.1* and *Split.2* curves, as shown by the isometric representation.

Figure 3.453 shows three curves: *Split.1*, *Split.2* and *Sketch.2*. At the top, the curves meet at the same point, as explained above. At the bottom, the endpoints are joined by an arc of radius R23.225 mm.

In the next step, the user creates a circle by clicking the *Circle* icon in the *Wireframe* toolbar and opens the *Circle Definition* box (Figure 3.454). From the *Circle type* drop-down list, the user chooses the *Three points* option, ticks the *Trimmed Circle* option in the *Circle Limitations* area and, then in the *Point* fields, chooses the three endpoints in the order shown in the figure (*Split.2*, *Sketch.2* and *Split.1*). In the *Support* field, the user can select the *XY Plane* or it remains the default set by the program.

Using the *Point* tool in the same *Wireframe* toolbar, the user adds two points representing the centres of the arcs of radii R17 and R24.5 mm, according to Figures 3.445 and/or 3.449. In the *Point Definition* selection box in Figure 3.455, the user chooses the *Circle/Sphere/Ellipse center* from the *Point type* drop-down list and then selects the circle arcs whose centres will be extracted. In fact, the *Split.1* and *Split.2* curves are chosen, which contain several arcs, each with its own centre. These are previewed, and it can be selected by the user.

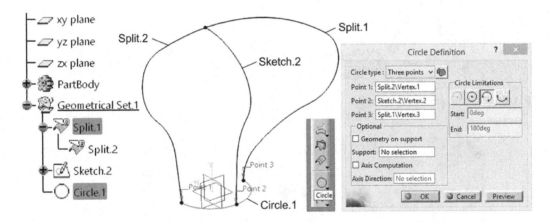

FIGURE 3.454   Creating a circle directly in the 3D environment using the *Three points* option.

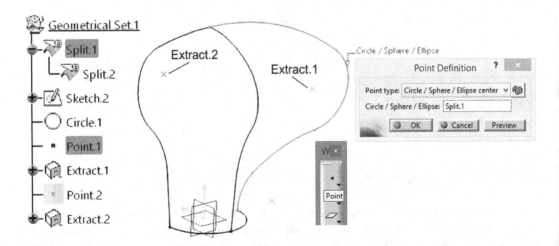

FIGURE 3.455   Creating two points in the centres of the already created arcs.

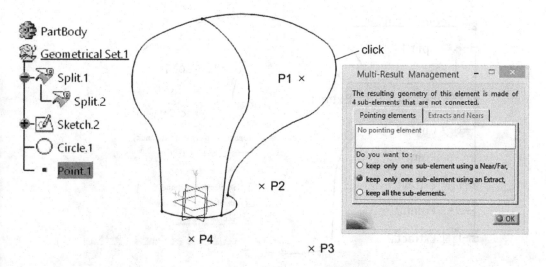

FIGURE 3.456 Dealing with *Multi-Result Management* selection box after applying the *Extract* tool.

The user clicks the *OK* button in the *Point Definition* selection box, and the *Multi-Result Management* selection box will open as in Figure 3.456. The feature *Point.1* is added to the specification tree. The user observes the four points marked *P1,…, P4* in the figure and ticks the *keep only one sub-element option using an Extract*.

This option allows the user to select only one item from a possible multiple selection of multiple items (points, in this case). The user clicks the *OK* button, and the *Extract Definition* options box presented in Figure 3.457 will open. The user specifies the point marked *P1* by clicking on it, and the *Extract.1* feature is added to the specification tree.

The *Extract.2* point is similarly defined (Figure 3.458). The user draws a line that joins these two points and is inclined 16.5° from the horizontal.

To create a line between *Extract.1* and *Extract.2*, the user clicks the *Line* icon in the *Wireframe* toolbar and the *Line Definition* dialog box will open as in Figure 3.459. In the *Line type* drop-down list, the user chooses the *Point-Point* option, selects the *Extract.1* and *Extract.2* features in both *Point* fields and clicks the *OK* button. *Line.1* is added to the model and to the specification tree.

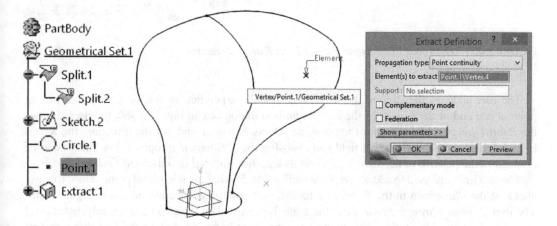

FIGURE 3.457 Using the *Extract* tool in a multi-result selection box.

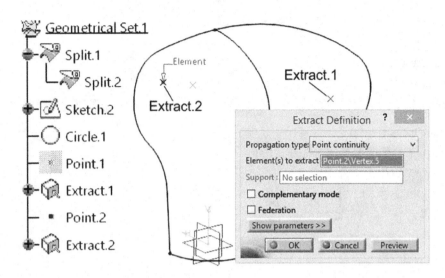

FIGURE 3.458   Using the *Extract* tool in a multi-result selection box to create the *Extract.2* feature.

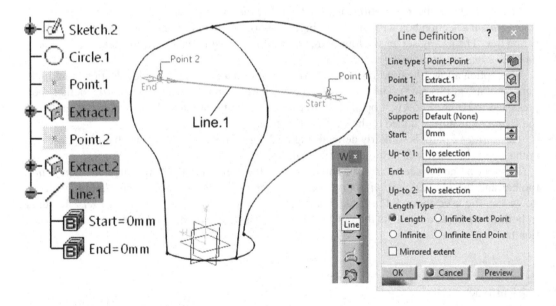

FIGURE 3.459   Creating a line between *Extract.1* and *Extract.2* features.

The user adds another line to the wireframe model, perpendicular to *Line.1*, with the starting point at one end of the *Line.1*. In the *Line Definition* dialog box in Figure 3.460, the user chooses the *Point-Direction* line type; for the point, he selects *Extract.1* and, for the direction, the *X* axis by right-clicking in the *Direction* field and choosing the *X Component* option. The *Line.2* can be oriented in the positive or negative *X* axis and its length is inserted into the field *End*.

The two lines are used to add a new plane inclined 16.5° from the horizontal plane. Thus, the user clicks on the *Plane* icon in the *Wireframe* toolbar and opens the selection box as in Figure 3.461. The user chooses *Through two lines* as the plane type, and then the two lines are selected, *Line.1* and *Line.2*, in the *Line* fields. The *Plane.1* is represented as a symbol on the line that was first selected, and it is added to the specification tree.

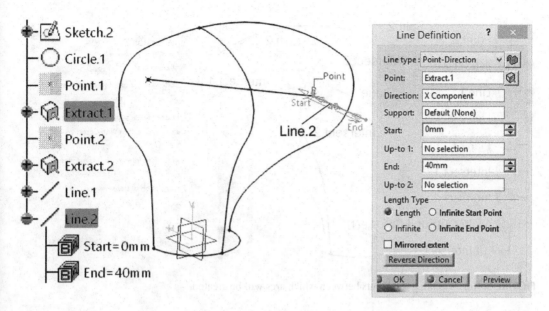

FIGURE 3.460 Creating a line with the option *Point-Direction*.

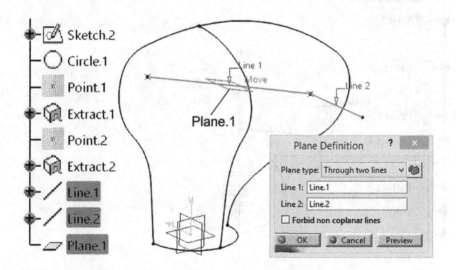

FIGURE 3.461 Creating a plane with the *Through two lines* option.

The lines used to position the plane can be hidden (*Hide/Show* option in the context menu of each). According to projection B in the 2D drawing shown in Figure 3.445, the user must draw a profile in this plane, consisting of arcs between certain points (Figure 3.462). Points are obtained by intersecting the *Plane.1* with the three curves: *Split.1*, *Split.2* and *Sketch.2* using the *Intersection* tool. As a result, the points *Intersect.1*, *Intersect.2* and *Intersect.3* are added to the specification tree.

Thus, in the *Sketch.3* of the *Plane.1*, five circles of radii R37 mm, R23 mm, R118 mm, R70 mm and R67 mm and a line are drawn, respecting coincidence and tangent constraints. The line is an auxiliary construction and coincides with the vertical axis *V* of the sketch coordinate system (Figure 3.463). The ends of the line must extend beyond the circles that will be drawn next because they will be edited relative to the vertical line.

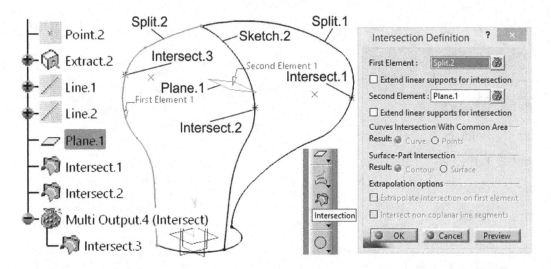

FIGURE 3.462   Creating the points between which arcs will be created.

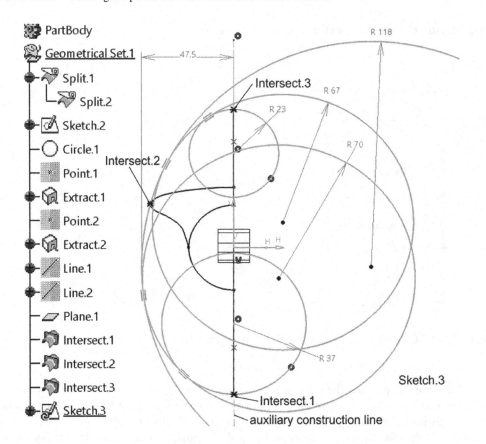

FIGURE 3.463   Creating the *Sketch.3*.

The first circle drawn is the R37 mm radius, with the centre positioned on the auxiliary construction line and obviously coinciding with the *V* axis. The circle is constrained to coincide with the point *Intersect.1*. The second circle inserted in the sketch is of radius R23 mm, with the centre on the auxiliary construction line. Point *Intersect.3* is constrained to coincide with this circle.

The third circle of radius R67 mm is determined to be tangent to the circle of radius R23 mm. Similarly, the fourth circle, of radius R70 mm, is drawn tangent to that of radius R37 mm. The last circle created by the user in the *Sketch.3* is of R118 mm; its constraint will transform the other circles to become completely and correctly constrained. Thus, the point *Intersect.2* belongs, by constraint, to this circle, and then two tangent constraints are imposed between the circle R118 mm and each of the circles R67 and R70 mm. The circumference of the circle is 47.5 mm (95/2 mm, section *A-A*) from the auxiliary construction line.

Each centre of the last three circles does not have a certain position, constrained by other geometric elements, but results from the other constraints set by the user.

Circles are edited using the *Quick Trim* tool in the *Operation* toolbar. The user double-clicks on the yellow eraser, which becomes and remains active, and then all the segments to the right of the vertical auxiliary construction line are deleted, according to Figure 3.464.

On the left side, only the outer contour is kept, shown in projection *B* in Figure 3.445. As can be seen, there are many differences between Figures 3.464 and 3.465, the latter missing many circle segments removed by the user. As the user deletes these arcs, the user may also notice the disappearance of certain constraints. Finally, after the profile has been obtained, the user can re-add these constraints (dimensions, tangents, etc.), or can completely constrain any geometric element by right-clicking on it and choosing the *Fix* option from the context submenu (example: *Circle.5 → Fix*).

The profile of the *Sketch.5* consists of five circular arcs, tangent to each other, and the free ends are at *Intersect.1* and *Intersect.3* (Figure 3.465).

Figure 3.466 shows the wireframe profiles obtained so far (the so-called wireframe model). The structure of the gearbox shifter knob consisting of these profiles in three isometric representations is observed.

Using these curves that can be accessed in the specification tree, the user creates a surface applying the *Multi-Sections Surface* tool in the *Surfaces* toolbar. In the *Multi-Sections Surface Definition* selection box in Figure 3.467, the user selects the sections through which the surface passes (*Circle.1* and *Sketch.3* curves) and, then in the field below, the guidance curves that delimit the surface between sections (*Split.1, Sketch.2* and *Split.2*).

Usually, these guiding curves are chosen in a specific order (from right to left, for example). The user then clicks the *Preview* button and gets the surface in Figure 3.467. It is noticeable that the surface is not complete at the top, but only in the area between the two sections.

In the *Relimitation* tab, there are two options (Figure 3.468) that limit the surface area between the two sections, even if the guidance curves exceed them. Thus, the user must uncheck *Relimited on start section* and *Relimited on end section* to remove this restriction. Sections are considered the start or the end depending on how the user selects them in the *Section* column. In this case, *Circle.1* is the start section and *Sketch.3* is the end section.

Figure 3.468 shows the complete *Multi-sections Surface.1* feature and its continuity between sections and between the guiding curves.

If these guiding curves had not been chosen in the order mentioned above, but, for example, *Split.1, Split.2* and *Sketch.2*, the surface would not have been possible, *CATIA* displaying an error message according to Figure 3.469, and the user must resume the selection of curves.

The part has an obvious symmetry around the *YZ Plane*, according to the representation in Figure 3.445 and the modelling strategy used in the application. From the *Operations* toolbar, the user can apply the *Symmetry* tool to mirror the existing surface as in Figure 3.470.

The user selects the surface *Multi-sections Surface.1* as the input in the *Element* field, and as a reference the *YZ Plane*. There is a possibility that the initial surface will become hidden and only the symmetrical one will be visible by pressing the *Hide/Show initial element* button. The *Symmetry* tool can also be used for both surfaces and solids by checking one of the *Surface* or *Solid* options.

The result, in this case, is an identical (symmetrically mirrored) surface to the original one, called *Symmetry.1* and added to the specification tree.

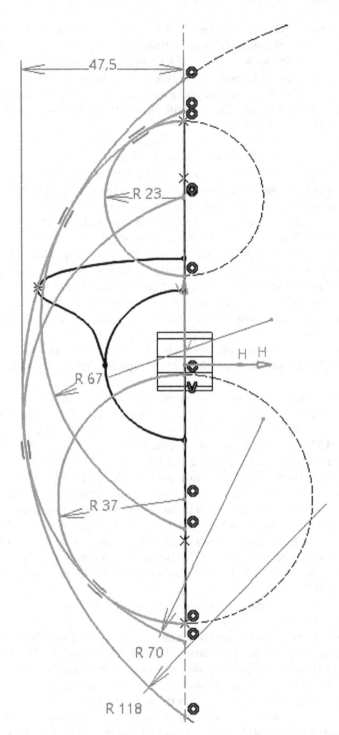

FIGURE 3.464 Editing by using the *Quick Trim* tool.

At the bottom of the two surfaces, there is an open area, bordered on one side by the *Circle.1* curve created according to the explanations related to Figure 3.454. On the other side of the *YZ Plane* is a curve, part of the *Symmetry.1* surface. The user has the ability to extract that curve/edge using the *Boundary* tool in the *Operations* toolbar.

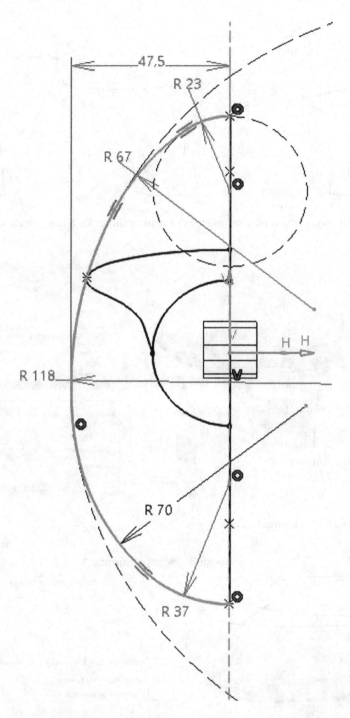

FIGURE 3.465 Displaying the final profile of *Sketch.3*.

Thus, in Figure 3.471, in the *Boundary Definition* selection box, the edge propagation type is chosen as *Point* continuity; the extracted curve is selected in the *Surface Edge* field and, in the *Limit* fields, two points belonging to the *Split.1* and *Split.2* curves. These two points are, in fact, the endpoints of the *Circle.1*. If the selection is correct and the edge can be removed, it turns green and *Boundary.1* is added to the specification tree.

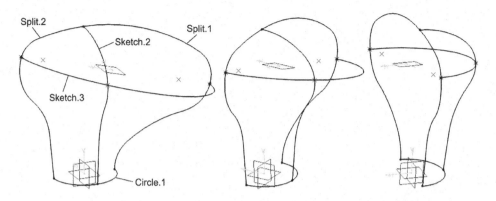

FIGURE 3.466   Displaying the wireframe geometry of gearbox shifter knob in isometric view.

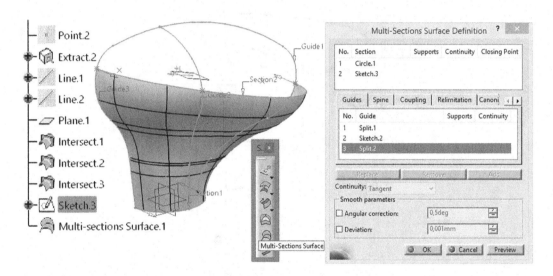

FIGURE 3.467   Displaying a partial surface.

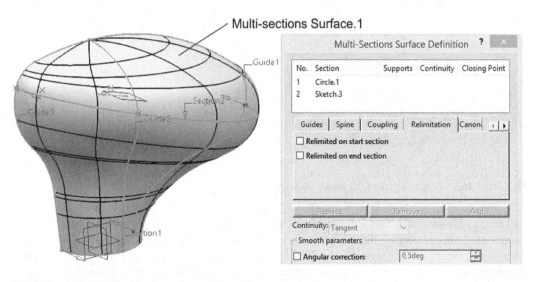

FIGURE 3.468   Displaying the entire surface without relimitation.

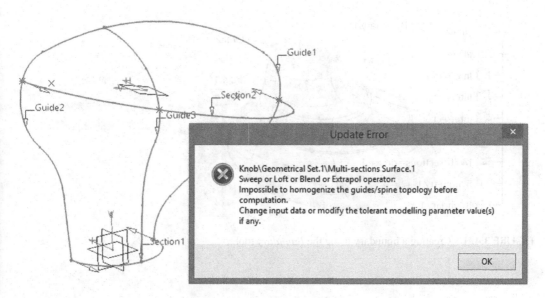

FIGURE 3.469 Displaying an error caused by the wrong order in selecting the guiding curves.

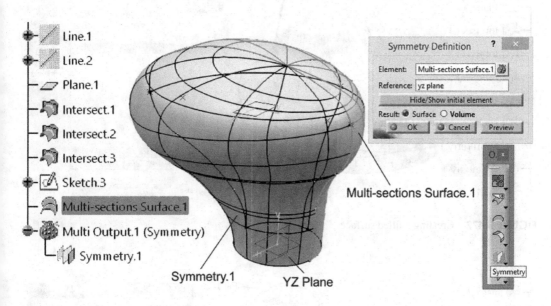

FIGURE 3.470 Applying the *Symmetry* tool.

The open area, enclosed by two edges *Circle.1* and *Boundary.1*, can be covered/filled by a surface with the *Fill* tool in the *Surfaces* toolbar. In the *Fill Surface Definition* selection box in Figure 3.472, in the *Boundary* field, the user chooses the two curves and then clicks the *Preview* button. As a result, if the surface is possible to be generated, it is represented on the 3D model, the message *Closed Contour* is displayed, and the feature *Fill.1* is added to the specification tree.

Up to this stage, the part model consists of three surfaces: *Multi-sections Surface.1*, *Symmetry.1* and *Fill.1*. For the surface part to be transformed into a solid model, a check of the connectivity and tangency of these three surfaces is required.

Thus, the *Connect Checker Analysis* option is used from the *Insert* → *Analysis* menu and the selection box in Figure 3.473 opens.

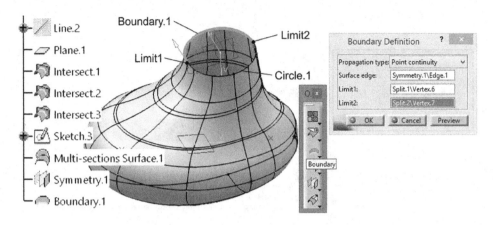

FIGURE 3.471 Creating a boundary using the *Boundary* tool.

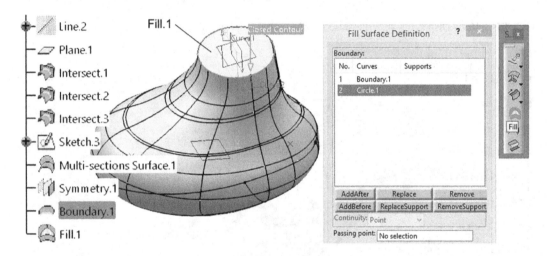

FIGURE 3.472 Creating a filled surface.

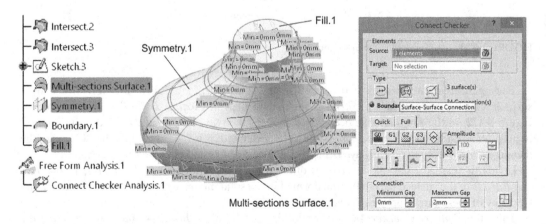

FIGURE 3.473 Checking the connectivity and tangency of the existing surfaces.

According to the selection of the three surfaces and for gaps between 0 and 2 mm (*Connection area*), the user also uses the analysis type *Surface-Surface Connection*, and the result is displayed graphically by numerous labels *Min = 0* mm. These are positioned along the sensitive edges of the 3D model of the surfaces: the edge joining the two surfaces *Multi-sections Surface.1* and *Symmetry.1*, the circular edge of the surface *Fill.1*, etc., especially where the surface suddenly changes its orientation.

In the *Join Definition* selection box opened by accessing the *Join* tool on the *Operations* toolbar, the user selects the three surfaces in the *Elements To Join* field (Figure 3.474). Checking the *Check tangency* and *Check connexity* options is no longer required as the check has been done before. However, the program may automatically reduce (and not user-controlled) the number of items (faces and edges) in the resulting *Join.1* area, if possible. The *Simplify the result* option is recommended for high complexity surfaces.

The *Federation* tab allows the user to create a group of patches that make up the surface and display *Join.1*. If by moving the mouse, it is positioned on the surface, no longer a patch or a small group of adjacent patch is selected, but the entire *Join.1* surface, like its selection in the specification tree. There are several options available to determine which areas to select at the same time, but for this application, *All* has been selected from the drop-down list.

The *Federation* option is useful when the user wants to click on an area to select not only a part of it (one or more patches), but the entire area. Also, certain geometric elements connected to this surface (example: the centre of a hole), if they have to change their position, will no longer refer to the patches where they were originally positioned, but to the entire surface.

If the *Federation* option was applied using the *Join* tool, its icon in the specification tree is completed with the yellow letter *F*.

Parallel to the *Plane.1* and at a distance of 50 mm above it, a new plane is inserted, *Plane.2*, using the *Plane* tool with the *Offset from plane* option in the *Plane Definition* dialog box (Figure 3.475). *Plane.1* is chosen as reference, and the value of 50 mm is entered in the *Offset* field.

In this plane, the *Sketch.4* is created by drawing an ellipse oriented with the large axis in a vertical position. To simplify explanations on how to constrain the ellipse, the surface *Join.1* will be hidden using the *Hide/Show* from the context menu shown in Figure 3.476. It is noted that to the surface in the figure, the *Federation* option was not applied because it was possible to select with the mouse the two patches of the *Join.1* surface, and then the user clicked the right button and opened the context menu.

The *Constraint* tool icon is activated to size the ellipse. Its major axis (in the vertical direction) is set to 114.5 mm, and the minor axis to 63.5 mm (in the horizontal direction). By default, the program would also add a value for the major axis, but after activating the constraint tool, the user presses the right button on the mouse and opens a context menu from which he can choose the option *Semiminor axis* and then position the dimensional constraint horizontally (Figure 3.477).

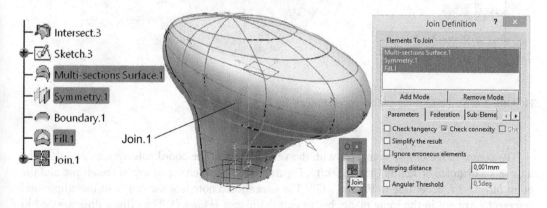

FIGURE 3.474   Joining multiple surfaces into a single one using the *Join* tool.

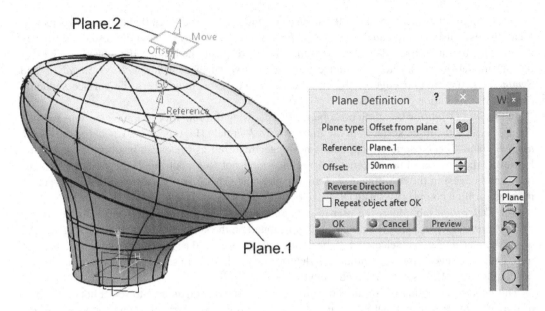

FIGURE 3.475   Creating a new plane parallel to the existing one at a specified offset.

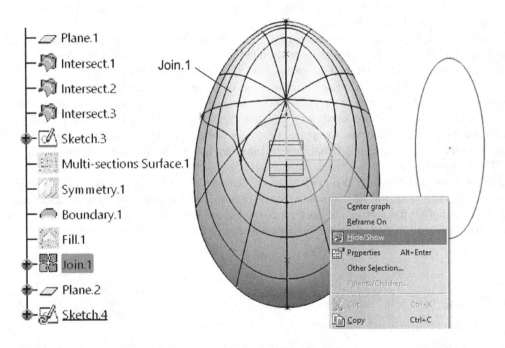

FIGURE 3.476   Performing a hiding operation on the *Join.1* surface.

The centre of the ellipse coincides with the vertical axis of the coordinate system, and then the user sets a distance of 72 mm (35 + R37 mm; Figure 3.445) between the centre of the ellipse and the point *Intersect.1*, according to Figure 3.478. The user should note that the centre of the ellipse and *Intersect.1* are not in the same plane, but in parallel planes (*Plane.1*). The ellipse dimensioned in this manner is represented and positioned in the sketch and in the specification tree.

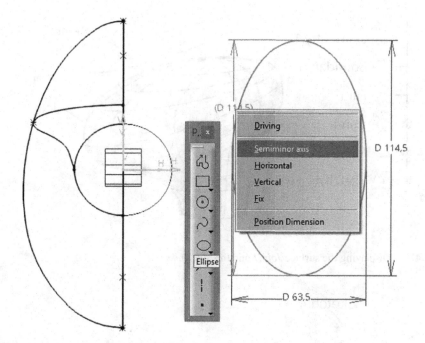

FIGURE 3.477 Dimensioning the minor axis of the ellipse.

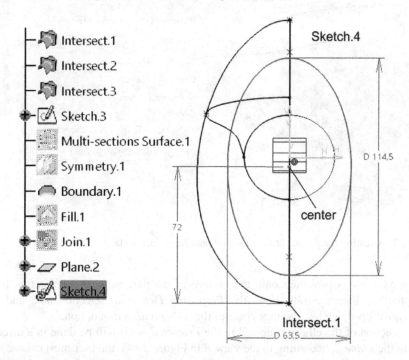

FIGURE 3.478 Placing a constraint for the centre of the ellipse.

The user restores the visibility of the *Join.1* surface, so Figure 3.479 displays it together with the *Sketch.4*. The inclined position of the sketch is observed according to the *Plane.2* that contains it. The ellipse was created above the surface *Join.1* because its dimensions and position are determined by the 2D drawing from the beginning. In fact, based on this ellipse, a decorative channel will be created at the top of the knob.

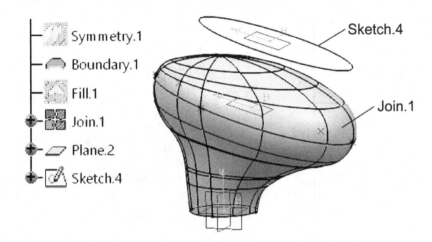

FIGURE 3.479 Displaying the surface *Join.1* and the *Sketch.4*.

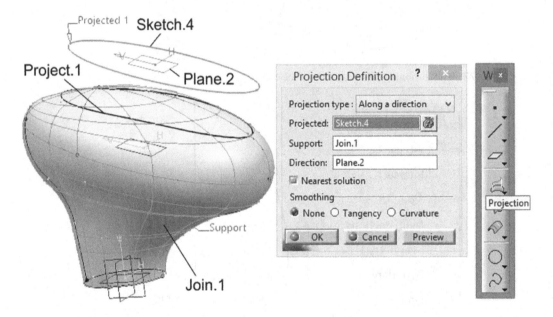

FIGURE 3.480 Creating the project of the *Sketch.4* onto the *Join.1* surface.

Thus, the *Sketch.4* is projected onto the surface of the part using the *Projection* tool in the *Wireframe* toolbar. Figure 3.480 shows the *Projection Definition* selection box, and from the *Projection type* drop-down list, the user chooses the *Along a direction* option.

As the projection of the ellipse (selected in the *Projected* field) will be done in a direction perpendicular to the *Plane.2*, according to the view *B* in Figure 3.445, the user must choose this plane in the *Direction* field. The surface on which the ellipse is projected is *Join.1*, selected in the *Support* field. As a result, a 3D curve, *Project.1*, is created on the surface of the previously generated surface part.

Once the projection is done, the ellipse and other features that were involved in creating the surfaces model can be hidden, such as *Split.1*, *Split.2*, *Sketch.2*, *Circle.1*, *Extract.1*, *Extract.2*, *Plane.1*, *Intersect.1*, *Intersect.2*, *Intersect.3*, *Sketch.3*, *Plane.2*, *Sketch.4* and *Boundary.1*, as defined above.

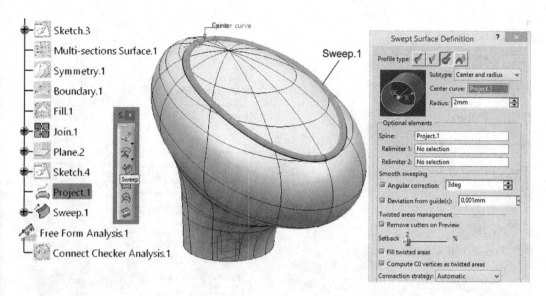

FIGURE 3.481   Creating a swept surface with the implicit profile *Circle*.

Along this curve, a circle moves to create a certain tubular surface, which will be extracted from the *Join.1* surface. There are several ways to get the surface from the top of the part, but the fastest one is to use the *Sweep* tool in the *Surfaces* toolbar. In the dialog box shown in Figure 3.481, the user can observe the *Circle* icon selected from the *Profile type* area. The user can also choose the *Center and radius* option from the *Subtype* drop-down list and then selects the *Project.1* feature for the guidance curve in the *Center curves* and *Spine* fields. The radius of the tubular surface is R2 mm according to the 2D drawing in Figure 3.445, section *A-A*.

The *Project.1* curve may seem relatively simple, but obtaining it on the *Join.1* surface causes some potential tangency problems between successive component segments of the curve. Thus, to avoid the error messages obtained by pressing the *OK* and/or *Preview* buttons, the user enters a value of 3° in the *Angular correction* field, the other options being checked according to Figure 3.481.

The *Sweep.1* surface is added to the specification tree, and it is observed that it follows correctly the shape and curvature of the *Join.1* surface. Editing is performed between the two surfaces using the *Trim* tool in the *Operations* toolbar. In the *Trim Definition* selection box, the user selects the surfaces and then presses the *Other side/next element/previous element* buttons until the correct cutting solution is obtained (Figure 3.482). This way the *Sweep.1* surface is extracted from *Join.1*.

The resulting surface, *Trim.1*, is added to the specification tree and displayed by three orientations in Figure 3.483. For checking purposes, the area of this surface is 41145.04 mm². The user can transform the *Trim.1* surface into a solid body (volume: 647,770.11 mm³) to complete the hybrid modelling of this part by using the *Close Surface* tool in the *Volumes* toolbar.

The 3D modelling of this gearbox shifter knob is of medium-high complexity, given by the initial profiles (*Sketch.1* and *Sketch.2*), the creation of the *Sketch.3* and, especially, the understanding of the part shape that required performing many working steps with wireframe and surface features.

The user can change the dimensions shown in Figure 3.445 to obtain other constructive solutions of this gearbox shifter knob.

A modelling version is also presented in the video solution: https://youtu.be/oH47PD9rp-g.

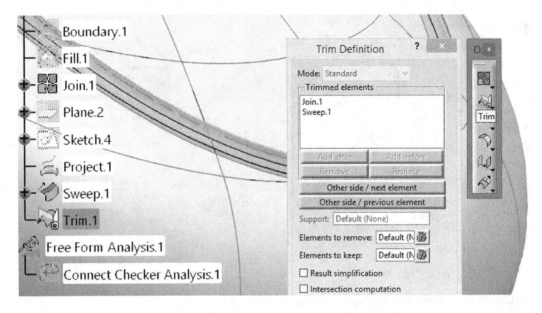

FIGURE 3.482   Editing the part by applying a *Trim* operation on selected surfaces.

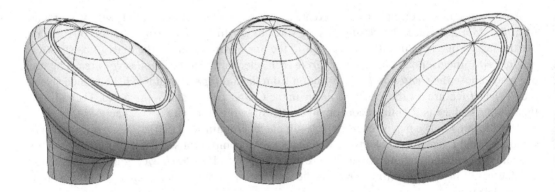

FIGURE 3.483   Displaying the result of the *Trim* operation, the final model of the gearbox shifter knob.

3.21   MODELLING OF A COMPLEX PLASTIC COVER

In this tutorial, a complex plastic part will be modelled combining *Part Design* and *GSD* workbenches and their tools.

Figure 3.484 shows the 2D drawing of the cover with a sufficient number of views, projections and section views. By observing its shape, it could be concluded that the main feature is a revolution surface with a couple of pockets. Especially interesting are side and back pockets. Also, in the section *A-A*, the constant value of thickness of 1.5 mm tells the user that this is a shell-like structure and that the *GSD* workbench could be a good starting choice to model this part, or *Shell* tool in *Part Design* may be a good replacement. After learning from this tutorial, the user may try another – his own approach.

In the *YZ Plane*, in the *Sketch.1*, the user begins to draw a profile consisting of two standard lines. By sketching, the user coincidently connects the first point of the first line with the *H* axis and finishes it by left-clicking somewhere inside of the upper right quadrant. The second line should be horizontal, and its second point is coincident with the *V* axis. This profile is open, but closed by axes, and for full constraining, some dimensions are needed, as shown in Figure 3.485.

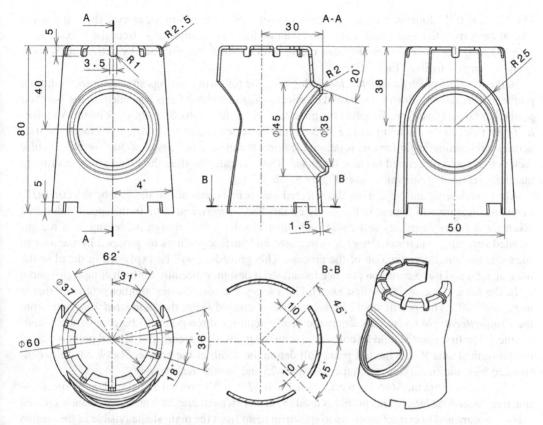

FIGURE 3.484   The 2D drawing of the complex plastic cover.

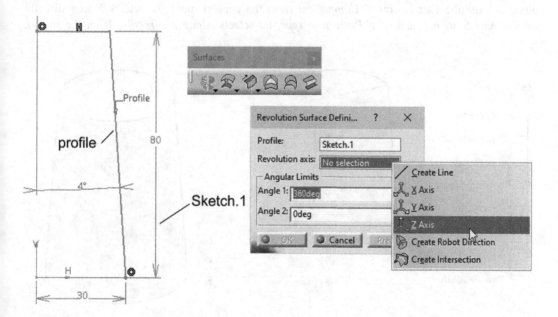

FIGURE 3.485   The first sketch and elements of the *Revolution Surface Definition* box.

The sketch is fully dimensioned by the height of profile of 80 mm from the *H* axis, the first point is 30 mm away from the coordinate origin, and the first line has an angle of 4° from the *V* axis.

By leaving the *Sketcher*, the user can create a rotational feature/surface by activating *Revolve* from the *Surface* toolbar (Figure 3.485).

In the *Revolution Surface Definition* dialog box, the following settings should be secured: (1) a profile to be revolved – *Sketch.1* – should be selected; (2) a revolution axis is needed, as the sketched profile has no axis line type. To define the missing axis, the user can do a right-click on the *No selection* blue field. In the context menu, *Z Axis* should be selected; and (3) In *Angular Limits*, the user defines a revolution angle for this rotational feature. It can be a full angle of 360° or two angular sectors that could be defined to close the shape. The user can also drag the green arrow in order to manually set the angular values as in Figure 3.486.

After previewing and approving the selected values and elements on the box by selecting *OK*, *Revolute.1* feature will appear in the newly established *Geometrical Set.1* in the specification tree (*Sketch.1* is under *PartBody* and could be hidden). In the hybrid design modelling, it is recommended that all geometrical elements, wireframe and surface features be grouped by the user in such sets for better organization of the process. This procedure will be explained in detail in the tutorial 3.25, and it is a common practice in industrial design, especially for high complexity parts.

In the same plane as in the first sketch (*YZ Plane*), the user creates another profile as shown in Figure 3.487. This profile could be automatically created using the *Elongated Hole* tool from the *Profile/Rectangle* tools. This automatic profile requires three points to be geometrically constrained. The first point should be coincident with 0,0 point (the origin). The second point is placed on the vertical axis *V,* and the last point will define the width of the profile. *Sketch.2* is fully constrained with additional dimensional constraints: 25 and 38 mm, respectively.

After exiting from the *Sketcher* workbench, a new *Sketch.2* is created and placed in the specification tree under *Revolute.1*. This profile is used in further steps to project it on the previously created surface feature and to extract contours to split/trim them from the main shape (visible in the section view *A-A* and back view in Figure 3.484).

To get a proper back shape of the cover, the user projects *Sketch.2* on the inner surface of the *Revolute.1* feature. To do so, the user activates *Projection* tool from the *Wireframe* toolbar and defines the following sets of elements as shown in Figure 3.488: (1) the projection should be done along *X* axis (the user selects *X Component* from the context menu, or selects *X* axis directly on the *Axis System.1*, and as a *Projection type*, he selects *Along a direction* from the menu);

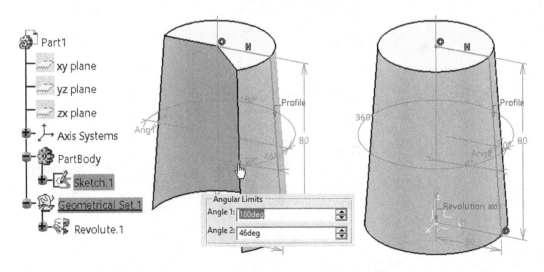

FIGURE 3.486   Defining angles of rotational surface features.

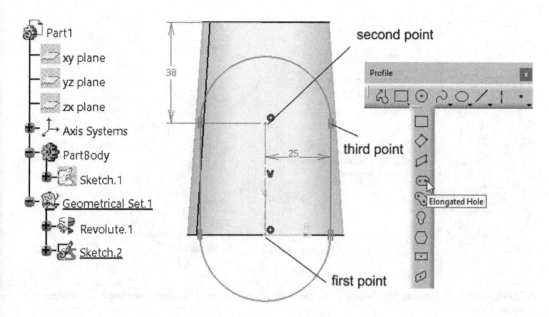

FIGURE 3.487 Using *Elongated Hole* tool to create the closed profile of the *Sketch.2*.

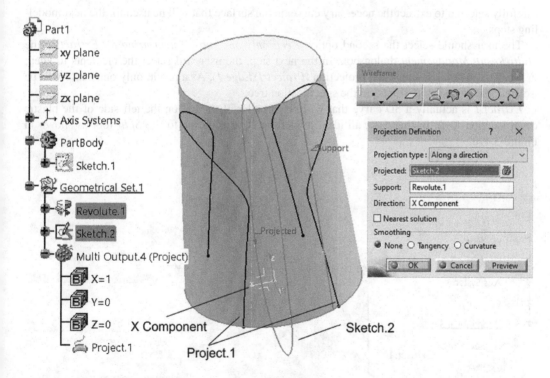

FIGURE 3.488 Projecting the *Sketch.2* on the wall of the revolved surface *Revolute.1*.

and (2) as an element to be projected, the user selects the *Sketch.2*, and as a *Support*, he chooses *Revolute.1* on which this profile will be projected. The *Nearest solution* option is unchecked to support projection along the *X* direction.

After approving these options, *CATIA* suggests to the user multiple options, as shown in Figure 3.489. As the only one side of the projected element is needed, some options should be

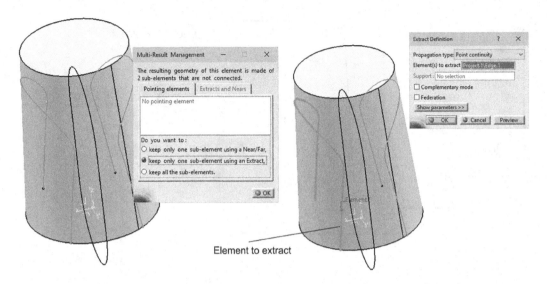

FIGURE 3.489 Projecting the *Sketch.2* on the wall of the revolved surface *Revolute.1* – isolating one element, *Extract.1*.

carefully selected to extract the necessary curve on the surface that will be used for the next modelling steps.

The user should select the second option *Keep only one sub-element using an Extract* in the *Multi-Result Management* dialog box. In the next step, the user will select the elements to keep; in this case, it is a left suggested projection (*Project.1\Edge.1*). As a result, only one curve will be isolated and placed as *Extract.1* in the specification tree.

Extract.1 is actually a 3D curve that will be a border of the cut on the left side of the plastic cover (section *A-A*). In this case, an ideal modelling step would be to use *Split* tool as shown in Figure 3.490.

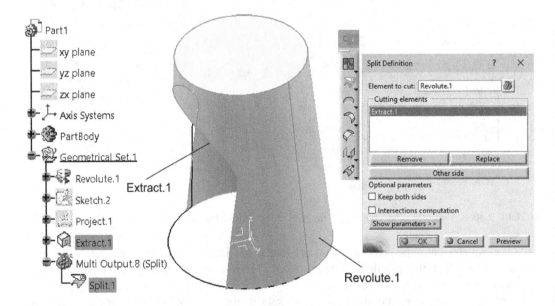

FIGURE 3.490 Splitting the *Extract.1* curve and removing its area from *Revolute.1*.

In the *Split Definition* dialog box, the user selects elements to cut, in this case *Revolute.1* and *Cutting elements*. CATIA suggests multiple options over which the user can intervene and select appropriate configuration or replace selection. After approval, *Split.1* feature will be added to the specification tree.

By observing the right side of the frontal view and section *A-A* in Figure 3.484, one circular element could be noticed with the diameter Ø45 mm. In that case, the user creates an additional sketch in the plane *YZ*, as shown in Figure 3.491. The centre of the circle is located on the *V* axis of the sketch and vertically shifted from the upper edge by the value of 40 mm.

Similarly, as with the first sketch, the *Sketch.3* will be projected from inner side to the closest surface using *Projection* tool. To set up a new feature *Project.2*, the user defines necessary elements as presented in Figure 3.492. As a result, a new 3D curve will be created on the existing surface, along to the *X* axis, and placed in specification tree.

Again, in this case, *Project.2* will be used to cut out/split the surface and to keep the patch or area outside of the circumference of the *Project.2*. For better visibility, features such as *Sketch.3* and *Extract.1* could be hidden from the specification tree. In the *Split Definition* box, the user defines which element should be trimmed/split and with by which element, as well as which elements to keep or remove, as shown in Figure 3.493. As a result, the new surface *Split.2* has been added to the specification tree, and the basic shape is closer to Figure 3.484.

The *Split.2* has one sharp edge on its top that should be filleted with the radius of R2.5mm. This could be done in the *GSD* workbench by using *Edge Fillet* tool as presented in Figure 3.494. In this simple procedure, the user defines which corner he wants to make rounded; in this case, he selects the edge of the *Split.2\Edge.1*. As a result, *EdgeFillet.1* feature will appear in the specification tree.

In the *ZX Plane*, the user creates another sketch with an open profile, as shown in Figure 3.495. This sketch is coincident with the *Split.2*, so the user should project some edges by standard procedures and use construction elements. The profile can be simply created using the *Profile* tool and fully constrained with the annotated dimensional constraints. The user should note that edges *1*, *2* and *3* are construction lines, out of which *1* and *2* are created with *Project 3D Elements* tool in *Sketcher*, and line *3* and sketch profile are created manually. The resulting *Sketch.4* is presented in 3D in the lower right corner.

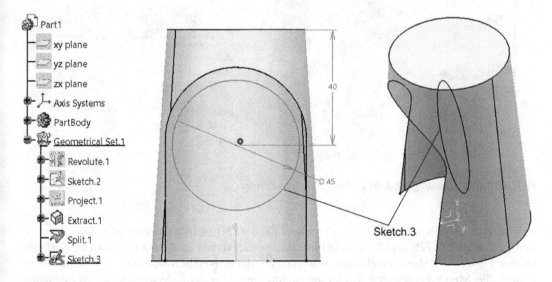

FIGURE 3.491 Sketching and constraining the circle in the *Sketch.3*.

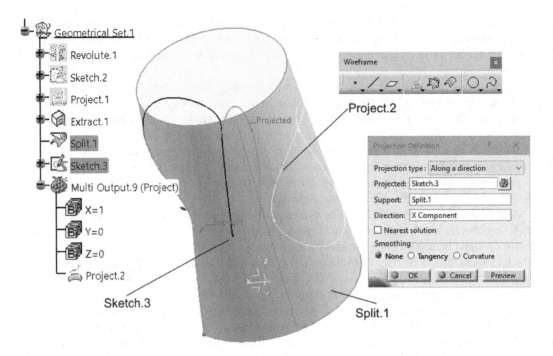

FIGURE 3.492 Projecting the *Sketch.3* on the wall of the *Split.1*.

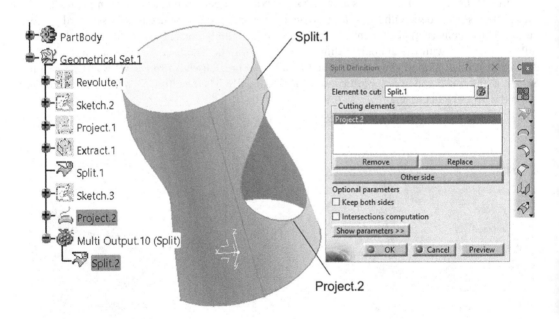

FIGURE 3.493 Splitting the *Split.1* with the 3D curve *Project.2*.

The previously created *Sketch.4* is a profile of the next 3D surface feature that follows the edge of the *EdgeFillet.1* or *Project.2* curve. In this case, the user should use *Sweep* tool from the *Surfaces* toolbar. Figure 3.496 shows all elements that should be defined on the *Swept Surface Definition* box. As a profile, *Sketch.4* is selected, and for the guiding curve, *Project.2*. As a result, *Sweep.1* complex surface feature is generated along the *X* axis direction and placed in the specification tree.

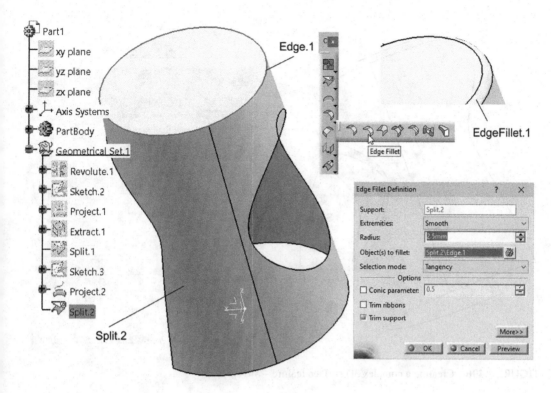

FIGURE 3.494 Creating an edge fillet using *Edge Fillet* tool.

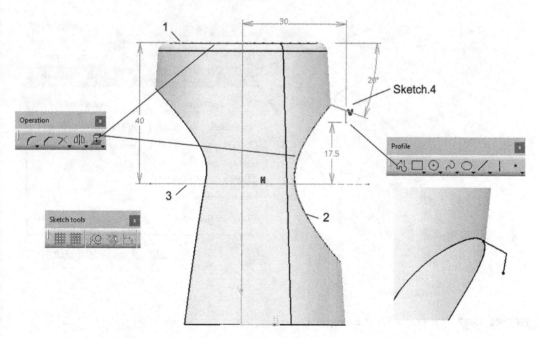

FIGURE 3.495 Creating a new sketch with open profile.

At this stage, the user should note that *EdgeFillet.1* and *Sweep.1* are independent surfaces, and thus, he must join them together in one surface using *Join* tool as shown in Figure 3.497. This is necessary as the following tool will be applied on the whole surface. This is especially important in making edges or corners filleting or chamfering.

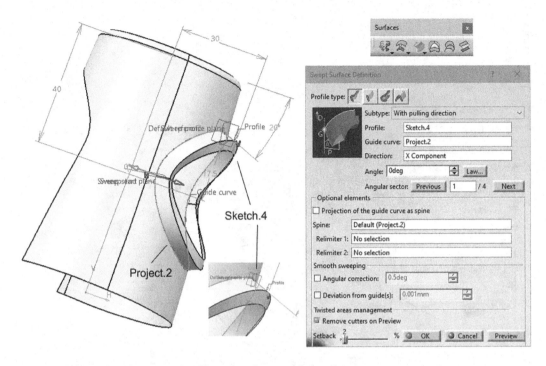

FIGURE 3.496   Creating a complex 3D surface feature – *Sweep.1*.

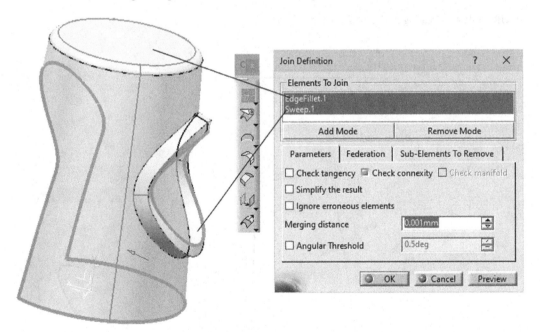

FIGURE 3.497   Creating a *Join.1*, a unique 3D surface.

After joining two surface features into one (*Join.1*), the user may round the edges of the sharp surface elements, as it is done in one of the previous steps using *Edge Fillet* tool. This procedure is explained in Figure 3.498, as well as which elements the user should select (*1* and *2*) in order to create *EdgeFillet.2*.

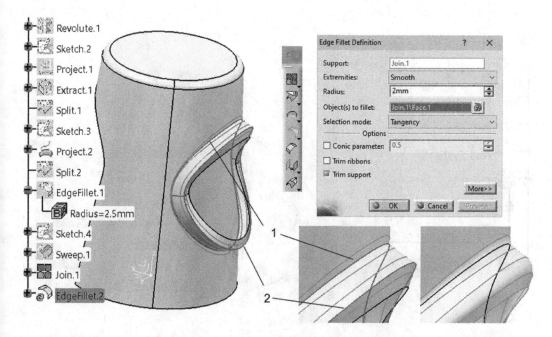

FIGURE 3.498 Creating an *EdgeFillet.2* over the sharp edges of the *Join.1*.

After this step, the user has fully modelled the outer silhouette of the plastic cover. In the following steps, the hybrid modelling procedure can continue in the *Part Design* workbench. There, the user can implement standard modelling tools to complete the shape with volumetric features, by adding, removing and multiplying them.

Usually, the first step after moving to *Part Design* from *GSD* is to activate the *PartBody* branch (right-click on the branch and select *Define In Work Object*), and then, the user can add a specific thickness of the wall to the surface with the *Thick Surface* tool as illustrated in Figure 3.499. Here, the user defines the surface (*EdgeFillet.2*) to be transformed in thick wall with 1.5 mm of thickness. The user may also notice that orange arrows are directed towards the inner area of the part, so the solid material will be added. This property can be regulated by pressing the arrow to change the direction of adding. As a result, the volumetric feature *ThickSurface.1* is created and placed in the specification tree. Other features could be hidden from the model, e.g. *EdgeFillet.2*, as it is not necessary in the next steps of modelling.

In the top view in the 2D drawing, one pierced pocket with the diameter Ø37 mm can be observed. This feature can be created with *Pocket* or *Hole* tool. In this approach, *Pocket* will be used, and, thus, a new *Sketch.5* will be created as presented in Figure 3.500. It is basically a circle with the centre located in 0,0 with diameter Ø37 mm, created with *Circle* profile.

By activating the *Pocket* tool, the user selects the profile of the volume to be removed. In this case, it is a previously created *Sketch.5*, and as a type, the user selects *Up To Next*. The resulting feature *Pocket.1* is illustrated in Figure 3.501.

Additionally, on the top side of the cover, another pocket could be identified that goes all along the part. Thus, a new sketch should be created on the same level as done previously (on the top plane/surface). This sketch has a triangular shape and is tangent to the yellow dashed line of the projected edge *1*. The tip of the triangle is located in the origin (0,0), and it is symmetrical over the *H* axis, as presented in Figure 3.502. The profile of the *Sketch.6* is also presented in a 3D view.

To cut the cover from top plane through all or next possible level, *Pocket* tool will be used again as in Figure 3.503. Here, *Up To Next* option should be selected and *Sketch.6*, as a profile of the cutting volume.

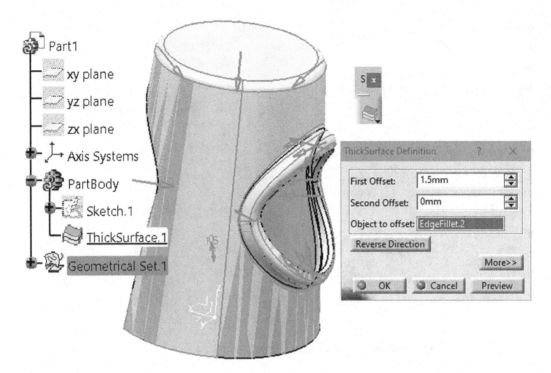

FIGURE 3.499 Creating a thick volumetric wall *ThickSurface.1* over the *EdgeFillet.2* in *PartBody*.

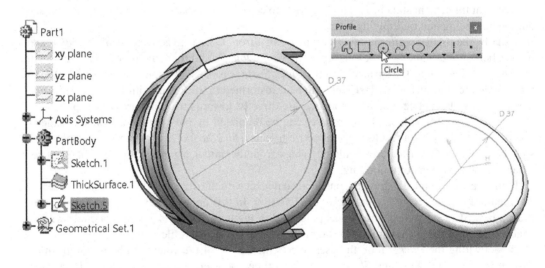

FIGURE 3.500 Creating a *Sketch.5* for circular pocket on the top of the cover.

By observing the 2D drawing of the cover, the user can note a small pocket on the upper side of the part. This pocket has a rectangular shape, it is symmetrical over *V* axis and has rounded edges on the bottom line. In the plane *YZ*, the user creates and constrains the profile of the *Sketch.7* presented in Figure 3.504. To speed up the process, the user may use *Centered Rectangle* tool located in the *Profile* toolbar. This tool requires a central point (1) that coincides the *V* axis of the sketch, and a second point located on the projected upper edge of the cover (2).

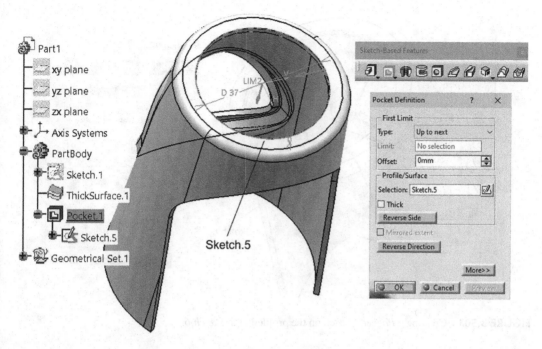

FIGURE 3.501 Creating a *Pocket.1* based on the profile in the *Sketch.5*.

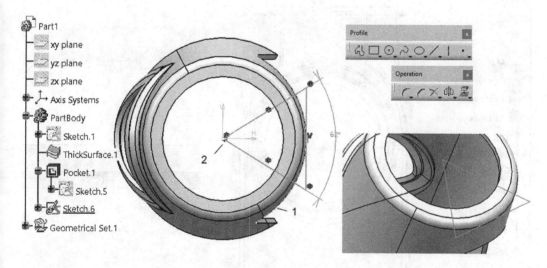

FIGURE 3.502 Creating a *Sketch.6* for triangular pocket on the top of the cover.

The main dimensions of the sketch are the following: width of rectangle of 3.5 mm and height 5 mm. The user can also make fillets of R1 mm in the sketch as it is done here, or later in *Part Design*.

To create the pocket up to the first limiting surface, the user selects *Up To Next*, based on the *Sketch.7*, in the *Pocket Definition* dialog box (Figure 3.505). Thus, *Pocket.3* will be created in the specification tree of the part.

This pocket should be circularly multiplied in two directions around the central rotational axis of the cover, and the most suitable way to do so is to implement *Circular Pattern* tool, in which case *Pocket.3* is selected for multiplying, as shown in Figure 3.506.

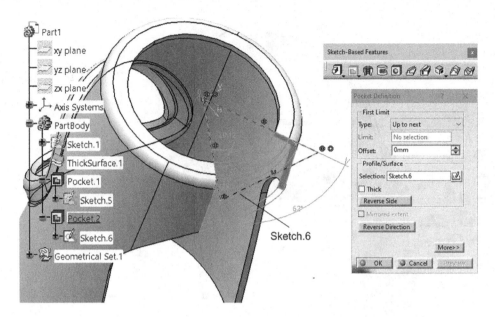

FIGURE 3.503   Creating a *Pocket.2* based on the profile in the *Sketch.6*.

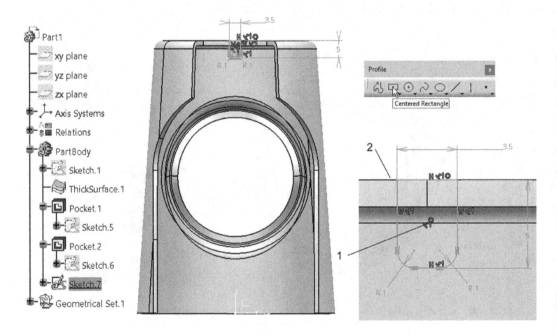

FIGURE 3.504   Creating a *Sketch.7* for a small side pocket, located on the *YZ Plane*.

    The user defines parameters of multiplying *Pocket.3* as follows: (1) *Instance(s) & angular spacing*, (2) *Number of instances*: 4, (3) *Angular spacing*: 36° and (4) as *Reference Direction*, the user may select any vertical axis that aligns with the rotational axis of the first revolved surface, or he can select the circular edge of the *Pocket.1* (*Pocket.1\Face.1*). The user should also be sure that the right object/feature for patterning is selected. In this case, it is *Pocket.3*. With option *Reverse*, the direction of multiplying could be changed (clockwise or counterclockwise).

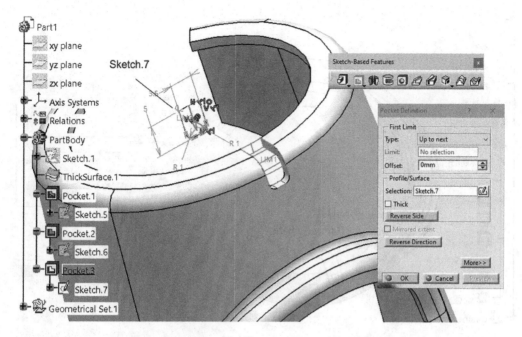

FIGURE 3.505 Creating a *Pocket.3*, *Up to next*, based on the profile of *Sketch.7*.

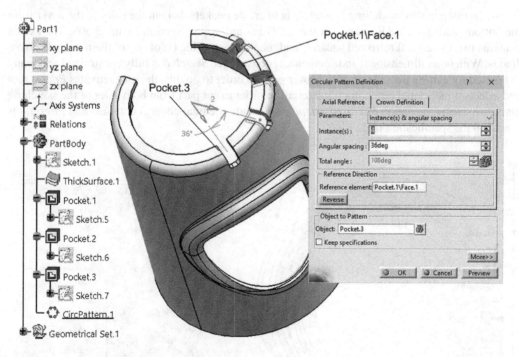

FIGURE 3.506 Multiplying a *Pocket.3* in circular order – direction 1.

As a result of this modelling step, *CircularPattern.1* feature is created and placed in the specification tree.

In a similar way, and using the same parameters, but in reverse direction, *CircularPatern.2* with *Pocket.3* is created and presented in Figure 3.507.

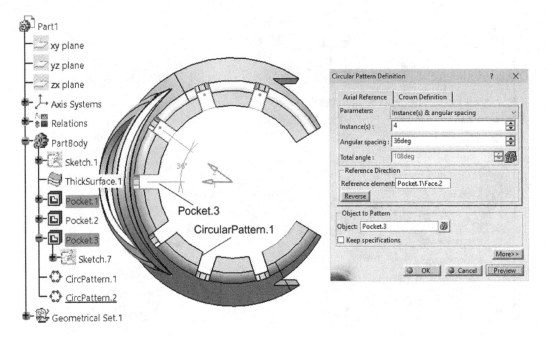

FIGURE 3.507   Multiplying a *Pocket.3* in circular order – direction 2.

The last step in this modelling procedure is to create pockets that cut the sides of the cover from the bottom side. *Sketch.8* is created in the *XY Plane* and is presented in Figure 3.508. The sketch contains two closed and mirrored squares, and the symmetry line (1) of one of them is on 45° from *H* axis. With other dimensional and geometrical constraints, *Sketch.8* is fully constrained and ready to be used for cutting the volume of the cover part. In order to do this, the user activates *Pocket* tool and selects dimensional removing of material. In order to cut part from both sides of the sketching plane, he selects the option *Mirrored extent*. As a result, the final cover is obtained and *Pocket.4* is placed in the specification tree.

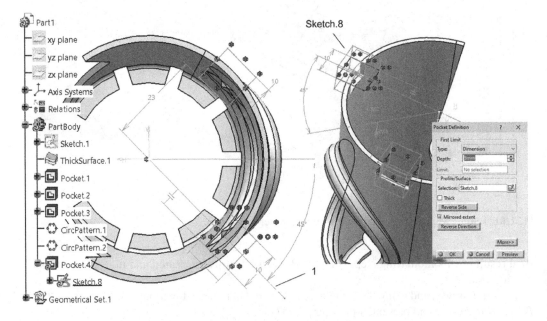

FIGURE 3.508   Creating a profile of the *Sketch.8* for removing volume – *Pocket.4*.

One modelling version of the same part is also presented in the video solution: https://youtu.be/_K5vTErh040.

3.22 MODELLING OF A WINDOW CRANK HANDLE

In hybrid design, the main bodies of the part are usually modelled with complex surface features and the additional ones are rather simple and they are created at the end of the modelling process. Some features could be organized in geometrical sets or bodies for *Boolean* operations, thus allowing creation of even more complicated shapes. The following crank handle model is a typical example of such kind of part that will be modelled by combining the *Part Design* and *GSD* workbenches.

Figure 3.509 shows the 2D drawing of the window crank handle with a complete number of views, projections and section views. By observing its shape, one transitional 3D shape ('neck') that connects lower and upper sides of the handle can be identified. Section *A-A* shows that the cross section of this handle is mainly elliptical and it changes its size to circle along guiding lines. In this case, the user has to define the cross-sectional profiles and guiding lines in *GSD* or *Wireframe & Surface Design* workbench and later continue the modelling process by creating rotational or pad 3D volumetric features in *Part Design*.

The first step in this tutorial is to create *Sketch.1* that contains the first profile of the handle – circle. In the *ZX Plane*, the user begins to draw a profile consisting in a circle of diameter Ø20 mm. By sketching, the user coincidently connects the centre of the circle to 0,0 and defines its diameter.

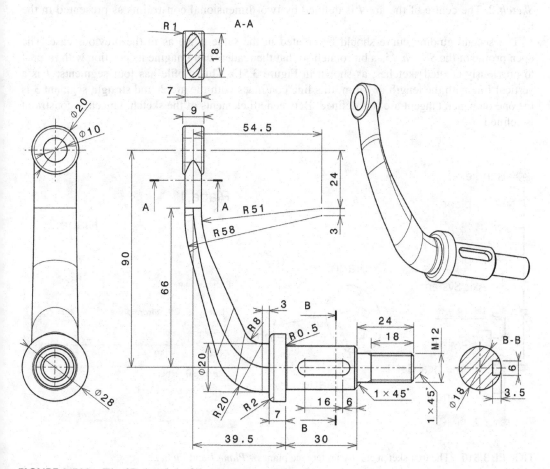

FIGURE 3.509   The 2D drawing of the window crank handle.

This *Sketch.1* is placed in the *PartBody* after leaving the *Sketcher* workbench. The user can observe two important dimensions that define the length of the handle neck: one is 66 mm and the other one is 90 mm. These dimensions are crucial for referencing, and in this case, the user can place two reference planes, parallel to plane *YZ* in reverse direction, as presented in Figure 3.510. The first reference plane (*Plane.1*) is parallel and defined by *Offset from plane* and the value 66 mm from *YZ Plane*, and another one (*Plane.2*) is created in the same manner, but with a distance of 90 mm from the same reference plane.

In the *Plane.1*, the user creates an elliptical shape profile using *Ellipse* tool from the *Profile* toolbar as shown in Figure 3.511. To create the ellipse, the user coincidently locates its centre on the *H* axis (1), the second point will define the height of ellipse (2), and the third will define the width (3). The centre of ellipse is located on the left side from origin at 32.5 mm, and the size of the ellipse is defined with the semiminor value of 7 mm and the semimajor value of 18 mm. The user can change these dimensional constraints with a right-click while dimensioning and selecting options *Semimajor/Semiminor axis* from contextual menu.

After defining two major profiles in *Sketch.1* and *Sketch.2*, the user can start to create the guiding curves. Thus, he first creates one open profile in *Sketch.3* located in the *XY Plane* as shown in Figure 3.512. This sketch profile connects the edges of the projected *Sketch.1* and *Sketch.2* (horizontal yellow lines, construction types). The easiest way to create this profile is to use the *Profile* tool, or to combine straight lines and circles/arcs. The first segment of the sketch is the vertical line with length of 3 mm starting on the right endpoint of the projected *Sketch.1*, on which the arc of R20 mm with tangency constraint continues up to line *3*. Arc *4* has a radius of R58 mm and is connected with a horizontal line that ends on the lower point of the projected *Sketch.2*. The centre of the arc *4* is defined by two-dimensional constraints as presented in the figure.

The second guiding curve should be created in the same plane as in the previous case. The open profile of the *Sketch.4* is a bit rotated so that the readers can imagine its position with respect to previously created sketches, as shown in Figure 3.513. This profile has four segments: *1* is a vertical line with the length of 3 mm, this line continues with the arc *2*, and straight segment *3* is the one on which tangent arc *4* is defined. Between all elements of the sketch, tangency constraint is defined.

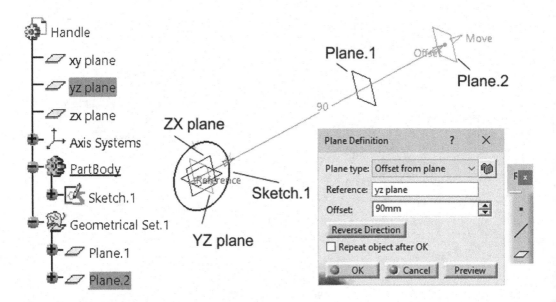

FIGURE 3.510　The first sketch and two reference planes – *Plane.1* and *Plane.2*.

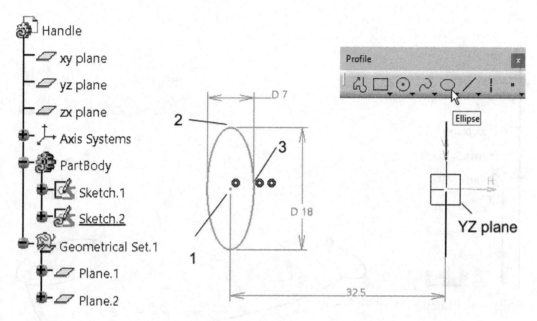

FIGURE 3.511   Defining closed elliptical profile in the *Sketch.2*.

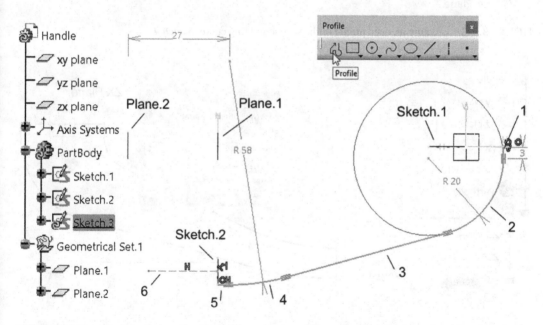

FIGURE 3.512   Creating an open profile of the first guiding curve – *Sketch.3*.

Besides these two horizontal guiding curves, two more should be created to follow the shape of the 'neck' vertically. These curves will be automatically created using *Spline* tool, and thus, sketching is not necessary. In order to create spline-type curve, the user needs to create starting and ending points, as well as reference lines over which spline will be tangent, thus defining its shape. For this, some small preparation steps are needed. First, the user creates an additional *Plane.3*, which will be parallel to *ZX Plane* (with an offset of 39.5–7 mm). This plane passes exactly through the middle of elliptical shape of the *Sketch.2*, as presented in Figure 3.514.

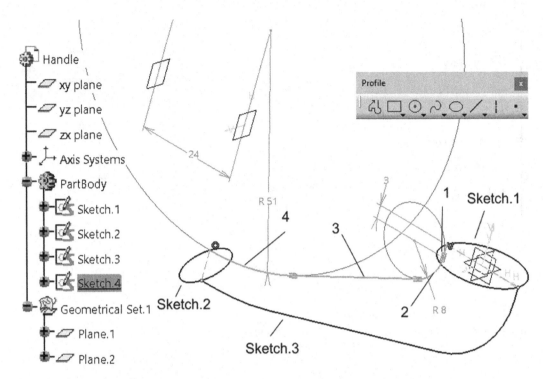

FIGURE 3.513   Creating an open profile of the second guiding curve *Sketch.4*.

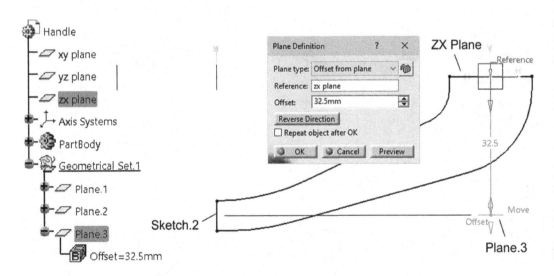

FIGURE 3.514   Creating an additional reference plane – *Plane.3*.

The user will now use *Plane.3* to make an intersection point with the profile of the *Sketch.2*, and to keep upper point only. To do so, he activates the *Intersection* tool and selects the elements between which he wants to make intersection (curve from profile and plane – as a result, two points will be created), as presented in Figures 3.515–3.517 (detail included).

CATIA generates two suggested points *P1* and *P2* as a result of intersection (Figure 3.516), and the user selects/keeps only one that will be extracted as a separate point – *Extract.1* – as shown in Figure 3.517.

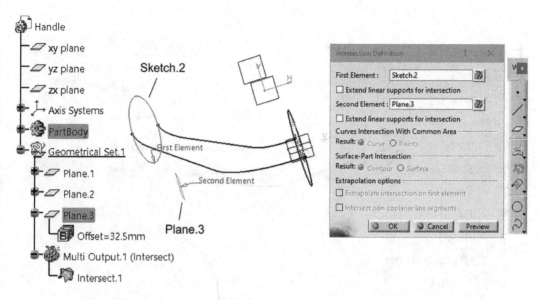

FIGURE 3.515 Intersecting *Sketch.2* with *Plane.3*.

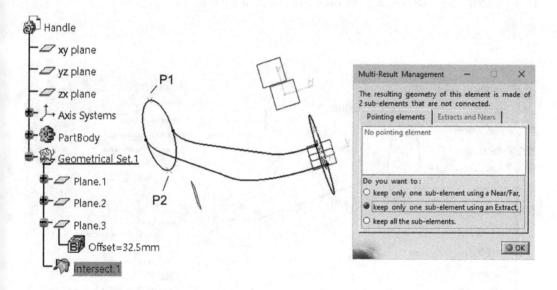

FIGURE 3.516 Suggested intersecting elements – points *P1* and *P2*.

Through the previously extracted point, a line that is parallel to the *X* axis should be created as defined in Figure 3.518. In the *Line Definition* dialog box, the user defines *Point-Direction* type of line definition, then a point from/through which the line will start/pass (*Extract.1*), direction (*X Component*) and the length of the line before and after the starting point (–20 to 20 mm). Thus, *Line.1* is created and placed under *Geometrical Set.1*.

Exactly the same procedure presented in Figures 3.515–3.518 will be applied over *Sketch.1*, in which case the reference point (*Extract.2*) will be extracted as intersection between *Sketch.1* and *YZ Plane* (*Intersect.2*) and the reference line (*Line.2*) will be created along *Y* axis through *Extract.2*, as presented in Figure 3.519.

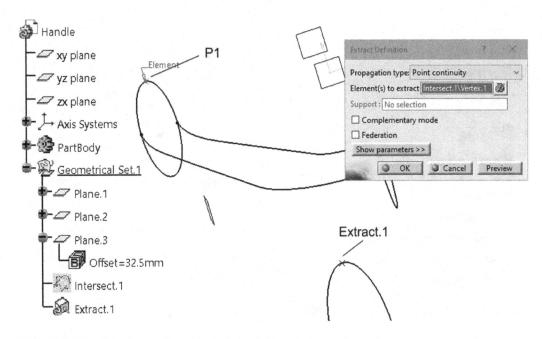

FIGURE 3.517   Selecting a point *P1* for isolation and extraction – *Extract.1*.

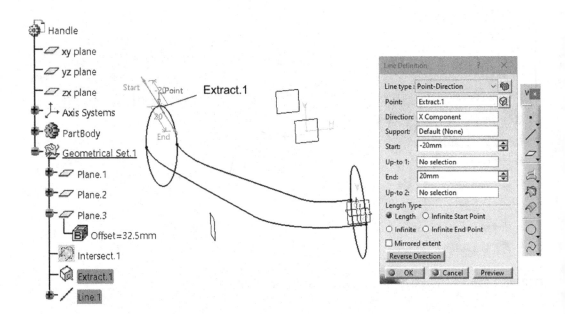

FIGURE 3.518   Creating a reference line in an exact point with known direction – *Line.1*.

Extracted points and created lines will be sufficient to construct the spline curve using *Spline* tool located in the *Wireframe* toolbar. By activating *Spline*, *Spline Definition* dialog box is opened. The user should select the starting point and tangent direction (*Extract.2* and *Line.2*) as well as end-point and tangent direction (*Extract.1* and *Line.1*). If *CATIA* suggests opposite tangency, the user can reverse it by clicking on *Reverse Tgt.*, as shown in Figure 3.520.

The curve with the same shape is located below the *Spline.1*, and instead of repeating the above-mentioned steps, the user can simply copy it symmetrically over *XY Plane*, as shown in Figure 3.521.

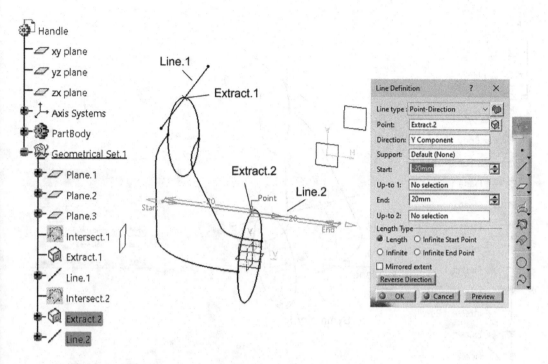

FIGURE 3.519 Generating a reference line – *Line.2*.

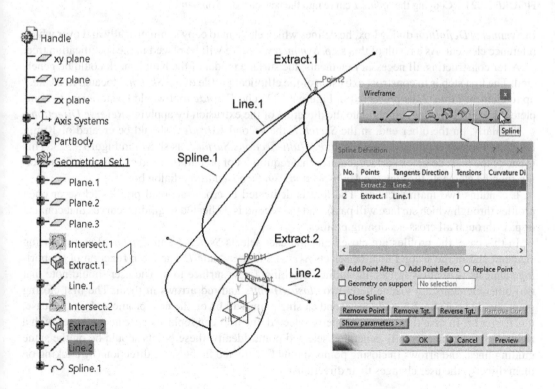

FIGURE 3.520 Constructing the *Spline.1* curve.

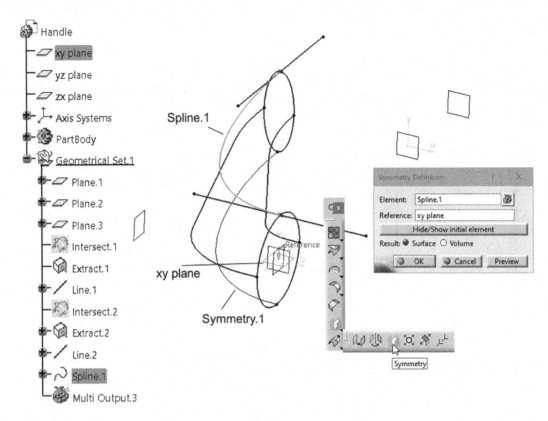

FIGURE 3.521   Copying the *Spline.1* curve into the new curve – *Symmetry.1*.

In *Symmetry Definition* dialog box, he defines which element to copy symmetrically and over which reference element. As a result of this step, *Symmetry.1* curve will be placed in the specification tree.

After constructing all necessary elements, the surface model of the handle neck could be generated. The first step is to construct extrusion of the elliptical profile of the *Sketch.2* located in *Plane.1* up to element *Plane.2*, as presented in Figure 3.522. The *Extrude* tool would be sufficient to complete this step. The user can regulate the direction of the extrusion by applying *Reverse Direction*.

Similarly, on the other end, in the *Sketch.1,* the second *Extrude.2* should be created of 10 mm length to support creating the next feature, *Multi-Sections Surface*, as shown in Figure 3.523. This is one of the most complex surface features that require lot of preparation and careful steps in selecting specific elements through the *Multi-Sections Surface Definition* dialog box.

It contains two main sections. The first is dedicated to cross-sectional profiles (close or open profiles through which surface will pass), and the second is dedicated to guiding curves that connect or pass through all cross-sectioning profiles.

In this case, the profiles are closed and the user selects *Sketch.1* and *Extrude.2* as supporting element, and in the second line, he selects *Sketch.2* and *Extrude.1*. The main function of supporting elements is to allow tangential adding/attaching a new surface to it. The user should note that two other elements are visible here: two *closing points* and red arrows in them. The first closing point should be *Extract.2*, and the second closing point will be on the next profile and, in this case, it is *Extract.1*. In case they are not properly selected, the user can replace or create new ones with a right-click over the currently generated/selected point. Ideally, these points should be on the same guiding lines, and arrows in closing points should be directed in the same direction (by clicking on them directly, the user changes their directions).

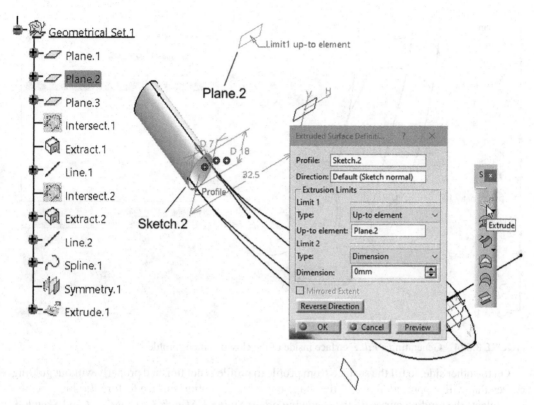

FIGURE 3.522 Generating the first surface feature – *Extrude.1.*

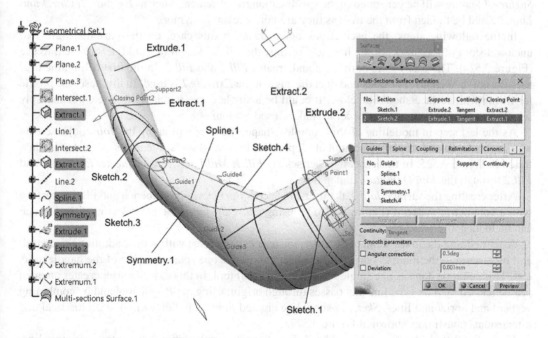

FIGURE 3.523 Generating complex multi-sectional surface feature – *Multi-sections Surface.1.*

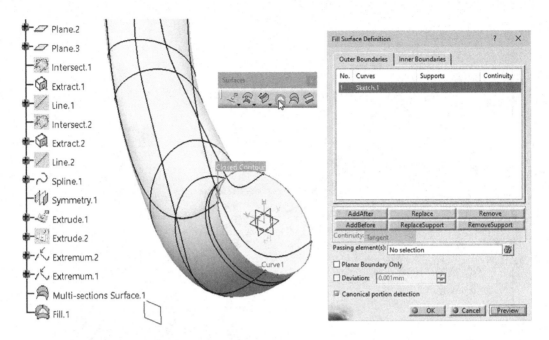

FIGURE 3.524   Generating a *Fill.1* surface inside of the closed contour/profile.

On the other side, to fill the surface from profile to profile is not finished properly without guiding curves that will support and 'guide' the shape of the multi-section surface feature. In this case, the user selects the guiding curves in the following order: *Spline.1*, *Sketch.3*, *Symmetry.1* and *Sketch.4*. If the user mixed lines (without the circular order), there is a big chance that the multi-section surface feature will not be created. By approving the selected elements in dialog box, *Multi-sections Surface.1* feature will be generated in the specification tree. Elements such as *Extrude.2*, *Line.1* and *Line.2* could be hidden from the tree, as they are not necessary anymore.

In the following steps, the user closes open areas; in this case, on the part there are two unclosed sides of the 'neck' of the handle. For this, the *Fill* surface tool would be sufficient to use (Figure 3.524). The user closes both sides and creates *Fill.1* and *Fill.2*. In the first filling, the user uses the curve *Sketch.1*; in the second one, the edge of the *Extrude.1* is used. In this case, the profile for filling with surface should be close. It could be a single curve (circle, ellipse, etc.) or multiply selected elements in appropriate order to make closed contour.

As the last step in modelling of this complex shape, the user will apply the *Join* tool to create watertight/closed and single surface out of all previously created surface features. This step is illustrated in Figure 3.525. In this case, the user selects *Fill.1, Multi-sections Surface.1, Extrude.1* and *Fill.2* through the *Join Definition* dialog box.

After creating the surface *Join.1*, the user can access *Part Design* workbench and fill this surface with a volumetric material (Figure 3.526) by using *Close Surface* tool and selecting *Join.1* which can be hidden afterwards.

After defining the first volumetric feature, the user can continue with adding additional material, and by observing the main 2D drawing, the user can notice one rotational part of the handle at its end. This could be done as one single feature, using *Shaft* tool. In this case, the user creates an open profile (closed by *Axis*-type line that passes through origin) using the *Profile* tool and by combining vertical and horizontal lines. *Sketch.5* should be created in the *XY Plane* with the geometrical and dimensional constraints shown in Figure 3.527.

Using *Shaft* tool, the user selects *Sketch.5* as the rotational profile that contains an axis line, and defines a rotational angle of 360° (Figure 3.464). After approving parameters defined in *Shaft Definition* dialog box, the *Shaft.1* feature is created and placed in the specification tree (Figure 3.528).

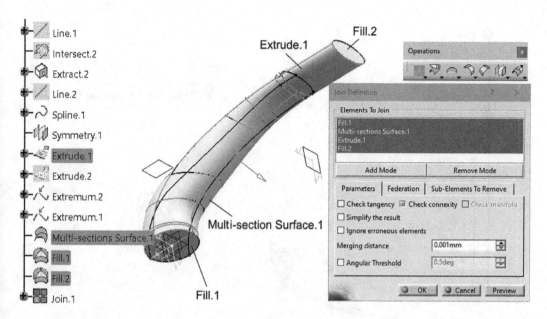

FIGURE 3.525 Creating a watertight surface – *Join.1.*

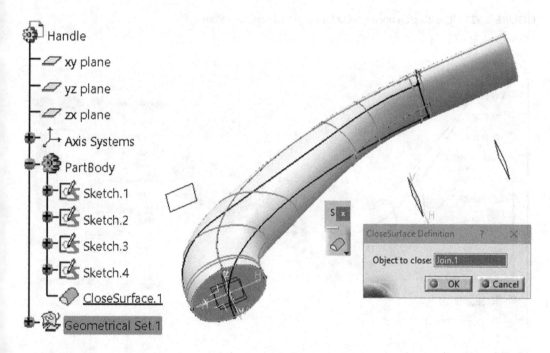

FIGURE 3.526 Creating a volumetric feature – *CloseSurface.1.*

On the other end of the handle, another prismatic feature with circular profile could be identified. As the new feature may clash/collide with the tip of the end, it is recommended to create a new body. The user creates a new body through *Insert → Body*, and the new *Body.2* will appear in the specification tree, as a separate branch. The user can also note that this body is now active/underlined.

Under this body, in *Plane.3* the user creates a circle with diameter Ø20 mm, of which the centre is located on *H* axis, and at the same time on the edge of *CloseSurface.1* (Figure 3.529).

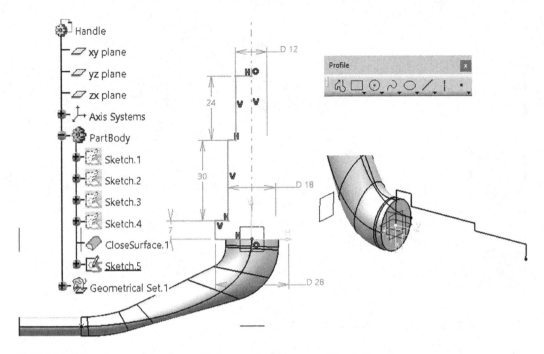

FIGURE 3.527   Creating an open profile for rotational feature – *Sketch.5.*

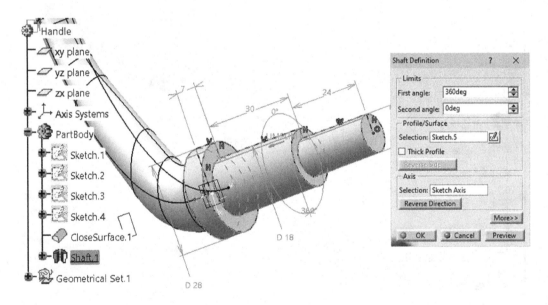

FIGURE 3.528   Creating a volumetric rotational feature – *Shaft.1.*

Sketch.6 is closed, and the *Pad* tool with mirrored extent of 4.5 mm on both sides is suitable for implementation. After approving these settings, the *Pad.1* is created. This feature is independent from the rest, as it is created in a separate body (Figure 3.530). Here, *Boolean* operation is a perfect solution for joining the *Body.2* with the main body, *PartBody*.

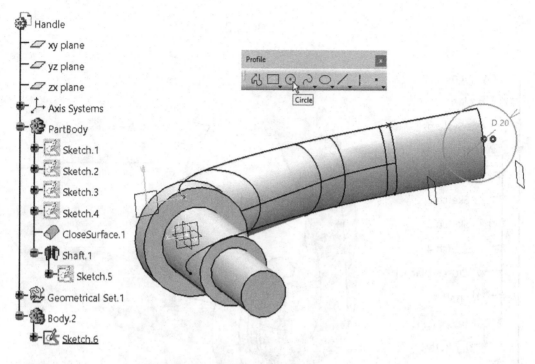

FIGURE 3.529 Creating a *Shaft.6* in a new *Body.2*.

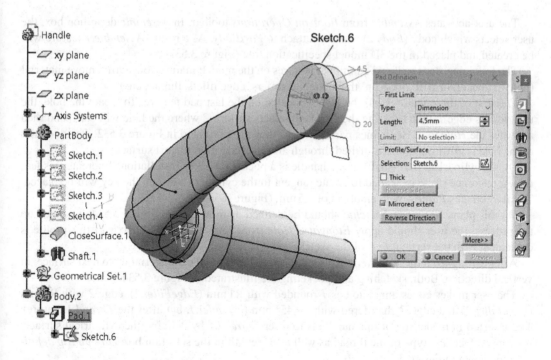

FIGURE 3.530 Creating a *Pad.1* in a new *Body.2*.

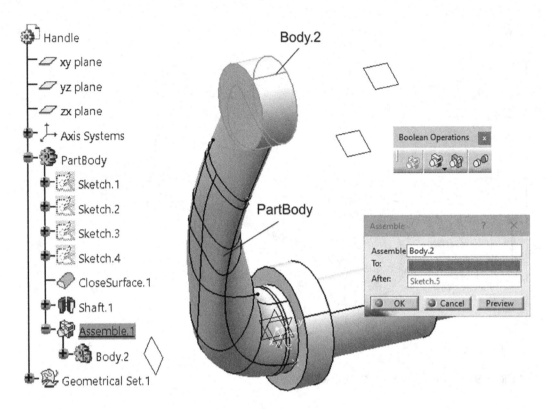

FIGURE 3.531 Assembling *Body.2* with the *PartBody*.

The user activates *Assemble* from *Boolean Operations* toolbar. In *Assemble* definition box, the user selects which body (*Body.2*) wants to attach to *PartBody*. As a result, *Assemble.1* feature will be created and placed in the 3D model specification tree (Figure 3.531).

The following steps will add additional features on the model; mainly, they will remove material (hole or pocket) or will 'dress up' the model (chamfers, edge fillets, thread, etc.).

The first of these features will be *Hole.1* located on the last pad feature. To create the hole, the user selects edge *1*, then (holding the *Ctrl* key) selects surface *2* where the hole will start, and then activates the *Hole* tool. He defines parameters as they are presented in Figure 3.532. The diameter of the hole is Ø10 mm, and it is drilled through all material, or up to next surface.

On the cylindrical part (shaft) of the handle is a pocket for a key installation. To create this feature, a reference plane is needed to create tangent to the cylinder on which the key will be placed. The user creates *Plane.4* with an offset of 5.5 mm (Figure 3.533).

On this plane (*Plane.4*), *Sketch.8* should be defined as presented in Figure 3.534. To speed up the process, the user should apply *Elongated Hole* from *Profile* toolbar, the axis of this profile is coincident with the vertical axis of the sketch, and it is fully constrained.

After exiting from the *Sketcher*, the user can apply *Pocket* and cut material up to next surface, in vertical direction. Both, sketching and pocketing are illustrated in Figure 3.534.

The user makes edges annotated as *1* rounded with R1 mm (*EdgeFillet.1*), edge *2* with R2 mm (*EdgeFillet.2*) and edge *3* chamfered with 1×45° mm (*Chamfer.1*, but after the *Thread.1* feature). The last step in modelling of this handle is to insert *Thread.1* by defining the cylindrical surface, starting surface and type of the thread as well as its length in the selection box *Thread/Tap definition*, as shown in Figure 3.535.

One modelling video solution of this part is also presented at: https://youtu.be/YvNi3Q-uHjk.

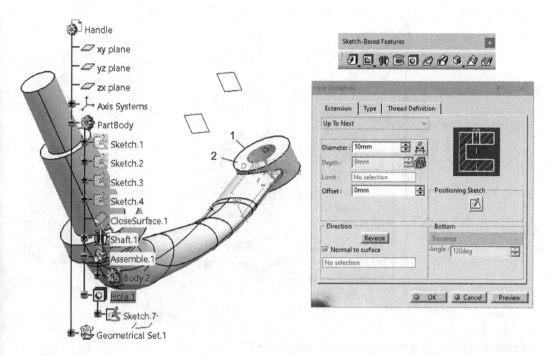

FIGURE 3.532 Defining parameters of the *Hole.1* feature.

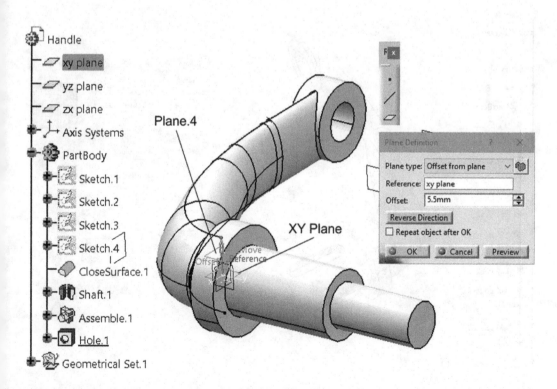

FIGURE 3.533 Defining reference *Plane.4* for key (pocket) feature.

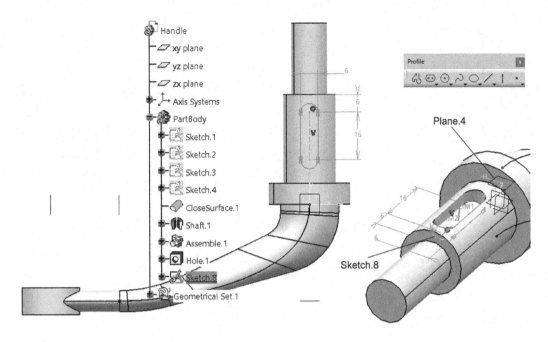

FIGURE 3.534   Drawing the *Sketch.8* and feature *Pocket.1*.

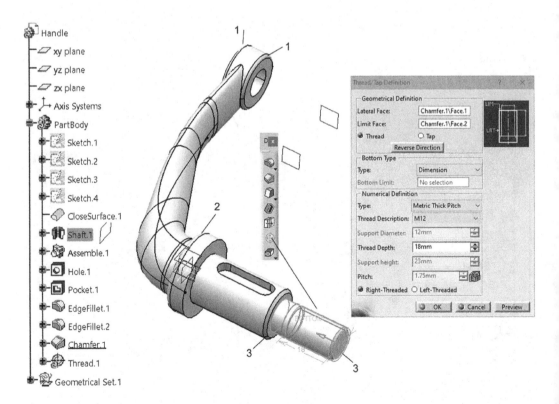

FIGURE 3.535   Dressing up the model with chamfer, edge fillets and thread.

3.23 MODELLING OF A SHIELD USING LAWS

The following tutorial covers an interesting part that looks less complicated, but some easy to follow, yet advanced modelling steps are necessary to complete the model.

Figure 3.536 shows the 2D drawing of the shield with all necessary views, details, and one section and isometric view. The user can observe that one surface cut area is repeated three times, which could lead to a circular pattern implementation. Similarly, as in previous examples, the surface model will be created in *GSD* workbench and a volumetric material will be added later in *Part Design* workbench.

The first step in this tutorial is to create *Sketch.1* in *YZ Plane* that contains a first open profile of the shield as shown in Figure 3.537. The user should note that the profile has one horizontal line that is coincident with the *H* axis of the sketch and then it tangentially continues with a long arc and finishes with a short vertical line. Also, the user must draw an axis-type line that is coincident with the *V* axis of the sketch. The profile of the sketch could be created with the *Profile* tool or by combining circles/arcs and standard lines.

After creating *Sketch.1*, the user activates the *Revolve* tool to rotate the sketch and to create a *Revolute.1* surface feature. In the *Revolution Surface Definition* (Figure 3.538), the user defines axis of rotation (*Z Axis*), profile to rotate (*Sketch.1*) and angle of rotation as 60°. The rotation is in counterclockwise direction if the user looks on the model from the top.

A sharp edge of the *Revolute.1* surface should be filleted by the *Edge Fillet* tool with a radius R10 mm. This feature appears in the model specification tree as *EdgeFillet.1*.

To continue with the modelling process, an additional reference plane should be created from the top of the model towards inside at 80 mm distance, using the *Plane* tool and the option *Offset from plane*, as shown in Figure 3.539. As a reference, the user selects the *XY Plane*.

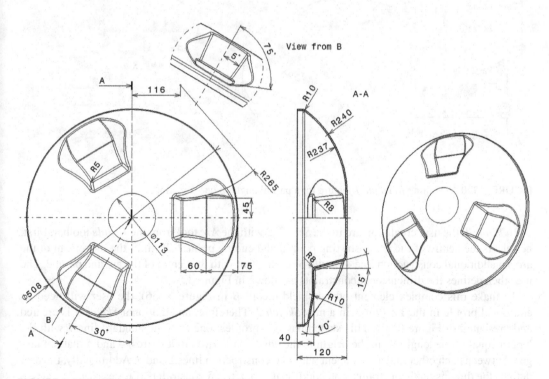

FIGURE 3.536   The 2D drawing of the shield.

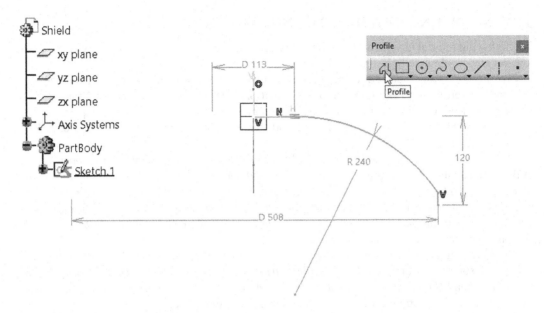

FIGURE 3.537   The first sketch that contains an open profile with axis-type line.

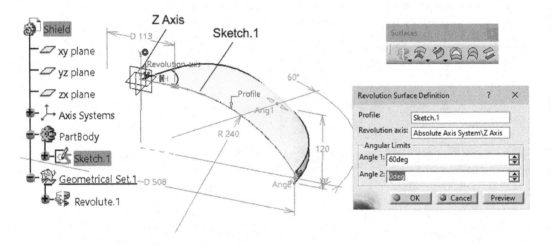

FIGURE 3.538   Defining *Revolute.1* feature for a partial angle in one direction.

In *Plane.1*, the user creates an arc in the *Sketch.2* with the *Arc* tool from the *Profile* toolbar. First, he selects the centre of the arc *1*, starting point *2* and ending point *3*. To limit the movement of the arc, an additional construction line *4* should be coincident with the centre of the arc and point *2*. The user then defines the dimensional constraints as shown in Figure 3.540.

To make this complex element of the shield (detail *B* in Figure 3.536), the user will need an additional profile in the *YZ Plane*, in a new *Sketch.3*. Therefore, an additional *Sketch.3* is created and presented in Figure 3.541. This sketch has two profiles: one straight standard line *1* with the length equal to the length from the origin to the edge *6*, and two circles (arcs) *4* and *5* that are tangent between each other and also tangent with two construction lines *2* and *3*. Additionally, the user defines the dimensional constraints and additional geometrical constraints if necessary.

After trimming the unnecessary elements, *Sketch.3* will have the profiles as presented in Figure 3.542.

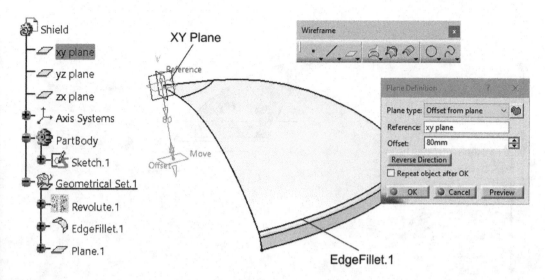

FIGURE 3.539 Defining a new *Plane.1* as a support for a new sketch.

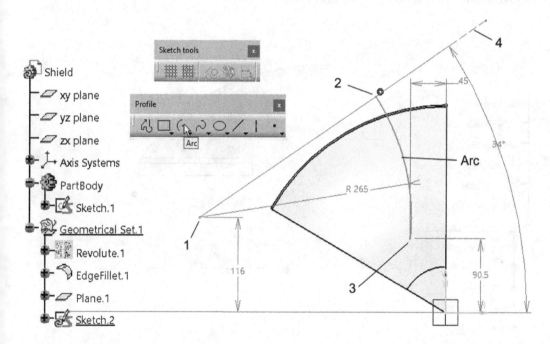

FIGURE 3.540 Drawing an open arc in the *Sketch.2*.

As it is already mentioned, there are two open profiles in the last sketch. Very often, the user may need to select only one profile from this sketch; in that case, he can extract (isolate) these profiles from the sketch and use them later separately. The best way to do this is to use the *Extract* tool. The user extracts separately two profiles from the sketch as *Extract.1* (straight line) and *Extract.2* (two arcs propagated by the *Point continuity*), as shown in Figure 3.543.

In the following step, the user will create a *law* (math function) based on the predefined profiles *Extract.1* and *Extract.2*. To do so, the *Law* command will be used and defined as presented in Figure 3.544. As a *Reference*, the straight line is selected, and as a *Definition*, the *Extract.2* is selected.

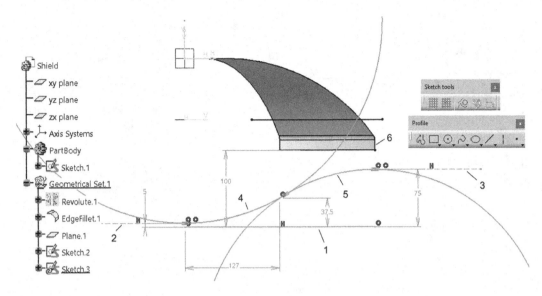

FIGURE 3.541   Drawing two open profiles in a *Sketch.3*.

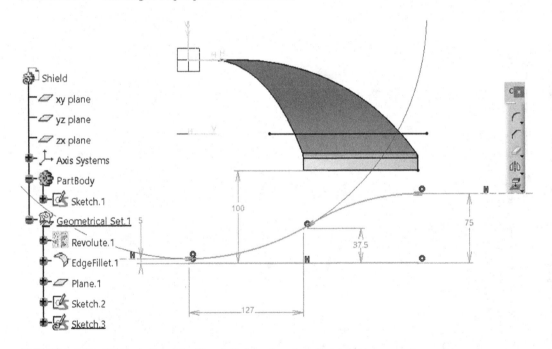

FIGURE 3.542   Final profile of the *Sketch.3* after trimming elements.

The *Law.1* is a function that can be implemented in various tools in *GSD* workbench. Here, this relation will be used to create *Sweep.1*, a very complex surface, as defined in Figure 3.545. The user selects the second option (*Line*) in the *Swept Surface Definition*. As a *Subtype*, *With draft direction* is selected, as a *Guiding curve 1*, *Sketch.2* is selected, and as a *Draft Direction*, the user selects the *Z* direction.

As *Length type 1* and *Length type 2*, the user selects *Standard* mode and defines lengths of 100 and 150 mm, respectively. The key step in defining the shape of the *Sweep.1* surface feature is to apply the *Law.1* function. To do so, the user selects *Law…* under *Wholly defined* tab.

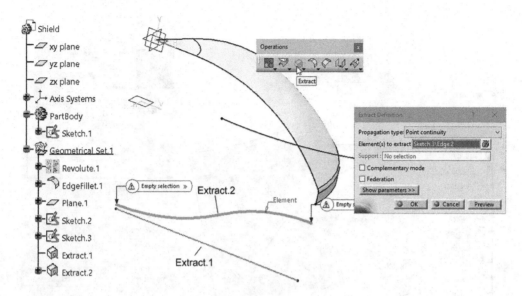

FIGURE 3.543 Isolation of the sketch profiles using the *Extract* tool.

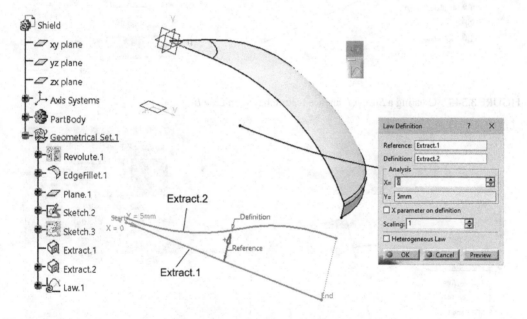

FIGURE 3.544 Making a law based on a line and a planar curve – *Law.1*.

By selecting the *Law.1*, directly from specification tree, the *Law Definition* graph will appear. After approving, *Sweep.1* is added to the specification tree (Figure 3.546).

After creating the second surface features based on law functionalities, the user creates one more sketch, named *Sketch.4,* on the *YZ Plane* by combining three straight lines, as presented in Figure 3.547. The first line covers the angle of 15° with the *V* axis of the sketch. Line *1* is horizontal and coincident with plane *2* (*Plane.1*), the last line is defined by another angle (10°), and its end is coincident with the edge *3*. The user also defines other dimensional constraints to make this sketch fully constrained.

The *Sketch.4* is then used to create bidirectional extrusion from the *YZ Plane* as presented in Figure 3.548. The first extrusion is performed at 20 mm, and the second at 225 mm.

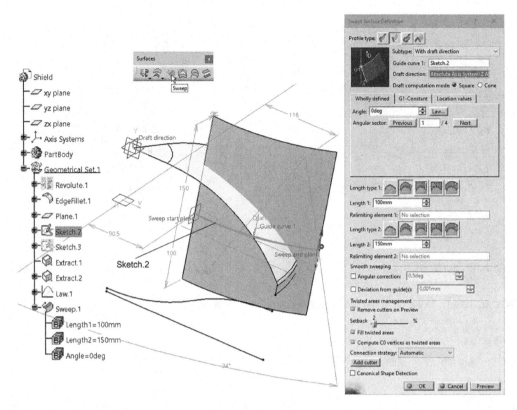

FIGURE 3.545   Creating a *Sweep.1* surface feature based on *Law.1*.

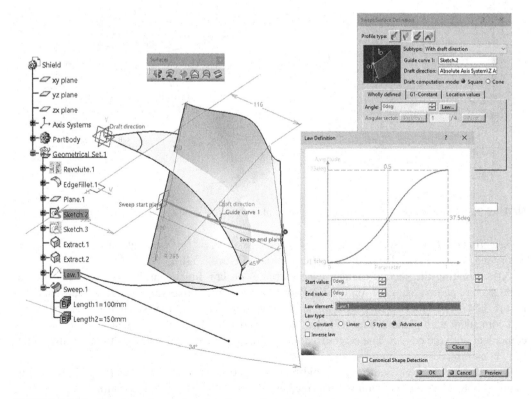

FIGURE 3.546   Creating a *Sweep.1* surface feature based on *Law.1* – after applying the law.

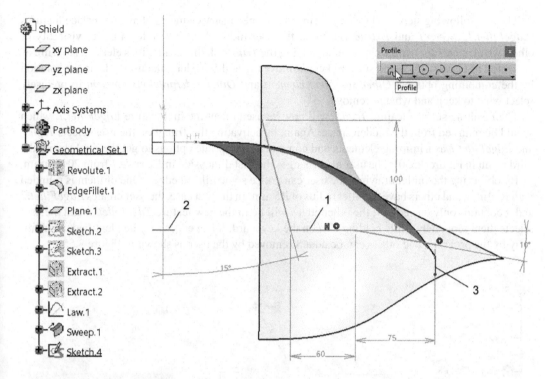

FIGURE 3.547 Drawing a *Sketch.4* with an open profile in *YZ Plane*.

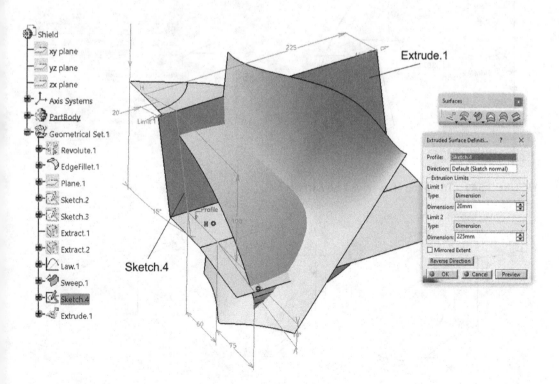

FIGURE 3.548 Extruding the *Sketch.4* in two directions – *Extrude.1*.

The two following steps will focus on trimming unnecessary elements of all the surface features (*EdgeFillet.1*, *Sweep.1* and *Extrude.1*). First, the user hides *EdgeFillet.1* for a better visibility of other two elements for trimming. After activating the *Trim* tool, the user defines elements to trim, as presented in Figure 3.549, and sides to keep/remove (*S1* and *S2*). Here, the user selects the solution by the combining options: *Other side/next element* and *Other side/previous element* to precisely select what to keep and what to remove.

The resulting surface feature *Trim.1* will now be used to cut the first feature *EdgeFillet.1*, which should be returned from the hidden space. Again, by activating the *Trim* tool, the user selects *Trim.1* and *EdgeFillet.1* as trimming elements and combines appropriate options to get the final shape (the third result in Figure 3.550). The user also changes the first dimension in *Extrude.1* from 20 to 0 mm.

By observing the initial drawing, the user can notice some filleted edges. One of them is rounded with R10 mm, and others have the fillet radius of R5 mm. In the first case, the user creates *EdgeFillet.2* and it contains only one edge *1*; the other edges will be in the new feature *EdgeFillet.3*, and the user selects them separately while holding *Ctrl* on the keyboard. Some edges may be identified automatically by the program, and others can be added/removed by the user as shown in Figure 3.551.

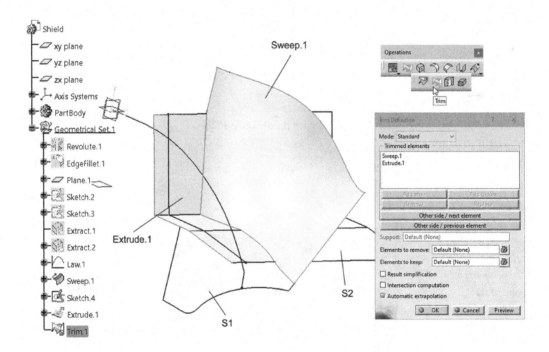

FIGURE 3.549   Trimming of *Sweep.1* with *Extrude.1*.

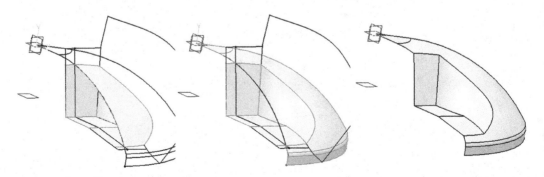

FIGURE 3.550   Trimming of the *Trim.1* with *EdgeFillet.1*.

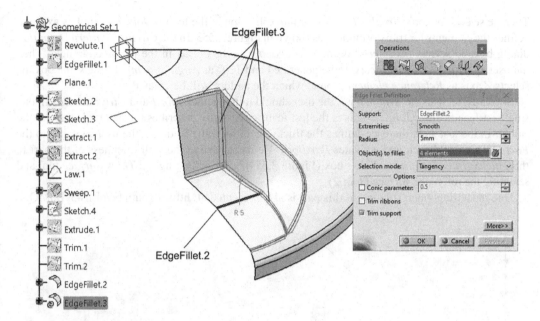

FIGURE 3.551 Creating rounded edges – *EdgeFillet.2* and *EdgeFillet.3*.

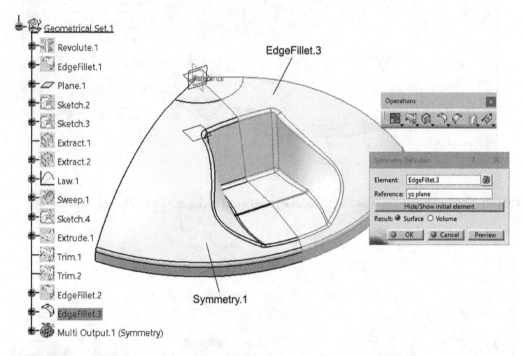

FIGURE 3.552 Transforming the *EdgeFillet.3* surface feature with the *Symmetry* tool.

After defining dress-up features, the final shape can be symmetrically transformed to make the full shape of the design. To do so, the user selects the *Symmetry* tool from the *Operations* toolbar and selects *EdgeFillet.3* and reference plane (*YZ Plane*) for symmetry transformation while keeping the initial object (Figure 3.552).

Here, the user should note that *Symmetry.1* and *EdgeFillet.3* are still independent features that should be joined together in order to be multiplied to a full surface of the shield. Thus, the user activates the *Join* tool and selects *Symmetry.1* and *EdgeFillet.3* for joining (*Join.1*).

Then he selects the tool *Circular Pattern* for multiplication of the feature *Join.1* around its axis and defines the elements for transformation, as shown in Figure 3.553. In the *Circular Pattern Definition* dialog box, he selects *Instance(s) & angular spacing* as the option in the *Parameters* field. Then the user specifies 3 for *Instance(s)* to be generated and 120° as *Angle spacing*. In the same box, he selects *Z* axis as *Reference element* around which the pattern will be created.

Similarly, as with the *Symmetry.1*, the user should join together *Join.1* and *Circ.Pattern.1* in one single surface feature (*Join.2*). This is the last feature that fully describes the external shape of the shield. In the last step, the user defines the thickness of its wall. To do so, the user switches to the *Part Design* workbench and activates *PartBody*. He has to define the wall thickness of the part in the *ThickSurface Definition* dialog box (Figure 3.554), after activating the *Thick Surface* tool and selecting the reference surface (*Join.2*).

One modelling video solution of this part is also presented at: https://youtu.be/oLus-3yBhMQ.

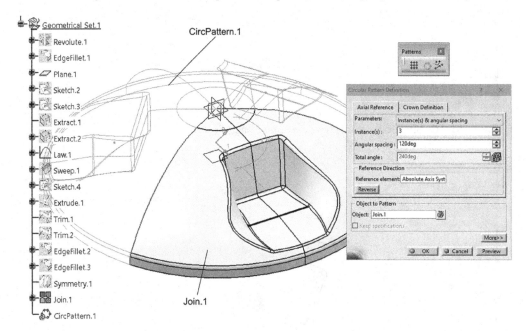

FIGURE 3.553   Multiplying *Join.1* with *Circular Pattern* tool to create the *Circ.Pattern.1* feature.

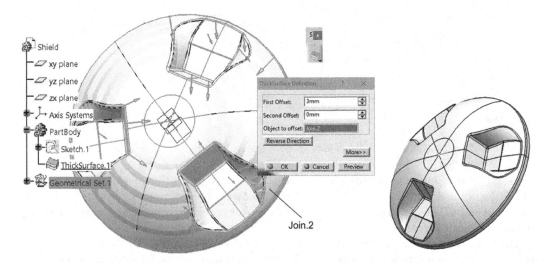

FIGURE 3.554   Creating a thick volumetric feature *ThickSurface.1* based on the *Join.2* surface feature.

3.24 MODELLING OF A CITRUS JUICER

The tutorial proposed for this chapter covers the modelling process of a citrus juicer that can be useful for every household. A representation of the juicer in four views (three orthogonal and one isometric) is presented in Figure 3.555. The dimensions of the features of the part will be mentioned as the modelling process continues. The thickness of the part wall is 1 mm. The workbenches used to create the part are *GSD* and *Part Design*.

To keep organized, the modelling process is divided into three major steps, as in Figure 3.556. The modelling starts with the creation of the cup where the juice will gather in, for example, citrus extraction. Next, the main active part of the juicer will be added and then the edge completes the part as a whole. The user should open a new file in the *GSD* workbench.

The first step is to create the wireframe geometry for the cup of the part analysed. For this purpose, a new *Geometrical Set* is added to the specification tree. The name of the set is *Step1_wireframe*, and it has as its father the part itself. It can be added from the *Insert* menu, *Geometrical Set,* as in Figure 3.557. The geometrical set is used to group the wireframe geometry for the first step.

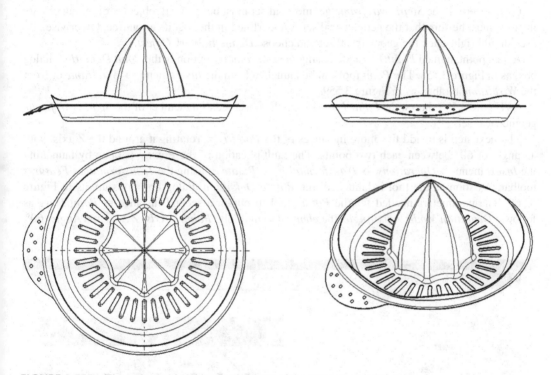

FIGURE 3.555 The orthogonal and isometric views of the part to be modelled.

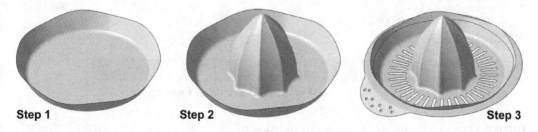

Step 1   Step 2   Step 3

FIGURE 3.556 The three major steps of the modelling process.

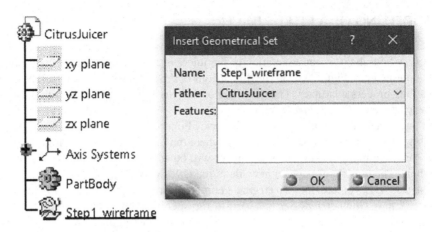

FIGURE 3.557   Displaying the geometrical set *Step1_wireframe.*

Once inserted, the *Step1_wireframe* geometrical set must be the work object for moving on. So, the user must be sure that the geometrical set is underlined in the specification tree. Otherwise, the user should right-click the geometrical set and choose *Define In Work Object.*

A new point, named *Point.1*, is added using the *Coordinates* option in the *Point Definition* dialog box, as in Figure 3.558. The *Point* tool can be launched from the *Insert* menu → *Wireframe* or from the *Wireframe* toolbar, as in Figure 3.559.

As Figure 3.558 shows, the *Point.1* feature can be observed as a member of the *Step1_wireframe* geometrical set.

The next step is to add five more instances of the *Point.1*, by rotating it around the *Z* axis, with an angle of 60° between each two points. The multiplication of the *Point.1* is done by launching the *Insert* menu → *Operations* → *Transformation* → *Rotate* or using the *Transformation Features* toolbar. As soon as the tool is launched, the *Rotate Definition* dialog box pops up, as in Figure 3.560. The parameters needed for the *Rotate* tool to obtain a second instance of *Point.1* are as follows: *Definition Mode*: *Axis-Angle*, the *element* to multiply: *Point.1*, *Axis: Z Axis* and *Angle: 60°.*

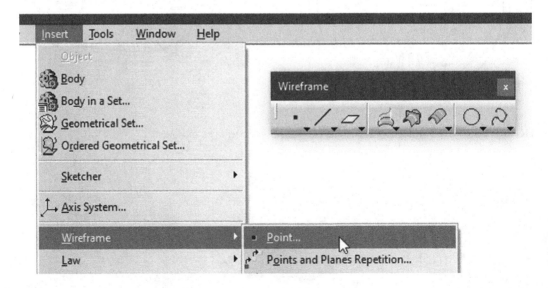

FIGURE 3.558   Displaying the coordinates for the *Point.1.*

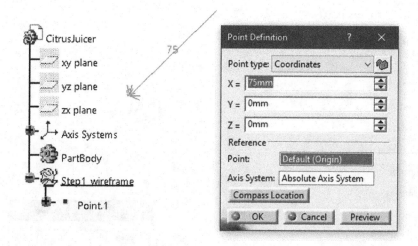

FIGURE 3.559 Displaying the location of the *Point* tool.

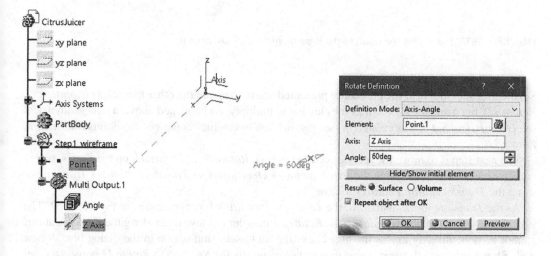

FIGURE 3.560 Displaying the *Rotate Definition* dialog box.

To create more than one copy of the *Point.1*, the user must check the option *Repeat object after OK*. After confirming and clicking the *OK* button, a new dialog box, *Object Repetition*, appears and the user is asked to enter the number of instances to be created. The value to be entered is 4, as in Figure 3.561. In the figure, the result of the *Rotate* tool can also be observed, which adds to the specification tree five more instances (*Rotate.1,..., Rotate.5*) of the *Point.1*. It should be noted that the last four of them are placed in the new *Geometrical Set.2*, which is subordinate to the *Step1_wireframe*, but at the same level with *Point.1* and *Rotate.1*.

These six points are the first series of points needed to create the wavy edge of the cup of the citrus juicer. A second series of six points are needed to complete the edge of the cup. To create this series, a new point is needed, which is obtained based on the *Point.1* also. For this, the *Rotate* tool is used again, but, this time, the *Angle* field is filled with the value of 30° and the option *Repeat object after OK* is not checked. The result consists in the appearance of point *Rotate.6* between *Point.1* and *Rotate.1* in the graphic area. For the second series of points also, six points are needed.

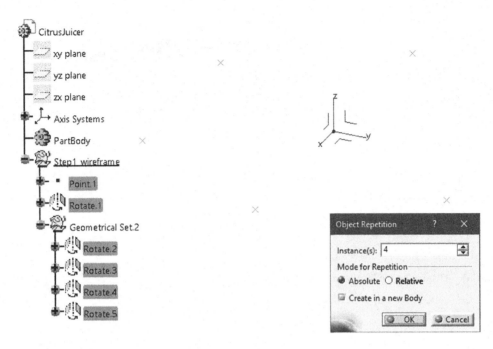

FIGURE 3.561 Displaying the result of the *Repeat object after OK* option.

So, the user should repeat the procedure presented above to create the other five points (*Rotate.7,...,* *Rotate.11*), but using the *Rotate.6* as the element to multiply. As happened above, a new geometrical set, *Geometrical Set.3,* was created in the specification tree, which contained the points *Rotate.8,...,* *Rotate.11*.

The next step is to move the second series of points, *Rotate.6,..., Rotate.11,* on the Z axis, with a negative value of −4 mm, using the *Insert* menu → *Operations* → *Transformation* → *Translate* or using the *Transformation Features* toolbar.

As the *Translate* tool is launched, the *Translate Definition* box appears, as in Figure 3.562. The user should select the points *Rotate.6,..., Rotate.11* in order to move them along the Z axis. But this cannot be done directly unless the user clicks the multi-selection option in the dialog box. A new dialog box opens, and the user is able to select sequentially the *Rotate.6* to *Rotate.11* points (six elements). The other parameters are the following: *Vector Definition: Direction, distance; Direction: Z Component;* and *Distance:* −4 mm. The *Translate* tool will copy the source elements (six points) to the specified distance, along the selected direction. In this case, the source points are no longer needed, so the user should click *Hide/Show* the initial element option. This way, the source points will be hidden. As Figure 3.562 shows, the specification tree is completed with the *Multi Output.4* feature, which contains the needed points, *Translate.1,..., Translate.6*.

The next step in the modelling procedure is to obtain the wavy feature of the cup by creating a spline which will go through the newly formed points. To create a spline, the user should launch the *Spline* tool, from the *Insert* menu → *Wireframe*, or from the *Wireframe* toolbar. In the *Spline Definition* dialog box, as in Figure 3.563, which appears when launching the tool, the user selects the following points (from the specification tree or in the graphic area, in this specific order): *Point.1, Translate.1, Rotate.1, Translate.2, Rotate.2, Translate.3, Rotate.3, Translate.4, Rotate.4, Translate.5, Rotate.5* and *Translate.6*. The final step, very important, in getting the spline curve done is to check the option *Close Spline* in the same dialog box. The user can now confirm with the *OK* button, and after that, he observes the wavy aspect of the *Spline.1*, which is the new feature added to the specification tree.

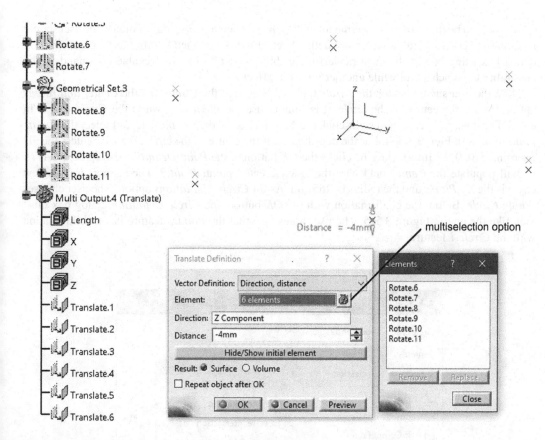

FIGURE 3.562 Displaying the *Translate* tool and its result.

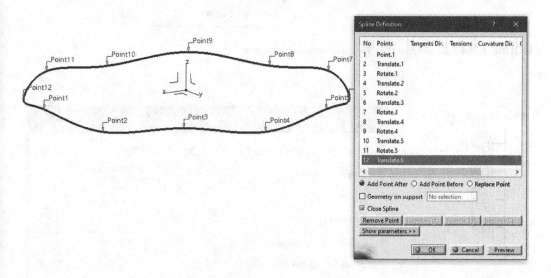

FIGURE 3.563 Displaying the *Spline Definition* dialog box.

The next step is to create the wireframe geometry for the bottom part of the cup, in the same *Step1_wireframe* geometrical set. To achieve this, the user should place a circle, *Circle.1*, 20 mm below the *XY Plane*, with the centre in the coordinates (0, 0, −20 mm). In the *GSD* workbench, the

Circle tool can be found in the *Insert* menu → *Wireframe* or in the *Wireframe* toolbar. The user should note that the centre of the circle is not defined yet; it will be defined within the *Circle Definition* dialog box using one of the most powerful capabilities of *CATIA v5*, stackable commands – the possibility to launch a tool while another tool is in effect.

Now, the user should launch the *Circle* tool and choose as the *Circle type* the *Center and radius* option. As for the centre of the circle, it is time to use the *Point* tool, while the *Circle* tool is in effect. For this, the user should right-click the *No selection* in the *Center* field and select the *Create Point* tool, as in Figure 3.564. For the coordinates of the centre of the circle, the user enters the 3D coordinate (0, 0, −20 mm). Once he clicks the *OK* button in the *Point Definition* dialog box, *CATIA v5* will populate the *Center* field with the newly created point, *Point.2*. The support for the new circle is the *XY Plane*, and the radius is R65 mm. As for *Circle Limitations* options, the user chooses *Whole Circle*. Before the confirmation with the *OK* button, the *Circle Definition* dialog box must look like the one in Figure 3.565. The user notes also that the *Point.2* feature has a child relation with the *Circle.1* feature.

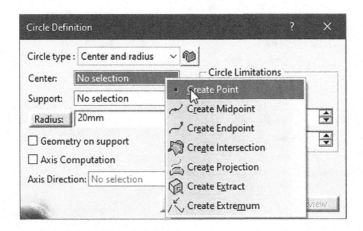

FIGURE 3.564   Displaying the *Circle Definition* dialog box.

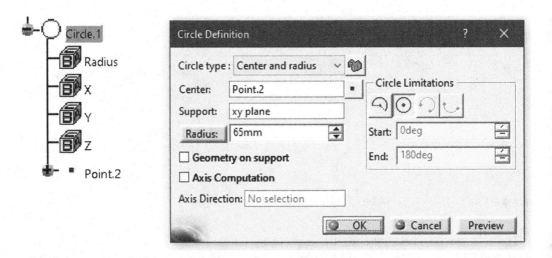

FIGURE 3.565   Displaying the parameters in the *Circle Definition* dialog box.

Now the specification tree contains *Spline.1* and *Circle.1* features, which are needed to create the shape of the cup by generating a surface between them.

To keep the work well organized, the user creates a new geometrical set, *Step1_shapes*, which will be used as a container for the surfaces generated for the cup. So, from the *Insert* menu, the *Geometrical Set* tool is launched and the user inserts the *Step1_shapes* geometrical set, as in Figure 3.566.

Once inserted, the *Step1_shapes* becomes the current work object and it is underlined in the specification tree.

It is time to create the wall of the cup by using the *Multi-Sections Surface* tool from the *Insert* menu → *Surfaces* or from the *Surfaces* toolbar. By default, the tool admits a number of sections through which a surface will be created. For the citrus juicer, these sections are *Spline.1* and *Circle.1*. So, in Figure 3.567, the user launches the *Multi-Sections Surface* and selects *Spline.1* and *Circle.1* as the input sections. Without any options selected or tweaking, the surface should look like the one in Figure 3.567.

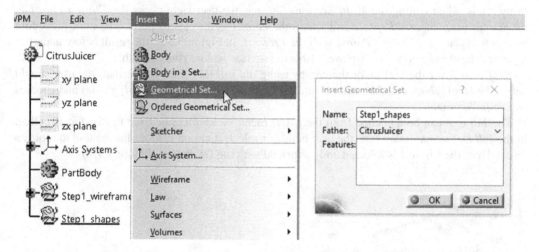

FIGURE 3.566 Inserting *Step1_shapes* geometrical set.

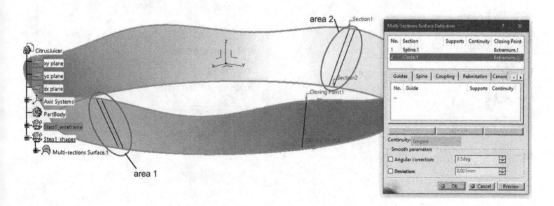

FIGURE 3.567 Displaying the *Multi-sections Surface.1* before tweaking.

In Figure 3.567, it can be observed that the *Multi-sections Surface.1* is not very smooth in the two areas, 1 and 2. To fix that, the closing points should be on the same side, having the tangents in the same direction. So, the user will apply the *Point.1* (as it belongs to the *Spline.1*) as the *Closing Point1* for the first section and the *Point.1* projection on the *Circle.1* as the second *Closing Point2* for the second section. To achieve this, the user must create the *Point.1* projection on the *Circle.1* as it is not present in the graphic area. The projection of a point is wireframe geometry, so it must be placed in the appropriate geometrical set. In the case of the citrus juicer, it is *Step1_wireframe*.

The user right-clicks the *Step1_wireframe* branch in the specification tree and chooses *Define In Work Object*. The feature gets underlined, and the user can launch the *Projection* tool from the *Insert* menu → *Wireframe* or from the *Wireframe* toolbar.

In the *Projection Definition* dialog box, the user should select the *Point.1* as the input for the *Projected* field and *Circle.1* as the input for the *Support* filed. The other options remain untouched. The *Projection Definition* dialog box is presented in Figure 3.568, and the results consist in the appearance in the specification tree of the feature *Project.1*.

The user returns to edit the *Multi-sections Surface.1* by double-clicking it in the specification tree. The user has to replace the existing closing points with the *Point.1* and *Project.1*, respectively. Thus, when editing the *Multi-Sections Surface*, the user right-clicks the *Closing Point1* in the graphic area, chooses *Replace* and selects the *Point.1* as the closing point. The same procedure applies to replace the *Closing Point2* with the *Project1*. In Figure 3.569, the result before and after replacing the closing points is displayed. Now the surface feature runs smooth.

Next is to cover the bottom of the cup by using the *Fill* tool. But, before that, the user should define the *Step1_shapes* as the work object. So, the user right-clicks the *Step1_shapes* and chooses *Define In Work Object*.

The *Fill* tool can be launched from the *Insert* menu → *Surfaces* or from the *Surfaces* toolbar. In the *Fill Definition* dialog box, the user builds or selects a closed contour for the tool to return a result. Thus, the *Circle.1* is selected and confirmed with the *OK* button (Figure 3.570).

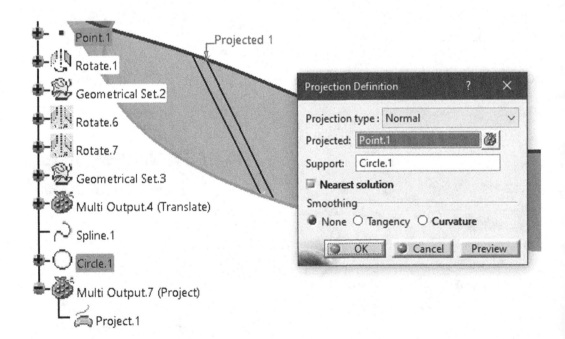

FIGURE 3.568 Displaying the *Projection Definition* dialog box.

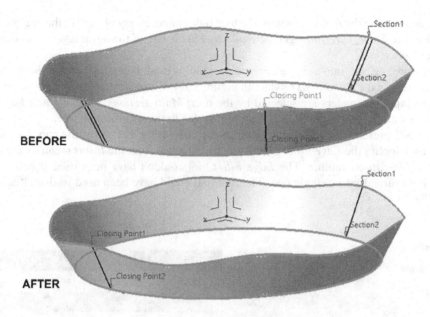

FIGURE 3.569 Displaying the result before and after replacing the closing points.

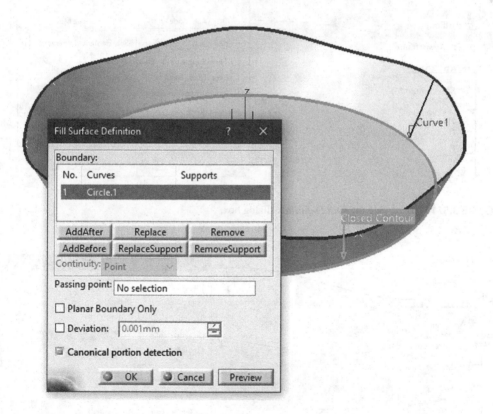

FIGURE 3.570 Displaying the *Fill Surface Definition* dialog box.

In the specification tree, in the *Step1_shapes* geometrical set, there are two surfaces, *Multi-sections Surface.1* and *Fill.1*. The last thing the user should do to complete the major first step is to join both surfaces into a single one and to apply a fillet radius on the edge at their intersection.

Before committing the next operations, the user hides unnecessary objects in the graphic area by hiding the entire *Step1_wireframe* geometrical set. This is one of the advantages of working with geometrical sets.

To turn more surfaces into a single one, the *Join* tool is used. It can be found in the *Insert* menu → *Operations* or in the *Operations* toolbar. In Figure 3.571, it is presented the *Join Definition* dialog box and the input parameters are selected by the user: *Multi-sections Surface.1* and *Fill.1*. After confirmation with the *OK* button, *Join.1* feature is added to the specification tree.

Figure 3.572 presents the fillet performed on the edge at the bottom of the cup. To create the fillet, the user applies the *Edge Fillet* tool. It can be launched from the *Insert* menu → *Operations* or from the *Operations* toolbar. The *Edge Fillet* tool couldn't have been used if both surfaces wouldn't have turned into a single one. The tool that could have been used in that situation was *Shape Fillet*.

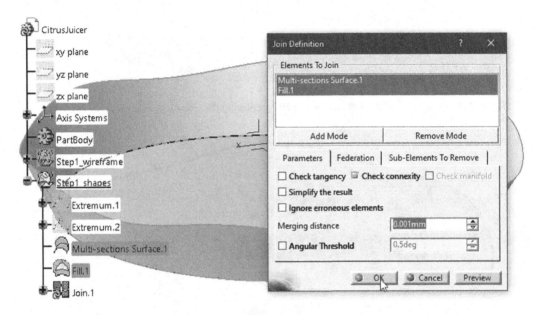

FIGURE 3.571   Displaying the *Join Definition* dialog box.

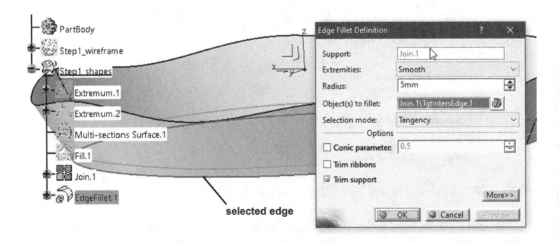

FIGURE 3.572   Displaying the *Edge Fillet Definition* dialog box.

Performing the *Edge Fillet* operation completes the major first step.

The next major step in the modelling process is to create the main active part of the citrus juicer, presented in Figure 3.573c. For this to be achieved, the user must create the wireframe geometry, to create the surface and to multiply it around an edge. These steps are presented in Figure 3.573.

To keep the work organized, the user inserts a new geometrical set, named *Step2_wireframe*, at the same level with the previously created *Step1_wirefame* geometrical set. All the wireframe geometry created for this step will be placed under this container.

The wireframe geometry for this step consists in two sketches (*Sketch.1* and *Sketch.2*), a rotated copy of one of *Sketch.2* (*Rotate.12*) and a circle, *Circle.2*, as illustrated in Figure 3.574.

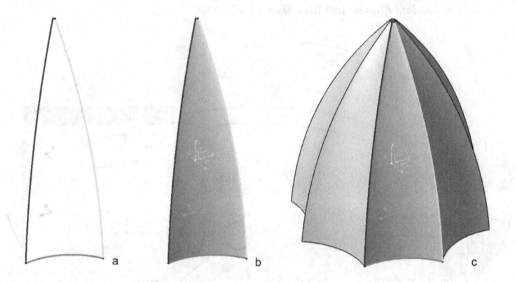

FIGURE 3.573 Displaying the steps for the main active part modelling.

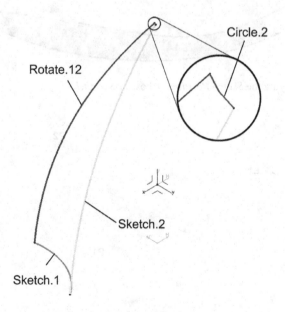

FIGURE 3.574 Displaying the wireframe geometry for major step 2.

To create *Sketch.1*, the user selects the bottom surface created with *Fill* in the first major step and then launches the *Positioned Sketch* tool from the *Sketch* toolbar. The user checks the appropriate option in the *Sketch Positioning* dialog box to switch between the direction of the *H* axis as in Figure 3.575 and confirms with the *OK* button. The *Sketch.1* feature is added to the specification tree.

At this moment, the geometrical set *Step1_shapes* can be hidden to clear the screen for the drawing of the *Sketch.1* objects. Thus, a line, *L1* in Figure 3.576a, is drawn using the *Line* tool from the *Profile* toolbar. It starts from the origin; its length is set to 35 mm; and an angle constraint is placed between the line *L1* and the *H* axis. The value of the angle is 45°, which was purposely chosen as it is a divisor of 360° (a full circle). This line was drawn only to help the user to completely constrain the next object, which is an arc. So, the line is converted to an auxiliary construction line using the *Construction/Standard Element* tool from *Sketch Tools* toolbar.

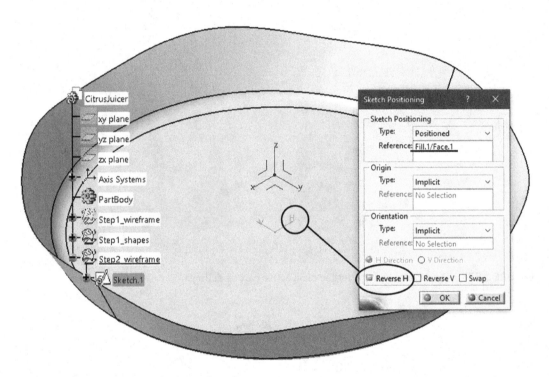

FIGURE 3.575   Displaying the *Sketch Positioning* box for *Sketch.1*.

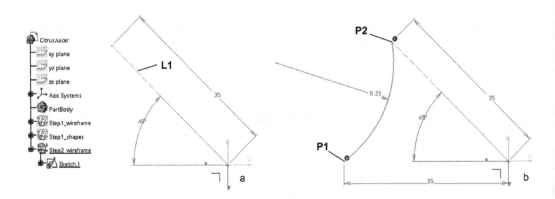

FIGURE 3.576   Displaying the *Sketch.1*.

The next step is to create the mentioned arc using the *Arc* tool from the *Profile* toolbar. The radius of the arc is set to R25 mm, and its endpoints (*P1* and *P2* in Figure 3.576b) are as follows: *P1* is coincident with the *H* axis and at 35 mm from the origin and *P2* is coincident with the endpoint of the line *L1*. As the objects in *Sketch.1* turn green, this means that the sketch is *ISO-constrained*.

The *Sketch.2* is created using again the *Positioned Sketch* tool from the *Sketch* toolbar. This time, the sketch support is *ZX Plane* and the directions of the *H* and *V* axes are as in Figure 3.577.

Using the *Spline* tool from the *Profile* toolbar, the user draws a spline controlled by three points (*P3*, *P4* and *P5* – arbitrarily picked using the mouse in the graphic area; Figure 3.578). After exiting the *Spline* tool, the three points (*P3*, *P4* and *P5*) are constrained as follows: *P3* is coincident with *P1* from *Sketch.1* (Figure 3.576b), *P4* has the coordinates (25 mm, 25 mm), and *P5* has the coordinates (1 mm, 55 mm), according to Figure 3.578. This way, the *Spline.1* object, which belongs to *Sketch.2*, is completely constrained (*ISO-constrained*).

Another step in continuing the creation of the wireframe geometry is to generate a rotated copy of *Sketch.2*. For this, the user launches the *Rotate* tool from the *Transformation Features* toolbar. This brings to the user the *Rotate Definition* dialog box, which takes as parameters the following: *Definition Mode: Axis-Angle, Element: Sketch.2* and *Angle: –45°*, leaving the other options untouched, as in Figure 3.579. The user confirms by clicking the *OK* button, and the *Rotate.12* feature is added to the specification tree.

The last step in completing the wireframe geometry is to create a circle at the upper part of the existing wireframe geometry using the *Circle* tool from the *Wireframe* toolbar. Launching the *Circle* tool, the *Circle Definition* dialog box opens and the user is required to enter/select the following parameters: *Circle Type: Two points and radius, Point 1* and *Point 2* – the user selects them in the graphic area as in Figure 3.580, *Support: XY Plane* and *Radius:* 2 mm. One important option

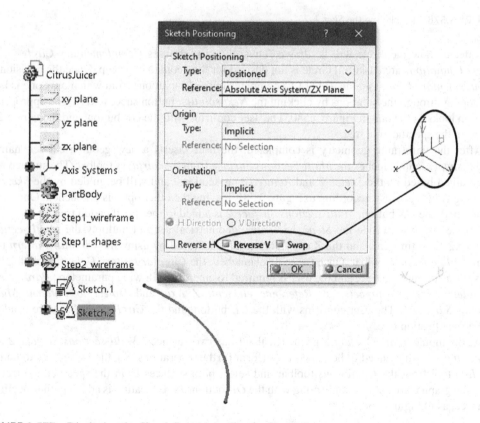

FIGURE 3.577 Displaying the *Sketch Positioning* box for *Sketch.2*.

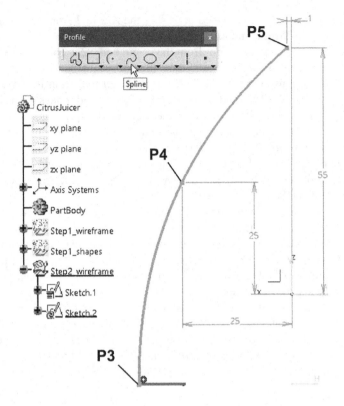

FIGURE 3.578   Displaying the *Sketch.2*.

is to decide how the circle will be drawn. Thus, the user chooses *Complementary Circle* in the *Circle Limitations* area as a full circle is not desired for the geometry to keep. Once the decision of *Complementary Circle* is done, *CATIA v5* computes multiple solutions from which only one is kept by looping through the solutions by clicking the *Next solution* button successively. The appropriate solution looks as the one in Figure 3.580. The user confirms with the *OK* button, and the *Circle.2* is added in the specification tree.

After the wireframe geometry is completed, the user inserts a new geometrical set, named *Step2_shapes*, and launches the *Multi-Sections Surface* from the *Surfaces* toolbar. The surface will be obtained based on the *Sketch.2* and *Rotate.12* (as sections) and will be guided by the *Sketch.1* and *Circle.2* (as guides), according to Figure 3.581, leaving the other options untouched. Once confirmed with the *OK* button, *Multi-sections Surface.2* is added to the specification tree.

The next step is to hide the *Step2_wireframe* geometrical set and multiply the *Multi-sections Surface.2* eight times around the *Z* axis. This can be achieved by using the *Circular Pattern* tool from the *Replication* toolbar. Once the tool is launched, the *Circular Pattern Definition* dialog box pops up, as in Figure 3.582, and the user is required to enter the following parameters: *Parameters: Complete crown, Instance(s):* 8, *Reference element: Z Axis* and *Object to Pattern: Multi-sections Surface.2*. The user confirms with the *OK* button, and the *CircPattern.1* feature is added to the specification tree.

At the moment, *Step2_shapes* geometrical set has two surfaces, *Multi-sections Surface.2* and *CircPattern.1*, which need to be joined to perform further operations. So, the user needs to launch the *Join* tool from the *Operations* toolbar and select both surfaces from the specification tree or from the graphic area. After confirming with the *OK* button, *Join.2* feature is added to the specification tree, as in Figure 3.583.

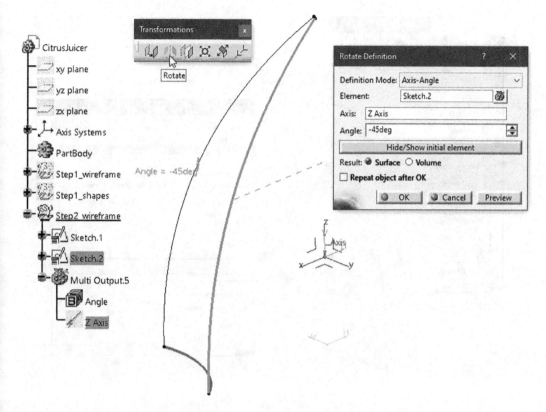

FIGURE 3.579   Displaying the *Rotate Definition* dialog box.

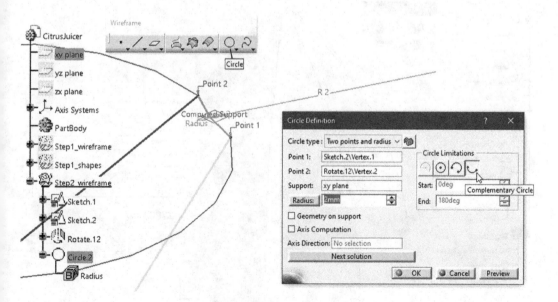

FIGURE 3.580   Displaying the *Circle Definition* dialog box.

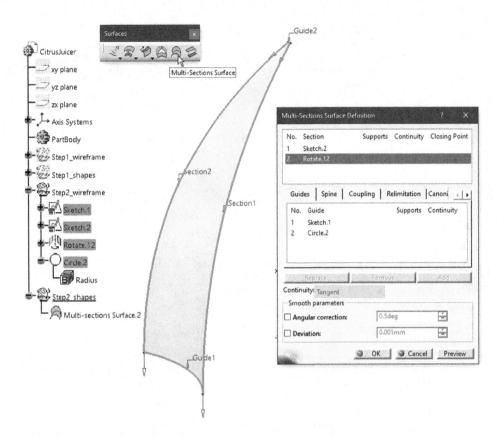

FIGURE 3.581   Displaying the *Multi-Sections Surface Definition* dialog box.

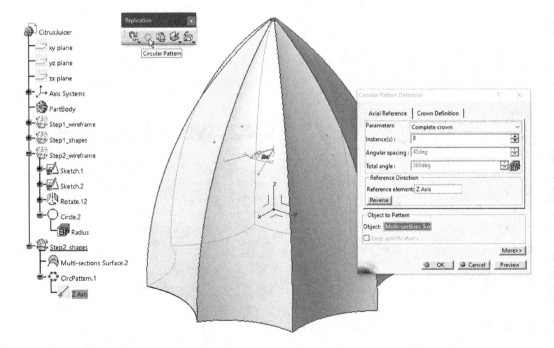

FIGURE 3.582   Displaying the *Circular Pattern Definition* dialog box.

It can be observed that at the upper part of the surface *Join.2*, there is a hole that needs to be filled. For this purpose, the user will use the *Face-Face Fillet* tool from the *Operations* toolbar. Before launching the *Face-Face Fillet* tool, the user will rotate the part to look at it from underneath. In the *Face-Face Fillet Definition* dialog box, presented in Figure 3.584, the user selects two opposite surfaces, then modifies the *Extremities* parameter to *Minimal* and enters the *Radius* of R3 mm. Confirming with the *OK* button makes the *FaceFillet.1* feature to be added to the specification tree. The user should apply the same procedure to the other three pairs of surfaces.

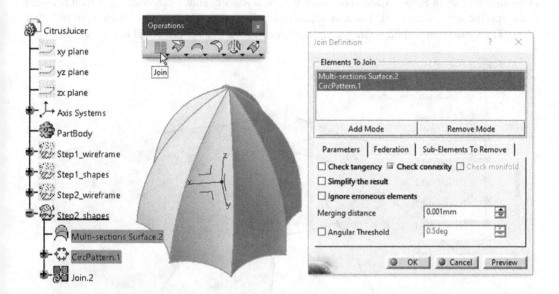

FIGURE 3.583   Displaying the *Join Definition* dialog box.

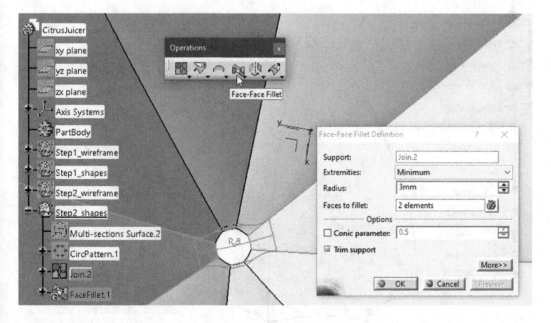

FIGURE 3.584   Displaying the *Face-Face Fillet Definition* dialog box.

When done, the surface and the specification tree look as in Figure 3.585, *FaceFillet.1,...,* *FaceFillet.4* features being parts of the specification tree.

That last thing to be done is to fillet the edges of the surface known as *FaceFillet.4*. For this purpose, the user will use the *Variable Fillet* tool from the same *Operations* toolbar. As the name speaks for itself, the tool allows the user to specify different values for the radius used to fillet the edges. In this situation, only two points (the endpoints of the edges) will be used to specify two values for the fillet radius, R3 and R0.5mm, respectively. In the *Variable Radius Fillet Definition* dialog box, the user selects from the graphic area the edge that is to be filleted and double-clicks the value that needs to be modified, as in Figure 3.586. The result will be a feature named *EdgeFillet.2*, which is added to the specification tree as well. The user should apply this procedure seven times more, to the other edges. The surface, named *EdgeFillet.9*, and the specification tree are presented in Figure 3.587.

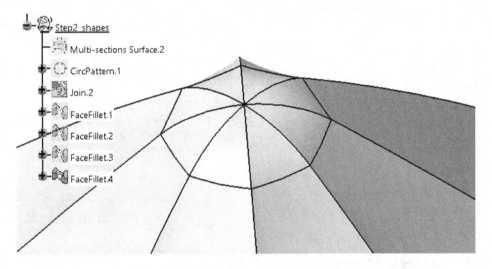

FIGURE 3.585   Displaying the result after applying four times the *Face-Face Fillet* tool.

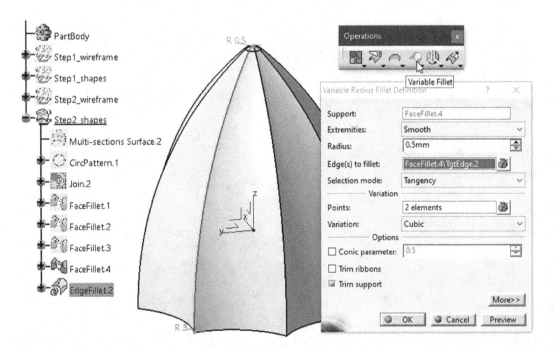

FIGURE 3.586   Displaying the *Variable Radius Fillet Definition* dialog box.

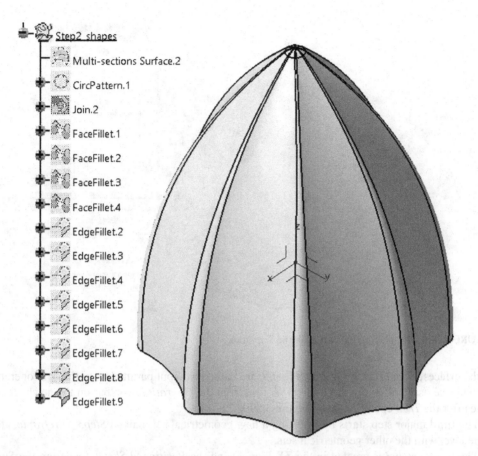

FIGURE 3.587 Displaying the result after applying eight times the *Variable Radius Fillet* tool.

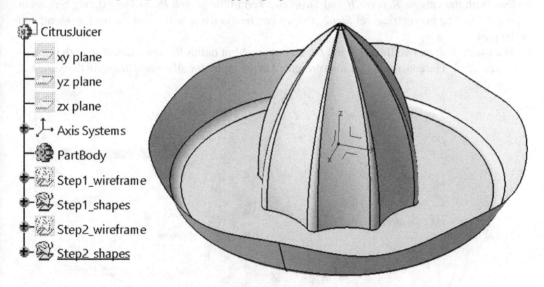

FIGURE 3.588 Displaying the part after the first and second major steps.

After the first two major steps were completed, the citrus juicer looks like that in Figure 3.588. Before moving to the third major step, two more operations need to be performed on both surfaces, *EdgeFillet.1* and *EdgeFillet.9*. First, the user will trim the surfaces by using the *Trim* tool from *Operations* toolbar. In Figure 3.589, the *Trim Definition* dialog box is presented.

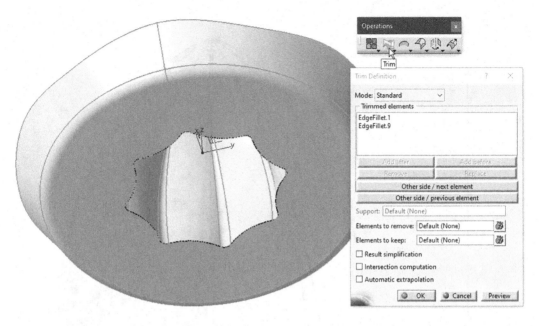

FIGURE 3.589   Displaying the part after the *Trim* tool.

Both surfaces, *EdgeFillet.1* and *EdgeFillet.9*, are passed as input parameters to the trim operation. Second, the user will apply the *Edge Fillet* tool from the *Operations* toolbar on the newly created edge after the *Trim* operation, as in Figure 3.590.

The third major step starts by inserting a new geometrical set, named *Step3_wireframe*, at the same level with the other geometrical sets.

The new *Sketch.3* is created in the *ZX Plane* using the *Positioned Sketch* tool from the *Sketch* toolbar, with the options *Reverse H* and *Swap* checked in the *Sketch Positioning* dialog box, as in Figure 3.591. The geometrical set *Step2_shapes* can be hidden now to simplify the representation of the part.

The user will draw two lines, *L2* and *L3*, and an arc *A1* of radius R5 mm, tangent to both lines, as in Figure 3.592. The constraints are also presented in the figure for all objects drawn in the *Sketch.3*.

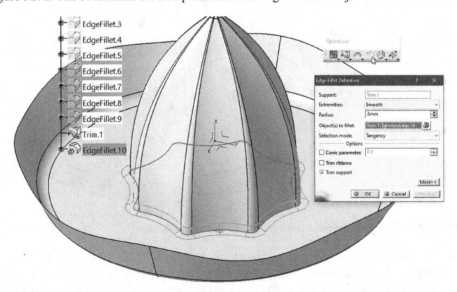

FIGURE 3.590   Displaying the part after applying the *Edge Fillet* tool.

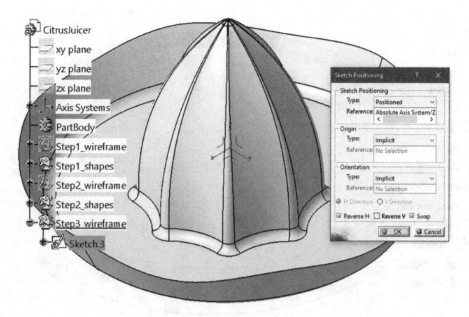

FIGURE 3.591   Displaying the *Sketch Positioning* box for *Sketch.3*.

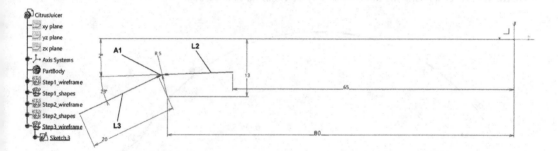

FIGURE 3.592   Displaying the *Sketch.3*.

The *Sketch.3* will be used to create a surface by using the *Revolve* tool from the *Surfaces* toolbar. Before using the *Revolve* tool, the user needs to insert a new geometrical set, named *Step3_shapes*.

Once the *Step3_shapes* set is inserted, the user launches the *Revolve* tool and enters the following parameters in the *Revolution Surface Definition* dialog box: *Profile: Sketch.3, Revolution Axis: Z Axis* and *Angular Limits:* 0 deg for *Angle 1* and 360 deg for *Angle 2*, as the surface should describe a full circle, as in Figure 3.593. The new feature *Revolute.1* is added to the specification tree.

To create the handle of the citrus juicer, the user will define the *Step3_wireframe* as the work object and creates a new sketch, *Sketch.4*, in the *XY Plane*. In the *Sketch.4*, the user draws a circle of diameter Ø170 mm, with the centre at origin of an arc of radius R40 mm, having the centre point coincident with the *H* axis and at 65 mm from the *V* axis and the start and the end of the arc on the circle of diameter Ø170 mm. The circle and the arc will be joined with two corners with the radii of R20 mm each, as in Figure 3.594.

The next step is to project the *Sketch.4* onto the *Revolute.1* surface to create the curve *Project.2* needed to obtain the handle. The projected curve is the result of the *Projection* tool from the *Wireframe* toolbar. In the *Projection Definition* dialog box, the input parameters are as follows: *Projection type: Normal, Projected: Sketch.4* and *Support: Revolute.1*, according to Figure 3.595. The result is the addition of the *Project.2* feature to the specification tree. *Sketch.4* can be hidden as it is not required anymore.

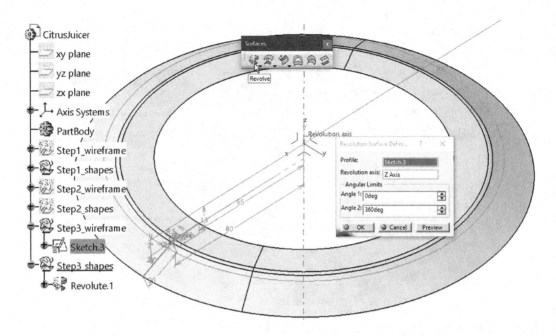

FIGURE 3.593 Displaying the *Revolution Surface Definition* dialog box.

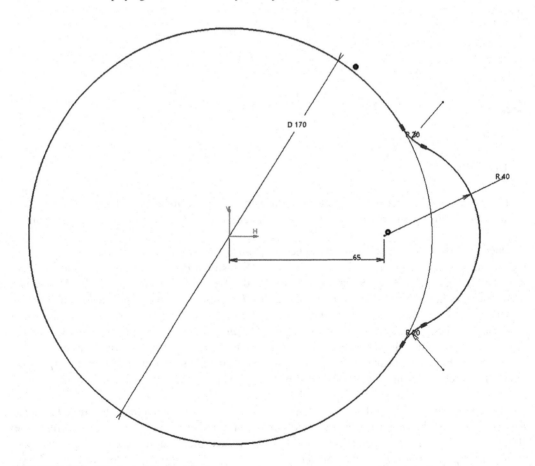

FIGURE 3.594 Displaying the *Sketch.4*.

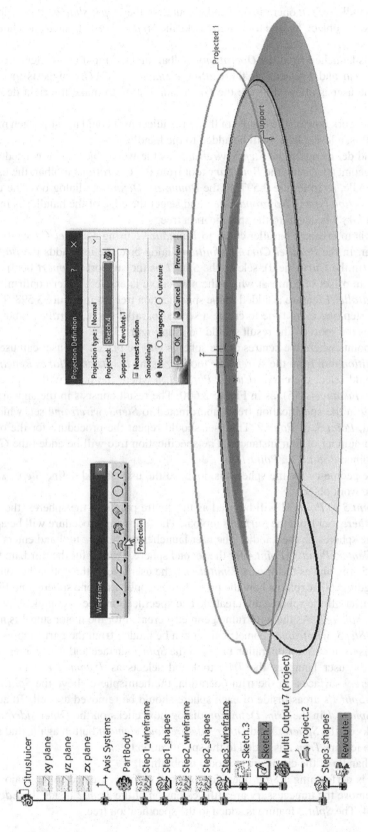

FIGURE 3.595 Displaying the *Projection Definition* dialog box.

To obtain the handle, a *Split* operation is needed, but first, the *Step3_shapes* geometrical set must be defined as the work object. So, the user right-clicks the *Step3_shapes* branch and chooses *Define In Work Object*.

The *Split* tool is launched from the *Operations* toolbar, and the input parameters are *Revolute.1* as the *Element to cut* and *Project.2* as the *Cutting element*. As *CATIA* suggests the wrong side to be removed, the user is allowed to use the *Other side* button to make the right decision, as in Figure 3.596.

The user should work now on the handle of the citrus juicer as it could not slip when manipulated by a human. For this, a few spheres will be added to the handle.

The user should define again the *Step3_wireframe* as the work object, as it is needed to create more wireframe geometry. Using the *Boundary* tool from the *Operations* toolbar, the user extracts an edge of the handle, as in Figure 3.597. In the *Boundary Definition* dialog box, the user should change the *Propagation type* to *No propagation* and select the edge of the handle, as in the figure. The feature *Boundary.1* is added to the specification tree.

The next step is to create a parallel curve to *Boundary.1* using *Parallel Curve* tool from the *Wireframe* toolbar. In the *Parallel Curve Definition* dialog box, the user adds the *Boundary.1* as the source curve (in the *Curve* field), selects the *Split.1* as the support (*Support* field) for the new curve and enters an offset of 5 mm at which the new curve is created. After confirming with the *OK* button, the *Parallel.1* feature is added to the specification tree, as in Figure 3.598. The user has to repeat the last step one more time to create a second parallel curve, *Parallel.2*, but having the *Parallel.1* curve as the source. The result should be as in Figure 3.599.

To create the points where the centres of the spheres will be placed, the user can use the *Points and Planes Repetition* tool from the *Wireframe* toolbar. In the *Points and Planes Repetition* dialog box, the user should select the curve (*Curve: Parallel.1*) where the point will be placed and the number of points (*Instance(s): 5*), as in Figure 3.600. The result consists in the appearance of the *Geometrical Set.9* in the specification tree, subordinated to *Step3_wireframe* set, which contains the points needed, *Point.3,..., Point.7*. The user should repeat the procedure for the other curve, *Parallel.2*, with a number of four instances. The specification tree will be added the *Geometrical Set.10* with the points: *Point.8,..., Point.11*.

The wireframe geometry for the spheres is done, so the user should define *Step3_shapes* geometrical set as the work object.

The points *Point.3* to *Point.11* will be used as the centre points of the spheres; the user places them using the *Sphere* tool from the *Surfaces* toolbar. The following procedure will be applied nine times to create the spheres on the handle. The user launches the *Sphere* tool and enters the following parameters: *Center: Point.3* (*Point.4* for the second sphere, *Point.5* for the third, and so on) and *Sphere radius:* 1.5 mm, and as the *Sphere Limitations*, the user selects the option/button *Create the whole sphere*. Figure 3.601 explains how the procedure is applied for one sphere, and Figure 3.602 shows the result after all the spheres are created. The specification tree is completed with the features *Sphere.1,..., Sphere.9*. As the wireframe geometry created for the major step 3 is not required anymore further, *Step3_wireframe* geometrical set can be hidden from the part's representation.

The next task is to run a *Trim* operation between the *Split.1* surface and the spheres *Sphere.1,..., Sphere.9*. Thus, the user launches the *Trim* tool and selects as *Trimmed elements* the *Split.1*, *Sphere.1,..., Sphere.9* surfaces. In the trim operation, the hemisphere above the *Split.1* should be removed and the *Split.1*'s areas inside of each sphere should be removed as well. To achieve that, the user selects *Sphere.1* in the *Trim Definition* dialog box, clicks on the *Other side/previous element* button, clicks on the *Sphere.2* and *Other side/previous element* button again, and so on, as in Figure 3.603. The feature *Trim.2* is added to the specification tree.

This way, the handle of the citrus juicer model is obtained.

The next task is performing a *Split* operation on both surfaces created in the major step 2 and major step 3 to remove the unnecessary area (the extension of the *Trim.2* inside the *EdgeFillet.10*), as in Figure 3.604. The *Split.2* feature is added to the specification tree.

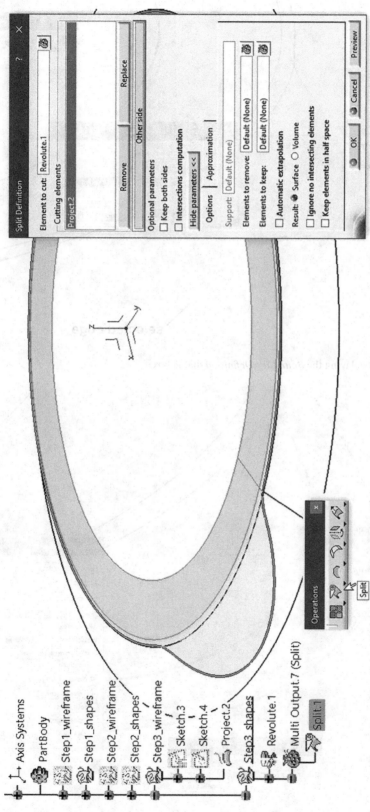

FIGURE 3.596 Displaying the *Split Definition* dialog box.

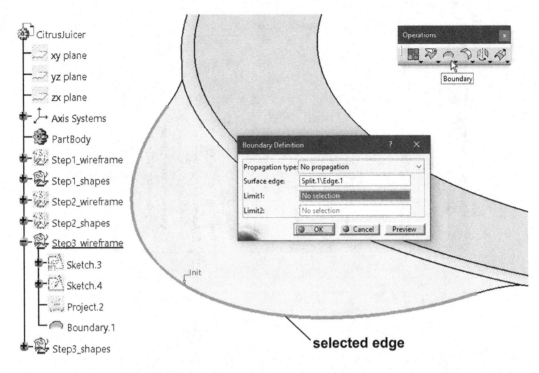

FIGURE 3.597 Displaying the *Boundary Definition* dialog box.

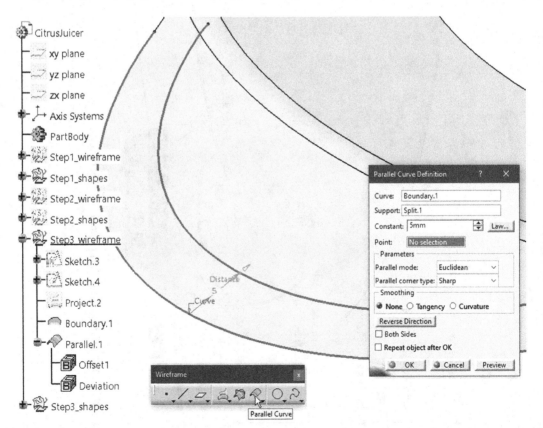

FIGURE 3.598 Displaying the *Parallel Curve Definition* dialog box.

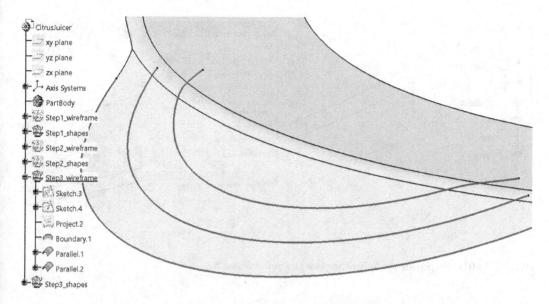

FIGURE 3.599 Displaying the result after applying the *Parallel Curve* tool twice.

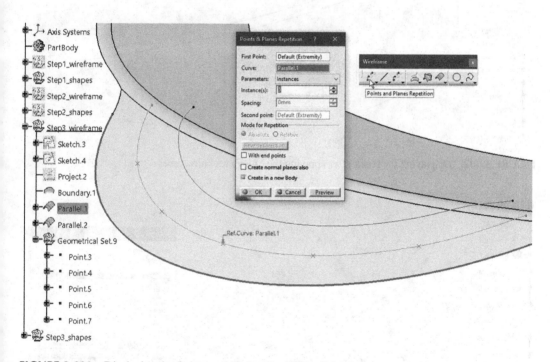

FIGURE 3.600 Displaying the *Points and Planes Repetition* dialog box.

The next step in the modelling process is to create the slots through which the juice should go to the pot below. To create the slots, the user needs to create the necessary wireframe geometry. So, he defines the *Step2_wireframe* as the work object and creates a new sketch, *Sketch.5*, on the bottom surface of the part. An *Elongated Hole* is placed with its centreline collinear to the *H* axis, and it is constrained as in Figure 3.605.

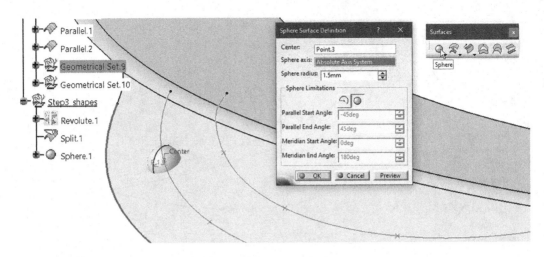

FIGURE 3.601   Displaying the *Sphere Surface Definition* dialog box.

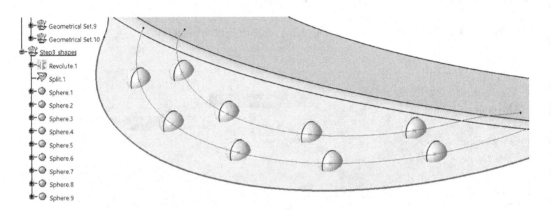

FIGURE 3.602   Displaying the result after applying the *Sphere* tool nine times.

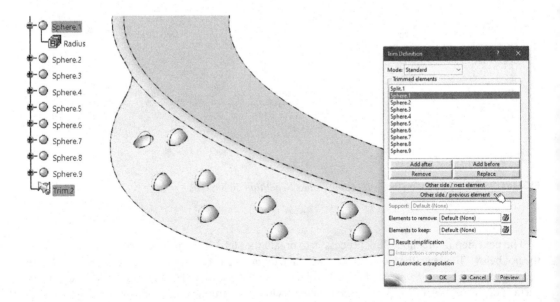

FIGURE 3.603   Displaying the *Trim Definition* dialog box.

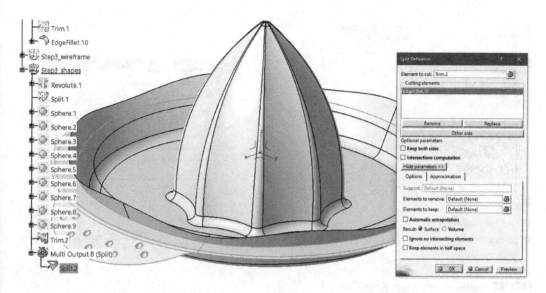

FIGURE 3.604 Displaying the *Split Definition* dialog box.

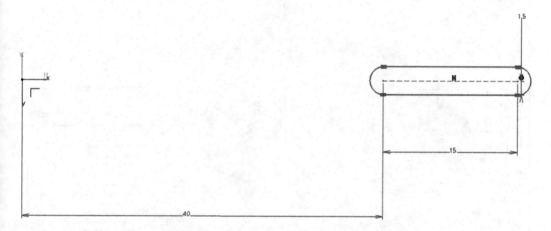

FIGURE 3.605 Displaying the constraints of *Sketch.5*.

The next step is to replicate the *Sketch.5* around the Z axis 36 times. The user launches the *Circular Pattern* tool from the *Replication* toolbar and enters the following parameters, as in Figure 3.606. A new feature, *CircPattern.2*, is added to the specification tree.

The next step is performing the *Split* operation on the *EdgeFillet.10* surface using the *Sketch.5* and the *CircPattern.2* as the cutting elements. But before doing that, the *Step2_shapes* geometrical set must be defined as the work object. Then the *Split* tool can be applied, as in Figure 3.607. The result, *Split.3*, is added to the specification tree, in the *Step2_shapes* geometrical set. Now, the geometrical set *Step2_wireframe* can be hidden as it is not used anymore.

Now, there are two surfaces, *Split.2* and *Split.3*, in the specification tree that need to be converted into a volumetric solid. To achieve that, the user switches to *Part Design* workbench, defines the *PartBody* as the work object and inserts a new body, named *Body.2*, using the *Body* tool from the *Insert* menu.

The user defines again the *PartBody* as the work object and launches the *Thick Surface* tool from the *Surface-Based Feature* toolbar or from the *Insert* menu → *Surface-Based Features*.

In the *ThickSurface Definition* dialog box, the user enters 0.25 mm either for the *First Offset* or for the *Second Offset* and selects the *Split.3* surface as the *Object to offset*, as in Figure 3.608.

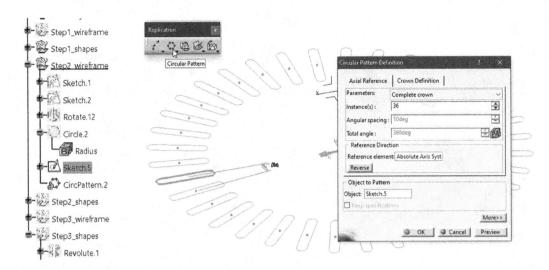

FIGURE 3.606 Replication of the *Sketch.5*.

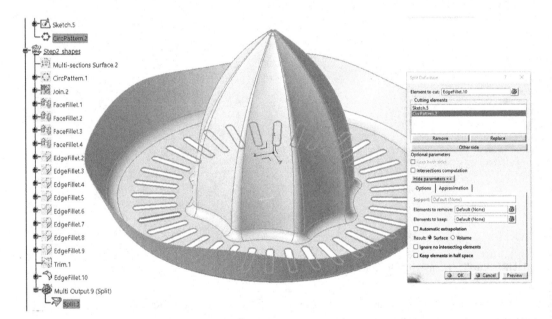

FIGURE 3.607 Creating the slots using the *Split* tool.

After confirming with the *OK* button, a new feature *ThickSurface.1* is added to the specification tree, under the *PartBody*.

The next task is to convert the other surface, *Split.2*, into a volumetric solid. Thus, the user defines the *Body.2* as the work object, then launches the same *Thick Surface* tool and applies the same offset to the surface, but the object to offset is *Split.2* now.

After these operations were performed, the surfaces *Split.2* and *Split.3* can be hidden as they are not used anymore.

Now the specification tree contains two solid bodies, *PartBody* and *Body.2*, that need to be assembled into a single one. The user should apply the *Assemble* tool from the *Insert* menu → *Boolean Operations*. The dialog box for the *Assemble* tool is presented in Figure 3.609, where the user selects the *Body.2* feature as the input and confirms with the *OK* button. The *To* field is automatically filled with the *PartBody* feature.

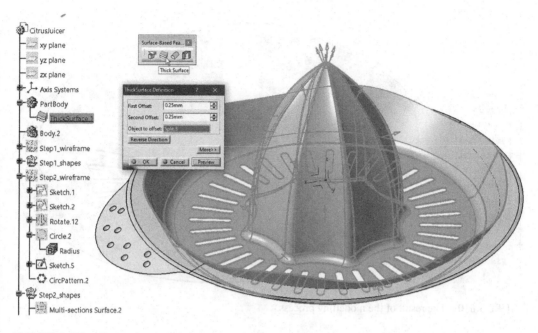

FIGURE 3.608   Converting a surface into a volumetric solid.

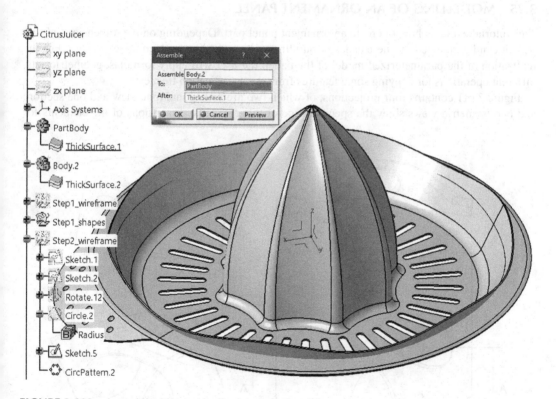

FIGURE 3.609   Assembling the part bodies into a single solid.

This operation was the last one in the modelling process of the citrus juicer. The result is presented in Figure 3.610.

In the modelling phases of the juicer, the user had to create several geometric sets. In this manner, he kept the 3D model organized in wireframe and shape features, easy to be edited later.

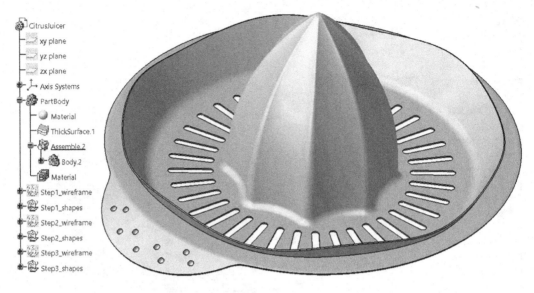

FIGURE 3.610   The result of the modelling process.

3.25   MODELLING OF AN ORNAMENT PANEL

This tutorial explains how to create an ornament panel part. Depending on the dimensions of this part, it can be mounted on the facades of buildings with a shading or decorative purpose. For the realization of the parameterized model of the panel, the user will apply formulas, geometric sets, different operations for copying some features from the specification tree, etc.

Figure 3.611 contains four projections, of which two are orthogonal (one view and one section) and two isometric views show the special shape of this part. The dimensions of the elements for

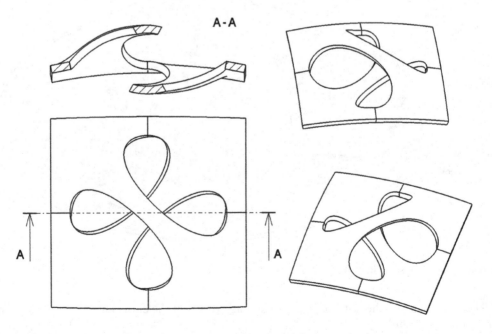

FIGURE 3.611   Representations of the ornament part to be modelled.

hybrid modelling are presented and used in the following explanations, but in general, it has a squared shape with sides of 100×100 mm.

The user opens the *CATIA GSD* workbench from the *Start → Shape* menu and, then in the *Sketch.1* of the *XY Plane*, draws a centred rectangle, positioned with the intersection of the diagonals (centre) at the origin of the *HV* coordinate system. The sides of this rectangle are constrained to the value of 100 mm, according to Figure 3.612.

The figure shows the equidistance of the vertical and horizontal sides, respectively, from the origin of the coordinate system.

In the *Sketch.2* of the *YZ Plane*, a spline curve is drawn through three points, as follows: the ends of the curve are symmetrical about the vertical axis *V* and the midpoint is constrained to be coincident with it. A distance of 100 mm is required between the ends of the spline curve, they are 10 mm above the horizontal axis *H*, and the midpoint is 15 mm (Figure 3.613). Note that the *H* axis is in the *XY Plane* of the rectangle in the *Sketch.1*.

The same spline curve is similarly drawn in the *ZX Plane*, or the user can copy the *Sketch.2* from the *YZ Plane* to the *ZX Plane*. Thus, in the specification tree, from the context menu of the sketch, the user chooses the *Copy* option, then right-clicks on the *PartBody* feature and, from

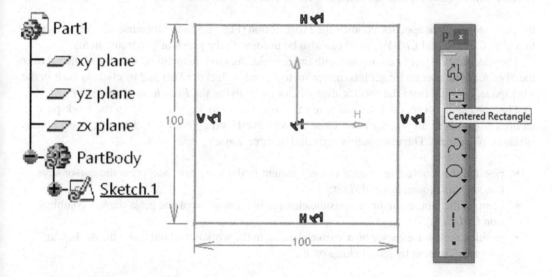

FIGURE 3.612   Drawing the first rectangle.

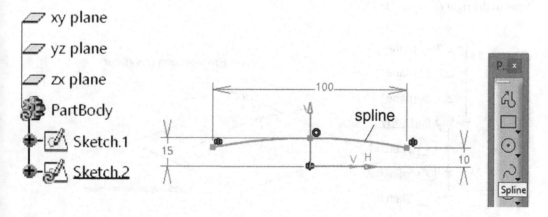

FIGURE 3.613   Drawing the first spline curve.

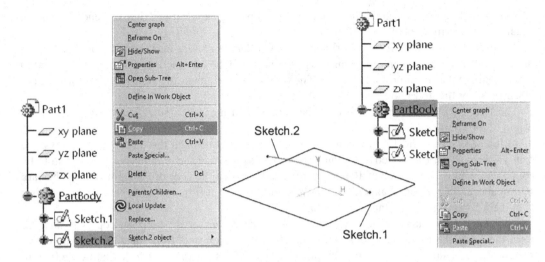

FIGURE 3.614   *Copy-Paste* of the first spline curve from one plane to another.

the context menu that appears, chooses the *Paste* option (Figure 3.614). Of course, the key combination *Ctrl-C* (*copy*) and *Ctrl-V* (*paste*) can also be used, with the previously selected items.

The specification tree is completed with the *Sketch.3* feature, an identical sketch and placed over the *Sketch.2*. The user no longer has the option to directly select the *Sketch.2* by clicking on it in the workspace, but only from the specification tree or by applying the *Preselection navigator*.

This navigator is especially useful when the user has to make a selection in the workspace in an area where there are two or more close or even overlapping geometric elements: lines, points, surfaces, planes, etc. The navigator is activated in three ways:

*   position the mouse cursor on a visible element in the work area and press the cursor keys (arrows) (*up*, *down*, *left* and *right*);
*   position the mouse cursor on a visible element in the work area and press the key combination *Ctrl-F11*;
*   position the mouse cursor on a visible element in the work area, hold down the *Alt* key, and select the element by left-clicking on it.

As a result, a red circle with four small arrows is displayed and the selected item appears inside it. Confirmation of the correct selection can be seen both in the specification tree and in the list of items to the right of Figure 3.615.

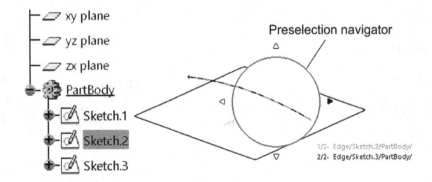

FIGURE 3.615   Selecting a curve using the *Preselection navigator*.

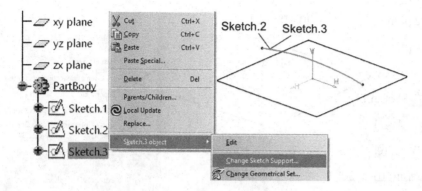

FIGURE 3.616 Changing the planar support for *Sketch.3*.

Sketch.3 must be moved to another plane perpendicular to the plane in which it was copied. Thus, from the context menu of the respective sketch, the user selects the option *Sketch.3 object →* *Change Sketch Support* (Figure 3.616).

In the *Sketch Positioning* selection box in Figure 3.617, in the *Type* list, the user chooses the *Sliding* option and the reference is the *ZX Plane* in which the *Sketch.3* is positioned. The user visually checks its correct placement and ticks the *Move geometry* option from the lower area of the selection box.

If their positioning is correct, the two sketches/spline curves must have one point in common, and this can be verified by applying the *Intersection* tool in the *Wireframe* toolbar. In the two fields *First Element* and *Second Element*, the user selects the two sketches and the result is a point, named *Intersect.1*, according to Figure 3.618. The appearance of a geometric set containing this point of intersection can be seen in the specification tree.

Currently, this intersection point is not required in the next steps of the application, and the user can remove it and return to the state of the part in Figure 3.617.

The two spline curves are used to create a surface using the *Sweep* tool in the *Surfaces* toolbar. In the *Swept Surface Definition* selection box, the user chooses the *Explicit* option/icon from the *Profile type* list and then the two sketches, *Sketch.3* and *Sketch.2* in the *Profile* and *Guide curve* fields (Figure 3.619), and the obtained surface, *Sweep.1*, is added to the specification tree in *Geometric Set.1*. The three sketches can be hidden using the *Hide/Show* option in the context menu of each one.

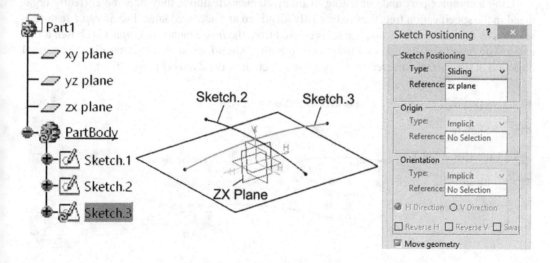

FIGURE 3.617 Moving the *Sketch.3* to *ZX Plane*.

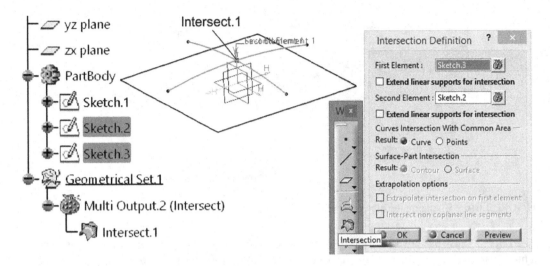

FIGURE 3.618 The two sketches have a point in common.

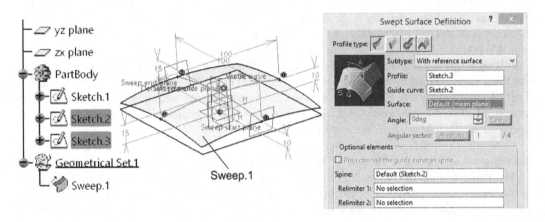

FIGURE 3.619 Creating the *Sweep.1* surface based on the two spline curves/sketches.

Being a complex part and consisting of many elements/features, they must be correctly organized in the specification tree, thus to be easily identified and accessed later. The *Sweep.1* feature is selected by the user; a new geometric set is created from the *Insert* menu, and *Inputs* is chosen for its name. The *Invert Orientation* tool is applied from the *Operations* toolbar to determine the normal direction of the surface directed in the positive direction of the Z axis (Figure 3.620).

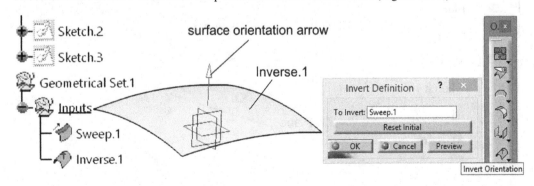

FIGURE 3.620 Choosing the surface orientation using the red arrow pointed upwards.

The *Sweep.1* surface becomes hidden in the specification tree; its place being taken by the *Inverse.1* surface. Its name is changed in *InputSurface* by the user using the *Properties → Feature Properties* tab in the context menu.

The most suggestive names of the features that are part of the hybrid model are important for a smooth go through the application, both for the user that makes it as a creator and for another user who, according to the changing specifications of a project, must subsequently modify it. The 3D model of the part should be very clear and easy to understand for this second user.

The contour of the *InputSurface* is extracted using the *Boundary* tool in the *Operations* toolbar (Figure 3.621).

Although it consists of four elements, they are not independent, but a single entity which can be divided into four curves (*Curve.1,…, Curve.4*) using the *Disassemble* tool. In the selection box with the same name (Figure 3.622), the user selects the *Boundary.1* feature from the specification tree (*Input elements* field becomes equal to 1) and then the *All Cells* option gets the value 4, equal to the number of curves.

The naming and numbering of the curves are done automatically by the program. The orientation in which these curves are considered to be drawn differs from one curve to another, so the user must set a certain orientation for all. Thus, the *Invert Orientation* tool is used again for each curve.

In the *Invert Definition* selection box in Figure 3.623, the user selects the curves, one by one, in the *To Invert* field and presses the *Click to Invert* button, if necessary, so that the red arrows point in the clockwise direction, as shown in the figure. As a result, the curves *Curve.1,…, Curve.4* are hidden in the specification tree, their place to delimit the *InputSurface* being taken by the features *Inverse.2,…, Inverse.5*.

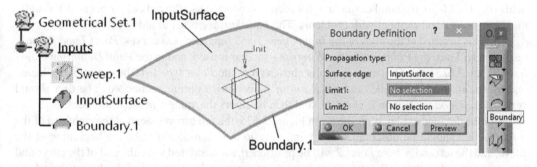

FIGURE 3.621   Extracting the contour of the *InputSurface*.

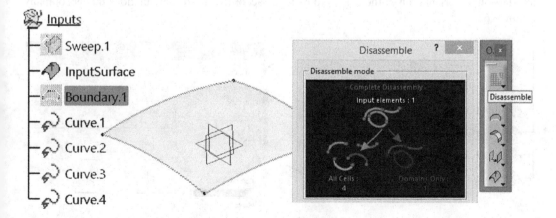

FIGURE 3.622   Disassembling the *Boundary.1* feature into four edge curves.

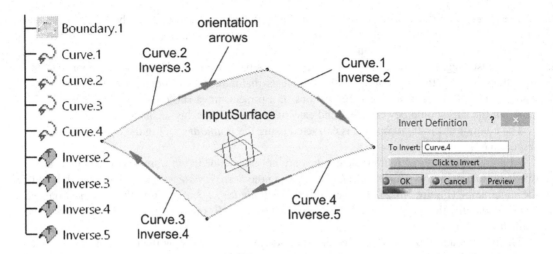

FIGURE 3.623 Changing the orientation of the extracted curves.

Although the user can keep these default names, it is still recommended to change them using the *Properties* → *Feature Properties* tab in the context menu in: *InputCurve.1,…, InputCurve.4* (example: *Inverse.2* becomes *InputCurve.1*; *Inverse.4* becomes *InputCurve.3*). These curves are further used in the modelling process and, for example, if many of such entities are present in a part, and in order to be more visible and easier to select, it is recommended to change their colour and thickness using the same context menu, the *Graphic* tab, *Lines and Curves*: *Colour* and *Thickness* fields (Figure 3.624). Remember that it is not recommended to use the red colour because it usually means a geometrical element with problems. The *InputSurface* is set to hidden.

A point is inserted at the ends of the *InputCurve.1* and *InputCurve.3* curves: *Point.1* and *Point.2*, respectively. Thus, the *Point* tool in the *Wireframe* toolbar is used, and in the *Point Definition* dialog box (Figure 3.625), the *On curve* option is chosen from the *Point type* list, the curves are selected one by one in the *Curve* field, and the *Ratio of curve length* option is checked. The user should verify if the *Geodesic* option is checked, and then he enters the value 1 in the *Ratio* field.

From the detailed explanation shown in Figure 3.625, the red arrow placed at the right end of the *InputCurve.3* should be observed. That point, of ratio 0, is considered to be the beginning of the curve, and the left end, where *Point.2* will be positioned, is considered to be the end of the curve and has a ratio of 1. Similarly, *Point.1* was previously inserted at the right end of the *InputCurve.1*. As a reference, the program chooses by default the end of ratio 0 of the curve. In this way, it is justified to establish the orientation of the curves to be made according to the explanations that accompany Figure 3.623.

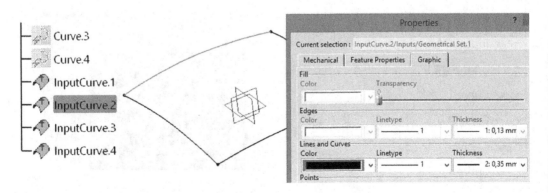

FIGURE 3.624 Changing the properties of the *InputCurve* curves.

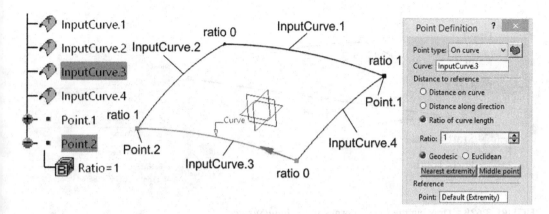

FIGURE 3.625   Creating points at the ends of *InputCurve.1* and *InputCurve.3*.

Between the two points, using the *Line* tool in the *Wireframe* toolbar, the user draws the *Line.1* and, then, in the middle of it, inserts a third point, *Point.3*, in a similar way to the definition of the previous points, but the ratio value is 0.5.

Figure 3.626 shows the line between *Point.1* and *Point.2*, its orientation by the red arrow, the *Point Definition* dialog box and *Point.3* located according to the ratio 0.5 on the line and also present with this parameter in the specification tree.

Point.3 is projected perpendicular to the *InputSurface*, according to Figure 3.627. In the *Projection Definition* selection box opened after applying the *Projection* tool, the user chooses the

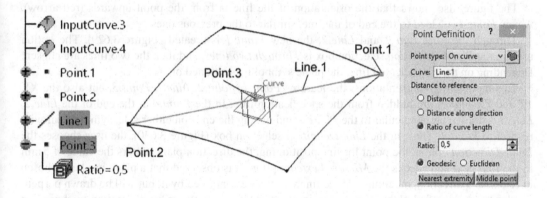

FIGURE 3.626   Creating the *Line.1* and the *Point.3* in its middle.

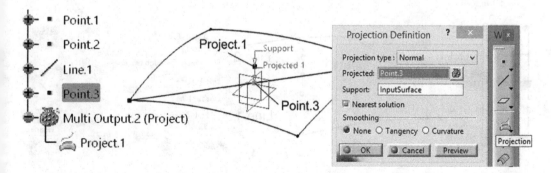

FIGURE 3.627   Projecting the *Point.3* on the *InputSurface*.

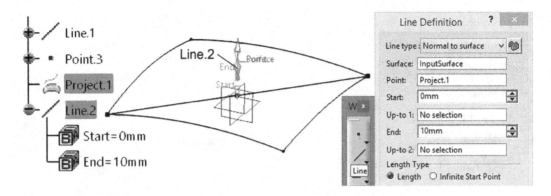

FIGURE 3.628 Drawing the *Line.2* normal to the *InputSurface*.

Normal type, the element to project and the projection support. As can it be observed, it is not necessary to display the *InputSurface*, but it can be selected from the specification tree.

The *Project.1* point represents the result of the projection, but also the lower end of a line perpendicular to the *InputSurface* and pointing upwards. Thus, the *Line* icon is activated, and in the *Line Definition* selection box (Figure 3.628), the *Normal to surface* type is chosen, then the *InputSurface* feature is selected from the specification tree in the *Surface* field, and the starting point is *Project.1*.

The two fields *Start* and *End* are completed with values of 0 and 10 mm, respectively. The significance of the value 0 mm in the *Start* field is that *Line.2* will start right from this point, and if the user had entered another value, then the line would have been drawn from a different position than *Project.1*, but in the same perpendicular direction on the surface.

The figure also shows that the orientation of the line is from the point upwards (red arrow). A new *Point.4* is added to the end of the line, similar to the previous ones.

Through the lines *Line.1* and *Line.2*, the new *Plane.1* is created (Figure 3.629). The option used in the *Plane Definition* selection box is *Through two lines*, and then the two lines are chosen. Depending on the first selected line, the plane symbol is positioned on it.

To simplify the representations, the features *Point.1*, *Point.2*, *Line.1*, *Point.3*, but also the *XY*, *YZ* and *ZX Planes* are hidden from the specification tree. In the *Point.4* at the end of the *Line.2*, *Line.3* is drawn, perpendicular to the *Plane.1* and having the ends on either side, symmetrical with respect to *Point.4*. Thus, in the *Line Definition* selection box (Figure 3.630), the user chooses the *Point-Direction* option, the point for line positioning, the direction plane, enters the value of 5 mm in the *End* field and checks the *Mirrored extent* option. It is observed that a plane has been chosen to establish a direction, meaning that the line will not be contained by it, but will be drawn perpendicular to the specified plane. If possible, in other situations, the user has the option to choose an existing line in the drawing or an *X*, *Y* or *Z* axis from the context menu of that field.

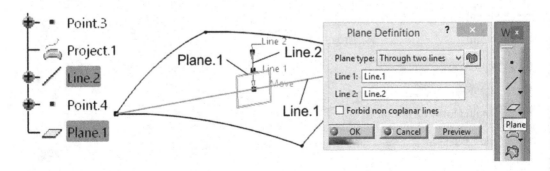

FIGURE 3.629 Inserting the *Plane.1* through two lines.

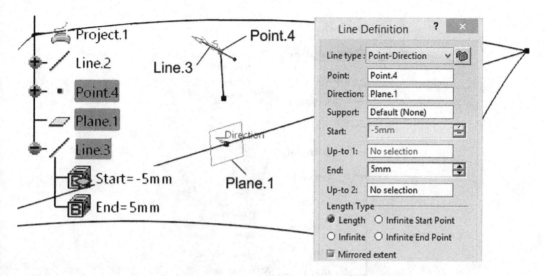

FIGURE 3.630 Drawing the *Line.3* in *Point.4* and perpendicular to *Plane.1*.

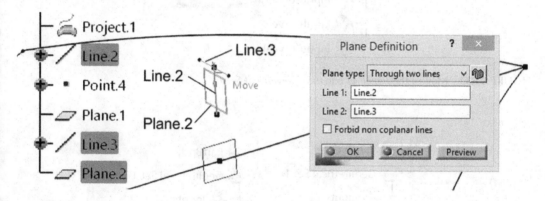

FIGURE 3.631 Defining the *Plane.2* through *Line.2* and *Line.3*.

The *Line.3* in Figure 3.631 is 10 mm long, according to the distances of 5 mm each on both sides of the *Point.4*. *Line.2* and *Line.3* are used to insert the *Plane.2* in a similar way to creating the *Plane.1*.

In the specification tree, the user selects the features *Point.1*, *Point.2*,…, *Line.3*, *Plane.2* by holding down the *Shift* key and choosing the first and last features from the list. With these ten features selected, the user accesses the *Geometrical Set* option from the *Insert* menu and enters the *Panel* name for this geometric set.

In the specification tree, the new set is placed after the *InputCurve.4* feature and it is initially a part of the *Inputs* geometric set. However, the user will move the *Panel* set to the *Geometrical Set.1* so that the sets *Inputs* and *Panel* are at the same level. To do this, from the context menu of the *Panel* set, he chooses the *Cut* option, *Panel* temporarily disappears from the specification tree, and then the user selects *Geometrical Set.1*, right-clicks and, from the context menu, chooses the *Paste* option (Figure 3.632). After moving, the components of the *Panel* set change their numbering; for example, *Project.1* becomes *Project.2*, *Line.2* becomes *Line.5*, *Plane.1* becomes *Plane.3*, and *Plane.2* becomes *Plane.4*.

It is easy to see how features of the specification tree can be defined, copied, arranged, and so on, with the help of context menus. At this stage of modelling, *Geometrical Set.1* contains the *Inputs* and *Panel* sets, according to Figure 3.633.

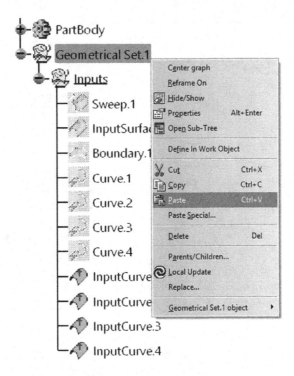

FIGURE 3.632 Moving the *Panel* set into the *Geometrical Set.1*.

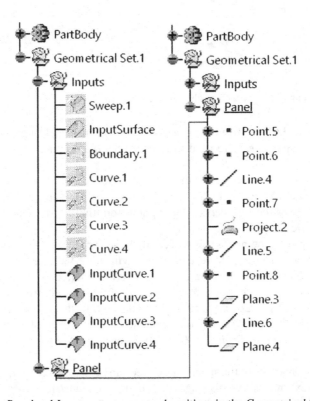

FIGURE 3.633 The *Panel* and *Inputs* sets are on equal positions in the *Geometrical Set.1*.

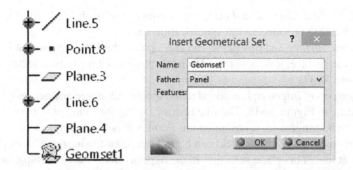

FIGURE 3.634 Placing a geometric set in the specification tree, part of another geometric set.

A new geometric set is inserted in the specification tree (Figure 3.634), as follows: in the *Insert Geometrical Set* selection box, the user enters its name *Geomset1* and then chooses the element/set where to be placed the new set in the *Father* list. In this case, *Panel* is selected and the new geometric set is positioned after the *Plane.4*. In this way, the geometric sets are placed directly and it is no longer necessary to move them later into the specification tree. Note that the set is underlined, so any newly inserted feature will be added to its content.

On the *InputCurve.3* curve, a *Point.9* is added to the middle of the distance between its ends. Thus, the ratio 0.5 is specified in the *Point Definition* selection box (Figure 3.635), and then the correct positioning of the point according to the orientation of the red arrow is observed.

Through this point, the user creates a plane perpendicular to the *InputCurve.3* using the *Normal to curve* option in the *Plane Definition* selection box (Figure 3.636).

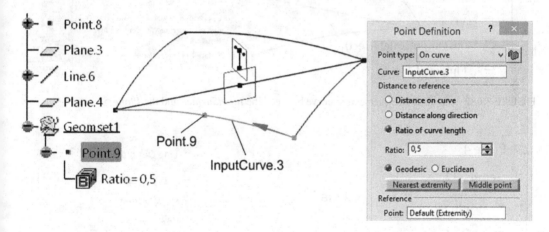

FIGURE 3.635 Inserting the *Point.9* in the middle of *InputCurve.3* using a ratio of 0.5.

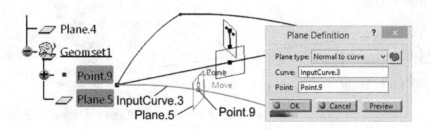

FIGURE 3.636 *Plane.5* is created in the middle of *InputCurve.3* through the *Point.9*.

The two available fields, *Curve* and *Point*, are completed with the features *InputCurve.3* and *Point.9*, and the *Plane.5* is added to the specification tree.

Between the *InputSurface* and this plane, the user applies the *Intersection* tool in the *Wireframe* toolbar. To select the surface, it does not need to be displayed and it can be selected from the specification tree (Figure 3.637).

The intersection curve, *Intersect.1*, is added to the specification tree and acts as a tangent element for *Line.7*, according to Figure 3.638. The line is drawn using the *Tangent to curve* option, the user selects the curve to be tangent to, it has an end at *Point.9*, it is 5 mm long, and it is directed inside the wireframe structure. The *Support* field can be left blank, or the user can choose the *Plane.5* to contain the tangent line. This plane selection in the *Support* field is also a simple way to check the correctness of drawing the curves, points and planes up to the current stage.

From the specification tree, the user can hide the *Intersect.1* curve, then copy the *Geomset1* set to the *Panel* set and change the copy name to *Geomset2* (right-*click* on *Geomset1*, *Copy* option, right-click on *Panel*, *Paste* option, right-click on the copied set, option *Properties*, *Feature Properties* tab, rename set).

The two geometric sets, *Geomset1* and *Geomset2*, are overlapping, so some changes are needed in the second for its repositioning. The definition of the *Geomset1* set began with choosing the *InputCurve.3* to position the *Point.9*. The user accesses the context menu of the set (Figure 3.639), chooses the option *Geomset2* object → *Edit Inputs* and opens the selection box *Definition of Geomset2*.

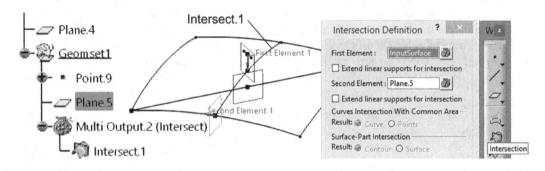

FIGURE 3.637   Creating the intersection curve between the *InputSurface* and the *Plane.5*.

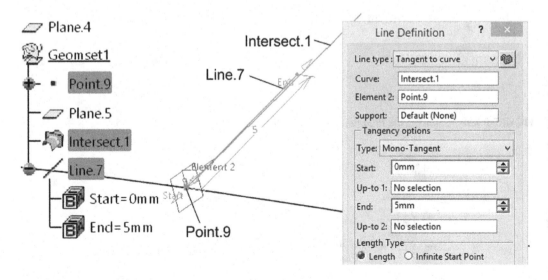

FIGURE 3.638   Drawing the *Line.7* tangent to the *Intersect.1* curve.

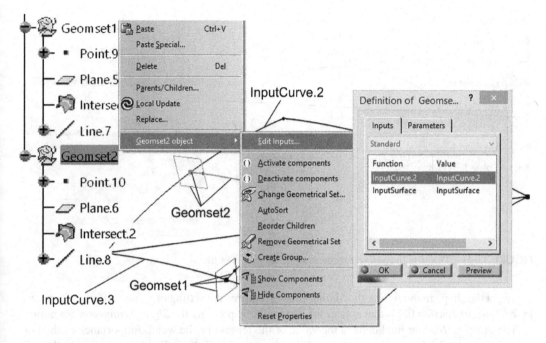

FIGURE 3.639 Editing the inputs of the *Geomset2*.

The *InputCurve.3* curve is present on the two *Function* and *Value* columns, but the user will choose *InputCurve.2* according to the figure.

It is observed that the set *Geomset2* moves on this curve and its component features have the numbering incremented by 1 for names: *Point.10, Plane.6*, etc. This is correct because the second geometric set, although obtained by copying, cannot contain the same features, and they are multiplied by copying for each set.

The user repeats the previous steps for copying geometric sets, changing the name and the supporting feature, to get *Geomset3* on the *InputCurve.1* and *Geomset4* on the *InputCurve.4*. Figure 3.640 shows the four geometric sets positioned on the boundary curves of the *InputSurface*.

To add the next features to the *Panel* geometric set, the user chooses the *Define In Work Object* option from its context menu. The *Panel* set, thus, becomes the current working set, and the user draws a curve, *Spline.1*, through the endpoints of the lines *Line.7, Line.6* and *Line.10*, according to Figure 3.641.

In the *Spline Definition* selection box, points 1, 2 and 3 are specified in the *Points* column, and in the *Tangents Dir.* Column, three planes: *Plane.5, Plane.4* and *Plane.8*, corresponding to the lines. Basically, the curve is perpendicular to these planes at those points.

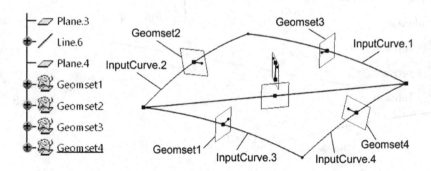

FIGURE 3.640 Placing the four geometric sets on the boundary curves of *InputSurface*.

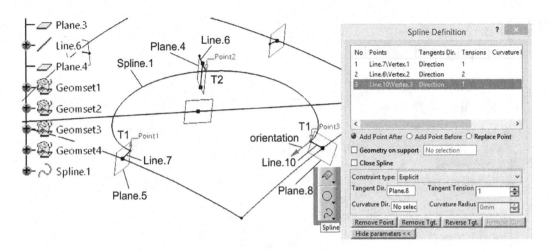

FIGURE 3.641   Drawing the *Spline.1* through some endpoints of lines.

To get the shape in the figure, the *Spline.1* curve requires the setting of values different from zero in the *Tangent Tension* fields that appear for each point by pressing the *Show Parameters* >> button.

The *Tangent Tension* field defines the value of the tension or the weight/importance of the tangent direction. A higher tension value results in the drawing of the *Spline.1* curve more in the tangent direction, which is, of course, always along the curve. Its ends receive the value 1 (noted as *T1*), and the midpoint, the value 2 (noted as *T2*). The tension values are listed in the *Tensions* column. The figure also shows the red arrow of the curve orientation.

The geometric sets that contain the features on which *Spline.1* was drawn are *Geomset1* (*Plane.5* and *Line.7*), *Panel* (*Plane.4* and *Line.6*) and *Geomset4* (*Plane.8* and *Line.10*), according to the explanations and the wireframe modelling method presented above.

Similarly, the *Spline.2* curve is drawn symmetrically with respect to the *Plane.3* (formerly *Plane.1*), using the features selected from Figure 3.642.

The geometric sets that contain the features specified in the figure are *Geomset3* (*Plane.7* and *Line.9*), *Panel* (*Plane.4* and *Line.6*) and *Geomset2* (*Plane.6* and *Line.8*).

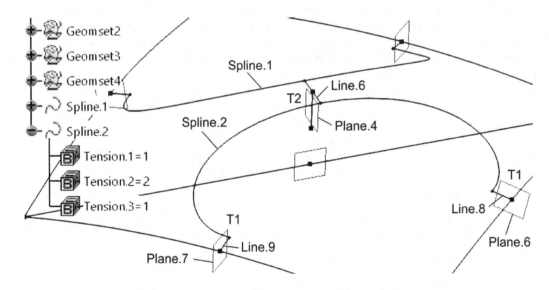

FIGURE 3.642   Drawing the *Spline.1* through some endpoints of lines.

Using the curves that delimit the part on its diagonal, a surface *Fill.1* is created, according to Figure 3643. The user opens the *Fill Surface Definition* selection box of the *Fill* tool and selects in order all the ten curves *InputCurve.1,..., Line.9*. On the *Supports* column, the user selects the *InputSurface* for each *InputCurve*. Thus, the surface obtained, *Fill.1*, becomes tangent to the surface *InputSurface* in the area of those curves.

The selection should be made very carefully to obtain a closed contour, and the confirmation is given by the message *Closed Contour* written in white on a light blue background that appears after the selection of the *Line.9* (last feature/curve in the list). Note that although the *InputCurve.1,..., InputCurve.4* are longer than necessary, *CATIA* uses only the portions needed to close the contour, and the user no longer has to edit/shorten those curves.

The surface *Fill.1* is added to the specification tree after the *Spline.2* curve and within the *Panel* geometric set. Figure 3.644 also shows how to transform the surface into a solid feature using the

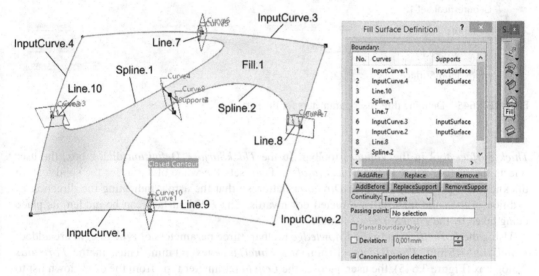

FIGURE 3.643   Creating the *Fill.1* surface by filling a closed contour delimited by curves.

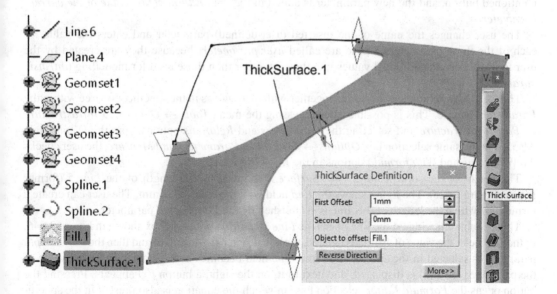

FIGURE 3.644   Adding thickness to the *Fill.1* surface.

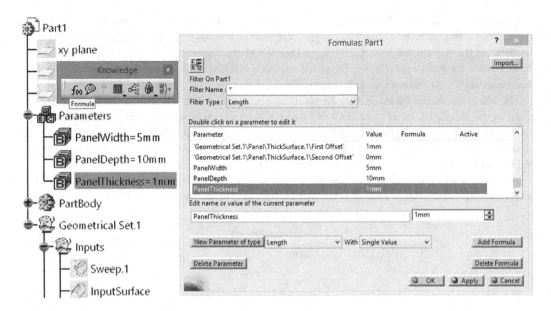

FIGURE 3.645   Defining three user parameters for the part.

Thick Surface tool in the *Volumes* toolbar. In the *ThickSurface Definition* dialog box, the user selects the surface *Fill.1* in the *Object to offset* field, sets the value of 1 mm for the solid feature thickness and presses the *Reverse Direction* button so that the arrow indicating the direction of addition of the solid material is directed downwards. The *Fill.1* surface can be hidden, its place being taken by the *ThickSurface.1* feature.

Using the *Formula* icon in the *Knowledge* toolbar, three parameters of type *Length* are added: *PanelWidth* = 5 mm, *PanelDepth* = 10 mm and *PanelThickness* = 1 mm. Thus, in the *Formulas* dialog box (Figure 3.645), the user chooses the *Length* parameter type from the drop-down list to the right of the *New Parameter of type* button and the *Single Value* option. The user presses the mentioned button, and the new parameter is added in the field *Edit name or value of the current parameter*.

The user changes the name of the inserted (user-defined) parameter and enters the value for each of the three parameters. These are called *user parameters* because they are created by the user to insert certain additional values into the part, other than those used for modelling (*intrinsic parameters*).

The three user parameters are added together with their values in the specification tree within the *Parameters* feature. This is possible after accessing the menu *Tools → Options → Infrastructure → Part Infrastructure* and selecting the *Parameters* and *Relations* options from the *Display* tab. Also, from the same selection box *Options → General → Parameters and Measure*, the user checks the *With value* and *With formula* options in the *Knowledge* tab.

The height of the solid feature, *ThickSurface.1*, is given by the length of the *Line.5* (former *Line.2*; Figure 3.628). The figure shows the *End* field with the value of 10 mm. The user can create a formula by which the length of this line is established by the *PanelDepth* parameter.

Thus, double-clicking on the line opens the *Line Definition* selection box shown in Figure 3.646. In the *End* field, the value of 10 mm is deleted, the equal sign (=) is inserted, and then the *PanelDepth* parameter is selected in the specification tree. The field becomes non-editable, the initial value of the parameter (10 mm) is displayed, and next to it, on the right, a button *f(x)* appears. Pressing the button opens the *Formula Editor* selection box, in which the equation is also present in the specification tree within the *Relations*: *Geometrical Set.1\Panel\Line.5\End = PanelDepth*.

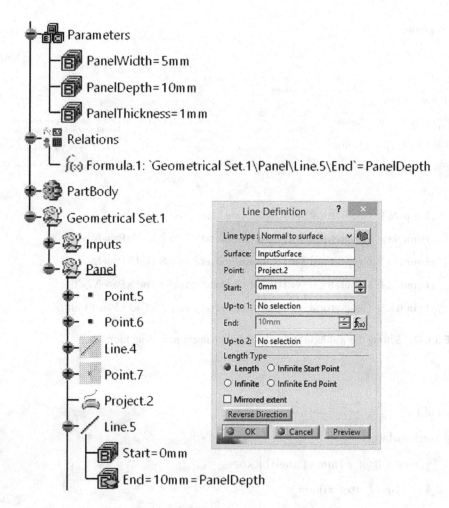

FIGURE 3.646 Defining the length of a line with a user parameter.

By modifying (double-clicking) the user parameter, the height of the *Line.5* and, implicitly, the surface *Fill.1* and the solid *ThickSurface.1* are edited. The line length can no longer be changed directly by the user by editing in the *Line Definition* selection box, but only via the *PanelDepth* parameter. If, however, at some moment, the user wants to free the line length from the influence of this parameter, he must disable the formula using its context menu in the specification tree.

Similarly, the lengths of the lines *Line.7* (*Geomset1*), *Line.8* (*Geomset2*), *Line.9* (*Geomset3*) and *Line.10* (*Geomset4*) are constrained by the parameter *PanelWidth*=5mm (Figure 3.647). The figure shows the four formulas (*Formula.2,...*, *Formula.5*) that determine the width of the ends of the ornament part. It is also observed that the formulas are active and the complex names of the intrinsic parameters of the lines contain their names, but also the geometric sets that contain the lines.

The thickness of the solid feature is given by the intrinsic parameter within *ThickSurface.1*. The value of this parameter is constrained to be equal to that of the user parameter *PanelThickness*=1 mm. In the *First Offset* field of the *ThickSurface Definition* dialog box (Figure 3.648), the user deletes the initial value of 1 mm, enters the equal sign (=) and selects the *PanelThickness* parameter in the specification tree.

The user also notices the locked field on the parameter value and the *f(x)* button that opens the formula: *Geometrical Set.1\Panel\ThickSurface.1\First Offset = PanelThickness*.

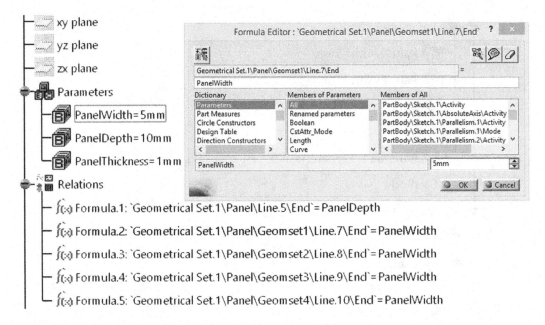

FIGURE 3.647 Setting the widths of the ends of the ornament part using formulas.

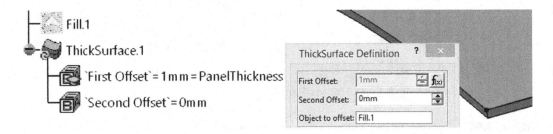

FIGURE 3.648 Setting the surface thickness of the part using a formula.

Figure 3.649 shows the six formulas that parameterize the ornament part, but also a representation of it for other values of the user parameters. Basically, with the help of the three parameters, the shape of the ornament part can be easily changed, even by other users who do not know the locations in the specification tree of the intrinsic parameters.

The solid feature shown in Figure 3.649 does not represent the final model of the ornament part, but only a part (half) of it. For the rest of the model, the user can resume wireframe and surface modelling in all stages up to this stage. Another option is to use and edit what already exists in the specification tree.

Thus, the user selects (by holding down the *Shift* key) the first and last feature within the *Panel* geometric set. According to Figures 3.646 and 3.648 above, the first feature is *Point.5* and the last is *ThickSurface.1*.

Having selected the 18 features, the user creates a new geometric set, called *HalfPanel1* (Figure 3.650). The *Insert Geometrical Set* selection box shows the name of the set, the features it contains (*Point.5*, *Point.6*, *Line.4*, etc.), and also the geometric set in which it is positioned: *Panel*.

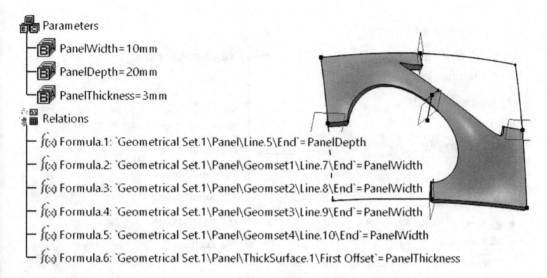

FIGURE 3.649 The six formulas that parameterize the part.

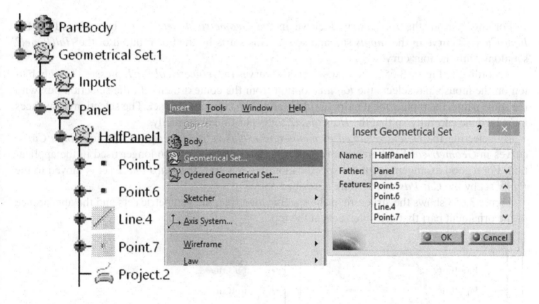

FIGURE 3.650 Inserting the *HalfPanel1* geometric set within *Panel* set.

To create by copying the rest of the solid part, it is necessary to multiply the *HalfPanel1* set, but also other curve features (*InputCurve.1,…, InputCurve.4*) from the *Inputs* set. This is because the *InputCurve* curves were used to create the features in *HalfPanel1*.

The user selects the four curves in the *Inputs* set and *HalfPanel1* in the *Panel* set by holding down the *Ctrl* key, and then he performs a right-click and, from the context menu, chooses the *Copy* option (Figure 3.651). From the context menu of *Geometric Set.1*, the user chooses the *Paste* option to finish copying the four curves and the *HalfPanel1* set, as shown in the figure.

The *HalfPanel1* set is renamed to *HalfPanel2*, and the four curves that accompany it must be redefined. At this point, the *InputCurve.1* curve in the *Inputs* set overlaps with the *InputCurve.1* in the *Geometrical Set.1*, as do the others, and *HalfPanel2* set is placed over the *HalfPanel1* set due to this overlap of the curves.

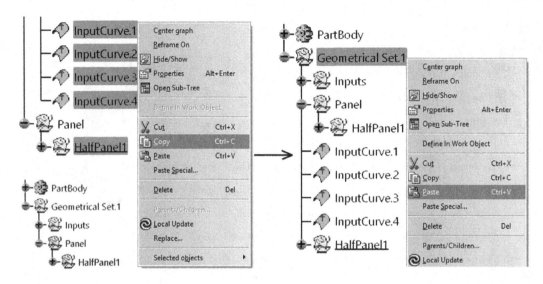

FIGURE 3.651 *Copy-Paste* of the *HalfPanel1* set and of the four *InputCurve* curves to the *Geometrical Set.1.*

For this reason, the *InputCurve.1* curve in the *Geometrical Set.1* must be replaced by the *InputCurve.2* curve in the *Inputs* set, and so on. This starts by the 90° rotation of the *HalfPanel2* set along with the four curves.

According to Figure 3.652, the user selects the curves in *Geometrical Set.1*, presses the right button on the mouse and selects the *Replace* option from the context menu. In the selection box with the same name, the replacement pairs (*Replace* and *With* columns) are set. The selection of surfaces must be done carefully so that the *HalfPanel2* set is positioned correctly.

By selecting the *Delete replaced elements and exclusive parents* option, the four *InputCurve* curves in *Geometrical Set.1* are removed after replacement and are no longer used in the application. For a good arrangement of the specification tree features, the *HalfPanel2* set is moved to the *Panel* set by the *Cut-Paste* procedure described above.

Figure 3.653 shows the structure of the specification tree, the geometric sets and the appearance of the ornament part that self-intersects at the top.

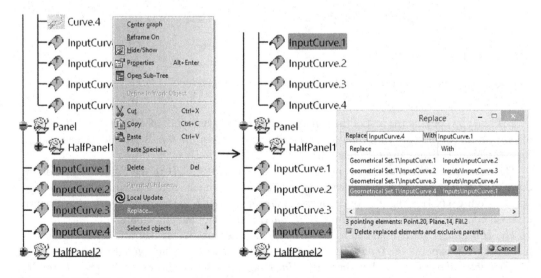

FIGURE 3.652 Replacing the *InputCurve* curves of *Geometrical Set.1* with the curves of the *Inputs* set.

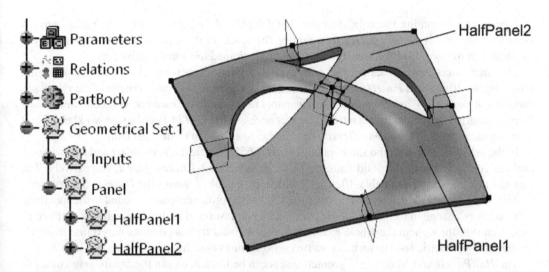

FIGURE 3.653 The ornament part self-intersects at the top.

To correct the solid *ThickSurface.3* feature of the *HalfPanel2* set, the user must edit two wire-frame features in this geometric set. Thus, the line constrained by the *PanelDepth* user parameter is identified in the set as *Line.19*.

The numbering and the line name were changed by copying *HalfPanel1* → *HalfPanel2* and moving the latter to the *Panel* set. Due to the association of the line with that parameter, the editing of *ThickSurface.3* is very simple. The user cannot change the line length (the *End* field is not editable), but he can change the drawing direction by pressing the *Reverse Direction* button in the *Line Definition* selection box (Figure 3.654) that appears by double-clicking on the line.

The *Point.24*, placed at the end of the line, can also be similarly edited by pressing the *Reverse Direction* button in its definition box.

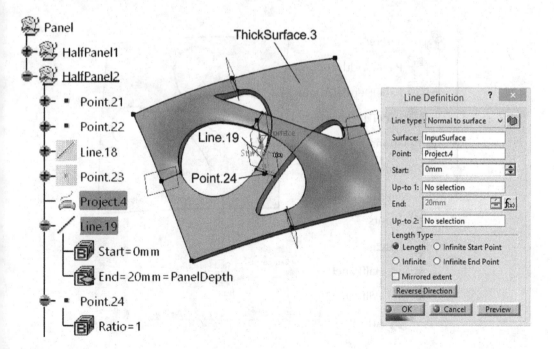

FIGURE 3.654 Changing the drawing direction of the *Line.19*.

Figure 3.654 contains the solid features *ThickSurface.1* and *ThickSurface.3*, parameterized using formulas involving user parameters. Editing the values of these parameters leads to changing the shapes of the two solid features, but they need to be joined to form the solid part.

The user enters new values for the ornament part parameters (example: *PanelWidth*=6 mm, *PanelDepth*=10 mm and *PanelThickness*=2 mm) and, then from the context menu of the *Panel* set, chooses the *Define In Work Object* option. The panel becomes underlined and will contain any new features inserted in the specification tree. The user accesses the *CATIA Part Design* workbench, and from the menu *Insert → Boolean Operations*, he chooses the icon of the *Add* tool.

In the selection box with the same name (Figure 3.655), the fields *Base object* and *Added operand* are completed with the solid features *ThickSurface.3* and *ThickSurface.1*, respectively. The user can also see the non-editable *After* field, which, by default, contains the *Panel* geometric set.

The *Add.1* feature, which represents the final solid of the ornament part, is added to the specification tree. Any change in the values of the parameters is transmitted to the change in the part geometry, even if it, throughout the whole modelling process, went from a complex wireframe structure divided into geometric sets, to surfaces, to two solid features and, finally, to a single solid.

The *HalfPanel1* and *HalfPanel2* geometric sets can be hidden, as can the *InputCurve* curves in the *Inputs* set.

Figure 3.656 shows three variants of the part for three different sets of parameters. It is observed that the shape and overall dimensions of the part do not change because they were not included in the parametric modelling, but only the loops that form the inner surfaces.

The example of parametric hybrid modelling presented in this application is of high complexity, requiring the user's attention in establishing the wireframe structure, copying/moving geometric sets and curves, etc. The names of geometric features (points, planes, lines, curves, etc.) often change automatically during such a modelling approach, so there may be some differences between the written text of this application and the part created by the user.

Also, the user can try to model the ornament part by combining other methods, in a variant with a round base shape, without copying geometric sets, etc.

One modelling solution can be found here: https://youtu.be/HMQyIJgMgJ8.

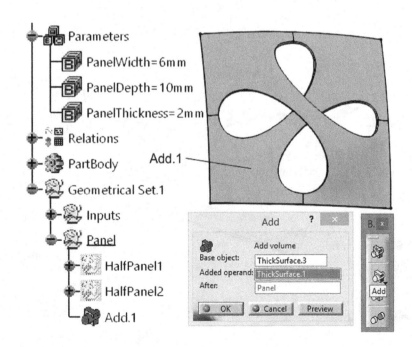

FIGURE 3.655   Joining two solid features to create the ornament part's final body.

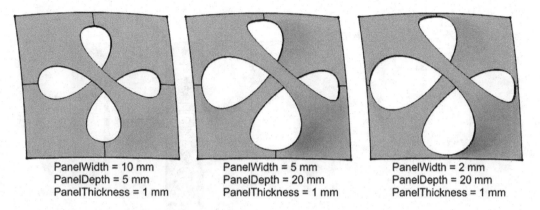

PanelWidth = 10 mm	PanelWidth = 5 mm	PanelWidth = 2 mm
PanelDepth = 5 mm	PanelDepth = 20 mm	PanelDepth = 20 mm
PanelThickness = 1 mm	PanelThickness = 1 mm	PanelThickness = 1 mm

FIGURE 3.656 Three variants of the part for three different sets of parameters.

3.26 MODELLING OF PARAMETRIC BELLOWS IN DIFFERENT CONSTRUCTIVE SOLUTIONS

The bellows is a flexible element made of rubber, leather or thick cloth, rubberized to protect and cover cables, hinges and moving parts or to connect moving or fixed elements, but through which a certain fluid or gas passes. The sizes and shapes of the bellows are very diverse depending on the functional role and applicability. There are, therefore, bellows elements that provide the moving parts between the wagons of a train and tram, bellows that connect two pipes of an engine, bellows that protect some axle of a car, etc. Also, few musical instruments use this type of flexible element.

Figure 3.657 shows three variants of bellows represented isometrically and without dimensions. Following the explanations and the modelling manner in this tutorial, the user will be able to 3D model any other shapes and types of bellows.

The user opens the *GSD* workbench and, then, in the *Sketch.1* of the *XY Plane*, draws a centred rectangle, with the intersection point of the diagonals at the origin of the coordinate system and the sides equal to 120 mm. The four corners of the square are rounded with R20 mm using the *Corner* tool in the *Operation* toolbar, according to Figure 3.658.

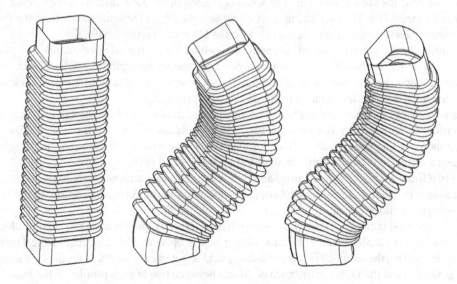

FIGURE 3.657 Representations of the bellows part to be modelled.

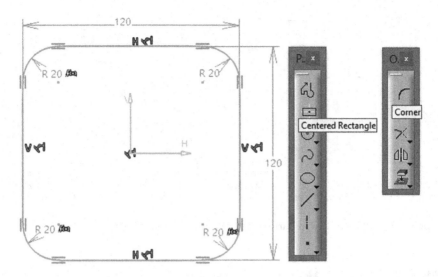

FIGURE 3.658   Centred rectangle with rounded corners.

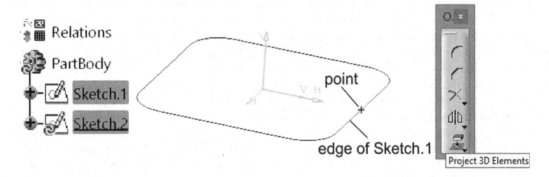

FIGURE 3.659   Projecting the edge of *Sketch.1* in the *YZ Plane* as a point of the *Sketch.2*.

The user exits the sketch, and then a new one is created in the *YZ Plane*, named *Sketch.2*. From an isometric view of the *Sketch.1*, the user chooses the right side of the square and clicks the *Project 3D Elements* icon to project the selected side into the *Sketch.2* (Figure 3.659).

According to the planes selected to create the two sketches, the edge of *Sketch.1* is perpendicular to *Sketch.2*, so the projection will be a point marked in the figure. By default, it is of standard construction, but the user turns it into auxiliary construction by selecting it and clicking the *Construction/Standard Element* icon in the *Sketch Tools* toolbar.

The user checks that the icon has been deactivated and draws from this point two oblique lines, symmetrical to an auxiliary construction line. This one is also drawn by the user, horizontally, through the right end of the bottom line. An angle of 45° is established between the two oblique lines, and a distance of 20 mm between their left ends (Figure 3.660).

The two lines are rounded to R5 mm, and then the lines and the resulted arc are selected (multiple selection by holding down the *Ctrl* key) and multiplied in the vertical direction, in the sketch, by pressing the *Translate* icon.

There is no field in the *Translation Definition* box to select the profile to be moved or multiplied, so the selection is usually made before the dialog box is opened. The user checks the *Duplicate mode* option, enters the value 20 in the *Instance(s)* field to get twenty identical profiles (clones) with the original one, and then sets the distance of 20 mm between two of such profiles in the *Value* field. This value must be equal to the one chosen by the user between the lines ends in Figure 3.661.

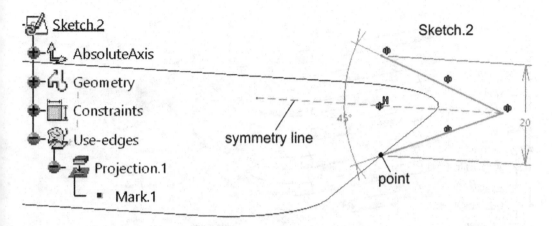

FIGURE 3.660 Drawing of two oblique lines, symmetrical to an auxiliary horizontal construction line.

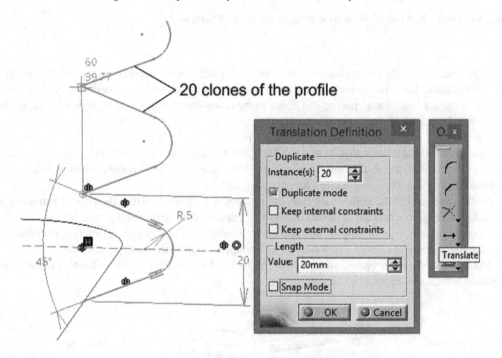

FIGURE 3.661 Creation of twenty more clones above the initial profile of *Sketch.2*.

The user clicks the *Exit workbench* icon and notices that the specification tree contains two sketches in perpendicular planes. The *Sweep* tool in the *Surfaces* toolbar is applied between the sketches, and the *Swept Surface Definition* dialog box opens (Figure 3.662). From the list of *Profile type* icons, the user chooses the first, *Explicit*, and then from the drop-down list with *Subtype* options, *With reference surface*. Thus, the surface is perpendicular to the *XY Plane* in which the *Sketch.1* is selected in the *Guide curve* field. The *With pulling direction* option can also be chosen, but must be followed by selecting the *Z* axis in the *Direction* field. In the *Profile* field, the user selects the profile *Sketch.2* and, pressing the *Preview* button, displays the *Sweep.1* surface in Figure 3.662.

Instead of choosing the *Z* axis as the surface creation direction, the user has the option to also select the *XY Plane* and thus specify to the program that the direction is, in fact, normal to the chosen plane. The user should be sure that the normal is the *Z* axis.

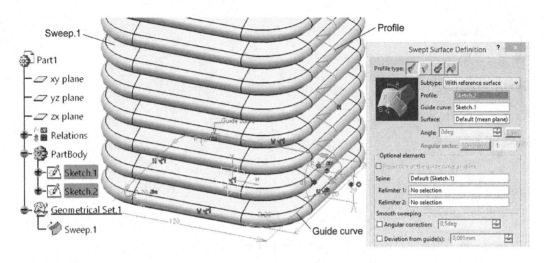

FIGURE 3.662   Preview of the swept surface in the shape of bellows.

In practice, there are often situations in which a surface, for example, must be created along a certain direction, but this one does not coincide with any of the axes of the coordinate system, nor it is easy to draw a line to materialize it. Thus, a plane or another surface that is perpendicular to the respective direction can be used for selection.

The *Sketch.2* can be hidden, but the *Sketch.1* profile at the base of the *Sweep.1* surface is kept visible. It is extruded downwards, below the *XY Plane* at a distance of 60 mm. The *Extrude* tool was used (Figure 3.663), the *Z* axis was chosen as the direction, and the *Reverse Direction* button was pressed because the implicit direction for extrusion was *Z+*.

At the upper end of the bellows surface, the procedure is similar, so that the edge is extracted as *Boundary.1* (Figure 3.664). It is observed that this curve is continuous and green. In this case, the user can choose any type of propagation (*Complete boundary*, *Point continuity* or *Tangent continuity*), the result being the same: extracting the entire upper edge of the *Sweep.1* surface.

There is also the *No propagation* option in the list, and its use leads to the extraction of only one segment, which was chosen by the user.

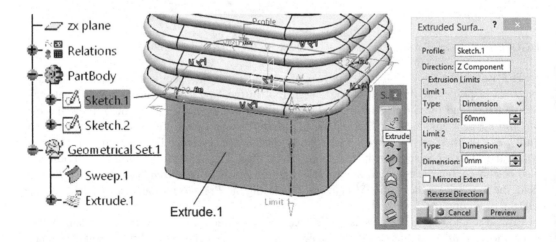

FIGURE 3.663   Extruding the *Sketch.1* below the *XY Plane*.

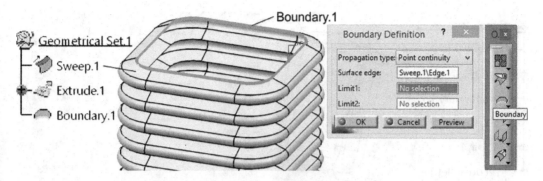

FIGURE 3.664 Extracting the upper edge of the *Sweep.1* surface as *Boundary.1* curve.

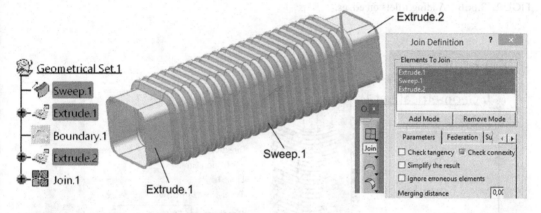

FIGURE 3.665 Joining three surfaces to create a single surface – *Join.1*.

The extracted edge is extruded over the same distance of 60 mm along the *Z* axis, and the *Extrude.2* surface is obtained. The specification tree contains three surfaces, according to Figure 3.665. By clicking the *Join* icon, *CATIA* opens the *Join Definition* selection box and the user adds those surfaces in the *Elements To Join* list. In the options list, he ticks the *Check connexity* option to verify if the *Join.1* surface is continuous.

The *Join.1* surface is added to the specification tree, but it is still necessary to fillet the edges between the former *Extrude* and *Sweep* surfaces. This fillet is possible only after grouping the surfaces into *Join.1*. From the *Operations* toolbar, the user selects the *Edge Fillet* tool and opens the selection box in Figure 3.666. In the *Radius* field, he enters the value of the radius R5 mm and selects the two edges of the *Join.1* surface. Selecting an edge can also be done by selecting only one element, line or arc, followed by choosing *Tangency* in the *Selection mode* drop-down list.

The first version of the bellows is completed and will be used to create a new one through the following explanations.

In the *Sketch.3* of the *YZ Plane*, the user draws a vertical line coinciding with the *V* axis of the bellows coordinate system. The line protrudes beyond its ends, and then the user exits the sketch and extrudes the line on either side of it with 120 mm to create the *Extrude.3* surface, according to Figure 3.667.

In the *Extruded Surface Definition* dialog box, the user selects the sketch that contains the line in the *Profile* field, right-clicks the extrusion direction in the *Direction* field, sets the value to 120 mm and checks the *Mirrored extent* option. It is recommended to hide the *Sketch.3* vertical line and change the colour of the *Extrude.3* surface from the context menu of each item.

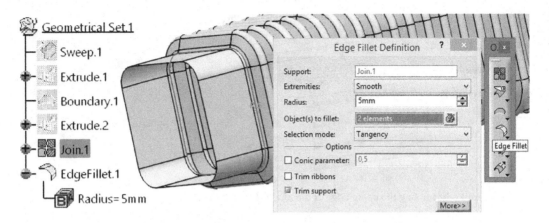

FIGURE 3.666   Adding fillets on edges.

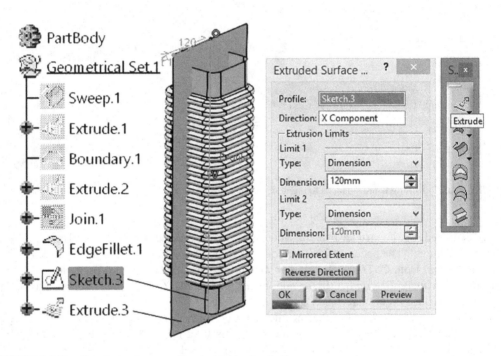

FIGURE 3.667   Creating a planar surface inside the first bellows.

In the same *YZ Plane*, in the *Sketch.4*, a spline curve is drawn, next to the surface *EdgeFillet.1* which represents the bellows. The curve should be similar in size to the *Sketch.3* line and relatively smooth, with no sudden changes in direction and small radii. Outside the *Sketch.4*, this spline curve is extruded and the *Extrude.4* surface is obtained in Figure 3.668. The extrusion is similar to that in which the *Extrude.3* surface was created (Figure 3.667), with the same values of 120 mm.

The user accesses the *Wrap Surface* icon on the *Advanced Surfaces* toolbar and opens the *Wrap Surface Deformation Definition* selection box shown in Figure 3.669. The three fields are filled, by selection, with existing surfaces in the specification tree, as follows: in the *Element to deform* field, the *EdgeFillet.1* surface of the first bellows variant is chosen, *Extrude.3* is selected as the reference surface, and *Extrude.4* is specified as the surface whose shape is to be taken.

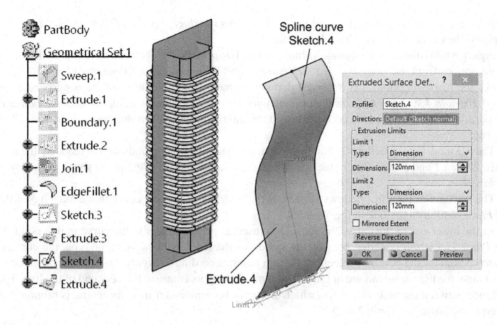

FIGURE 3.668 Extrusion of the spline curve from *Sketch.4*.

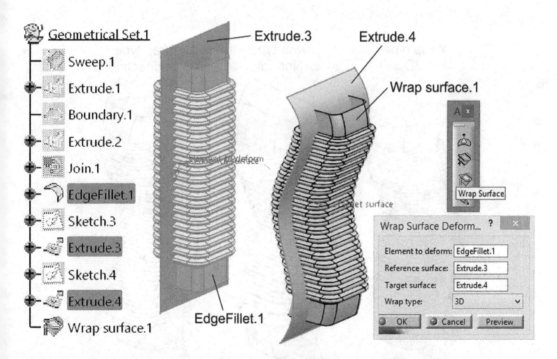

FIGURE 3.669 Applying the *Wrap Surface* tool to obtain the second bellows variant.

A first solution, corresponding to the *3D* option in the *Wrap type* list, is displayed by pressing the *Preview* button. Both *EdgeFillet.1* and *Wrap surface.1* are visible, but by pressing the *OK* button and confirming the options presented, the *EdgeFillet.1* (first bellows) surface is hidden, but *Extrude.3* remains visible.

It is observed that, in fact, the resulting surface, *Wrap surface.1* is *EdgeFillet.1*, which has been imposed the curvature of the *Extrude.4* surface.

Figure 3.670 shows three possible variants of the *Wrap surface.1* obtained by applying three different options from the *Wrap type* list: *3D*, *Normal* and *With direction: Y Component*.

For the *Normal* option, it is observed that any element of the bellows surface is perpendicular to the *Extrude.4* surface. Also, in the case of *With Direction: Y Component*, the bellows elements are parallel to the *Y* axis of the coordinate system, although they follow the shape of the *Extrude.4* surface.

The 3D model of the bellows is completely parametric, so that any modification made by the user to the *Sketch.1* and *Sketch.2* or to the spline curve in *Sketch.4* is reflected in the shape of the *Wrap surface.1*.

The creation of the third bellows variant starts by drawing a spline curve in the *Sketch.5* of the *YZ Plane* similar to the one in the *Sketch.4*.

At the lower end of the curve, the *Plane.1* is inserted perpendicular to the curve in that point. In the *Plane Definition* selection box, the user chooses the *Normal to curve* plane type, the curve is *Sketch.5*, and for the point, he selects its bottom point, according to Figure 3.671. In the *Sketch.6* of this plane, the user draws an arc of radius R100 mm having its centre at the very end of the line. The distance between the ends of the arc, which are set to be equidistant from its centre, is proposed to be equal to 140 mm (Figure 3.672).

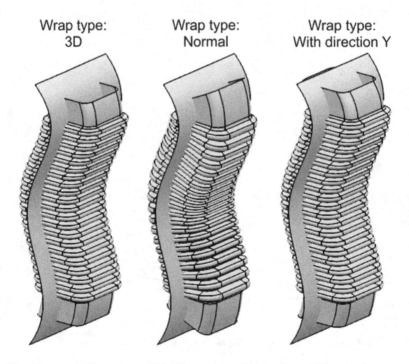

Wrap type: **3D**   **Wrap type:** **Normal**   **Wrap type:** **With direction Y**

FIGURE 3.670   Three possible variants of the *Wrap surface.1* using three different options.

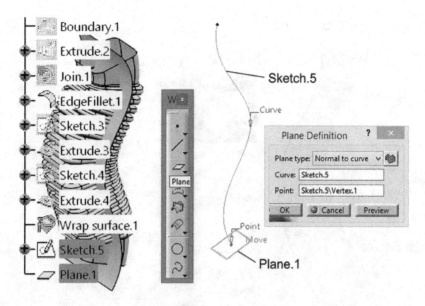

FIGURE 3.671 Creating a new curve in *Sketch.5* and inserting a plane normal to this curve.

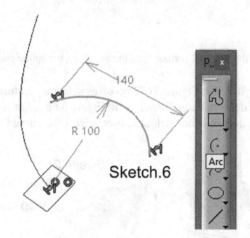

FIGURE 3.672 Drawing an arc in *Sketch.6* with the centre in the endpoint of the curve.

The arc is extruded along the spline curve using the *Sweep* tool. Thus, in the *Swept Surface Definition* dialog box in Figure 3.673, the user chooses the *Profile type*: *Explicit*, and then in the *Profile* field, the sketch containing the arc is selected. As a guide curve, the *Sketch.5* is used. The result, the *Sweep.2* surface, is added to the specification tree, and the two sketches can be hidden from the 3D representation.

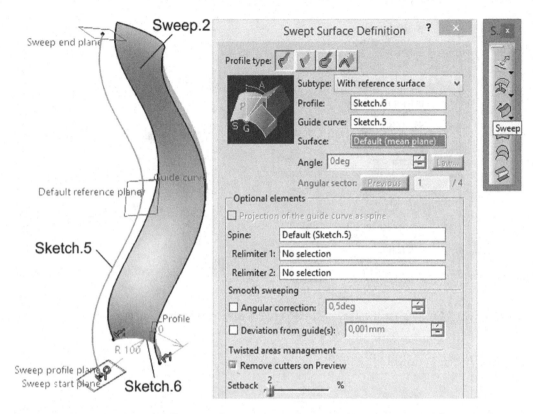

FIGURE 3.673   Creation of the *Sweep.2* surface with the arc along the spline curve.

Obtaining the *Wrap surface.2* in Figure 3.674 involves selecting features from the specification tree as follows: *Element to deform: EdgeFillet.1*, as in Figure 3.669; *Extrude.3* is chosen as the reference surface; and in the *Target surface* field, the user selects the curved surface *Sweep.2*.

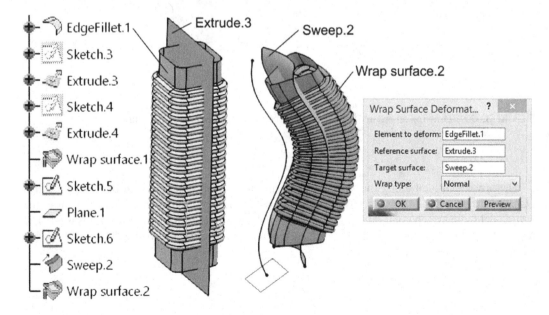

FIGURE 3.674   Creation of the *Wrap surface.2* based on the curved surface *Sweep.2*.

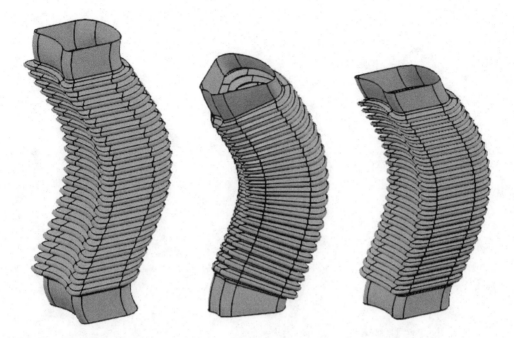

FIGURE 3.675   Three possible variants of the *Wrap surface.2* using three different options.

As in the case of the previous version of obtaining the bellows surface (Figures 3.669 and 3.670), by establishing other options in the *Wrap type* list, different shapes can be created for the *Wrap surface.2*, shown in Figure 3.675.

One modelling solution can be found here: https://youtu.be/uXrFScT8oJU.

4 Parametric Modelling and Sheetmetal Design

4.1 INTRODUCTION TO PARAMETRIC MODELLING OF PARTS AND FAMILIES OF PARTS

Parametric modelling is a process that contributes to the ability to change the shape of a model geometry as soon as a dimension value that defines the model or its features is modified. Parametric modelling is implemented through the design of computer programming code such as a script to define the dimension and the complex shape of the model. This model can be visualized in programs with 3D capabilities to resemble the behaviour characteristics of a real object. Every parametric model uses feature-based modelling tools to manipulate the attributes of the model.

The main advantage of the parametric modelling is, when setting up a 3D geometric model, the shape of its geometry can be changed as soon as the parameters such as the dimensions or curvatures are modified; therefore, there is no need to re-model the part whenever it needs a change. This greatly saves time for engineers, in all the design phases, especially if the model is prone to be changed frequently and needs to be robust. Thus, before the application and development of the parametric 3D modelling programs, changing the shape of a 3D model was very difficult.

Also, in his design activity, the *CATIA v5* user often encounters situations in which he has to create families of similar parts. These parts have, in principle, the same constructive shape, but differ in size or in some specific features that can be deactivated or activated. Also, in the design of the assemblies, it is useful to create a dimensional link between two or more components, so that the modification of one of them will automatically lead to the modification of the others due to the direct association between various *CATIA v5* workbenches. This design manner requires advanced knowledge of modelling, programming and managing parameters and relations, called: parametric-aided modelling (or design).

Parametric modelling can be considered as a process based on algorithmic thinking that enables the expression of parameters, formulas and rules that, together, define, encode and clarify the geometrical or dimensional relation between design intent and design response. It is a paradigm in design where the relations between elements are used to manipulate and assist the design of complex geometries and structures.

The term *parametric* originates from mathematics (*parametric equation*) and refers to the use of certain parameters that can be edited to manipulate or alter the result of an equation or system. While today the term is used in reference to computational design systems, there are precedents for these modern systems in the works of architects, one of the first being Antoni Gaudí (1852–1926).

Currently, there are several software solutions on the *CAD* market that allow parametric modelling of parts and assemblies, but the *CATIA v5* program has a workbench dedicated to identifying, creating, using and verifying the parameters involved in the construction of the 3D models, from sketch to the 3D objects and assemblies. Thus, the *CATIA Knowledge Advisor* workbench allows users to integrate and efficiently use all available data in the design, facilitating technical decisions, reducing the number and influence of possible errors or automating the design with the highest possible productivity.

The user can integrate various data in the design of his products through rules, parameters, formulas, reactions and checks, elements that are considered and used together in a given context.

Parameters and relations are used in the modelling activities to create certain connections/links between the dimensions of the part or assembly at all stages of a new product design.

Parameters contain the properties of a project, and when used in relations, they act as arguments. Parameters are defined by a name, a type and a value, but a relation can be used instead of a value.

DOI: 10.1201/9781003281153-4

In this case, the parameter is constrained by that relation (formula) and the user cannot change its value directly.

The parameters are divided in two categories:

- intrinsic parameters – they belong to the project and depend on the application (and are not visible in the specification tree). The purpose of a *CATIA Knowledge Advisor* application is to create, use and indicate how these parameters can be constrained by relations, rules and reactions;
- user parameters – they are created and inserted by the users to add and define additional information in parts and assemblies (the user parameters are visible in the specification tree).

In most cases, in the modelling process to create a part, the user starts by drawing a sketch, from which he obtains a solid body or a surface (features combined in part), to which various parametric data can be added, which can be modified geometrically, whose material properties are changed, etc. Modifying, adding or deleting a parameter has an effect on the project.

Examples of user-defined parameter types are: *Real, Integer, String, Boolean, Length, Angle, Time, Mass, Volume, Density, Area, Moment of Inertia, Energy, Force, Inertia, Moment, Pressure, Temperature, Frequency, Electric Power*, etc. These parameters can be assigned with single or multiple values, or they can have a range of values.

The parameters are displayed in the specification tree if the *Parameters* option is checked in the list of the menu *Tools → Options → Infrastructure → Part Infrastructure → Display* tab (Figure 4.1). The *CATIA Knowledge Advisor* workbench manages the following:

- formulas, which define how a parameter should be calculated based on other parameters. The formulas use mathematical operators to define the parameters;
- design/parametric tables, which contain data in an ordered form, used to control the parameters of a project;
- rules, a set of instructions that conditionally execute a group of formulas and conditions in a context or depending on the values of certain parameters;
- checks, a set of instructions by which the user is warned whether certain conditions in the rules are met or not. The checks do not change the parameter values.

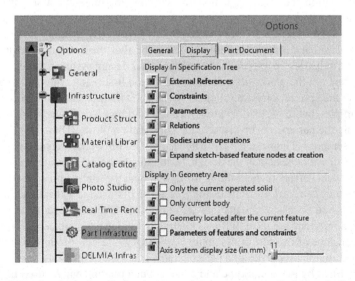

FIGURE 4.1 Options to be checked to display the constraints, parameters and relations in the specification tree.

Formulas are elements used to define or constrain parameters. To write a formula, the user uses parameters, mathematical operators and functions. Once created, a formula can be manipulated like any other feature of the specification tree through its context menu.

The formulas are displayed in the specification tree if the *Relations* option in the menu *Tools → Options → Infrastructure → Part Infrastructure → Display* tab is checked (Figure 4.1).

If a user parameter is constrained by a formula, it is displayed together with the parameter it constrains, if the *With formula* option in the menu *Tools → Options → General → Parameters and Measures → Knowledge* tab is checked (Figure 4.2). Also, in the same menu, the *With value* option is checked to display the value of the respective parameter defined by the formula.

Each formula has an associated parameter named *Activity*. Its value is of *Boolean* type, being defined by *True/False* (active/inactive, 1/0, as it is known in programming), as shown in Figure 4.4.

It is observed that the second formula is marked by a pair of red round parentheses, meaning that it is inactive. The formula can be activated or deactivated in the *Formula object* dialog box of the context menu (Figure 4.3) or by executing a rule.

The reaction is a set of operations related to a component of the project, being performed in response to an event. Any reaction is defined by: the component to which it is applied, a set of operations, and an event that triggers the list of operations.

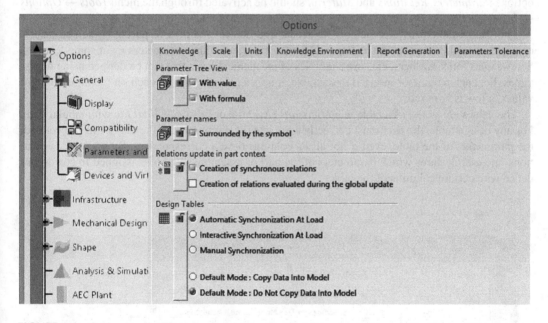

FIGURE 4.2 Options to display parameters with values and formulas in the specification tree.

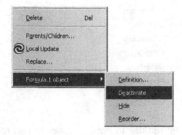

FIGURE 4.3 Deactivation of a formula.

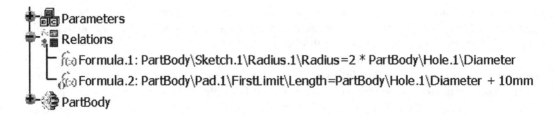

FIGURE 4.4 Displaying of two formulas: the first is active and the second is deactivated.

Each reaction is written in the *Visual Basic Application (VBA)* programming language.

The rules and checks are also relations that can only be created and controlled with the *CATIA Knowledge Advisor* workbench.

Once created, the rule and check can be accessed and used in a similar way to any other feature of the specification tree: by double-clicking, an edit box is displayed. Other options such as *Delete*, *Cut*, *Copy* and *Paste* are also available in the context menu.

The settings in Figures 4.1 and 4.2 are often used for a single part, but for an assembly, the options *Parameters*, *Relations* and *Material* should be activated through the menu *Tools → Options → Infrastructure → Product Structure → Tree Customization* tab (Figure 4.5).

The design table provides methods to create and control families of parts. These can be, for example, mechanical parts that differ only by the values of some parameters or in some features. Screws, nuts, pins, washers, etc., are examples of mechanical parts that can be defined using design tables. For each parameter inserted in the table, it uses a column, and for each set (configuration) of values, a row is required.

The values are stored as a table in a *Microsoft Excel* file or in an *ASCII txt* file with tabular data. For any design table, the user must correctly associate the parameters of the *CATIA v5* project with the parameters of the table, even if not all its columns have a correspondent. By creating associations, the user declares which parameters of the part or assembly should be assigned in correspondence with certain columns of the table.

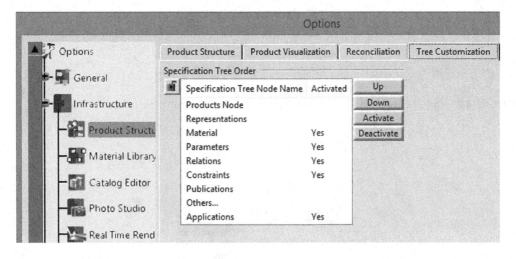

FIGURE 4.5 Activation of important options to have parameters and relations for assemblies displayed in the specification tree.

	A	B	C	D	E
1	Index	D (mm)	d (mm)	A (mm)	B (mm)
2	1	10	5	14	12
3	2	12,056	6,015	13,1	14,75
4	3	8,2	4,1	21,35	15,11
5					
6					

FIGURE 4.6 Example of a design table shown as an *Excel* file.

Figure 4.6 shows the format of the Microsoft Excel file. It is very important that the values mentioned in its cells are expressed in units of measurement (e.g. mm and deg); otherwise, they cannot be correctly associated with the parameters of the *CATIA v5* project.

To demonstrate how the user may set parameters, apply and use relations, rules and design tables, several examples will be presented in this chapter.

4.2 PARAMETRIC MODELLING OF A PART USING FORMULAS AND RULES

In the following tutorial, a simple prismatic part is considered. Its 2D drawing is represented in Figure 4.7. Numerous parameters can be observed in the projection and section views, such as the overall dimensions of the part in section *A-A*, the diameter of the central hole, the dimensions of the side cuts and the length of the part.

Certain relations will be established between these parameters, so that, when modifying some of them, others may be modified according to certain formulas and rules created by the user.

The part is created combining the *CATIA Sketcher* and *CATIA Part Design* workbenches, as follows: in a sketch in the *XY Plane*, the user draws a square with the side of 20 mm, symmetrical with respect to the origin of the coordinate system, and a circle with a diameter of Ø7 mm, with the centre in this origin (Figure 4.8). The profile is extruded with the *Pad* tool and converted into a solid feature over a distance of 50 mm (Figure 4.9). The four parameters that represent the basic of the creation of this solid have already been identified: the dimensions of the sides, the diameter of the circle and the distance of extrusion.

A small rectangle is then drawn on one of the frontal flat faces of the extruded solid, constrained according to Figure 4.10: the length of the rectangle is not very important, but its left side must be inside the flat face at a distance of 4.8 mm from the right side. The width is 6.2 mm, and the two horizontal sides are symmetrical about the horizontal axis of the coordinate system.

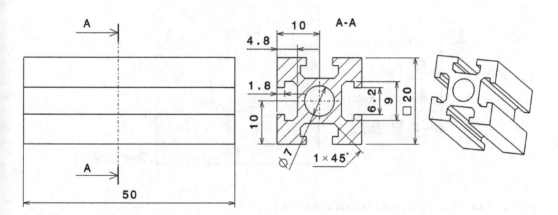

FIGURE 4.7 The 2D drawing of the parametric part.

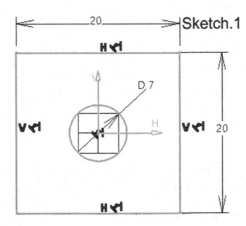

FIGURE 4.8   The first sketch of the part, a square of side 20 mm and a central circle.

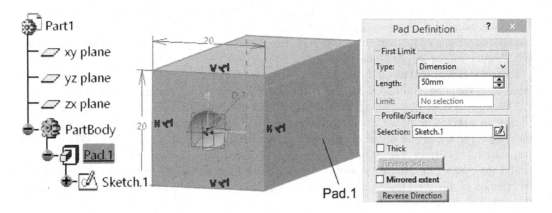

FIGURE 4.9   Extruding the profile from the *Sketch.1*.

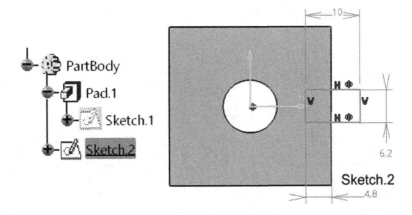

FIGURE 4.10   Drawing a small rectangle in the *Sketch.2*.

Using this profile, a *Pocket* cut is applied to remove a volume of the solid to a depth of 50 mm. It was preferred to use the *Dimension* type in the *First Limit* drop-down list (Figure 4.11), and the user enters this value to keep another parameter available later in the process of creating relations.

The *Pocket.1* cut is multiplied in four copies on the other flat side faces of the solid part. Thus, the *Circular Pattern* tool is applied by the user, according to Figure 4.12.

In the *Circular Pattern Definition* dialog box, in the *Axial Reference* tab, in the *Parameters* field, the user chooses the type *Complete crown* and, in the *Instance(s)* field, enters the value 4, so that a cut will be created on each side of the part. In the *Crown Definition* tab, it is observed that the four cuts are, in fact, arranged on a single circle of radius 20 mm (the *Circle Spacing* field in Figure 4.13). The *CircPattern.1* feature is added in the specification tree.

A rectangular profile with 1.8 mm from the right side of the part is drawn on the same front face of the part, according to Figure 4.14. The left side of the rectangle coincides with the vertical inner edge of a cut, and the horizontal sides are symmetrical about the *H* axis of the coordinate system. The created profile (3×9 mm in size) is also involved in a *Pocket.2* cut to a depth of 50 mm (Figure 4.15).

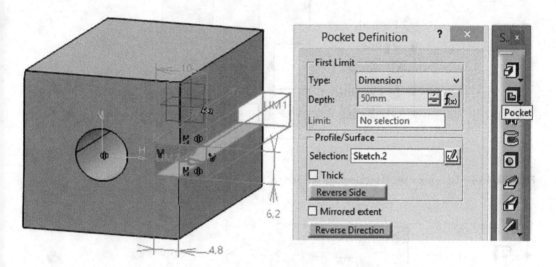

FIGURE 4.11 Creating the first *Pocket* cut on one side of the solid part.

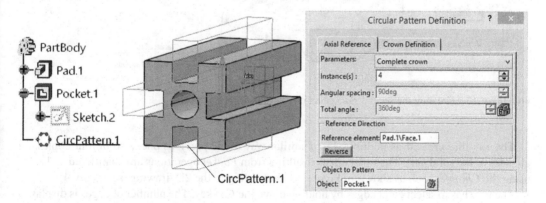

FIGURE 4.12 Multiplication of the *Pocket.1* cut on all four faces of the part.

Circular Pattern Definition   ?   ×

Axial Reference	Crown Definition

Parameters: Circle(s) & circle spacing

Circle(s) : 1

Circle spacing : 20mm

Crown thickness : 0mm

FIGURE 4.13 Parameters of the *Crown Definition* tab.

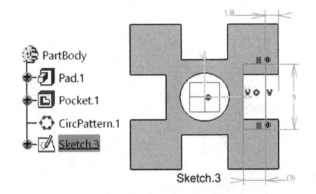

FIGURE 4.14 Drawing a small rectangle in the *Sketch.3*.

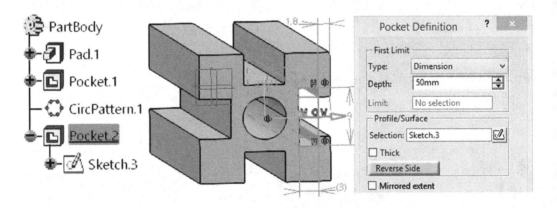

FIGURE 4.15 Creating the second *Pocket* cut on one side of the solid part.

The cut, thus created, is multiplied in a similar way to the previous one, also in four copies, on all sides of the solid part. The inner edges resulting from *Pocket* operations are chamfered as 1×45° using the *Chamfer* tool, as shown in Figure 4.16 according to the 2D drawing, section *A-A*.

The user has to select eight edges by holding down the *Ctrl* key. The number of edges is displayed in the *Object(s) to chamfer* field, and clicking on the icon next to it opens the list of their names.

Following all these operations results in the 3D model of the part, and in Figure 4.17, its specification tree is presented. Certain intrinsic parameters were used during the modelling process.

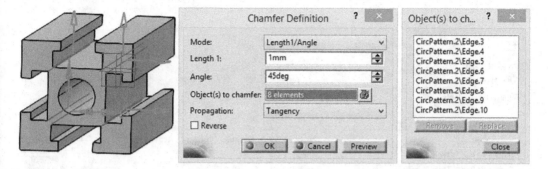

FIGURE 4.16 Chamfering certain inner edges of the part.

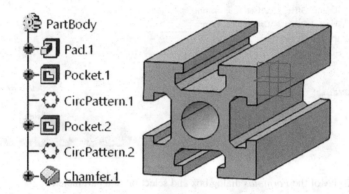

FIGURE 4.17 Specification tree and the 3D model of the parametric part.

The 3D model of the part is quite simple with a reduced number of parameters; this will facilitate the following explanations.

In order to enter the relations between the parameters, it is advisable to follow the order in which they appear in the specification tree.

For example, the depth of the two *Pocket* cuts is set to be equal to the distance of the *Pad* extrusion (*Pad* operation precedes *Pocket* operations). To do this, the user clicks the *Formula f(x)* icon in the *Knowledge* toolbar, and the *Formulas* dialog box, shown in Figure 4.18, is displayed.

In the list of parameters from the left, the one that imposes the cutting depth of the first *Pocket* operation is identified and selected. Its full name (*PartBody\Pocket.1\FirstLimit\Depth*) contains all data for an easy identification. By its complex name, the user is informed that it is of type *Depth* and belongs to the *Pocket.1* feature. Its value of 50 mm is also displayed.

The user clicks the *Add Formula* button to open the *Formula Editor* dialog box in Figure 4.19. The previously selected parameter from the list is present in a non-editable field, and to its right is the equal sign (=), which means the possibility of creating a formula in the editable field below.

The user has at his disposal some filters in search of the parameters involved in the equation. Thus, in the *Members of Parameters* list, the *Length* type is chosen, and from the *Members of Length* list, the parameter that contains the *Pad* extrusion distance is selected (with a double-click). The name of this parameter (*PartBody\Pad.1\FirstLimit\Length*) is intuitive. Its value of 50 mm is displayed in the corresponding field at the bottom of the dialog box. The displayed value often represents a check/confirmation for the user that he has chosen the correct parameter from the list.

```
PartBody\Pocket.1\FirstLimit\Depth=PartBody\Pad.1\FirstLimit\Length
```

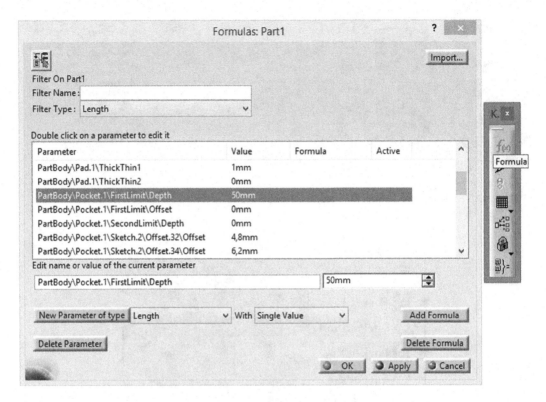

FIGURE 4.18 Display of the *Formulas* dialog box and selection of a parameter.

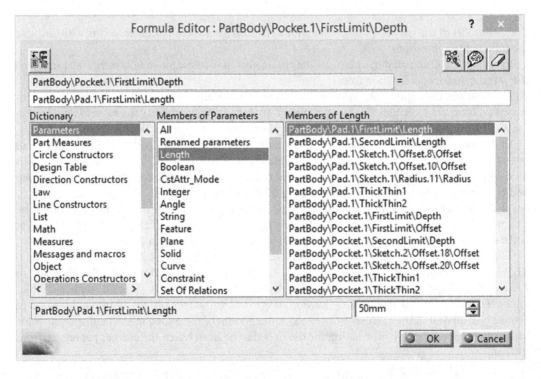

FIGURE 4.19 Formula that establishes an equality between two length parameters.

The user confirms the formula by pressing the *OK* button and similarly sets a second formula for the depth of the second cut (*Pocket.2*).

```
PartBody\Pocket.2\FirstLimit\Depth=PartBody\Pad.1\FirstLimit\Length
```

Figure 4.20 shows the *Formulas* dialog box, and the two depth parameters of the *Pocket.1* and *Pocket.2* are observed to be equal to the same *Pad.1* extrusion distance/length (on the *Formula* column). The status of the two formulas is *yes* on the *Active* column, meaning that both are active and have an influence on the part's geometry.

Next, a formula will be created to specify that the hole radius is 1/6 of the square profile side. The user identifies and selects the radius parameter in the list shown in Figure 4.18 (*PartBody\ Pad.1\Sketch.1\Radius.11\Radius*), clicks the *Add Formula* button and, in the *Formula Editor* dialog box, double-clicks on the parameter that contains one of the side values of the square profile (*PartBody\Pad.1\Sketch.1\Offset.8\Offset*; Figure 4.21). Finally, to complete the equation, the user adds the ratio condition: */6*.

```
PartBody\Pad.1\Sketch.1\Radius.11\Radius= PartBody\Pad.1\Sketch.1\Offset.8\
Offset /6
```

As a result of the above operations, the specification tree will contain three formulas (Figure 4.22), as previously established. The formulas are active until the user decides to intervene on one of them to deactivate it (see Figure 4.3) or to create another formula or rule to define a parameter between the three involved in the formulas.

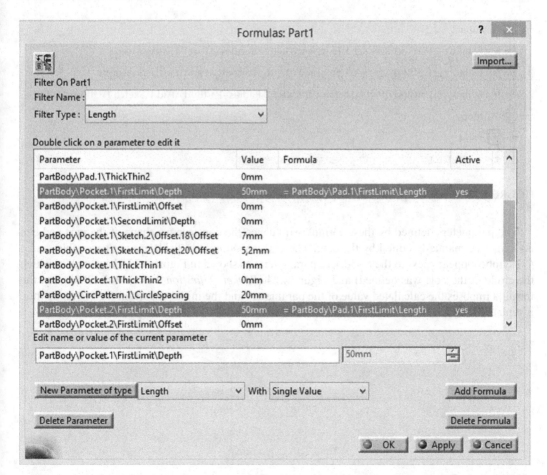

FIGURE 4.20 Formulas presented as active in the dialog box with their values and parameters.

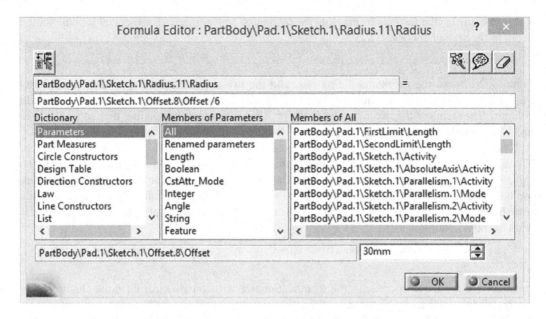

FIGURE 4.21   Formula that establishes a ratio between two length parameters.

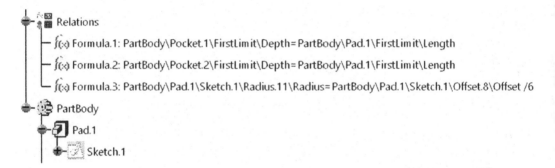

FIGURE 4.22   Three active formulas displayed in the specification tree of the parametric part.

The parameters defined by these formulas influence the constructive shape of the part and can no longer be manually edited by the user. The corresponding edit fields are not accessible, and a *f(x)* symbol appears next to the respective parameters, as shown in Figure 4.23 (the sketch in which the circle of the hole was defined) and Figure 4.24 (*Pocket Definition* dialog box). Obviously, the formula imposes the calculated value of the parameters and the initial one, from the 3D modelling, is replaced.

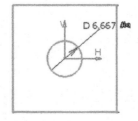

FIGURE 4.23   The *f(x)* symbol displayed near a value of a parameter defined by a formula.

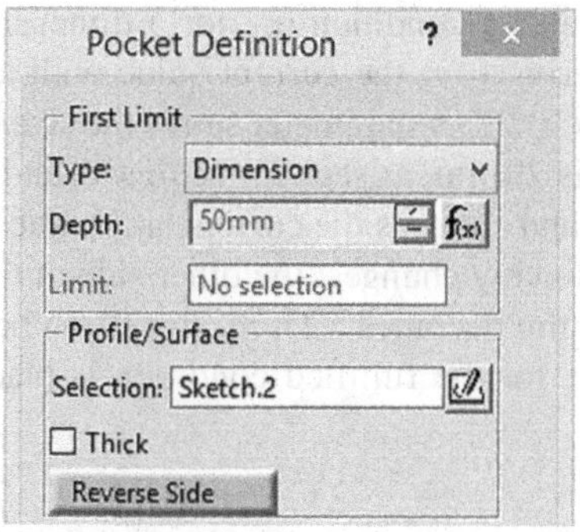

FIGURE 4.24 A parameter defined by a formula cannot receive a value from the user.

The parametric set-up of the part continues by creating a rule that changes the dimensions of the *Pocket* cuts according to the dimensions of the square profile sides. The *CATIA Knowledge Advisor* workbench can be accessed to enter the rule's code. From the *Reactive Features* toolbar, the user clicks the *Rule* icon to open the *Rule Editor* dialog box, shown in Figure 4.25.

It is noticed that the rule receives a name, an information regarding the user who wrote it and the date of creation, and also its placement in the specification tree. After completing them, the user presses the *OK* button and he may enter the *Visual Basic* code sequence:

```
/*Rule created by User 21.12.2021*/
if PartBody\Pad.1\Sketch.1\Offset.8\Offset == 30 mm
{PartBody\Pad.1\Sketch.1\Offset.10\Offset = 30 mm
PartBody\Pocket.1\Sketch.2\Offset.18\Offset = 7 mm
PartBody\Pocket.1\Sketch.2\Offset.20\Offset = 5.2 mm
PartBody\Pocket.2\Sketch.3\Offset.29\Offset = 11 mm
PartBody\Pocket.2\Sketch.3\Offset.30\Offset = 2.2 mm}
```

The meaning of the above relations is as follows: the first line contains the identifying data, placed between the special characters /* – to open the comment – and */ – to close the comment. This first line containing the comment may be missing from the rule code, but its presence is recommended at the beginning of the rule. Also, to add important explanations in the *Visual Basic* code, the user can enter any text/information between the special characters, anywhere inside the code sequence. What is entered as a comment is not taken into account by the rule.

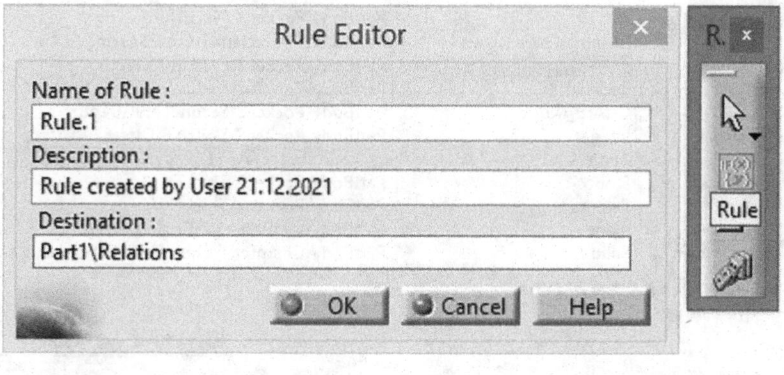

FIGURE 4.25 Several details for a rule created by the user.

The following/second line states a condition (*if*) for a dimension. The rule is waiting for the condition to be fulfilled, comparing (==) the current value with a possible value of 30 mm. The *PartBody\Pad.1\Sketch.1\Offset.8\Offset* parameter stores the size of the right side of the square in Figure 4.8. By default, its value is 20 mm, as seen in the figure, and also in the 2D drawing.

If the user edits the *Sketch.1* and changes the current size value from 20 mm to 30 mm, the rule will automatically make the following changes: the other side of the square becomes 30 mm, size 6.2 mm becomes 7 mm, size 4.8 mm becomes 5.2 mm, size 9 mm becomes 11 mm, and size 1.8 mm becomes 2.2 mm. The rule code for the fulfilled condition is placed between curly brackets, as follows:

```
{
code line 1
code line 2
....
}
```

These curly brackets can also be followed on the same line by relations, as in the sample rule above.

Of course, these dimensional changes will modify the shape of the part. The square profile becomes larger, and the *Pocket* cuts also change their dimensions. The user must follow exactly the syntax presented in the code sequence (spaces, double equals in the first relation, units of measurement, brackets, etc.).

The way of inserting the parameters in the rule is very similar to the one applied in the case of formulas (double-click on the parameter in the list to add it to the rule code), to which the equals, spaces, mm and brackets are manually added.

The *Rule Editor* dialog box in Figure 4.26 presents how the parameters are inserted, and Figure 4.27 shows the specification tree that contains the active *Rule.1*.

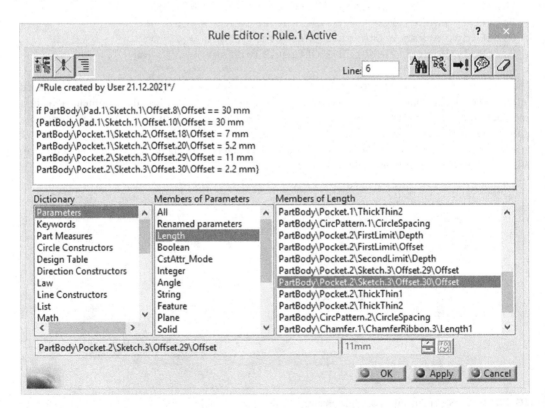

FIGURE 4.26   Adding a parameter from the list to the rule code sequence.

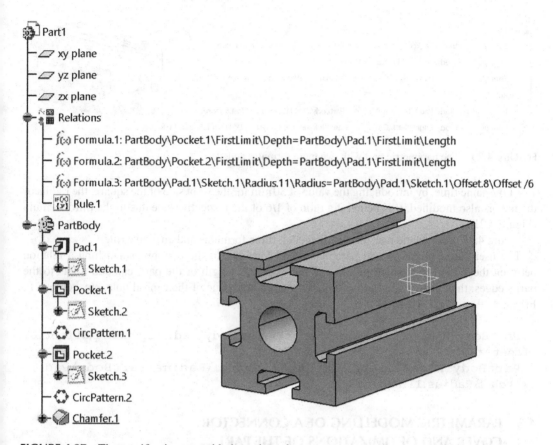

FIGURE 4.27 The specification tree with the active rule.

To test the rule, the user double-clicks the *Sketch.1* of the *Pad.1* feature, showing the constrained square (Figure 4.28). It is noticed that the dimension of the right side of the square is editable, and the other one, from above, has an icon *f(x)* next to it, due to the constraint imposed by the rule.

The figure also shows the *Constraint Definition* dialog box obtained by double-clicking on the editable dimension. In the *Value* field, the user enters the new value, 30 mm, and if the *More >>* button is clicked, the name of the respective dimensional constraint, *Offset.8*, will be displayed. This name is, of course, present in the list of relations that are part of the rule.

After confirming the change in the dimension value (pressing the *OK* button of the dialog box), the rule, being active, changes the dimensions constrained by the relations in the code sequence.

Obviously, the rule only applies when the user enters exactly the value of 30 mm, present in the rule condition, and any other values do not affect or launch the rule. Disabling the rule from its context menu allows the user to change the dimensions of the square to their initial values.

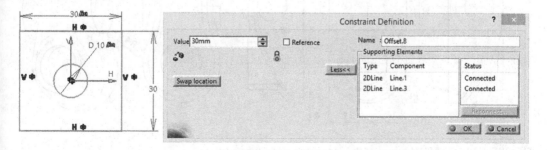

FIGURE 4.28 Fulfilling the condition of running the rule by entering the value of 30 mm.

⊞ Relations

— ƒ(x) Formula.1: PartBody\Pocket.1\FirstLimit\Depth= PartBody\Pad.1\FirstLimit\Length

— ƒ(x) Formula.2: PartBody\Pocket.2\FirstLimit\Depth= PartBody\Pad.1\FirstLimit\Length

— ƒ(x) Formula.3: PartBody\Pad.1\Sketch.1\Radius.11\Radius= PartBody\Pad.1\Sketch.1\Offset.8\Offset /6

— ⊞ Rule.1

— ƒ(x) Formula.4: PartBody\Pad.1\FirstLimit\Length= PartBody\Pad.1\Sketch.1\Offset.8\Offset *2

— ƒ(x) Formula.5: PartBody\EdgeFillet.1\CstEdgeRibbon.1\Radius= PartBody\Pad.1\Sketch.1\Radius.11\Radius

FIGURE 4.29   The parametric part with two more formulas added.

At the same time, by introducing the value of 30 mm for the right side of the square, the radius of the hole is also modified, this being of a ratio of 1/6 of the respective side due to the third formula (Figure 4.21).

Figure 4.27 presents the parametric 3D model, three formulas and an active rule.

The user can continue to add parameters and formulas to the part by establishing a relation between the side of the square profile and the extrusion length of the part, can add fillets to the part's edges, their radius values to be equal to the radius value of the central hole, etc., as seen in Figure 4.29.

```
PartBody\Pad.1\FirstLimit\Length=PartBody\Pad.1\Sketch.1\Offset.8\
Offset *2
    PartBody\EdgeFillet.1\CstEdgeRibbon.1\Radius=PartBody\Pad.1\
Sketch.1\Radius.11\Radius
```

4.3   PARAMETRIC MODELLING OF A CONNECTOR COVER AND OPTIMIZATIONS OF THE PART

The following tutorial briefly presents the steps in the solid parametric modelling of a cover-type part of an electrical connector represented by the 2D drawing in Figure 4.30. The part was proposed as phase II in the Model Mania contest of 2011: *https://blogs.solidworks.com/tech/wp-content/uploads/sites/4/Model-Mania-2011-Phase-2.jpg.*

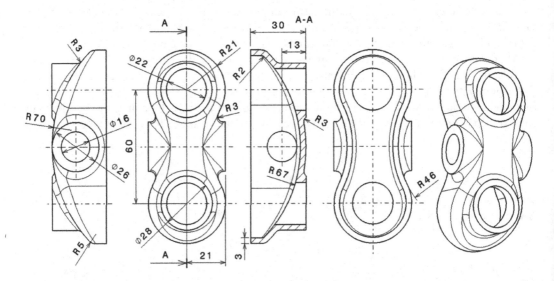

FIGURE 4.30   The 2D drawing of the parametric cover-type part.

Depending on certain dimensions and some features of the connector, the cover may have different construction variants, but very similar shape as a 3D model. This example will also present steps for parametric design and optimization of some dimensions in order to achieve a condition related to the mass of the part. The names of the sketches and, in general, of the features that appear in the specification tree as the part is created are very important because they belong to the complex names of the parameters considered later during the application.

In the *XY Plane*, the user draws the sketch shown in Figure 4.31. The two circles with diameters of Ø42 mm are symmetrical and have their centres at distances of 30 mm from the vertical axis *V* of the *Sketch.1*. Using the other two circles with diameters of Ø92 mm, the user connects the first circles (by tangent constraints) and then removes certain elements of the sketch profile (*Trim*) to obtain a closed contour, that is *Pad* extruded to a height of 30 mm (Figure 4.32).

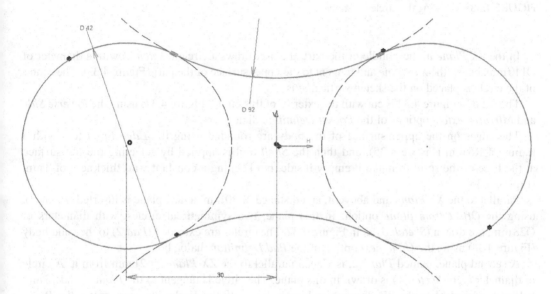

FIGURE 4.31   Drawing the profile of *Sketch.1*.

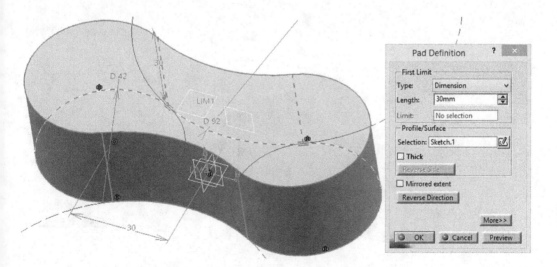

FIGURE 4.32   Extrusion of the first profile to obtain the *Pad.1* feature.

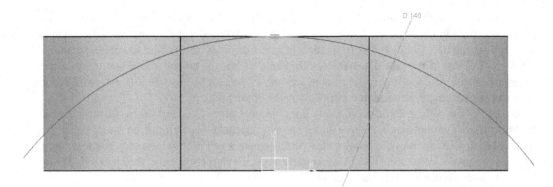

FIGURE 4.33 Drawing the circle of *Sketch.2*.

In the *ZX Plane* in the middle of the part, the user draws a circle (*Sketch.2*) with a diameter of Ø140 mm, below the *XY Plane* and tangent to the upper surface of the part (Figure 4.33). The centre of the circle is placed on the sketch vertical axis.

The solid in Figure 4.32 is cut with the exterior of this circle (Figure 4.34) using the *Reverse Side* and *Mirrored extent* options of the *Pocket Definition* dialog box.

The edges on the upper surface of the body are rounded using the *Edge Fillet* tool with a radius of R5 mm (Figure 4.35), and then the *Shell* tool is applied by selecting the flat surface at the base of the solid to make it empty inside, but keeping a constant wall thickness of 3 mm (Figure 4.36).

Parallel to the *XY Plane* and above it, at a distance of 30 mm, a new plane is inserted (*Plane.1*), using the *Offset from plane* option. In this plane, two symmetrical circles, with diameters of Ø28 mm, are drawn (*Sketch.3*) as in Figure 4.37. The circles are extruded (*Pad.2*) to the solid body (Figure 4.38) using the *Up to next* option of the *Pad Definition* dialog box.

A second plane, named *Plane.2*, is placed parallel to the *ZX Plane*, at 21 mm from it. A circle of diameter Ø26 (*Sketch.4*) is drawn in this plane. The circle is tangent to the *Plane.1* and 13 mm below it, according to the 2D drawing in Figure 4.3. With the circle, the user creates the *Pad.3* extrusion.

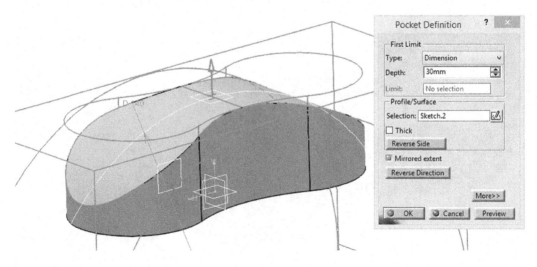

FIGURE 4.34 Cutting the solid to obtain the *Pocket.1* feature.

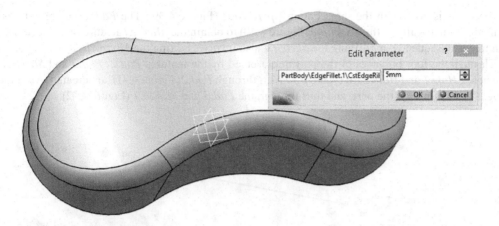

FIGURE 4.35   Rounding the upper edges of the solid to create the *EdgeFillet.1* feature.

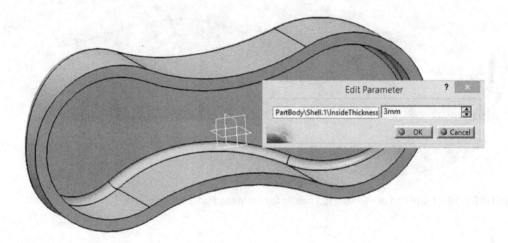

FIGURE 4.36   Applying the *Shell* tool to make the solid empty inside.

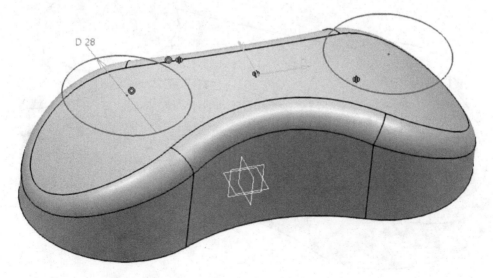

FIGURE 4.37   Drawing two circles of *Sketch.3* in *Plane.1*.

Extrusion is done with the same option *Up To Next* (Figure 4.39). The *Pad.3* feature is transformed symmetrically to the *ZX Plane* (Figure 4.40) to obtain the *Mirror.1* feature in the specification tree. Consequently, modifying the *Sketch.4* also involves editing the *Mirror.1*.

Through the four bossages of the part, represented by the features *Pad.2*, *Pad.3* and *Mirror.1*, two holes, *Hole.1* and *Hole.2* (of type *Up To Next* through *Pad.2*; Figure 4.41), are created. Also, the *Hole.3* passes through the part, and thus through the *Pad.3* and *Mirror.1* (Figure 4.42).

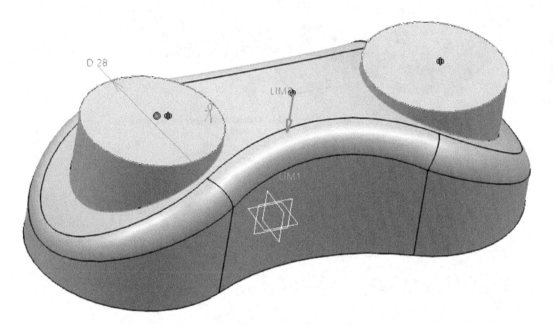

FIGURE 4.38 Extrusion of the *Sketch.3* profile to obtain the *Pad.2* feature.

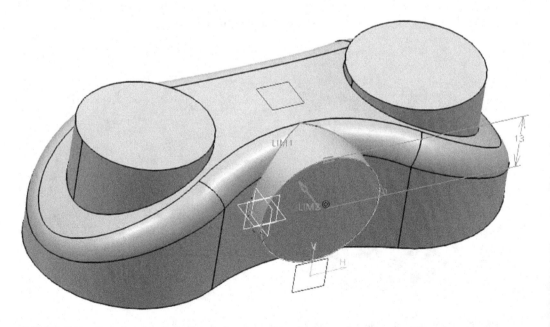

FIGURE 4.39 Drawing the circle *Sketch.4* and extrusion *Pad.3* starting from the *Plane.2*.

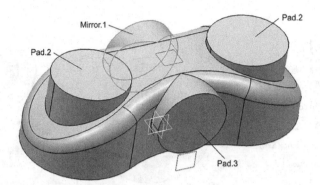

FIGURE 4.40   Symmetry of the *Pad.3* feature with respect to the *ZX Plane*.

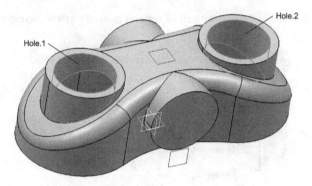

FIGURE 4.41   *Hole.1* and *Hole.2* features created through *Pad.2*.

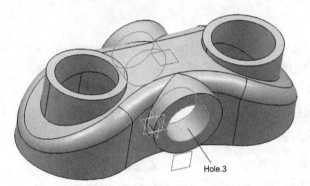

FIGURE 4.42   *Hole.3* feature created through *Pad.3*.

According to the 2D drawing, *Hole.1* and *Hole.2* have diameters of Ø22 mm and *Hole.3* of Ø16 mm. Also, the depth of the *Hole.3* is not specified to be pierced (*Up To Last*), but is required to be 50 mm. This value will be used in the next parameterization step of the part modelling.

On several edges of the part, the *Edge Fillet* tool is applied, with radii of R3 mm. Individual fillets were chosen in order to keep the corresponding radius parameters.

Figure 4.43 shows the specification of the cover part and its 3D model on which the names of all the features are marked. It is noted that although the modelling process is relatively simple, the number of these features is quite large.

In order to make parameters accessible from the specification tree, the user checks the options related to their visibility, as presented in this chapter, according to Figures 4.44 and 4.45.

The first parameters inserted by the user in the specification tree are automatically measured on the part by the program.

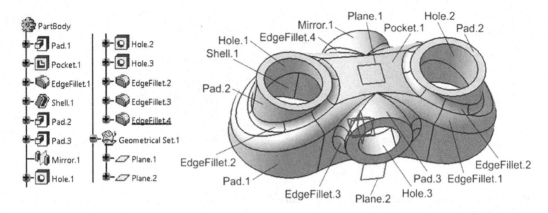

FIGURE 4.43 The 3D model of the part with all its features marked on their corresponding surfaces.

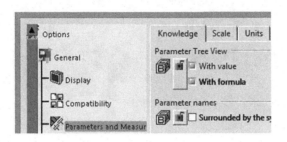

FIGURE 4.44 Checking the *With value* and *With formula* options.

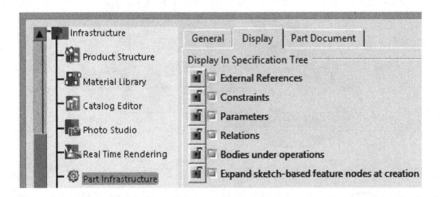

FIGURE 4.45 Checking the *Parameters* and *Relations* options.

Thus, from the *Measure* toolbar, the *Measure Inertia* icon is selected by the user (Figure 4.46). The dialog box with the same name is opened and the user presses the *Customize* button, and then from the available list, he selects the options *Area*, *Volume* and *Mass* (Figure 4.47). Returning to the dialog box, the *Keep measure* option is also selected to add and keep the measured parameters in the specification tree. It is observed that the measured parameters have blocked values, because they resulted based on the value of the density (part's material) and the volume determined by the program.

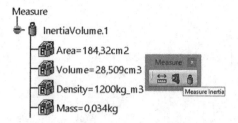

FIGURE 4.46 Displaying the area, volume, density and mass information in the specification tree.

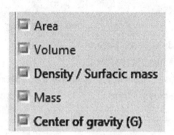

FIGURE 4.47 Checking the corresponding options in the *Measure Inertia* selection box.

It is observed that for the density of 1200 kg/m³ (density of a plastic material, which can be changed by establishing another material for the part, using the tool *Apply Material*), the part has a mass of 0.034 kg. The other geometric properties of the part are also presented: the area of 184.32 cm² (useful value in the calculation of the necessary amount of paint if the user decides to paint a batch of such parts) and the volume of 28.509 cm³. Some of these values are dependent on each other, and changing the volume of the part by editing certain parameters leads to updating the whole solid model and, of course, to changing the values for area and mass, while keeping the characteristics of the material.

If a value for the part mass is imposed (in the case of large batches, the value of the mass is very important for handling and transport), a procedure can be initiated in a few steps to optimize the geometry of its solid model.

The user opens the *Product Engineering Optimizer* workbench in the *Start → Knowledgeware* menu, as shown in Figure 4.48.

In this case, the user clicks on the *Optimization* icon and opens the dialog box with the same name, from Figure 4.49. The *InertiaVolume.1\Mass* field contains the current value of the part mass (0.034 kg).

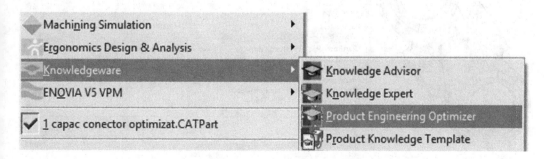

FIGURE 4.48 Opening the *Product Engineering Optimizer* workbench.

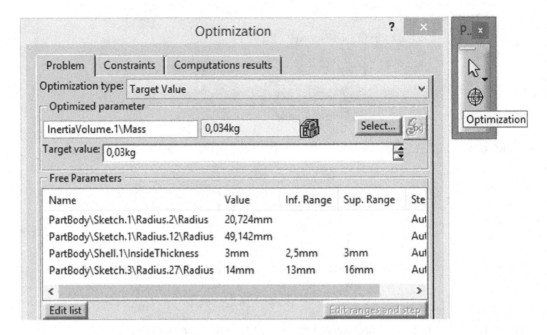

FIGURE 4.49 Opening the *Optimization* dialog box to enter the target value of the mass parameter.

The user enters a new smaller value (0.03 kg) in the *Target value* field. Then, he must specify the parameters whose values are allowed to be modified in several iterations by the program. The user clicks the *Edit list* button placed below the *Free Parameters* area and opens the selection box (*Select the free parameters*) shown in Figure 4.50. Practically, any parameter can be selected by the user in the *Parameters* field (left side) and added to the list on the right side.

The four selected parameters are as follows: from the *Sketch.1*, radii of the circles (Figure 4.31), the constant thickness of the part wall (inside) established using the *Shell* tool (Figure 4.36) and the radius of the circles in the *Sketch.3* (Figure 4.37).

From Figure 4.49, it is observed that the first two parameters are free, without the imposed value; thus, the program has the possibility to identify and try any value. However, the values proposed by the optimization analysis are close to the initial ones.

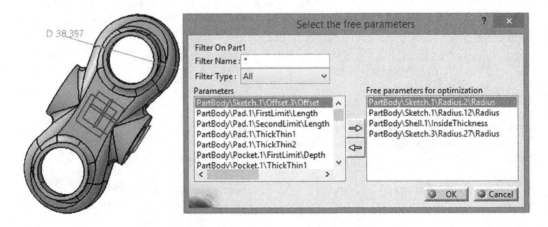

FIGURE 4.50 Selection of the parameters to be considered in the optimization process.

For the other two values, it is possible to set a range and the number of steps, values in mm. For example, for the wall thickness of the part, the minimum value is chosen to be 2.5 mm and the maximum 3 mm. The step value helps the program reducing the number of variants that should be computed. Implicitly, the analysis will have a shorter time of execution.

The *Problem* tab (Figure 4.51) contains other options, such as:

- *Algorithm type*: different computation variants, with and without constraints imposed, in the *Constraints* tab (Figure 4.49);
- *Termination Criteria*: it establishes the maximum number of updates/computation iterations, but also the maximum time allocated for them;
- *Save optimization data*: the option is checked to keep and save the analysis results;
- *Update Mode*: three variants are possible while moving the slide left-right:

  1. *Global Update*: for the entire *CATPart* file, the analysis is slow, but accurate and with many data saved in an *Excel* file;
  2. *Mixed Variational Update*: the analysis is slightly simplified and faster, but does not update the whole part;
  3. *Local Update*: a simple analysis.

If the part proposed for optimization belongs to an assembly or to a simulation with finite elements and its dimensions are connected in a parametric manner with the dimensions of other components, the *Global Update* variant is recommended.

The more complex is the part and the more parameters and conditions are involved (*Constraints*), depending on the chosen algorithm and the update variant, the more accurate will be the optimization analysis, and it will also last longer.

Launching the optimization is done by pressing the *Run optimization* button, then the user is asked to choose the location where to save the *Excel* file with the analysis results (in several variants), and an information box is displayed (Figure 4.52) with the stage of computations.

The file with the *.xls* extension (Figure 4.53) contains the columns with the parameters chosen by the user to be involved in the optimization process, according to the selection in Figure 4.50, and the rows with the different variants identified by the program. It chooses an optimal variant, but after

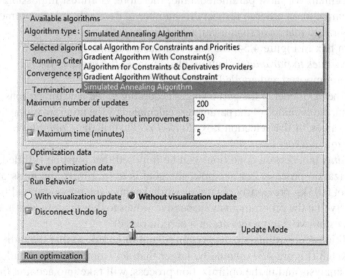

FIGURE 4.51   Setting the options for analysis and optimization.

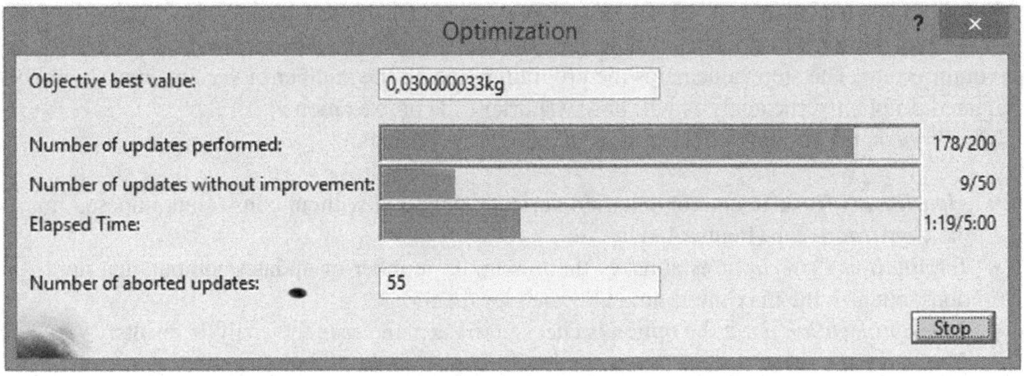

FIGURE 4.52　Launching the optimization to achieve the target value of the part's mass.

	A	B	C	D	E	F	G
1	Nb	Best (kg)	Part1\InertiaVc	Part1\PartBod	Part1\PartBod	Part1\PartBod\	Part1\PartBod
2	0	0,03	0,029999999	19,19867338	76,17736691	2,836591344	13,01338469
3	1	0,03	0,029986688	19,05780282	72,3185464	2,82374187	13
4	2	0,03	0,030266328	19,78635417	72,97636922	2,832582058	13,06038465
5	3	0,03	0,030421676	20,15600153	75,03832171	2,840717631	13,11855667
6	4	0,03	0,030298687	20,21336471	73,75153262	2,840973195	13
7	5	0,03	0,030455212	20,1272057	75,42695082	2,842364486	13,14504402
8	6	0,03	0,030180799	19,25443756	71,58050725	2,842636963	13
9	8	0,03	0,030260651	19,94067721	74,7633066	2,840942215	13,03374779
10	9	0,03	0,030047547	20,70377945	77,3237218	2,80783413	13

FIGURE 4.53　An *Excel* file with different variants identified by the program after parametric optimization.

viewing the data in the file, if the user does not agree with the variant proposed by the program, he can extract and apply the values of a certain row.

Thus, a specific option is selected from the *Computations results* tab. The display order of the available variants is chronological (in the order in which they were computed by the program) or from the best solution to the worst.

The solution identified by the program is not always the most suitable because it computes values with many decimals for most parameters (and, therefore, is almost impossible to process by a manufacturing company). In order to restrict this number of decimals, it is recommended to impose a step value in the range that a parameter can take (Figure 4.49).

In the selection box in Figure 4.54, the user chooses a variant, checks the dimensions and then presses the *Apply values to parameters* button, placed under the variants list. Thus, these values on the selected row are imposed and applied to the part.

Consequently, the 3D model of the part changes according to the new dimensions. By pressing the *Show curves* button, the evolution of the parameters calculated by the program is graphically displayed.

Note that the values in the selection box (*Sorted results*) of the *Computations results* tab are identical to those in the *Excel* file.

In the *Constraints* tab, certain conditions can be imposed that must be fulfilled by a parameter in the part optimization process. For example, the goal is to minimize the mass of the cover part (below the value of 0.03 kg previously obtained) while keeping the previous conditions and parameters (the possibility that the program gives successive values to these parameters and the conditions related to the values of two parameters are in a certain range).

Two new conditions/constraints are also required and imposed, so that the radii of the circles at the ends of the sketch (Figure 4.31) should be larger than R19 mm, but also smaller than R21 mm. *CATIA v5*, in the analysis and in the optimization process, will take into account these constraints, and if for certain solutions/variants, it is not be able to respect them, it will inform the user.

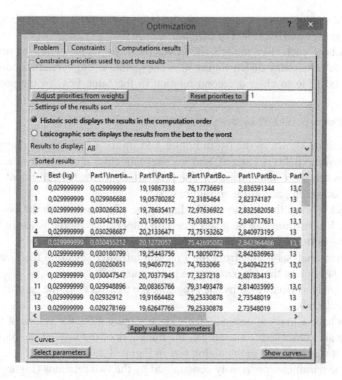

FIGURE 4.54 Selection of a set with values to be applied to the part parameters.

Thus, in the *Constraints* tab of the *Optimization* dialog box (Figure 4.55), the user presses the *New...* button to enter the first constraint. In the *Optimization Constraints Editor* box, he enters the following formula:

```
PartBody\Sketch.1\Radius.2\Radius>19mm
```

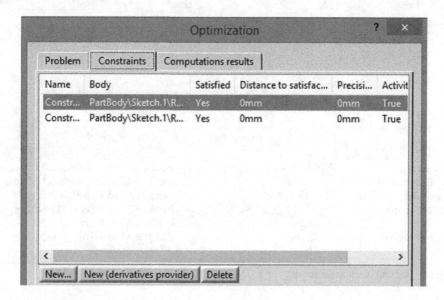

FIGURE 4.55 Constraints defined for a parameter of type radius.

The selection of this parameter is easier with the help of the three filters/columns (*Dictionary*, *Members of Parameters* and *Members of Length*).

Depending on how each user models the cover part, the name of the parameter *PartBody\Sketch.1\Radius.2\Radius* may differ. Figure 4.56 shows that a single relation/constraint has been entered. It is not possible to enter a second relation in the same box, so the procedure must be repeated for this:

```
PartBody\Sketch.1\Radius.2\Radius<21mm
```

The two constraints are shown in the list in Figure 4.55. In the *Optimization* dialog box, the name of the constraint (*Name* field) is entered, and the user can enter a comment or an explanation for the formula he has set (the *Comment* field). The program confirms the fulfilment of the constraint (*Satisfied* symbol on green), according to Figure 4.57.

The *Activity* field allows the user to enable/disable the relation through the two *Boolean* values *True* or *False*. Each constraint can be given a certain importance by the value in the *Weight* field. Below is the *Edit...* button to edit the relation.

The optimization analysis is run from the *Problem* tab after selecting the *Gradient Algorithm With Constraint(s)* option in the *Algorithm* type list (Figure 4.58).

Following the analysis performed by the program, the user receives the resulting variants in the *Computations results* tab.

These variants have a mass of around 0.0275 kg; in Figure 4.59, the values of the other parameters involved are also observed (for example, the thickness of the wall of the part varies between 2.5 and 3 mm, as it was imposed in the dialog box in Figure 4.49).

The value of the parameter previously constrained by the two relations (Figure 4.56) is close to the minimum, slightly above the value of 19 mm. Any change in conditions, constraints, etc. must be followed by the resumption of the analysis.

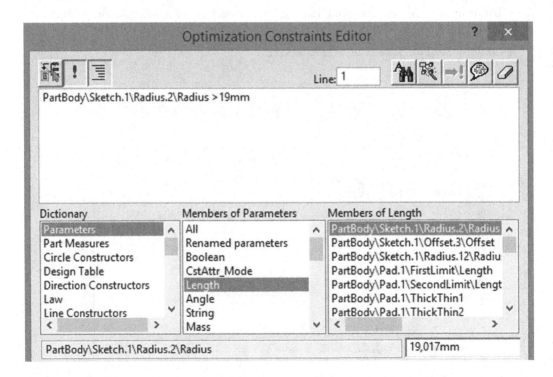

FIGURE 4.56 Writing the first constraint for a parameter to be larger than 19 mm.

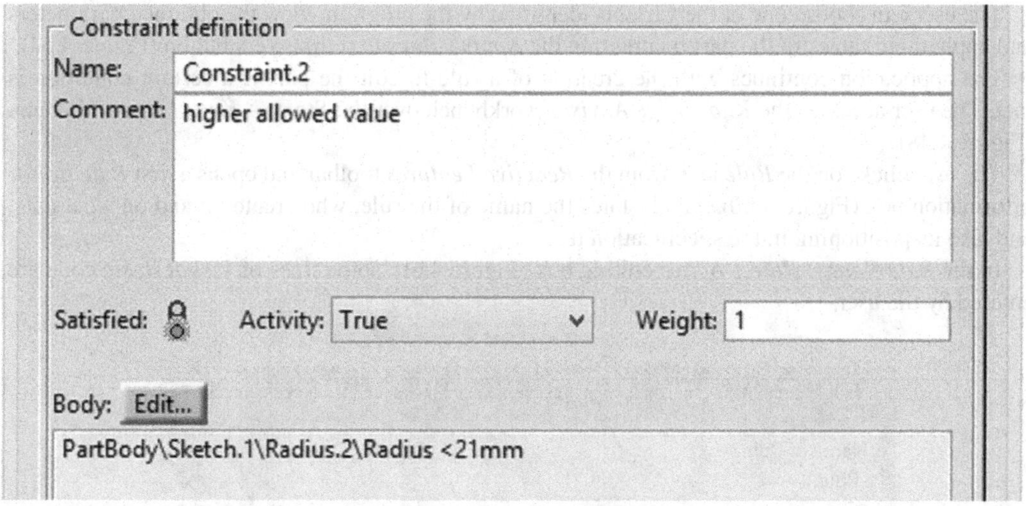

FIGURE 4.57 Adding name, comment and importance for a constraint.

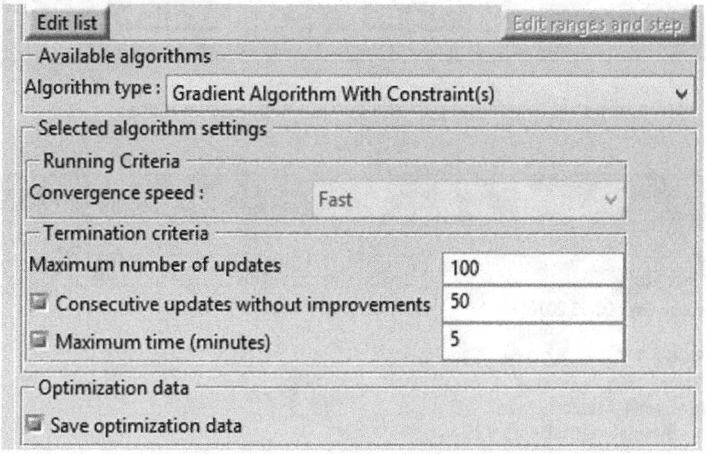

FIGURE 4.58 Selection of the analysis algorithm to run the optimization analysis.

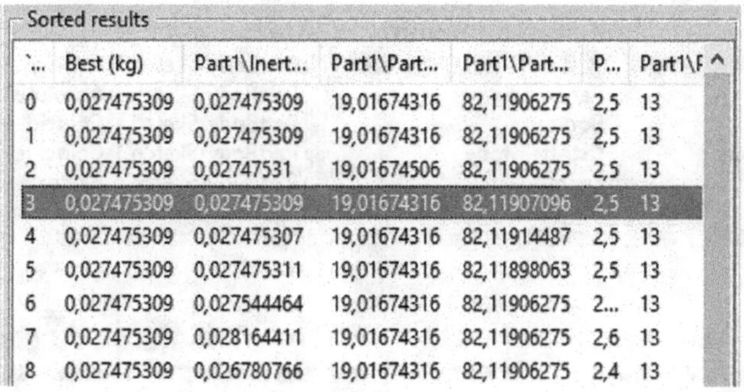
Sorted results

`...	Best (kg)	Part1\Inert...	Part1\Part...	Part1\Part...	P...	Part1\F
0	0,027475309	0,027475309	19,01674316	82,11906275	2,5	13
1	0,027475309	0,027475309	19,01674316	82,11906275	2,5	13
2	0,027475309	0,02747531	19,01674506	82,11906275	2,5	13
3	0,027475309	0,027475309	19,01674316	82,11907096	2,5	13
4	0,027475309	0,027475307	19,01674316	82,11914487	2,5	13
5	0,027475309	0,027475311	19,01674316	82,11898063	2,5	13
6	0,027475309	0,027544464	19,01674316	82,11906275	2...	13
7	0,027475309	0,028164411	19,01674316	82,11906275	2,6	13
8	0,027475309	0,026780766	19,01674316	82,11906275	2,4	13

FIGURE 4.59 Results after the analysis is performed.

The user can choose one of the variants identified by the program, view the obtained parameters and apply their values to the part by pressing the *Apply values to parameters* button (Figure 4.54).

The application continues with the creation of a rule to edit the part if a certain condition is met. The user accesses the *Knowledge Advisor* workbench over the *Start → Knowledgeware* menu (Figure 4.48).

The user clicks on the *Rule* icon from the *Reactive Features* toolbar and opens a first *Rule Editor* information box (Figure 4.60), which states the name of the rule, who created it and on what date, and also its positioning in the specification tree.

In the *Rule Editor: Rule.1 Active* editing box (Figure 4.61), some lines of *Visual Basic* code are entered by the user.

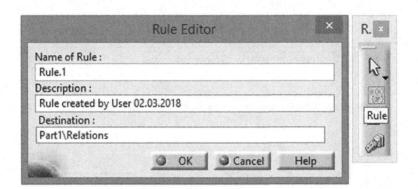

FIGURE 4.60   Creation of a rule to edit the part if a certain condition is met.

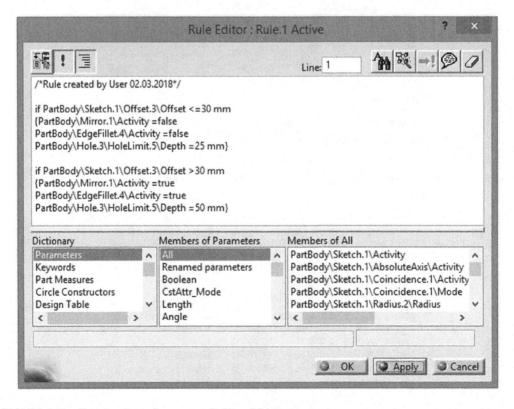

FIGURE 4.61   Entering the code sequence in *Visual Basic*.

The code sequence compares a parameter with a certain value and, if the condition is fulfilled, one of the two computation variants of the other parameters is executed. The syntax *if... {....}* is used to control the distance (initially, it was 30 mm) between the two circles of the *Sketch.1* and the vertical axis *V* (Figure 4.31), which is compared to be less than or equal to (<=) 30 mm. If this condition is fulfilled by the respective parameter, then the first variant is executed run.

Otherwise, when the distance is greater (>) than 30 mm, the second variant is run. The two variants are briefly explained below.

`if PartBody\Sketch.1\Offset.3\Offset <=30 mm`	the distance is compared with the value of 30 mm
`{PartBody\Mirror.1\Activity =false`	the *Mirror.1* feature is disabled
`PartBody\EdgeFillet.4\Activity =false`	the *EdgeFillet.4* feature is disabled
`PartBody\Hole.3\HoleLimit.5\Depth =25 mm}`	the depth of the *Hole.3* becomes 25 mm
`if PartBody\Sketch.1\Offset.3\Offset >30 mm`	the distance is compared with the value of 30 mm
`{PartBody\Mirror.1\Activity =true`	the *Mirror.1* feature is enabled
`PartBody\EdgeFillet.4\Activity =true`	the *EdgeFillet.4* feature is enabled
`PartBody\Hole.3\HoleLimit.5\Depth =50 mm}`	the depth of the *Hole.3* becomes 50 mm

In the first variant (Figure 4.62), when the value of the distance is 28 mm, the features *Mirror.1* and *EdgeFillet.4* (also marked in Figure 4.43) are deactivated with the help of the *Boolean* parameter *Activity* of each one. For each disabled feature, the program adds a set of parentheses *()* next to its icon in the specification tree.

By deactivation, the geometries of the two elements disappear from the 3D model of the part. Also, the depth of the *Hole.3* decreases to 25 mm (the part is no longer pierced). As an example, in this variant, the plastic cover can be used for a smaller and simpler connector with a single input (the remaining features *Pad.3* and *Hole.3* in the specification tree).

If the distance between the circles' centres of the *Sketch.1* is greater than 30 mm (31 mm in the example in Figure 4.63), the features *Mirror.1* and *EdgeFillet.4* become active and the depth of the hole *Hole.3* receives the value of 25 mm (is pierced). In this variant, the part can be mounted as a protection cover for a connector with two inputs.

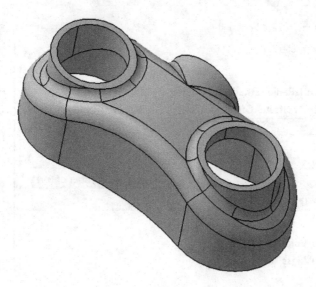

FIGURE 4.62   The cover part represented in the first simpler variant with a single input.

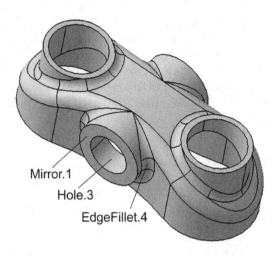

Mirror.1

Hole.3

EdgeFillet.4

FIGURE 4.63   The *cover* part represented in the second variant with two inputs.

For both variants, the syntax of the *Visual Basic* language must be observed and respected, the code sequence after the comparison (<= or >) begins and ends with curly brackets (*{ }*), each value is followed by the unit of measurement, the parameter names are complete, etc.

The user also has an option at his disposal through which he can try different construction solutions for the part in progress. Thus, from the *Product Engineering Optimizer* workbench, the user can access the *Design Of Experiments* (*DOE*) tool that allows to vary some parameters and track their influence on other parameters.

Accessing the icon with the same name opens the *DOE* selection box shown in Figure 4.64. In the upper field, the user can select the input parameters, which can be modified between two limits (*Inf. Range* and *Sup. Range*) and with a certain step (*Nb of Levels*), and in the lower field, the output parameters.

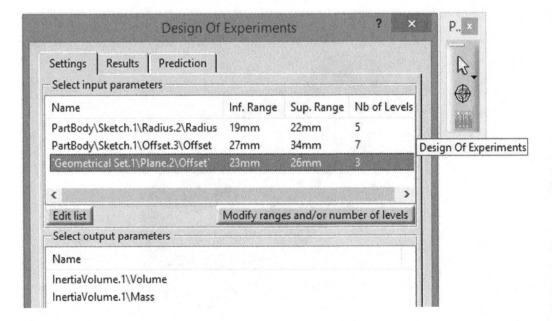

FIGURE 4.64   The *Design Of Experiments* selection box to try different dimensional solutions.

As input parameters that can be modified, the user may select:

- the radius of the circles at the ends of the *Sketch.1* (Figure 4.31), between 19 mm and 22 mm, with five steps;
- the distance between the centres of the two circles of the *Sketch.1* and the vertical axis *V* (Figure 4.31), between 27 mm and 34 mm, with seven steps;
- the distance from the *ZX Plane* to the *Plane.2* (Figure 4.39), between 23 and 26 mm, with three steps.

The parameters are selected using the *Edit list* button of the *Design Of Experiments* selection box (Figure 4.64).

The limits within the selected parameters can take values, and the number of steps are set by pressing the *Modify ranges and/or number of levels* button.

The volume and mass of the part have been selected in the output parameters field. Running the simulation is done by pressing the *Run DOE* button – the procedure is similar to the one in the case of *Run optimization*: the user is asked to establish a location to save the *Excel* file with the results (multiple variants). Depending on these results, the program may display an information box with some warnings (example in Figure 4.65; the *EdgeFillet.3* feature could no longer be created on the solid model and has been disabled).

The *Results* tab shows the computed values, and the user can try the variants obtained by selecting a specific row (Figure 4.66) and pressing the *Apply values* button.

The part's solid changes according to the chosen set of parameters, taking into account the *Rule.1* previously created and active (Figure 4.61).

In the *Prediction* tab, the user can try a certain variant of the part, with parameters entered manually (in the range of possible values). It is no longer necessary to run the entire simulation according to the options in the *Settings* tab, but only the respective model is created by pressing the *Run prediction* button. Figure 4.67 shows such a variant – the user enters values in the *Selected parameter value* field for each parameter from the *Valuate input parameters* list.

From the 3D model represented next to the *DOE* selection box, it can be seen that the part is in the simpler version, a cover for a single input electrical connector, due to the *Rule.1*, and the parameter *PartBody\Sketch.1\Offset.3\Offset* is less than 30 mm (Figure 4.61).

For the values of parameters of a certain variant, the volume and mass results are calculated and displayed in the *Predicted response for each output* field.

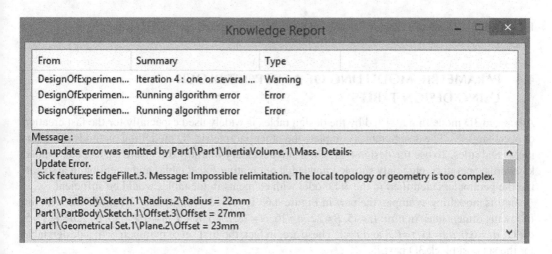

FIGURE 4.65 A *Knowledge Report* information box with warning and error messages.

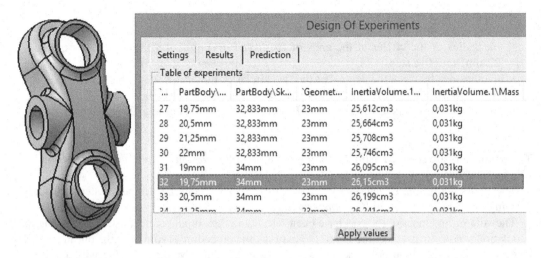

FIGURE 4.66 A table with different sets of values for the part's parameters.

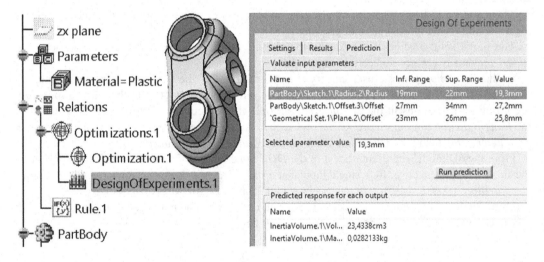

FIGURE 4.67 Creating a custom 3D model of the cover part by entering values for parameters.

4.4 PARAMETRIC MODELLING OF A SUPPORT BLOCK USING DESIGN TABLES

Advanced 3D modelling assisted by the design tables is widely used especially for the fast creation of parts families. The design tables complete the other parametric 3D modelling methods using formulas and rules. To use the design tables, it is not necessary to know the *Visual Basic* code syntax, keywords, operators, etc.; only simple knowledge of creating *Excel* tables and attention in associating the parameters identified in the 3D model with elements in the tables would be sufficient.

In this modelling example, the part in Figure 4.68 is considered, initially 3D modelled with the following dimensions, in mm: $L=45$, $B=22$, $h=16$, $N=18$, $d=5$, $K=9$, $m=6$, $P=15$, $f=7$, $b=8$, $s=5$, $c=12$, $d1=6.6$, $d2=11$, $t=6.8$ and $r=1$. These are, in fact, the first set of parameters of a design table for the support block 3D part.

The figure also contains the specification tree of the part.

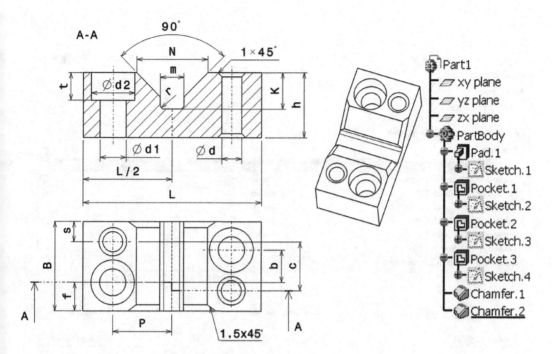

FIGURE 4.68   The 2D drawing of the parametric support block part.

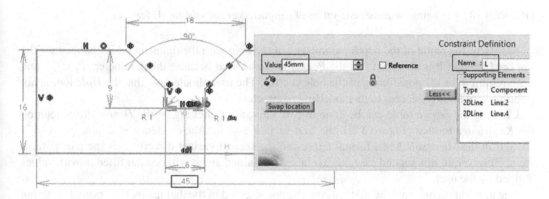

FIGURE 4.69   Changing the names and values of parameters in the *Sketch.1*.

In order to make the modelling process using a design table easier, the respective parameters will be identified and renamed from the first phase of creating the sketches.

For example, during the drawing of the first sketch that gives the shape of the part, the user double-clicks on each dimensional constraint, changes its value according to the set of parameters considered and clicks the *More >>* button in the *Constraint Definition* dialog box (Figure 4.69) to rename the parameter. The size of 45 mm corresponds to the parameter *L*.

Similarly, for all the parameters in all sketches, the user changes their values and names. The parameters of the 3D modelling tools (*Pad, Pocket,* etc.) are edited separately, using the *f(x)* icon on the *Knowledge* toolbar.

In the *Formulas* dialog box, shown in Figure 4.70, the *PartBody\Pad.1\FirstLimit\Length* parameter is selected in the list to appear in the *Edit name or value of the current parameter* field. Its name is changed by the user in *B*.

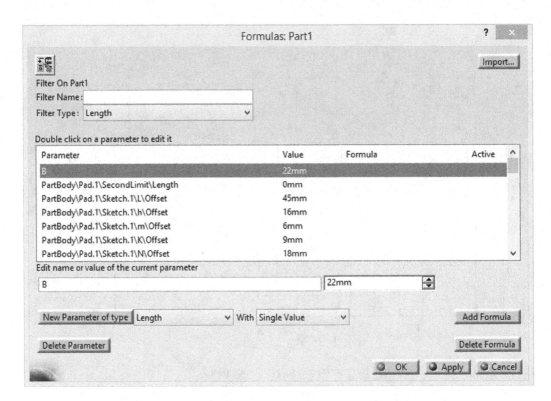

FIGURE 4.70   Changing the names and values of parameters created by the 3D features.

Similarly, the name of the depth parameter *t* of the hole with the diameter *d2* is changed for the *Pocket.1* feature. It is observed that the *Pocket* tool was used because the parameters *P, s, f, b* and *c* are required for the positioning of the hole centres. The user should note that the *Hole* tool is not offering very intuitive access to its modelling parameters.

Creating the design table can be done in two ways, after clicking on the *Design Table* icon on the *Knowledge* toolbar (Figure 4.71): the first variant assumes the existence of a table previously created in the *Microsoft Excel* format (*.xls*), and the second variant directly uses the part parameters. However, in this second case, an *.xls* table is obtained and it will be later filled in with values defined by the user.

The application presents the first variant, the one selected in the dialog box *Creation of a Design Table: Create a design table from a pre-existing file*, which involves creating a table file in *Microsoft Excel*. Thus, as shown in Figure 4.72, the table contains several columns, named after the part parameters in Figure 4.71. Each parameter must be followed by the unit of measurement, between round brackets. The parameter values are on four rows, each set being a unique configuration. For parameters *d, d1* and *d2* in the table, half of their values were used because they represent radius values (*Radius*; Figure 4.74).

In the *File Selection* dialog box in Figure 4.73, the user selects the *.xls* file previously saved as '*support table.xls*' and clicks the *OK* button. The user is asked in the *Automatic associations* dialog box if he wants the association between the table columns and the part parameters to be done automatically. It is recommended to press the *No* button to perform manual pairing, thus increasing the correctness of the pairing.

In the *Design Table* dialog box, in the *Associations* tab, there are two selection lists and one association list. The *Parameters* list contains the parameters of the part with their modified names. The *Columns* list contains the columns of the table in Figure 4.72. The user will select a parameter and a column from the respective lists, and then, when pressing the *Associate* button, a pair appears in the association list *Associations between parameters and columns* (Figure 4.74).

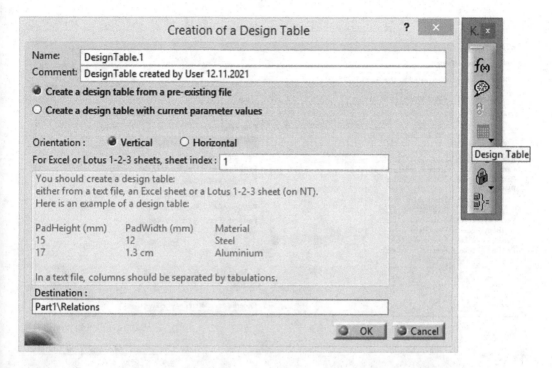

FIGURE 4.71 Creation of a design table from a pre-existing file or using the current values.

	A	B	C	D	E	F	G	H	I	J	K	L	M	N	O	P
1	L (mm)	B (mm)	h (mm)	N (mm)	d/2 (mm)	K (mm)	m (mm)	P (mm)	f (mm)	b (mm)	s (mm)	c (mm)	d1/2 (mm)	d2/2 (mm)	t (mm)	r (mm)
2	45	22	16	18	2.5	9	6	15	7	8	5	12	3.3	5.5	6.8	1
3	55	22	20	24	2.5	11	8	20	7	8	5	12	3.3	5.5	6.8	1
4	70	28	25	30	3	14	12	25	10	8	6	16	4.5	5.5	9	1
5	85	28	32	37	3	18	16	32.5	10	8	6	16	4.5	7	9	1
6																

FIGURE 4.72 Four sets of values for the part parameters in an *Excel* file.

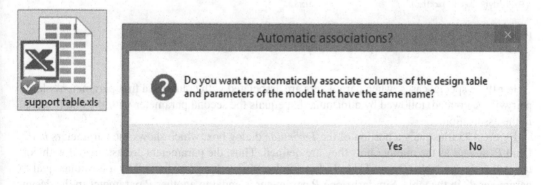

FIGURE 4.73 Dialog box asking the user to associate the table columns with the part parameters.

The pairing/association operation is repeated several times for all the involved parameters of the part. The figure shows that some parameters have already been associated with the corresponding columns, and also that there are two or more parameters of the part, with the same name: *d1*, *d2*, *d*, *P* and *r*. This is mainly due to certain symmetries of the 3D model and the fact that it has some similar holes (for example, the parameter *P* is used four times, for each hole; *r* is used twice in fillets; *d1* and *d2* are used twice for the bored holes).

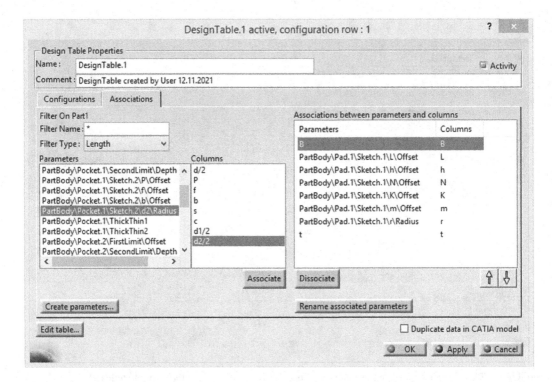

FIGURE 4.74   Associations between the table columns and the part parameters.

Parameter	Value	Formula	Active	▲
PartBody\Pocket.1\Sketch.2\b\Offset	8mm	DesignTable.1	yes	
PartBody\Pocket.1\Sketch.2\d2\Radius	5.5mm	= PartBody\Pocket.1\Sketch.2\d2\Radius	yes	
PartBody\Pocket.1\Sketch.2\d2\Radius	5.5mm	DesignTable.1	yes	
PartBody\Pocket.1\Sketch.2\f\Offset	7mm	DesignTable.1	yes	
PartBody\Pocket.1\Sketch.2\P\Offset	15mm	= PartBody\Pocket.1\Sketch.2\P\Offset	yes	
PartBody\Pocket.1\Sketch.2\P\Offset	15mm	DesignTable.1	yes	
PartBody\Pocket.1\ThickThin1	1mm			▼

Double click on a parameter to edit it

FIGURE 4.75   Parameters set by formulas and design table.

For these special cases, when parameters are associated with columns, a first parameter-column pair will be created, followed by a formula that equals the second parameter with the one involved in the association.

Figure 4.75 illustrates a portion of the *Formulas* dialog box, which shows the parameters *b*, *d2*, *f* and *P*, as well as the manner how they are defined. Thus, the parameters are associated with columns in a design table (*DesignTable.1*), but another parameter *d2* is defined by a formula equal to parameter *d2* in the table. Similarly, one *P* parameter is equal to another *P* parameter in the design table. The user should note that the values, set in mm, are equal.

The design table, also called family table, and formulas are present in the specification tree (*Relations*).

Figure 4.76 also shows the current configuration (1) of the design table. If the user wants the current part to be modified according to the dimensions set in the *Microsoft Excel* table (Figure 4.72), he will double-click on the *Configuration* feature to open the *Edit Parameter* dialog box (Figure 4.77).

Relations
DesignTable.1
Configuration=1
Sheet.1
Formula.1: PartBody\Pocket.1\Sketch.2\P\Offset=PartBody\Pocket.1\Sketch.2\P\Offset
Formula.2: PartBody\Pocket.3\Sketch.4\P\Offset=PartBody\Pocket.1\Sketch.2\P\Offset
Formula.3: PartBody\Pocket.3\Sketch.4\P\Offset=PartBody\Pocket.1\Sketch.2\P\Offset
Formula.4: PartBody\Pad.1\Sketch.1\r\Radius=PartBody\Pad.1\Sketch.1\r\Radius
Formula.5: PartBody\Pocket.1\Sketch.2\d2\Radius=PartBody\Pocket.1\Sketch.2\d2\Radius
Formula.6: PartBody\Pocket.2\Sketch.3\d1\Radius=PartBody\Pocket.2\Sketch.3\d1\Radius
Formula.7: PartBody\Pocket.3\Sketch.4\d\Radius=PartBody\Pocket.3\Sketch.4\d\Radius

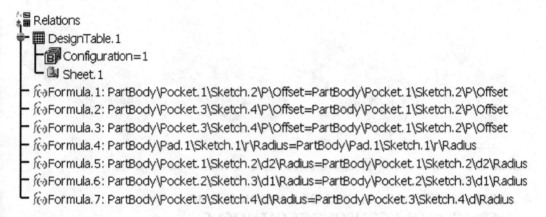

FIGURE 4.76 Formulas and the design table in the specification tree.

Relations
DesignTable.1
Configuration=1
Sheet.1
Edit Parameter
Relations\DesignTable.1\Configu
1
OK
Cancel

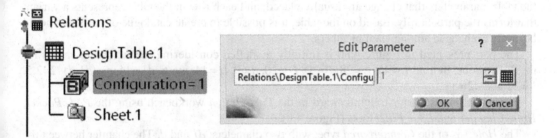

FIGURE 4.77 Opening the configurations list.

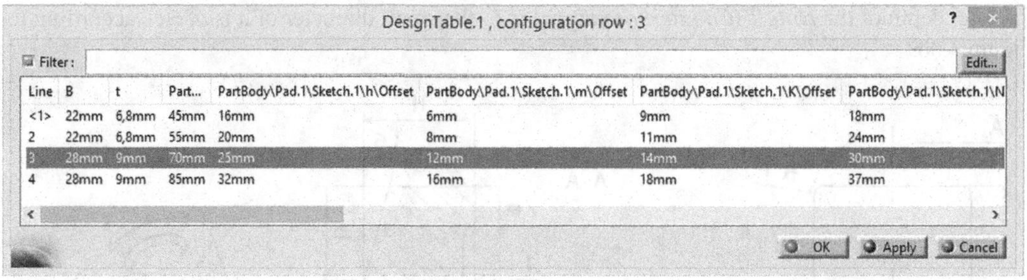

DesignTable.1 , configuration row : 3
Filter :
Edit...
Line B t Part... PartBody\Pad.1\Sketch.1\h\Offset PartBody\Pad.1\Sketch.1\m\Offset PartBody\Pad.1\Sketch.1\K\Offset PartBody\Pad.1\Sketch.1\N
<1> 22mm 6,8mm 45mm 16mm 6mm 9mm 18mm
2 22mm 6,8mm 55mm 20mm 8mm 11mm 24mm
3 28mm 9mm 70mm 25mm 12mm 14mm 30mm
4 28mm 9mm 85mm 32mm 16mm 18mm 37mm
OK
Apply
Cancel

FIGURE 4.78 Choosing a configuration/parameters set from the list to update the part geometry.

The configuration (parameters values set) is chosen from the *Design Table* selection box, which is available after clicking on the table icon to the right of the current configuration number. Figure 4.78 shows the selection of the third configuration/set of values.

Figure 4.79 compares the four variants of the parametrically designed part. The differences between the part variants are obvious. The choice of a configuration has, therefore, the effect of the complete transformation of the considered part, thus forming a various family of parts.

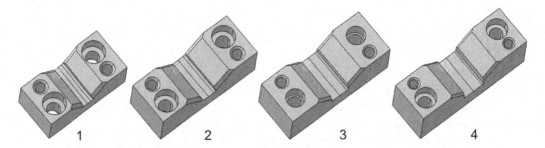

FIGURE 4.79   Four variants of the same part according to the values of the design table.

4.5   PARAMETRIC MODELLING OF A HOOK CLAMP. CREATION OF A COMPONENTS CATALOGUE

This *CAD* application uses a design table to quickly create families of 3D parts. They contain information about 3D models with similar shapes. Some values of parameters may be the same for all variants, but others change depending on the configuration selected in the table. In the columns of the table, parameters that change are usually placed, and each row in the table represents a variant that forms the parts family. Based on the table, it is possible to create catalogue of parts that could be easily exchanged among team members.

The part presented in Figure 4.80 is initially modelled considering the following values for the dimensions, in mm: $L=45$, $E=28$, $l=20$, $D=20$, $d=9$, $H=14$, $m=6$, $d1=14$, $c=(d1-d)/2=2.5$, $b=12$ and $d2=19$.

Part modelling is fairly straightforward in the *Part Design* workbench using the *Pad, Pocket, Hole, Chamfer* and *Edge Fillet* tools (Figure 4.81).

The *Hole.1* is of the *Counterbored* type, with two diameters: $d1$ and d. The chamfer between the two diameters is the *Chamfer.1* feature, the value c being given by the half of the difference of the two diameters. The part also has other chamfers; their value is $1\times45°$, two R0.5 mm fillets and two R1 mm.

To help the user in the modelling process, the steps to be followed are explained below. Figures 4.82–4.84 show some of the dimensions used to model the part. For example, the circle from which the feature *Pad.1* is extruded has a diameter D, the length of the extrusion is $L-H+1.5$–4 mm, the total depth of the *Hole.1* (*Counterbored* type) is L, the small diameter of it is d, etc., according to Figure 4.80.

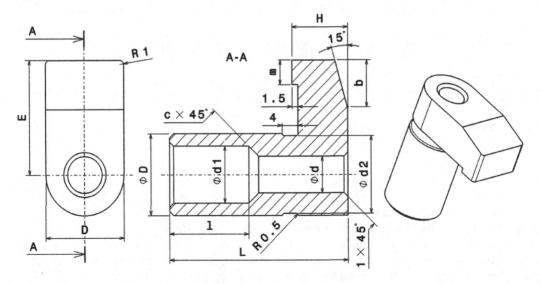

FIGURE 4.80   The 2D drawing of the part to create the catalogue for.

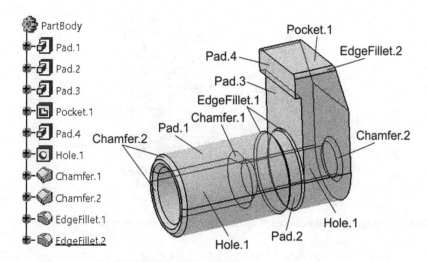

FIGURE 4.81 Displaying the features of the modelled part.

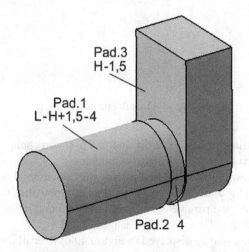

FIGURE 4.82 Displaying the dimensions of some features of the modelled part.

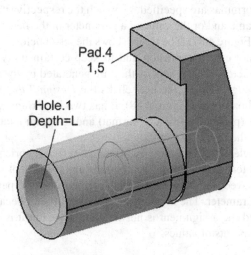

FIGURE 4.83 Displaying the dimensions of some features of the modelled part.

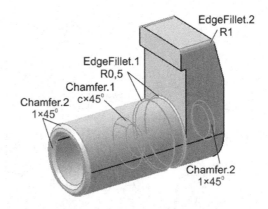

FIGURE 4.84   Displaying the dimensions of some features of the modelled part.

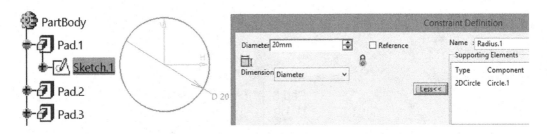

FIGURE 4.85   Displaying the *Constraint Definition* dialog box.

Also, during the modelling process or at the end, the user can change the names of the parameters using the *Formula f(x)* tool in the *Knowledge* toolbar.

In the *Sketch.1*, after drawing the circle and placing the dimensional constraints, the user should double-click on the value of the diameter, enter the value of 20 mm in the *Diameter* field and then press the *More >>* button to access the parameter's name. It can be found in the *Name:* field (Figure 4.85) and has the default name *Radius.1*, but it can be changed to *D/2*.

The *Formulas* dialog box (partly displayed in Figure 4.86) lists all the dimensional parameters used to create the 3D model of the part in Figure 4.80. The identification of the parameters can be done either by their names or by the values (*Parameter* and *Value* fields). Next to the *Value* column, for example, some formulas are specified, by which the respective parameters are calculated.

The user can edit the name and/or the value of a parameter in the field below *Edit name or value of the current parameter* (Figure 4.87). The figure shows the result before and after editing the name.

However, in order to not create confusion and to keep certain unity in the explanations, the default names of the parameters will be kept as they are generated by the program.

From the same *Knowledge* toolbar, the user clicks the *Design Table* icon and the *Creation of a Design Table* selection box opens (Figure 4.88). It has two important options: to create a design table from an existing file (preferably in *Excel* format) and to create a table with the values of the parameters used to model the part.

In the case of the first selection, the user must prepare a vertically oriented table (option selected below), so that the parameters names are in the columns and the sets of values for them are in the rows. Once created and saved, the table must be integrated into the part model by successively assigning a column to a parameter. The method is a bit more difficult because the design table must be in a certain format, and the assignment is not very intuitive, but it is applicable if such a table already exists and has many sets of values.

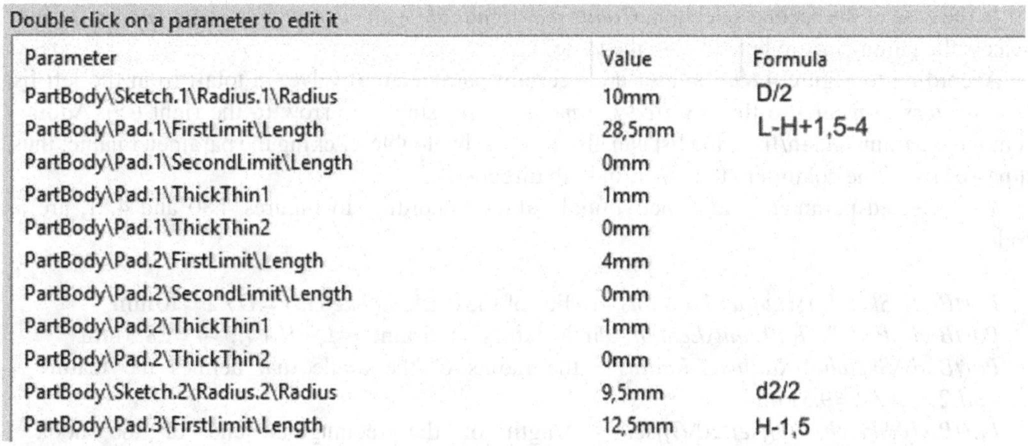

Double click on a parameter to edit it

Parameter	Value	Formula
PartBody\Sketch.1\Radius.1\Radius	10mm	D/2
PartBody\Pad.1\FirstLimit\Length	28,5mm	L-H+1,5-4
PartBody\Pad.1\SecondLimit\Length	0mm	
PartBody\Pad.1\ThickThin1	1mm	
PartBody\Pad.1\ThickThin2	0mm	
PartBody\Pad.2\FirstLimit\Length	4mm	
PartBody\Pad.2\SecondLimit\Length	0mm	
PartBody\Pad.2\ThickThin1	1mm	
PartBody\Pad.2\ThickThin2	0mm	
PartBody\Sketch.2\Radius.2\Radius	9,5mm	d2/2
PartBody\Pad.3\FirstLimit\Length	12,5mm	H-1,5

FIGURE 4.86   Displaying the list of the parameters.

Edit name or value of the current parameter

PartBody\Sketch.1\Radius.1\Radius	10mm

Edit name or value of the current parameter

D/2	10mm

FIGURE 4.87   Before and after editing a parameter's name or value.

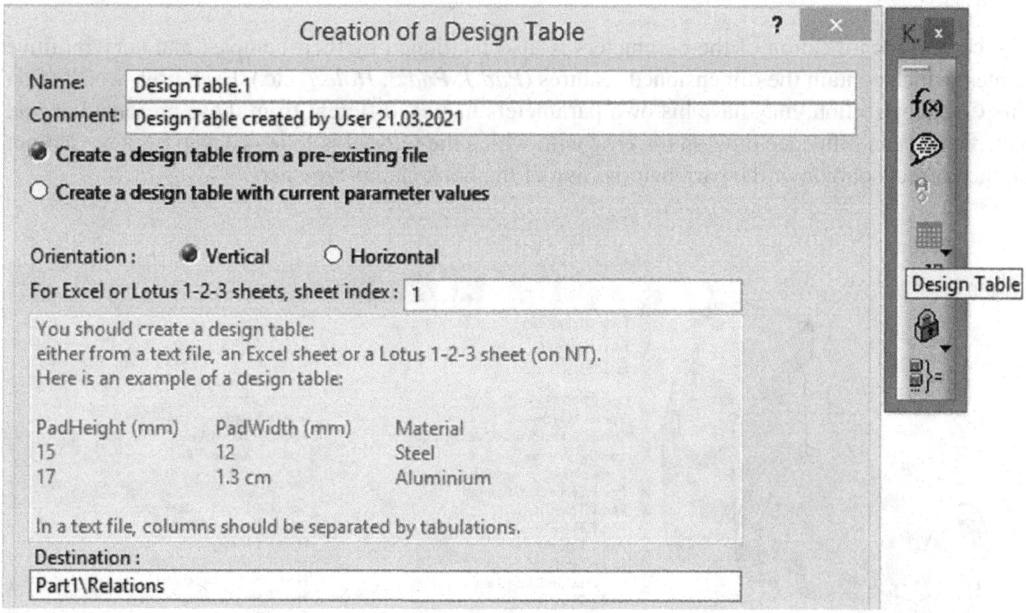

Creation of a Design Table   ?   ×

Name: DesignTable.1
Comment: DesignTable created by User 21.03.2021

● Create a design table from a pre-existing file
○ Create a design table with current parameter values

Orientation :   ● Vertical   ○ Horizontal
For Excel or Lotus 1-2-3 sheets, sheet index : 1

You should create a design table:
either from a text file, an Excel sheet or a Lotus 1-2-3 sheet (on NT).
Here is an example of a design table:

PadHeight (mm)	PadWidth (mm)	Material
15	12	Steel
17	1.3 cm	Aluminium

In a text file, columns should be separated by tabulations.

Destination :

Part1\Relations

Design Table

FIGURE 4.88   *Creation of a Design Table* definition dialog box.

In the case of the second selection, *Create a design table with current parameter values*, the user selects the parameters to be added to the table.

According to Figure 4.89, the user adds certain parameters (twelve, in total) from the left list (*Parameters to insert*) to the *Inserted parameters* list using the arrow to the right (->). Adding/removing parameters in/from the list can also be done by double-clicking the parameter name; thus, it passes from one column to the other, in both directions.

The selected parameters and their initial values, according to Figures 4.80 and 4.81, are as follows:

PartBody\Sketch.1\Radius.1\Radius - radius of the circle in *Sketch.1* $= D/2 = 10$ mm

PartBody\Pad.1\FirstLimit\Length - circle extrusion distance $= L - H + 1.5 - 4 = 28.5$ mm

PartBody\Sketch.2\Radius.2\Radius - the radius of the circle that defines the feature *Pad.2* $= d2/2 = 9.5$ mm

PartBody\Sketch.3\Offset.20\Offset - length of the rectangular end of the hook clamp $= E = 28$ mm

PartBody\Pad.3\FirstLimit\Length - width of the rectangular end of the hook clamp $= H - 1.5 = 12.5$ mm

PartBody\Sketch.4\Length.31\Length - the large side of the chamfer created by *Pocket.1* $= b = 12$ mm

PartBody\Sketch.5\Offset.46\Offset - width of the rectangular area *(Pad.4)* of the hook clamp $= m = 6$ mm

PartBody\Hole.1\Diameter - small diameter of the hook clamp hole $= d = 9$ mm

PartBody\Hole.1\HoleLimit.1\Depth - hook clamp hole depth $= L = 45$ mm

PartBody\Hole.1\HoleCounterBoredType.1\Diameter - hole large diameter of the hook clamp $= d1 = 14$ mm

PartBody\Hole.1\HoleCounterBoredType.1\Depth - bored hole depth of the hook clamp $= l = 20$ mm

PartBody\Chamfer.1\ChamferRibbon.1\Length1 - the length of the chamfer *(Chamfer.1)* between the two diameters of the hole $= c = (d1-d)/2 = 2.5$ mm

The correct identification of the parameters is also facilitated by their complex and very intuitive names, which contain the dimensioned features (*Pad.1, Pad.3, Hole.1*, etc.). Each user, working on this *CAD* application, may have his own parameters names, different from those presented above. This fact has an influence only on the ease with which the tutorial is followed step by step, and not on the correct solution of the parameterization of the hook clamp-type part.

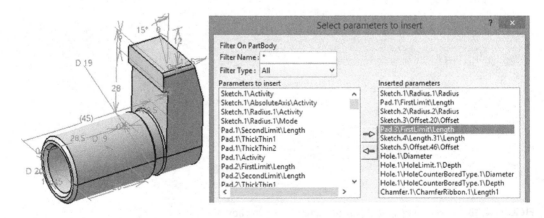

FIGURE 4.89 Selecting the parameters to be added to the design table.

When the parameter selection is completed, *CATIA v5* proposes to save an *Excel* file (but it can also be saved in *.txt* format) with the default *DesignTable.1.xls* name. It is recommended that the file to be kept in the same folder as the *CATPart* file of the hook clamp, or any other part that is parameterized in this manner.

After saving the file, the *Design Table active* dialog box opens (Figure 4.90), which offers, in the first two fields, the possibility to change the name of the design table (*Name:*) and to add a short comment (*Comment:*) about the author and the time of creation. These fields are, however, filled in by default by the program.

In general, if done correctly, in the *Configurations* tab, a row is displayed consisting of the values and their units of measurement (in this case, mm) assigned to a parameter. It is recommended to check the correct parameter-value combination.

The *Associations* tab (Figure 4.91) shows three lists of parameters that can be picked up by the user using the filters (by name – *Filter Name* – and/or type – *Filter Type*).

The *Parameters* list contains the parameters by which the part was created. Next to it, the *Columns* list contains the columns of the design table, but in this case, the association was made automatically by the program and its result completes the third list, *Associations between parameters and columns*. The figure shows the correctness of this association.

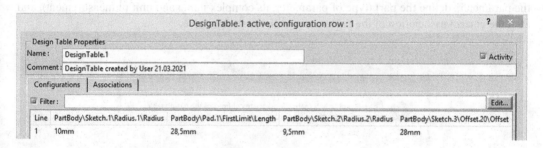

FIGURE 4.90 Saving a *Design Table* definition dialog box.

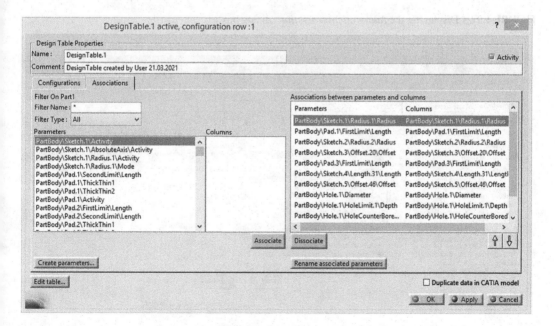

FIGURE 4.91 Displaying the *Associations* tab of a *Design Table*.

If in the selection window *Creation of a Design Table* in Figure 4.88 the user had chosen the first option, *Create a design table from a pre-existing file,* he would have been asked which file to upload, and a message is displayed (Figure 4.92) confirming the automatic association between the columns of the table and the parameters of the part model.

To support explanations, the entire *DesignTable.1.xls* file (or another, if available) will be uploaded. If the parameters have exactly the same name as the columns of the table, the user can press the *Yes* button, the result being similar to the one in Figure 4.91. Otherwise, the user can press the *No* button, the parameters' names in the table will appear in the *Columns* list, and he will manually associate it with the part parameters. For a small number of parameters and intuitive table column names, the association is relatively easy. Thus, as shown in Figure 4.93, the user can start creating (for testing) a new design table, *DesignTable.2,* by pressing the *No* button. Simplifying the parameter list is done by displaying only the *Length* parameters in the *Filter Type* drop-down list. The user should identify the *Parameters-Columns* pair and press the *Associate* button. The association is easy, being displayed in the list *Associations between parameters and columns.* In case of an incorrect pairing, the pair can be removed by selecting it and pressing the *Dissociate* button. Finally, the user presses the *OK* button to confirm the final list of associations between the part parameters and the columns of the design table.

Regardless of the method used, it requires a lot of attention and a perfect understanding of how the parameters define the part (type of parameter, its complex name and unit of measurement), and then the correct association of the parameters with the design table.

FIGURE 4.92　Displaying the *Automatic Associations* dialog box.

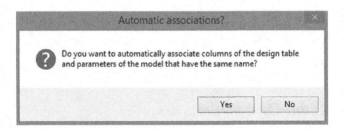

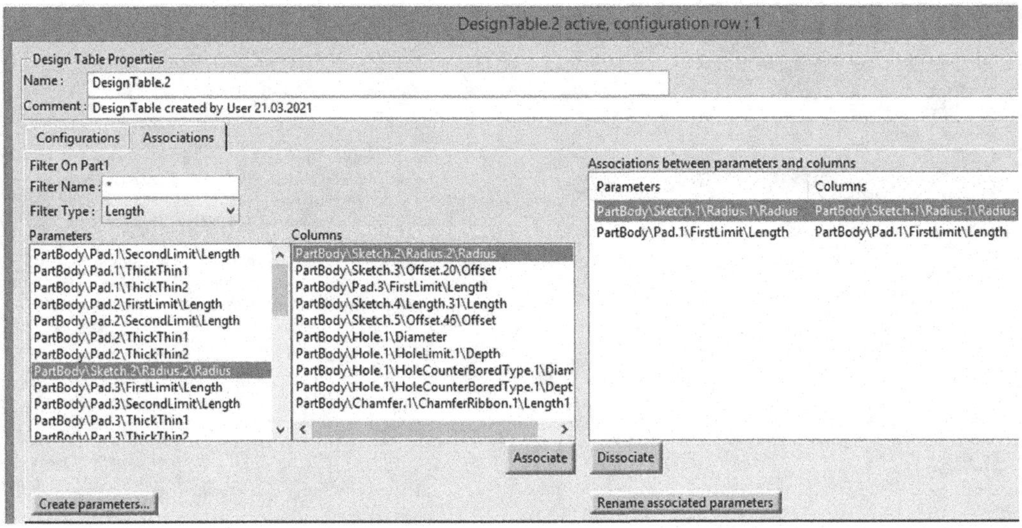

FIGURE 4.93　Manually creating the associations between the parameter's names and columns.

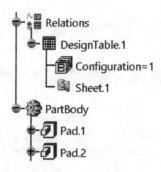

FIGURE 4.94 Specification tree after creating the *Design Table*.

The specification tree of the hook clamp part in Figure 3.81 is completed with the *Relations* feature (Figure 4.94), which contains the design table named *DesignTable.1* previously created and saved (Figure 4.90), and the first configuration is observed (*Configuration=1*) with the parameters taken from the 3D model of the part.

The user can open the *Excel* file and add other lines with values for the table parameters. Figure 4.95 shows the table completed with three more lines (configurations). The names of the parameters are not fully visible due to the narrowing of the columns in the *Excel* file, but by selecting the respective cell, the name is displayed, followed by the unit of measurement in round brackets.

The information enclosed in parentheses, although implied, must exist for the values of that parameter to make sense. In this case, where the table was created by the program and the user only completed it with other configurations, the units of measurement (mm) are added by default. However, if the user had used his own table, these units of measurement would have had to be added manually, in parentheses, next to the parameter name, as they are required by the program.

In the table in the *Excel* file shown in Figure 4.95, the columns of the table correspond to the order in which the parameters are associated with them according to Figure 4.93.

The table is saved (on the hard disk or other storage media), and the *Excel* program can close. Returning to *CATIA*, the user receives an information message (Figure 4.96) that the *Excel* file has been modified and that a successful synchronization has taken place between the file and the design table, with immediate effect on the 3D model of the part.

If something wrong is entered in the *Excel* table (for example, a letter instead of a digit in the parameter values), the synchronization takes place, but the association will be removed, in which case the information message is according to Figure 4.97.

Also, entering an inappropriate value (for example, the value of 12 mm was entered for the chamfer length (*Chamfer.1*) between the two diameters of the hole (parameter *c* in Figure 4.80)) leads to the impossibility of updating the 3D model of the part, and the diagnostic message in Figure 4.98 is generated and displayed. It specifies the feature (of the 3D model) that generates the error and two possible solutions. In this case, the user must return to the value of 2 mm for that parameter of the inner chamfer.

| F1 | ▾ | : | ✕ | ✓ | *fx* | PartBody\Sketch.4\Length.31\Length (mm) | —— unit of measurement |

	A	B	C	D	E	F	G	H	I	J	K	L
1	PartBody\	PartBody\	PartBody\	PartBody\	PartBody\	PartBody\	PartBody\	PartBody\	PartBody\	PartBody\	PartBody\	PartBody\
2	10	28,5	9,5	28	12,5	12	6	9	45	14	20	2,5
3	12	30	10,5	31	14	13	6	10	48	14	20	2
4	14	32	11,5	33	16	14	6	11	52	15	20	2
5	16	36	12	34	17	15	7	12	57	16	20	2

FIGURE 4.95 Adding more configurations in the *Design Table*.

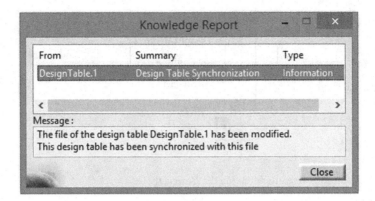

FIGURE 4.96   Displaying the synchronization report of a *Design Table*.

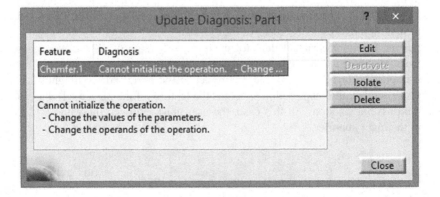

FIGURE 4.97   Displaying the synchronization report of a *Design Table*, but the association failed.

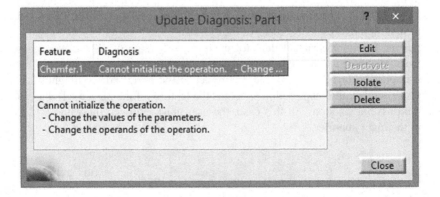

FIGURE 4.98   Displaying the error window when an inappropriate value is used for the parameter.

Once the association of the part parameters-table columns is completed and the file is saved, the user can choose another part configuration by double-clicking the *Configuration = 1* feature in the specification tree (Figure 4.94).

The *Edit Parameter* selection box opens as in Figure 4.99. In the grey field (non-editable), the value 1 can be found, which corresponds to the first configuration, on which the original hook clamp model was created. Choosing another configuration is done by pressing the table button, and placing the mouse cursor on it leads to the appearance of an informative message: *Opens a dialog that allows you to change driving design table configuration.*

From the *DesignTable.1 configuration row* selection box (Figure 4.100), another configuration is actually chosen for the parametric 3D model created in this application. The user can click the *Apply* button and see how the parameters on that line influence the shape and size of the part.

Clicking the *OK* button confirms the selected configuration. The part, thus parameterized, can be saved (*CATPart* file) and used in assemblies, for creating the 2D drawing, etc., the parameters in the selected configuration remaining active until a new selection is done.

Figure 4.101 shows the 3D model of the hook clamp in each of the four variants (configurations). The differences between the respective parts are clearly visible.

If a parametric model is applied to a family of standardized parts, then it is recommended to create a catalogue containing all the dimensional variants. The *CATIA v5* program, by default, contains catalogues (*EN, ISO, JIS* and *US standards*) with common standard parts such as screws, pins and nuts (stored, for example, in *C:\Program Files (x86)\Dassault Systemes\B21\intel\startup\components\MechanicalStandardParts*). These components are used in assemblies (*CATIA Assembly Design* workbench) and can also be opened individually in *CATIA Part Design.*

The standardized parts are accessed using the *Catalog Browser* option in the *Tools* menu, according to Figure 4.102 (for the *ISO* standards).

To create a new catalogue, in addition to those already existing in the program's library, a parametrically modelled part is required using a design table. Only one configuration can be accepted, but the role of the catalogue is to manage and use parts with a large and very large number of configurations. Also, part parameters can be controlled by two or more design tables, but only a single table can be used to create a catalogue.

For example, the actual 3D model of the hook clamp together with its table will be considered. The user needs to be sure that the associations are correct, the part is updated to a certain configuration, the files are saved, etc. Additionally, a new column (last position) with the name *PartNumber*

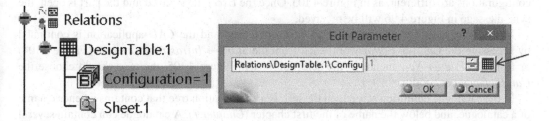

FIGURE 4.99 Displaying the *Edit Parameter* box to modify the configurations.

Line	PartBody\Sketch.1\Radius.1\Radius	PartBody\Pad.1\FirstLimit\Length	PartBody\Sketch.2\Radius.2\Radius	PartBody\Sketch.3\Offset.20\Offset	PartBoc
<1>	10mm	28,5mm	9,5mm	28mm	12,5mm
2	12mm	30mm	10,5mm	31mm	14mm
3	14mm	32mm	11,5mm	33mm	16mm
4	16mm	36mm	12mm	34mm	17mm

FIGURE 4.100 Displaying the dialog box for choosing another configuration.

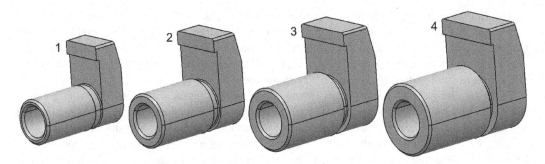

FIGURE 4.101   Displaying the hook clamp for all four configurations.

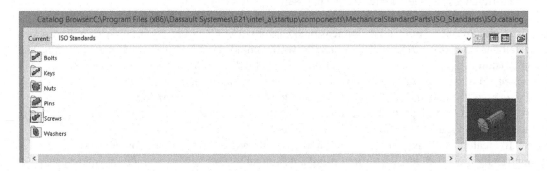

FIGURE 4.102   Displaying the *Catalog Browser* for ISO parts.

must be added to the *Excel* file, containing the name of the configuration on each line (*Hook clamp 1,..., Hook clamp 4*), as shown in Figure 4.103. This *PartNumber* column should have been in the first position in the table for an easier choice of configuration, but the program finds and recognizes it, no matter where it is created and inserted in the table. It is very important that the names of the configurations are different, as in Figure 4.103. Once the *Excel* file is saved and the part is open, the sync message in Figure 4.96 will be received.

In this step, both files (*Excel* and *CATPart*) are closed and the *CAD* application is continued by accessing the *Catalog Editor* workbench from the *Start → Infrastructure* menu (Figure 4.104) or from the *File → New* menu. In the selection box in Figure 4.105, the user should choose the *CatalogDocument* option.

In the catalogue definition screen, on the left, is a specification tree that contains a generic name of a catalogue, and below the name of the first chapter (*Chapter.1*). A catalogue can contain several chapters, an example being the catalogue with *ISO* standardized parts/components, its structure being presented in Figure 4.106.

It is noted that the parts are grouped in sections/categories: bolts, keys, nuts, pins, screws and washers. Each category contains, of course, its own family of standardized parts, structured by size and type.

	A	B	C	D	E	F	G	H	I	J	K	L	M
1	PartBody\	PartBody\	PartBody\	PartBody\	PartBody\	PartBody\	PartBody\	PartBody\	PartBody\	PartBody\	PartBody\	PartBody\	PartNumber
2	10	28,5	9,5	28	12,5	12	6	9	45	14	20	2,5	Hook clamp 1
3	12	30	10,5	31	14	13	6	10	48	14	20	2	Hook clamp 2
4	14	32	11,5	33	16	14	6	11	52	15	20	2	Hook clamp 3
5	16	36	12	34	17	15	7	12	57	16	20	2	Hook clamp 4

FIGURE 4.103   Adding a new column, *PartNumber*, to an existing design table.

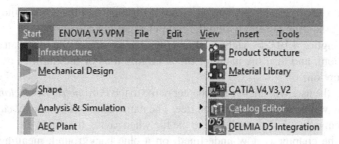

FIGURE 4.104 Creating a new catalogue from the *Start* menu.

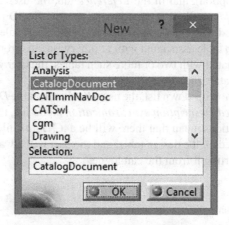

FIGURE 4.105 Creating a new catalogue from the *File → New* menu.

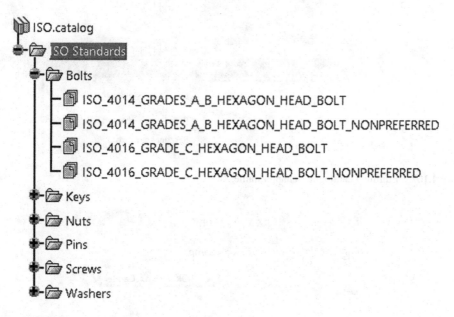

FIGURE 4.106 Displaying the *Catalog Editor* screen for the *ISO* parts.

The names of the chapters are very suggestive, also the names of the parts in the families are well established so that the user can easily find the right element in an assembly.

According to Figure 4.107, by right-clicking the chapter name, the user can change its name (*Definition…* option), and also the name of the configurations of the families of parts (*Keywords default values…* option).

In this example, the user will change the chapter name to the *Parametric Hook clamp* (Figure 4.108), which will become visible in the specification tree. The name *CatalogDocument2.catalog* will change according to the name that the user will give to the file when saving the catalogue.

The name of the chapter is now underlined, on a blue background, meaning that it is active and can be assigned families of parts. Thus, the *Add Part Family* icon (Figure 4.109) is clicked in the *Chapter* toolbar and the *Part Family Definition* dialog box opens. In the *Name* field, the user enters the name *Hook clamp* (it could be followed/supplemented, if necessary, by/with a standard or an additional brief description), and in the *Reference* tab, the user clicks the *Select Document* button and uploads the *.CATPart* file that contains the parametric 3D model previously created in the application and associated with the *DesignTable.1*. As a general rule, the selected *.CATPart* file must have at least one design table associated with it; otherwise, an error message will be displayed. If the *.CATPart* file is associated with two or more such tables, the user will be asked which of them will be transformed into a catalogue.

From the *Resolution mode* drop-down list, the user should select the *Descriptions will be resolved* option, and from the *Resolved Description synchronization mode* list, the user must choose *Always regenerate the Part*. The options mean that there will be a *.CATPart* file for each configuration the user chooses, generated based on the design table, and the part will always be updated with the new parameter values when inserting it from the catalogue.

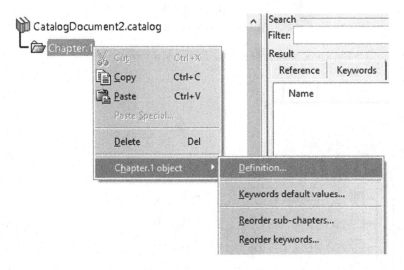

FIGURE 4.107   Editing a catalogue's chapter name.

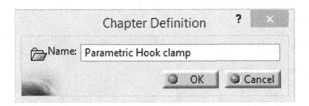

FIGURE 4.108   Changing the name of the chapter to *Parametric Hook clamp*.

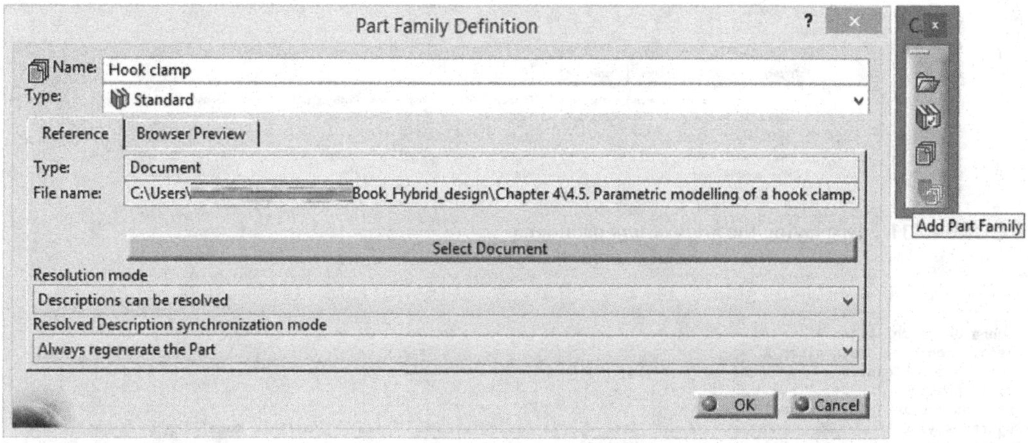

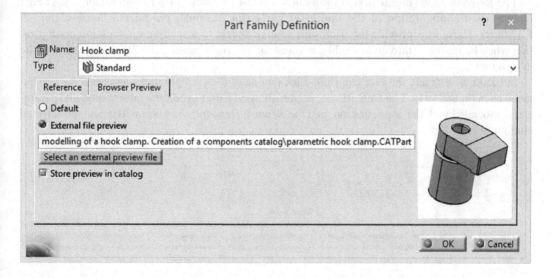

FIGURE 4.109 Displaying the *Reference* tab in the *Part Family Definition* dialog box.

In the *Browser Preview* tab (Figure 4.110), *External file preview* and *Store preview in the catalog* options are checked for the program to automatically create an image for the families of parts and attach it to the file where the catalogue is saved. The user also has the option to upload an image by clicking the *Select an external preview file* button.

For example, Figure 4.111 shows the *hook clamp* catalogue with four configurations. The part's parameters (the *Keywords* tab), their values in the table for each configuration number (1 to 4) and the name of each variant (*Hook clamp 1,..., Hook clamp 4*, as shown in Figure 4.103) are noted. The *Hook clamp* family of parts is highlighted in blue and underlined in the specification tree, and next to it, in the area on the right, its details. If the *Parametric Hook clamp* chapter contained several families of parts, the user's access to their details would have been made by double-clicking the name of that family, which would be underlined and highlighted in blue.

Also, the new name of the catalogue is observed, *hook clamp.catalog*, because the file was saved with the name *hook clamp* and the default extension *.catalog*. The user can also save the catalogue with the *.txt* and *.csv* extensions, a sample content for each extension being shown in Figure 4.112.

FIGURE 4.110 Displaying the *Browser Preview* tab in the *Part Family Definition* dialog box.

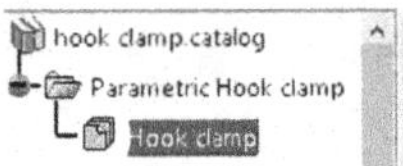

FIGURE 4.111 Displaying the *hook clamp* catalogue.

```
Nombre de description: 4
Description Name: Hook clamp 1
It's not a subChapter
Stored preview.
This Description has no objectFilter!
Object Name: C:_____Book_Hybrid_design\Chapter 4\4.5. Parametric modelling of a hook clamp.
Creation of a components catalog\parametric hook clamp.CATPart
It's a Family Instance.
This description is not computed.
Keyword Name (value = Hook clamp 1 )
Keyword (value = 10mm )
Keyword (value = 28,5mm )
Keyword (value = 9,5mm )
Keyword (value = 28mm )
Keyword (value = 12,5mm )
Keyword (value = 12mm )
Keyword (value = 6mm )
Keyword (value = 9mm )
Keyword (value = 45mm )
Keyword (value = 14mm )
Keyword (value = 20mm )
Keyword (value = 2,5mm )
Keyword PartNumber (value = Hook clamp 1)
Resolution Mode Hook clamp 1 (value = 0 )
Synchronization Mode Hook clamp 1 (value = 0 )
```

FIGURE 4.112 Displaying the *.txt* and *.csv* versions of the *hook clamp.catalog*.

The syntax used by each file type is different, but the information stored is the same, with the user intuitively accessing the names and parameter values. This information comes mainly from the *DesignTable.1*.

The *Reference* tab (Figure 4.113) shows how *CATIA v5* saves in *CATTemp* folder a *.CATPart* file for each configuration of the family of parts (for example, *parametric hook clamp_2. CATPart*). These files provide the user with a quick access to a specific *hook clamp* configuration when he inserts it into an assembly, without the program passing the 3D model through the design table.

In the *Keywords* tab, the user can right-click on one of the configurations and open the context menu as in Figure 4.114. The *Definition…* option allows the user to edit the information and parameters associated with the configuration, and the *Remove Description* option will delete the selected configuration.

FIGURE 4.113 Displaying the location where *CATIA* saves one *.CATPart* file per configuration.

Reference	Keywords	Preview	Generative Data							
Name	PartBody\Sketc...	PartBody\Pad.1\Fi...	PartBody\Sketch.2...	PartBody\Sketch.3...	PartBody\Pad.3\Firs...	PartBody\Sketch.4...	PartBody\Sketch...	PartBody\Hol...	PartBody\
1 Hook clamp 1	10mm	28,5mm	9,5mm	28mm	12,5mm	12mm	6mm	9mm	45mm
2 Hook clamp 2	12mm	30mm	10,5mm	31mm	14mm	13mm	6mm	10mm	48mm
3 Hook clamp 3	14mm	32mm	11,5mm	33mm	16mm	14mm	6mm	11mm	52mm
4 Hook clamp 4	16mm			34mm	17mm	15mm	7mm	12mm	57mm

Copy
Resolve Description

Definition...
Remove Description

Open Document
Open As New Document
Open Reference Document
Open As New Reference Document

FIGURE 4.114 Displaying the context menu for a configuration.

The *Open Document* option allows the user to open the program-saved file in the *CATTemp* folder, but the user does not have access to all *hook clamp* 3D model configurations, but only to parameters in the current configuration (for example, *Hook clamp 4*). The specification tree contains the name of the configuration as established in the table in Figure 4.103, and the presence of the *Relations* element is observed, but without parameters (Figure 4.115).

The *Open As New Document* option is similar, except that it opens a part in the selected configuration, but the file name is *CATPart1.CATPart*. The specification tree is identical to the one in the previous option.

The other two options, *Open Reference Document* and *Open As New Reference Document*, are also similar as the open file is the one saved after the parameterization of the hook clamp through the design table. The difference is that this file or a new reference of it with another name is open (for example, *CATPart2.CATPart*).

The specification tree in Figure 4.116 is identical to the one in Figure 4.94, being the same part, and contains under the *Relations* feature the design table and the configurations created by the user. The fourth configuration is selected because in Figure 4.114, the context menu with options was opened by right-clicking it.

These options allow the user to open the 3D model of the hook clamp in different configurations, with or without the possibility to choose another set of parameters, the user having access to the original file or its clones, created with certain dimensions.

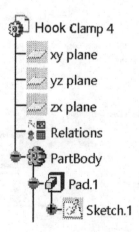

FIGURE 4.115 Displaying the specification tree of a part open from the context menu.

Part1
xy plane
yz plane
zx plane
Relations
DesignTable.1
Configuration=4
Sheet.1
PartBody
Pad.1
Sketch.1

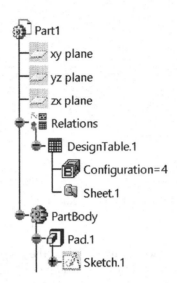

FIGURE 4.116 Displaying the specification tree of a part open as a reference from the context menu.

In practice, the methodology presented is particularly useful because it is possible to insert parameterized parts in an assembly without providing the access of some persons to the table, how the parameters of the part were associated, all the existing configurations, the catalogue created, etc.

A catalogue can contain, along with 3D geometry, other important information for its parts. If a material (for example, *Steel*; density 7860 kg/m^3) is applied to the hook clamp model, each configuration of the hook clamp, according to the design table, will have its own volume and, implicitly, its own mass. The user may need this information when choosing a part from the catalogue.

To insert a new column in the catalogue, the user should press the *Add Keyword* icon in the *Data* toolbar (Figure 4.117). In the *Keyword Definition* dialog box, the *Name* field contains the name of the column, the new property, which is *Mass* that has been chosen from the *Type* dropdown list.

A default value can be set for all configurations in the catalogue in the *Default Value* field, but since different volumes of parts are involved, each configuration will have its own mass written in the column. From the toolbar *Measure,* using the icon *Measure inertia,* the mass for each configuration can be found, which are (1) 110 g, (2) 179 g, (3) 266 g and (4) 370 g, respectively.

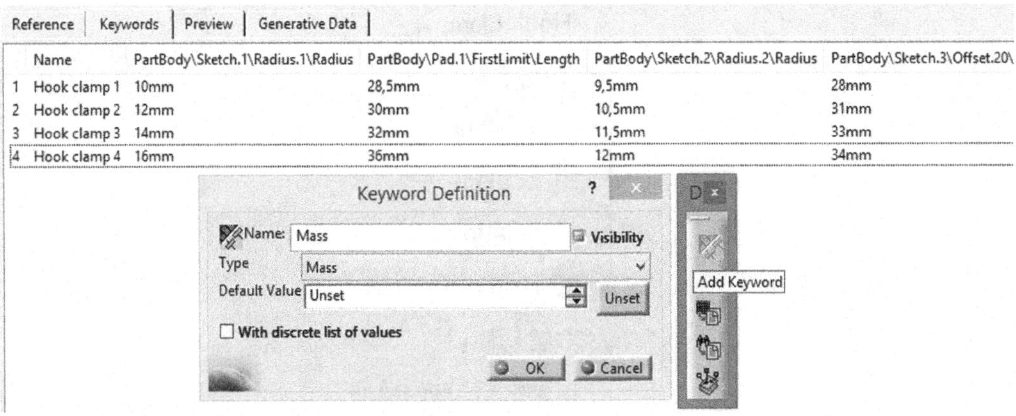

Reference
Keywords
Preview
Generative Data
Name
PartBody\Sketch.1\Radius.1\Radius
PartBody\Pad.1\FirstLimit\Length
PartBody\Sketch.2\Radius.2\Radius
PartBody\Sketch.3\Offset.20\
1 Hook clamp 1 10mm 28,5mm 9,5mm 28mm
2 Hook clamp 2 12mm 30mm 10,5mm 31mm
3 Hook clamp 3 14mm 32mm 11,5mm 33mm
4 Hook clamp 4 16mm 36mm 12mm 34mm
Keyword Definition
Name: Mass
Visibility
Type
Mass
Default Value
Unset
Unset
With discrete list of values
OK
Cancel
Add Keyword

FIGURE 4.117 Adding a new keyword in the catalogue.

In the list of the catalogue configurations, the user can right-click on one of them and choose the *Definition...* option from the context menu, as in Figure 4.118. The last column of the catalogue, the *Mass*, with three values already entered, can be seen.

In the *Description Definition* dialog box (Figure 4.119), in the *Keyword values* tab, a list with available parameters is displayed. At the bottom of the list, for the keyword *Mass*, the user enters the value of 370 g (for the fourth configuration, for example). The unit of measurement may or may not be specified, as it is automatically added by *CATIA v5*.

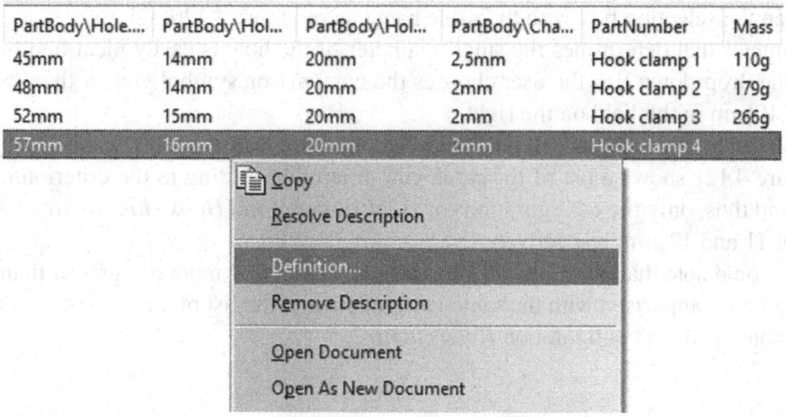

PartBody\Hole....	PartBody\Hol...	PartBody\Hol...	PartBody\Cha...	PartNumber	Mass
45mm	14mm	20mm	2,5mm	Hook clamp 1	110g
48mm	14mm	20mm	2mm	Hook clamp 2	179g
52mm	15mm	20mm	2mm	Hook clamp 3	266g
57mm	16mm	20mm	2mm	Hook clamp 4	

- Copy
- Resolve Description
- Definition...
- Remove Description
- Open Document
- Open As New Document

FIGURE 4.118 Editing a configuration from the context menu.

PartBody\Hole.1\HoleCounterBoredType.1\Depth	PartBody\Chamfer.1\ChamferRibbon.1\Length1	PartNumber	Mass
20mm	2,5mm	Hook clamp 1	110g
20mm	2mm	Hook clamp 2	179g
20mm	2mm	Hook clamp 3	266g
20mm	2mm	Hook clamp 4	370g

Description Definition — ☐ ✕

Name: Hook clamp 4

| Reference | Keyword values | Preview |

Keyword name	Value	^
PartBody\Hole.1\Diameter	12mm	
PartBody\Hole.1\HoleLimit.1\Depth	57mm	
PartBody\Hole.1\HoleCounterBoredType.1\Diameter	16mm	
PartBody\Hole.1\HoleCounterBoredType.1\Depth	20mm	
PartBody\Chamfer.1\ChamferRibbon.1\Length1	2mm	
PartNumber	Hook clamp 4	
Mass	370g	v

Value: 370g ⬍ Unset

OK   Cancel

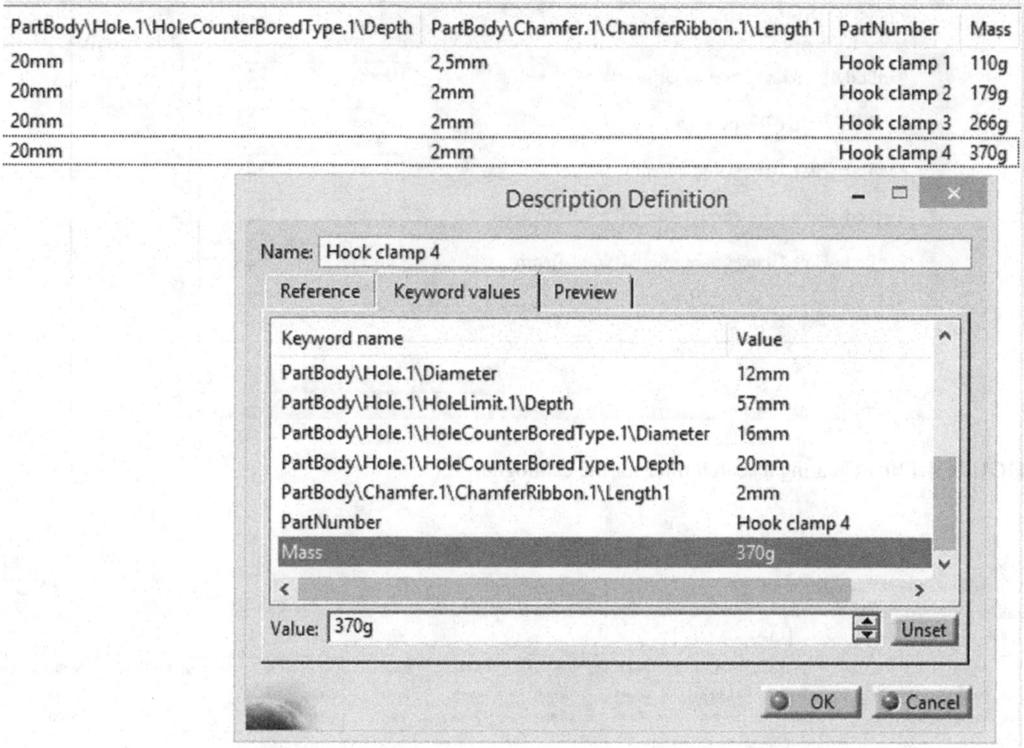

FIGURE 4.119 Adding a new value for a keyword of the selected configuration.

With this information entered, the user saves the *hook clamp.catalog* again. Containing only four configurations, the catalogue is relatively simple, but through the design table defined during the application, it can even contain dozens of configurations. To easily find a specific configuration based on a criterion, it is useful to define a search filter.

Thus, for example, the user will display from the catalogue only the configurations in which the small diameter of the *Hole.1* is larger than Ø10 mm. In the 2D drawing in Figure 4.80, the parameter that defines this diameter is *d*, and in the 3D model of the hook clamp, it is called *PartBody\Hole.1\ Diameter* (the list explained with parameters above Figure 4.89).

In the list of configurations, at the top of the screen, the user can click on the *Filter* icon (binoculars) and open the selection box with the same name as in Figure 4.120.

The parameter that determines the small diameter of the hole is easily identified by name, and then, from the drop-down list, the user chooses the comparison symbol greater than (>) and enters the value of 10 mm in the field on the right.

As a result, all configurations will be displayed containing values of the parameter *d* greater than 10 mm. Figure 4.121 shows a list of the catalogue, filtered according to the criterion of parameter *d* > 10 mm, and thus, only the configurations of *Hook clamp 3* and *Hook clamp 4* are displayed, with the values of 11 and 12 mm, respectively.

The user should note that if in Figure 4.120, the mathematical symbol >=(greater than or equal to) had been used for comparison with the same value of 10 mm, the list of the catalogue in Figure 4.121 would have contained the configuration *Hook clamp 2*.

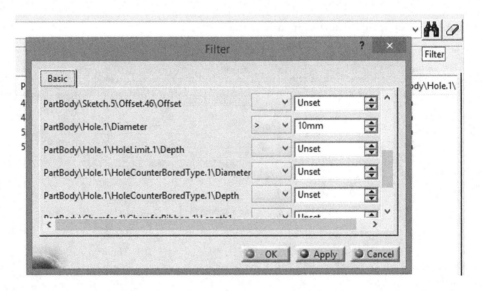

FIGURE 4.120   Creating a search filter for the catalogue.

Search
Filter: (x."PartBody\Hole.1\Diameter">10mm)
Result

	Name	PartBody...	PartBod...	PartBody\S...	PartBody...	PartBody\...	PartBody...	PartBod...	PartBody\Hole.1\Diameter	PartBody\...	PartBody\...
										Reference	Keywords	Preview	Generative Data
1	Hook clamp 3	14mm	32mm	11,5mm	33mm	16mm	14mm	6mm	11mm	52mm	15mm
2	Hook clamp 4	16mm	36mm	12mm	34mm	17mm	15mm	7mm	12mm	57mm	16mm

FIGURE 4.121   Displaying a filtered list of configurations based on the criterion defined.

The selection filter, once created and applied, is displayed in the *Filter* field. Even if the user saves the catalogue with a filter applied (so with fewer configurations), the next time the saved file is opened, the integrity of the catalogue is observed and all configurations are available and displayed. The filter only displays certain configurations and does not exclude the others from the catalogue.

A selection filter is active and influences the contents of the configuration list until the catalogue is reopened/loaded in the program or until the user decides to remove it (the filter) using the *Clear filter* icon (yellow eraser to the right of the *Filter* icon; Figure 4.121). According to Figure 4.120, there is a possibility of establishing complex filters, which compare several parameters with certain values; between each filtering sequence, the program inserts logical operators, written in the *Filter* field:

```
(x."PartBody\Sketch.1\Radius.1\Radius" >= 30mm) AND (x."PartBody\
   Hole.1\Diameter"> 10mm)
```

The AND operator can be replaced by OR operator, the filtering results being, of course, different.

The configuration list of this catalogue contains parameters organized in columns, as defined in the design table. When inserting a part from the catalogue, some columns that are important for the user's decision may not be in the top positions, so columns need to be rearranged.

In the specification tree of the catalogue, the user can right-click on the *Hook clamp* family of parts, and then from the context menu, he can choose the *Reorder keywords* option (Figure 4.122). Thus, the *Reorder Children* selection box opens (Figure 4.123), which displays a list of the parameters that define the variants of the hook clamp and four buttons for sorting: *Move up/down* the selected element and *Sort Ascending/Descending*. The *Mass* parameter (inserted according to Figure 4.117) was on the last column, and the diameter *d* of the hook clamp's hole (*PartBody\ Hole.1\Diameter*) was on the ninth column. Figure 4.123 shows the new positions of the respective columns after rearrangement.

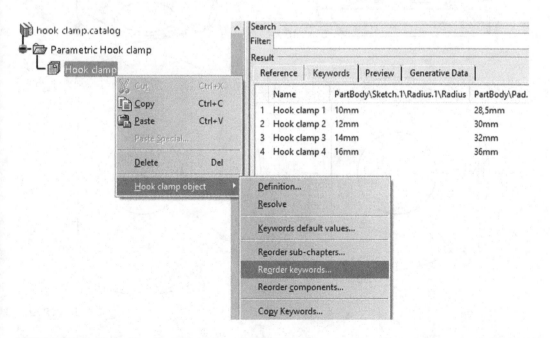

FIGURE 4.122 Launching the *Reorder keywords* from the context menu.

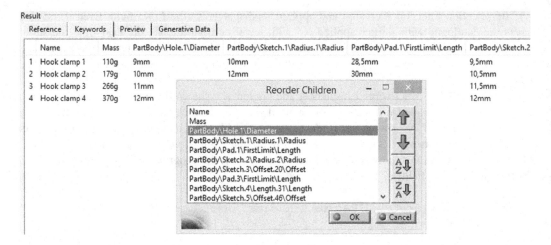

FIGURE 4.123   Rearranging the parameters of the parts family.

4.6   MODELLING OF A SHEET METAL COVER

This *CAD* application presents the steps of modelling a sheet metal part, obtained by cold blanking, piercing, drawing and forming processes, having the 2D drawing in Figure 4.124. Among the four projections in the 2D drawing, there is also an unfolded view, useful in the blanking stage. The application uses the *CATIA Generative Sheetmetal Design* workbench, which can be accessed from the *Start → Mechanical Design* menu.

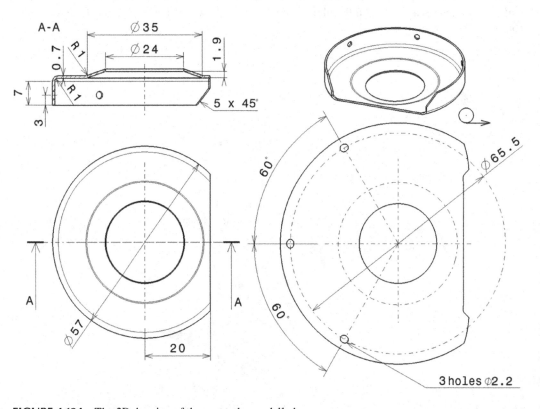

FIGURE 4.124   The 2D drawing of the part to be modelled.

The modelling of sheet metal parts cannot begin without establishing general parameters of the material of which the part will be made by cold-forming processes. Note that after accessing the workbench, all toolbars' icons are greyed out, except for *Sheet Metal Parameters*, which is in the *Walls* toolbar or in the *Insert* menu. The dialog box in Figure 4.125 shows the sheet thickness of 0.7 mm and the default bending radius of 1 mm (*Thickness* and *Default Bend Radius* fields in the *Parameters* tab). The specification tree is updated with the *Sheet Metal Parameter.1* feature.

A circle with a diameter of Ø53.6 mm (Figure 4.126) is drawn in a sketch of the *XY Plane*, with the centre at the origin of the coordinate system. Its value is obtained by subtracting from the size of Ø57 mm twice the thickness of the sheet (2×0.7 mm) and twice the value of the bending radius (2×1 mm).

To create the first flat surface of the part, the user can launch the *Wall* tool in the same *Walls* toolbar (Figure 4.127), with the *Sketch at extreme position* option. Using the red arrow, perpendicular to the plane, or the *Invert Material Side* button, the user can determine the direction in which the circle is extruded to obtain the disc of the sheet metal part. Thus, it becomes a reference wall, its thickness being obviously given by the value of the *Thickness* parameter, previously established (Figure 4.125).

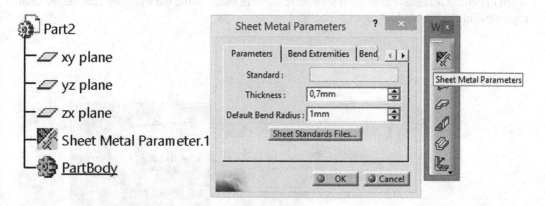

FIGURE 4.125 Displaying the *Sheet Metal Parameters* dialog box.

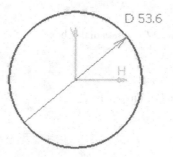

FIGURE 4.126 Displaying the first sketch of the sheet metal part.

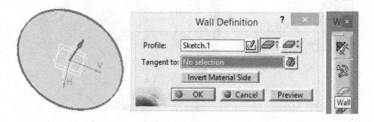

FIGURE 4.127 Displaying the *Wall Definition* dialog box.

The *Flange* tool bends the edge of the disc at a certain angle (90°), chosen by the user, creating, actually, a new wall with a R1 connection, the dimensions being established in the *Flange Definition* dialog box (Figure 4.128). The user has the possibility to change the connection radius, the value in the *Radius* field being initially taken from the *Default Bend Radius* field in the *Sheet Metal Parameters* dialog box (Figure 4.125).

The part has a cut at a distance of 20 mm from its centre, according to its 2D drawing. Cutting is done using the *Cut Out* tool in the *Cutting/Stamping* toolbar, which is similar in many ways (icon, application and options) to the *Pocket* tool in the *CATIA Part Design* workbench.

Thus, in a sketch of the *XY Plane*, the user draws a rectangle at 20 mm from the horizontal axis *H* (Figure 4.129). Its dimensions are not important, but in length and width, it must exceed the circumference of the part.

Cutout Definition dialog box shows the cutting options (*CutOut.1*, Figure 4.130). Thus, in the *Type* field in the *Cutout Type* area, the user can set the cutting mode, such as *Sheetmetal Standard* or *Sheetmetal Pocket*, the main difference being that the *pocket* is created only on a flat surface, the depth of cutting being less than the thickness of the wall sheet, while the *standard* cuts the material, regardless of the established depth.

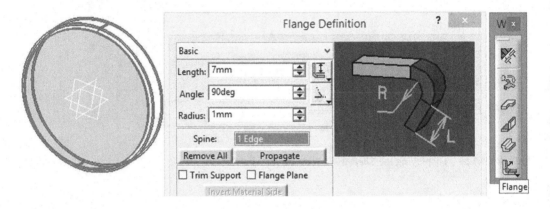

FIGURE 4.128 Displaying the *Flange Definition* dialog box.

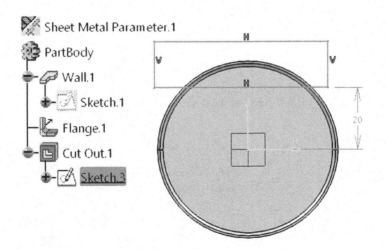

FIGURE 4.129 Displaying the sketch drawn for cutting the part.

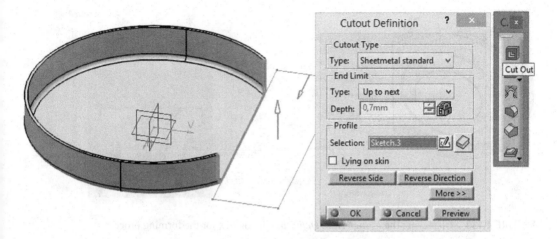

FIGURE 4.130 Displaying the *Cutout Definition* dialog box.

Figure 4.131 shows a very small edge in the area of the previous cut. Two such edges are selected (multiple selection using the *Ctrl* key on the keyboard), and by applying the *Chamfer* tool, similar to the tool with the same name in the *CATIA Part Design* workbench, two chamfers of 5×45° are created (Figure 4.132).

By applying the *Flanged Hole* tool, a circular perforation of the disc is obtained, combined with an outward reflection of the edges. The user should click on the icon with the same name and select the inner flat face at the base of the part (Figure 4.133).

In the *Flanged Hole Definition* dialog box, in the drop-down list in the *Parameters choice* field, the user can select the *Punch & Die* option out of the four available: *Major Diameter, Minor Diameter, Two diameters* and *Punch & Die*. Some of these options require entering values for two more diameters, one being in the *Diameter D* field and the other taking the place of the *Angle A* field, as the case may be, this field becoming *Diameter d* (*Two diameters* and *Punch & Die* cases).

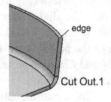

FIGURE 4.131 Displaying the part after the cutting process.

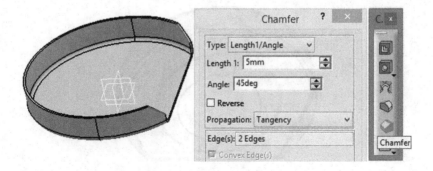

FIGURE 4.132 Displaying the *Chamfer Definition* dialog box.

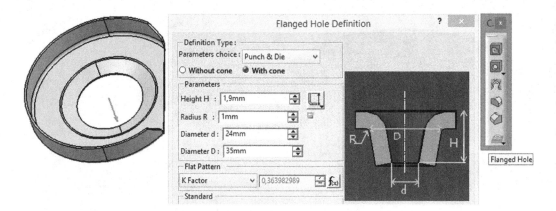

FIGURE 4.133 Displaying the *Flange Hole Definition* dialog box for the forming process.

In the *Height H* field, the user enters the value of the height of the perforated surface (1.9 mm) and, in the *Radius R* field, the value of the radius of reflection of the edges (1 mm). The value in the *Diameter D* (35 mm) field represents the large diameter, which is the diameter created when the punch enters the disc, and the value in the *Diameter d* field (24 mm) is the small diameter at the exit of the punch from the disc.

In the next stage, the three perforated holes will be created, arranged in a circle on the circumference of the part. Thus, in the *ZX Plane*, a circle with a diameter of Ø2.2 mm is drawn, located at 3 mm from the bent edge of the cover (Figure 4.134), according to the initial 2D drawing.

The circle (which has the centre point on the axis of the part) is involved in a cutting operation using the *Cut Out* tool. Thus, in the dialog box in Figure 4.135, the user should choose the type of the cut as *Sheetmetal Standard*, and the limit *Up to next*. Thus, a first hole (*Cut Out.2*) is obtained, on the bent wall. The hole should be multiplied twice on the left and the right of the newly created hole by using *Circular Pattern* tool in the *Transformations* toolbar. In the *Circular Pattern Definition* dialog box, in the *Reference element* field, the user chooses the inner cylindrical surface of the part, previously obtained with the *Flange tool* (Figure 4.128).

Clicking the *More >>* button expands the dialog box and makes the *Radial Alignment of Instance(s)* option available, which must be checked so that the axes of the multiplied holes are perpendicular to the cylindrical surface of the part (Figure 4.136).

Thus, two features *Circular pattern.1* and *Circular pattern.2* appear in the specification tree for the two holes on the left and right of the *Cut Out.2* hole.

As another working variant, already having a multiplication (*Circular pattern.1*), the second one can be obtained using the *Mirror* tool in the *Transformations* toolbar. Thus, in the *Mirror Definition*

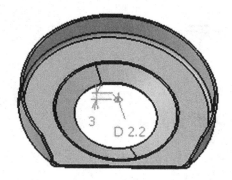

FIGURE 4.134 Displaying the sketch for perforated holes.

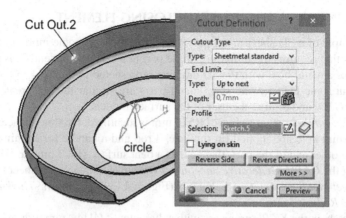

FIGURE 4.135 Displaying the *Cutout Definition* dialog box for creating the holes.

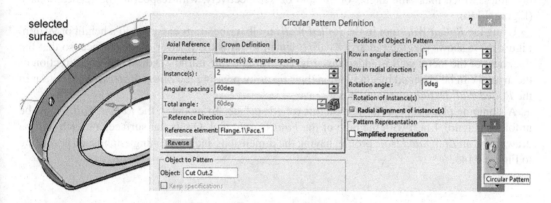

FIGURE 4.136 Displaying the *Circular Pattern Definition* dialog box.

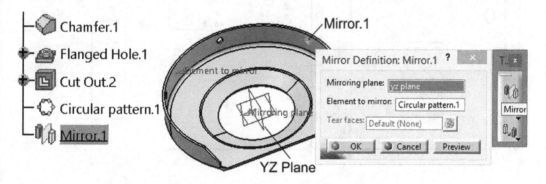

FIGURE 4.137 Displaying the *Circular Pattern Definition* dialog box with more options.

dialog box (Figure 4.137), in the *Element to mirror* field, the user selects the multiplied feature from the specification tree, and in the field above, *Mirroring plane*, the *YZ Plane*. The specification tree is completed, in this case, with the features *Circular pattern.1* and *Mirror.1*.

It should be noted that in the *CATIA Generative Sheetmetal Design* workbench, it is possible to use the *Mirror* tool of a previous element also obtained using another transformation tool (*Circular pattern* in the *Transformations* toolbar). This alternative way of working would not have been possible with the tools of the *CATIA Part Design* workbench.

4.7   MODELLING OF A SHEET METAL CLOSING ELEMENT

The second example of a 3D sheet metal part modelling presents some more advanced tools and also an approach that requires the unfolding of the part at a certain moment. This part is used as a closing element of a metal box, with the 2D drawing represented in Figure 4.138. It is not classical hybrid design, but some features and sheet metal parts are similar to ones created in *Generative Surface Design (GSD)* and remind the user of the surface models.

A rolling surface with an outer radius of R57.2 mm is identified. On this surface, a stamping and a piercing of Ø9 mm hole are applied (in this one, a rod and a handle for opening are fixed in an assembly). At the other end of the part, there exists a bent surface (R4.3 mm) as a hinge.

After opening the *CATIA Generative Sheetmetal Design* workbench, the sheet metal parameters are set in the *Sheet Metal Parameters* dialog box (Figure 4.139), as follows: *Thickness*: 1.3 mm and the *Default Bend Radius*: 3 mm.

In the first sketch, in the *XY Plane*, an arc with a diameter of Ø114.4 mm is drawn, with the centre at the origin of the coordinate system. The ends of the arc are delimited by two auxiliary construction lines, which make the angles of 51° and 62°, respectively, with respect to the vertical axis *V* (Figure 4.140).

Using the *Rolled Wall* tool in the *Rolled Walls* toolbar, the arc is extruded to a height of 94 mm. Figure 4.141 shows that the red arrows are oriented towards the interior of the surface, so that the thickness of the metal wall is added inwards, according to the initial 2D drawing. The direction of the arrows can be changed by clicking on their tip or by pressing the *Invert Material Side* button in the *Rolled Wall Definition* dialog box.

A stamping process is performed on this rolled surface. To model it, the *Rolled Wall.1* must be unfolded (Figure 4.142) with the help of the *Fold/Unfold* tool. On its flat surface, a sketch will be drawn with an *Elongated Hole* profile, having the dimensions and positioning established according to Figure 4.143.

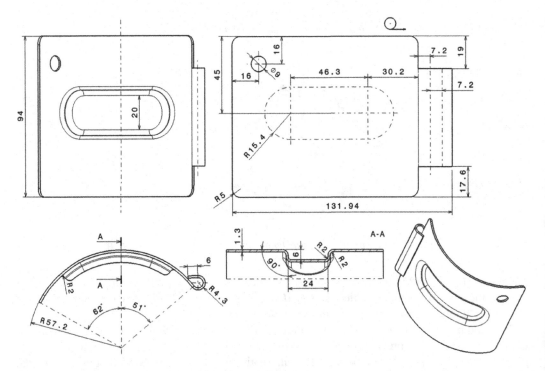

FIGURE 4.138   The 2D drawing of the sheet metal closing element.

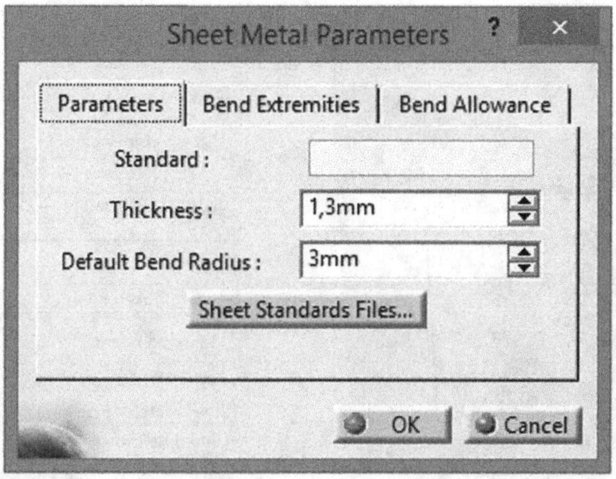

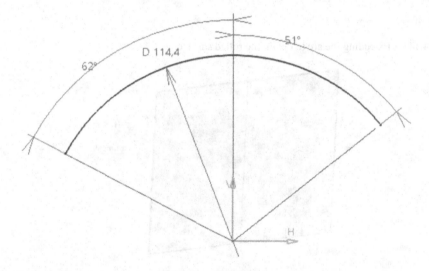

FIGURE 4.139   Setting the sheet metal parameters.

FIGURE 4.140   Drawing the first arc to create the rolled surface.

With the wall unfolded and the profile drawn on its flat surface, the *Point or Curve Mapping* tool in the *Bending* toolbar is applied to display the *Fold object definition* dialog box (Figure 4.144). In the *Object(s) list* field, the user selects the *Sketch.2* feature and the result is its projection on the wall surface, as if it were in rolled, working shape representation.

The specification tree is completed with the *Folded curve.1* feature, which also contains the *Sketch.2* (see Figure 4.145). After obtaining the profile projected on the surface, the *Sketch.2* is automatically hidden. For the representation in the figure, the user re-displays the sketch using the *Hide/Show* option in the context menu that appears by right-clicking on the *Sketch.2* feature in the specification tree.

Using the *Surface Stamp* tool in the *Cutting/Stamping* toolbar, the stamping is created on the rolled surface in the shape of the *Folded curve.1* profile. In the *Surface Stamp Definition* dialog box (Figure 4.145), in the *Profile* field, the user selects the *Folded curve.1* feature and, in the *Height H* field, enters the value of 6 mm, representing the embossing depth (Figure 4.138, section *A-A*). The values (2 mm) in the *Radius* fields indicate the inside and outside fillet radii.

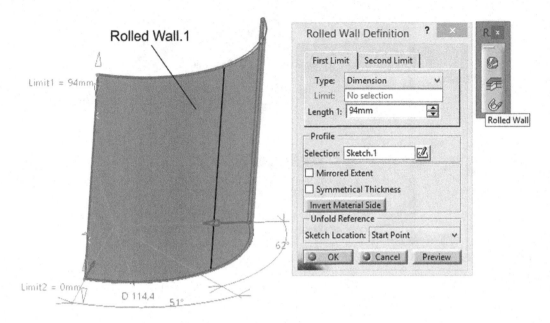

FIGURE 4.141 Extruding the arc to obtain the rolled surface.

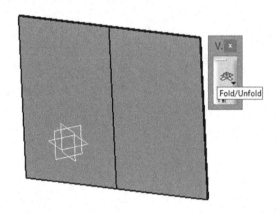

FIGURE 4.142 Unfolding the rolled surface.

The figure shows the other parameters of the stamping, and also a working view on it. It is also noticed that the *Sketch.2* (it is hidden) is the basis of the *Folded curve.1* profile, and this one determines the *Surface Stamp.1* stamping, according to the specification tree. Editing the dimensions of the sketch (Figure 4.143) leads to the change in the dimensions of the stamped surface. The feature *Folded curve.1* is an intermediary object between the sketch and the 3D shape of the resulting surface.

The part also has a perforation diameter of Ø9 mm at distances of 16 mm from the top and left edges of the sheet metal. To obtain the hole in the rolled surface of the part, the user applies the *Circular Cutout* tool on the same *Cutting/Stamping* toolbar. Its application can also be done directly on the rolled surface, but for a correct positioning of the hole centre, it is necessary to refer to either a previously created and positioned point, or a flat surface. Thus, the user will once again apply the *Fold/Unfold* tool to unfold the sheet metal part.

After activating the *Circular Cutout*, the user clicks on the flat/unfolded surface and starts positioning the centre of the hole. In the *Circular Cutout Definition* dialog box (Figure 4.146), he enters the diameter value in the *Diameter* field and notes that there is no option to set the distances from the centre to the part edges. The selection point represents the initial centre of the hole.

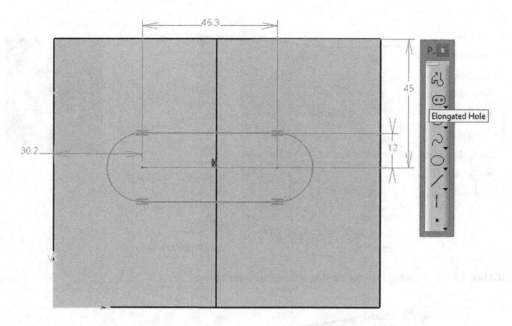

FIGURE 4.143 Drawing an elongated hole profile on the unfolded surface.

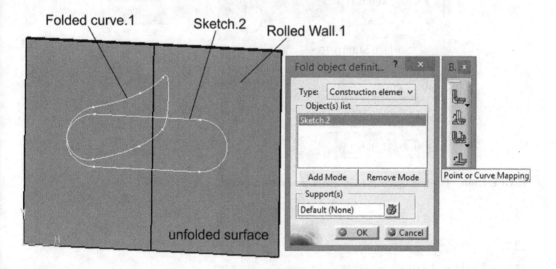

FIGURE 4.144 Representation of the folded and unfolded curves on the sheet metal surface.

Changing its position is done by editing (double-click) the *Sketch.3*, which is available in the specification tree under the *Circular Cutout.1* feature (Figure 4.146).

The user then returns to the rolled part (*Fold/Unfold*), a portion of which is visible in Figure 4.7.9. Obviously, because the hole was placed on the unfolded surface of the part (as it is presented in the 2D drawing), it does not have a circular shape anymore. The check can be done with the *Measure Item* tool on the *Measure* toolbar (Figure 4.147). Only the surface and perimeter information is available, but not the radius.

If the hole size (diameter) was given on one of the projections of the rolled surface in the 2D drawing, then the user should have used the *Hole* tool.

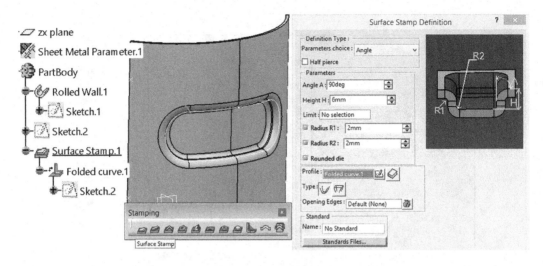

FIGURE 4.145   Creating the surface stamp using the folded curve.

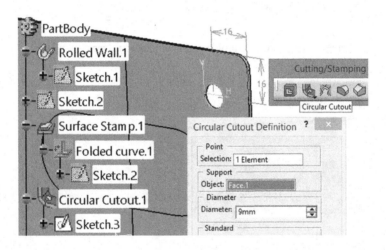

FIGURE 4.146   Positioning the hole in one of the part's corners.

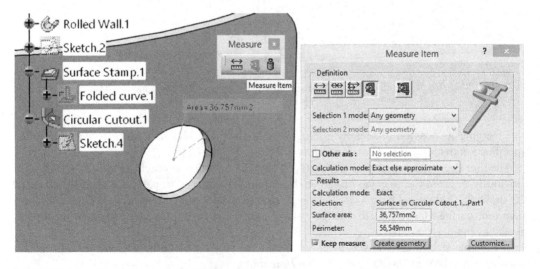

FIGURE 4.147   Checking the hole by measurement.

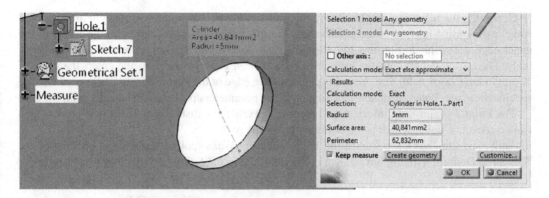

FIGURE 4.148 Checking the cylindrical hole by measurement.

Thus, for example, the *Hole.1* feature is inserted, and the application point of the centre is chosen on the rolled surface and can be completely defined with the help of the *Sketch.4* from Figure 4.148. The user applies the *Measure Item* tool on the hole surface to display the radius, surface and perimeter properties, so the surface is of *Cylinder* type.

Even if at this moment the hole has a circular shape on the rolled surface, on the unfolded surface, it loses this property. In conclusion, the choice of *Hole* and *Circular Cutout* tools is also made according to the way in which the holes/perforations are presented and dimensioned in the 2D drawings. Thus, after reading and understanding these explanations above, the user may remove the *Hole.1* feature because it is not indicated on the 2D drawing of the part.

The 3D model also has a flange-type bend, created using the *Tear Drop* tool, on the right side edge of the rolled wall. The bending, however, is not applied along the entire length of the edge, and it must be delimited at distances of 19 and 17.6 mm from the horizontal edges of the part, according to its 2D drawing (Figure 4.138).

Thus, the user clicks on the *Point* tool icon in the *Reference elements* toolbar, and in the *Point Definition* selection box (Figure 4.149), in the *Point type* options list, he chooses *On curve* and selects the edge on which the bend will be placed. The edge is represented in red, a reference point is chosen by default, and the user can specify another point by mouse click.

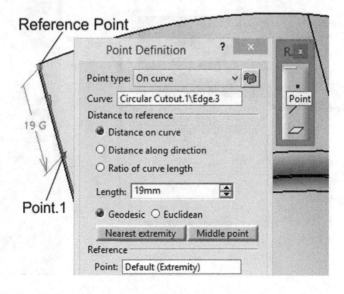

FIGURE 4.149 Placing the first point on the part's edge.

Generally, the reference point is set by the user as a particular point belonging to an edge over which a dimension is given in the 2D drawing. For this part, the point is chosen in the corner of the upper edge, according to Figure 4.149; thus, the *Point* field in the *Reference* area is filled in accordingly. Then, from the options, the user checks the *Distance on curve* option and enters the value of 19 mm in the *Length* field. The *Point.1* is placed on the edge of the part and becomes available in the specification tree. Later, by double-clicking on it, the parameters/position can be changed if needed.

The *Point.2* is similarly inserted, and the user specifies the distance of 17.6 mm (Figure 4.150) from a reference point, which is the bottom edge of the part.

Activating the *Tear Drop* tool in the *Walls → Swept Walls* toolbar leads to the opening of the *Tear Drop Definition* dialog box, shown in Figure 4.151. In the *Spine* field, the user selects the edge

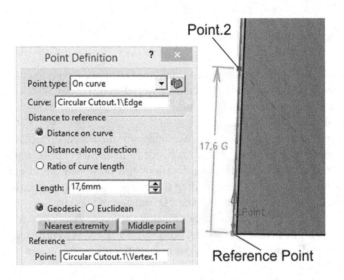

FIGURE 4.150 Placing the second point on the part's edge.

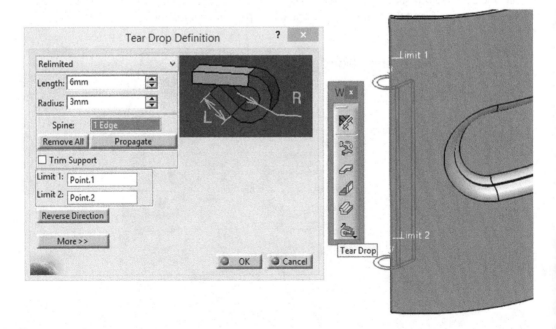

FIGURE 4.151 Choosing the bend limits points on the part's edge.

on which the two points *Point.1* and *Point.2* were inserted and then chooses the *Relimited* option from the drop-down list. The two fields *Limit 1* and *Limit 2* become available, and the user chooses the points by clicking on them in the workspace or in the specification tree. By default, if the *Basic* option is used from the drop-down list, the *Limit* fields are not available, and the bend is created along the whole length of the edge.

In the *Length* field, the user sets the length *L* of the right bending wall (6 mm), and, in the *Radius* field, the radius *R* of the bend (R3 mm), according to the representation next to the option fields. It is noticed that no value is required for the bending angle, but its wall is straightened and positioned close to (but without touching) the rolled wall of the part.

The four corners (multiple selection using the *Ctrl* key pressed) of the sheet metal part are rounded with a R5 mm radius applying the *Corner* tool on the *Cutting/Stamping* toolbar (Figure 4.152), and the final 3D model of the sheet metal part is obtained.

A tool similar to *Tear Drop* is *Hem* in the same toolbar *Walls* → *Swept Walls*. The options (Figure 4.153) are similar, but it is observed that the sheet metal wall of length *L* does not head towards the rolled wall of the part, but it is bent at an angle of 180°.

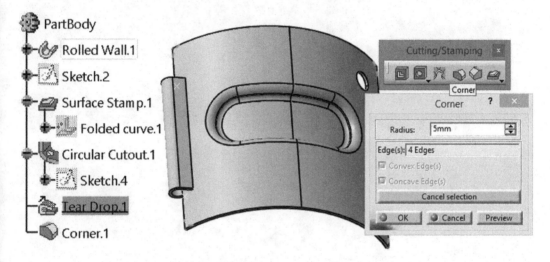

FIGURE 4.152 Selection of the *Corner* tool to round the part's corners.

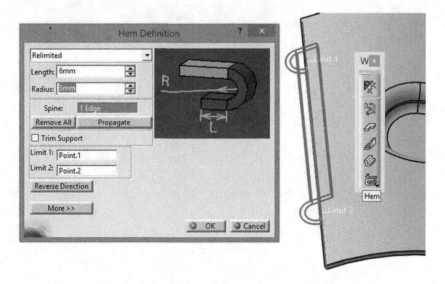

FIGURE 4.153 Additional example of application of the *Hem* tool and its options.

The difference between the two 3D models is the length of the sheet metal wall of R radius, which leads to a small change in the volumes, as follows: the part with the wall obtained by the *Tear Drop* tool has a volume of 15643.731 mm^3, and the part with the wall obtained by the *Hem* tool has a smaller volume, only 15391.17 mm^3. The measurements were performed using the *Measure Item* tool on the *Measure* toolbar (Figure 4.147).

Besides interesting features in designing sheet metal parts, in this *CAD* model, the user could also deal with various parameters: (1) general, defined at the beginning of the modelling process, and (2) specific, defined while using specific commands.

5 Programming, Automation and Scripting

5.1 INTRODUCTION TO AUTOMATION AND SCRIPTING IN CATIA v5

In the modelling process using *CATIA v5*, there are a number of necessary events that are repeated and thus time-consuming. It may be that mouse clicks are repeated in the same sequence a large number of times, or maybe there are many drawings that need to be converted to *pdf* format, or the user needs to import points to create geometry (e.g. spline). Hence, there is a need to find a way to automate such processes. And the answer to this need is automation and scripting in *CATIA v5*, done by creating or registering *Visual Basic (VB)* macros and scripts.

Macros are a sequence of functions, written in a scripting language, that are grouped into a single command and perform a predefined task automatically. *Macros* use notions of programming, but the user does not need to be a programmer or knowledgeable about it, although it would be useful. Thus, a macro reduces the time and possibilities of human errors by automating repetitive processes, standardizing, expanding *CATIA v5* capabilities and improving efficiency.

In *CATIA v5*, automation can be done with any application that can connect to the *Windows Component Object Model (COM)* interface, which can be *VBA, VBScript, JavaScript, .NET, Python*, etc. The *COM* is a *Microsoft* technology that allows binary code sharing between various applications and programming languages, and *CATIA v5* is one such application. *COM* object or component codes can be called, initialized or created at any time, being stored in *DLL (dynamic-link library)* files and registered in the *Windows* registry. When *CATIA v5* calls the *Excel* application, *CATIA v5* is the client and *Excel* is the server or component that provides the service to the client.

The most used programming languages, supported by *CATIA v5*, for creating macros are *CATScript* and *VBScript*.

CATScript is the *Dassault Systemes* version *of VBScript*, which are very similar. *CATScript* is a sequential programming language and is not graphically user oriented. Common text editors (*Notepad, Sublime Text, Atom*, etc.) can be used to write code. The advantages of using *CATScript* to write code include free macro registration, fast development and easy deployment. The disadvantages are low flexibility and difficulty in troubleshooting programs. The extension for such a file is *.CATScript*.

VBScript is a subset of *Visual Basic for Applications (VBA)*. Being a subset, all *VBScript* elements are found in *VBA*, but not all *VBA* elements are implemented in *VBScript*. The extension of a file written in *VBScript* is *.catvbs*.

In *CATIA v5*, there are two main ways to create a macro:

- recording a macro;
- using a text editor to write a new code.

In *CATIA v5*, macros are organized into libraries, which are of two types: *directories* and *VBA projects*. To create a library, the user should follow the following steps:

1. In the *Tools* menu, access *Macro → Macros*, as in Figure 5.1.
2. Click the *Macro Libraries* button.
3. Select the type of library as *Directories* and click *Add existing library...*, as in Figure 5.2.

DOI: 10.1201/9781003281153-5

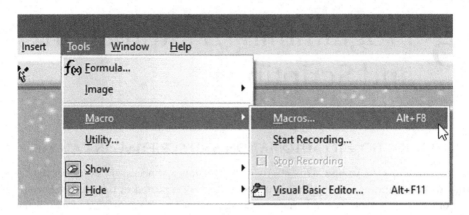

FIGURE 5.1 Displaying the *Tools* menu for accessing the *Macros* tool.

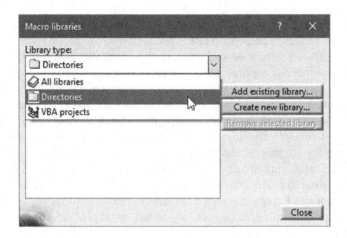

FIGURE 5.2 Displaying the selection box for adding a new or an existing library.

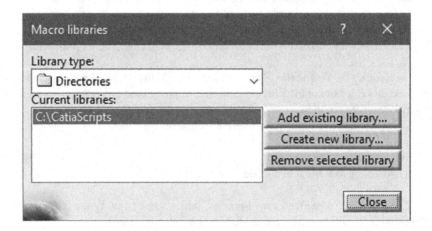

FIGURE 5.3 Displaying the result of adding an existing library.

4. Select a previously created directory (e.g. *C:\CatiaScripts*), where the *.CATScript* files are/ will be stored. The result will be as in Figure 5.3.
5. Close the *Macro Libraries* selection box.

Once the *Macro* library is created, as long as there are scripts with the *.CATScript* extension in the selected directory, they can be run. To open the macro window, the user can do one of the following:

1. *Tools* menu is accessed → *Macro* → *Macros*, as in Figure 5.1.
2. *Macros* toolbar is used.
3. Access the shortcut *Alt + F8*.
4. An icon can be created for a macro.

5.1.1 RECORDING A MACRO

As specified above, recording is one of the methods by which a macro is created. Recording a macro involves memorizing the actions that the user does by moving the mouse on the screen. For *macros* saved in a directory or in a *CATPart* or *CATProduct*, Dim statements will only be recorded for *CATScript*, not for *VBScript*. For *macros* registered in a *.catvba* library, *MS VBA* is the only option.

When recording a macro, a number of restrictions and obligations, given in Table 5.1, must be considered.

Also, when working with *macros*, it is recommended to consider some of the good practices, established over time, as follows:

1. Always *Undo* when a *Macro* has finished recording and run the *Macro* – this way, it checks if it has the expected results or not. If the *Macro* works in *CATIA v5* with good results, then the *Macro* code can be placed in a *VBA application*. Otherwise, the code needs to be revised.
2. After the *Macro* is registered, the user must verify the code. Sometimes, lines of code are added to the record, but they are not required. For example, a recorded macro for zoom in/zoom out may look like this:

```
Dim viewpoint3D As Viewpoint3D
Set viewpoint3D = viewer3D.Viewpoint3D
Viewer3D.ZoomIn
Set viewpoint3D = viewer3D.Viewpoint3D
Viewer3D.ZoomOut
Set viewpoint3D = viewer3D.Viewpoint3D
```

Note that 'Set viewpoint3D' appears several times. This occurrence is not necessary in this situation. It is enough that 'viewpoint3D' is set only once, after the statement Dim.
3. A *Macro* is often recorded in a window where a *CATPart* file is open. In testing, everything goes as planned, with the *Macro* having the expected results. In other situations, the *Macro* may not run for various reasons: there are several documents open and the *Macro* is not running in the right window or the *Macro* is running in a *CATProduct*. That is why it is necessary for the user to add error handling code to the macro.

TABLE 5.1

Restrictions and obligations when recording a macro

The User Should NOT...	The User MUST...
Switch between *CATIA v5* workbenches during recording!	Consider the *CATSettings* settings!
Record more than necessary!	Exit the workbench before recording is stopped!
Use *Undo* during recording!	Check each *macro* after it has been recorded!

5.1.2 Getting Started with Custom Code Writing

A *VBA* program in *CATIA v5* or a *Macro* is a subroutine called *CATMain ()*. In this function, the variables (using the Dim statement) are declared as type and a value (using the Set statement) is associated with them. The data types used can be *single, double, integer, string*, etc., or even *object*. The fact that the object can be a data type makes *CATIA v5* object-oriented programming (*OOP*). Understanding the principles of the *OOP* is critical to the success of macro programming. Thus, in the *OOP*, programmers define not only the type of data, but also the types of operations or functions that can be applied to the data structure. Everything in *CATIA v5* is an object; data fields are *properties*; and functions and subroutines are called *methods*. In *CATIA v5*, the user can distinguish:

- Object – an entity (in *CATIA v5* or *VB*): points, pads, ribs and parameters;
- *Property* – a characteristic of an object: *name, mass* and *color*;
- *Method* – an action performed by an object: *PartDocument.SaveAs ()*;
- *Collection* – a group or list of objects, grouped for a specific purpose.

Programming languages have a more or less rigid syntax, the knowledge of which depends on the successful creation of a script. In *CATIA v5*, the basic features of the syntax are as follows:

- Comments are added using the apostrophe ('), either at the beginning of a line or at the end of a line of code. It is recommended that comments be added as many times as necessary to document the code;
- Indentation is used to reflect the logical structure and make the code easier to read;
- Parentheses are used to get the expected result, e.g. order of operations;
- Colon (:) is used when you want to add multiple commands on the same line of code;
- Spaces are ignored by *VBA*, but are added to improve the clarity of the code;
- Underscore (_) – a character that is used to concatenate lines of code.

In a *Macro*, in *CATIA v5*, the code is run sequentially, from the first to the last line, except for repetitive structures, when the execution of the code must return to a certain line in the program. *Macros* can also be event driven, in the sense that execution depends on an event – the user presses a button, an object is selected in the graphics area, a certain text is entered in a form, etc.

5.1.3 CATIA Automation Documentation

CATIA v5 Automation documentation is installed with the program and can be accessed in the following two ways:

1. < *CATIA_installation_folder* >\intel_a\code\bin\v5Automation.chm;
2. *VBA* editor in *CATIA v5*, by accessing the menu *View* → *Object Browser* or *F2* on the keyboard.

The object architecture in *CATIA v5* presents a colour code that distinguishes between objects, abstract objects and object collections, as follows, according to Figure 5.4:

- Blue designates a single object.
- Purple designates an abstract object.
- Yellow designates a collection.
- A red arrow opens the next level of object structure.

The architecture of the objects, as in the documentation, is highlighted in Figure 5.5.

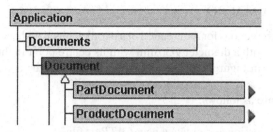

FIGURE 5.4 Displaying the colour coding for objects in *CATIA v5 Automation*.

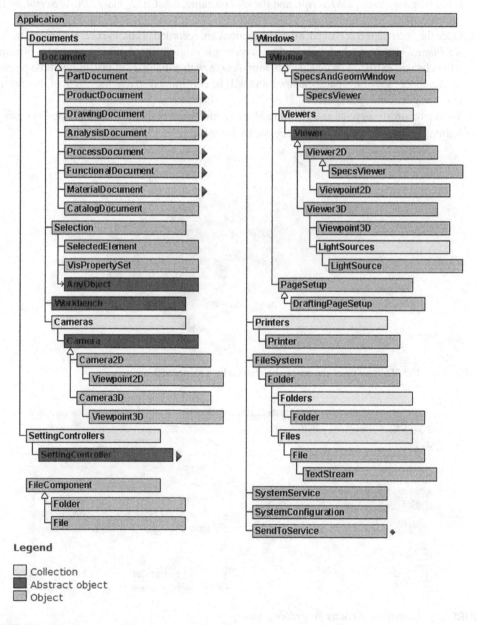

FIGURE 5.5 Displaying the infrastructure automation objects.

5.2 RECORDING A MACRO

It is proposed to record the actions for the creation of a parallelepiped with measures 100×60×20 mm and then making a hole with a diameter Ø30 mm with its centre point in the middle of the top face of the parallelepiped, as in Figure 5.6. The following steps should be completed in the process of recording a macro.

The steps to follow are as follows:

1. Open a *CATIA v5* session and create a new *CATPart* file.
2. Access the recording start command from the *Tools* menu → *Macro* → *Start recording...*, as in Figure 5.7.
3. Define the directory where the script will be saved, e.g. *C:\CatiaScripts*, the programming language used, *CATScript*, and the *macro* name, e.g. *Ch52_macro1.CATScript*, as in Figure 5.8.
4. Once the *Start* button is clicked, all mouse actions are recorded. Thus, the user will create in the *XY Plane* a sketch in which he will draw a rectangle origin-centred, measuring 100 × 60 mm, which he will extrude to a height of 20 mm. A hole with a diameter of Ø30 mm will be placed on the upper face of the parallelepiped and will be constrained so that the origin of the hole is at the origin of the axis system.
5. To stop recording actions and generate *Macro* code, the user should click the *Stop Recording* button, ▣ The first macro was successfully recorded.

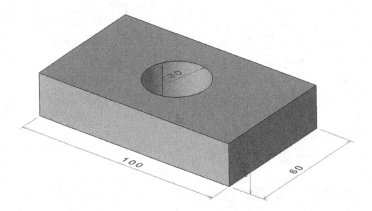

FIGURE 5.6   The 3D drawing of the part to be created by a recorded *Macro*.

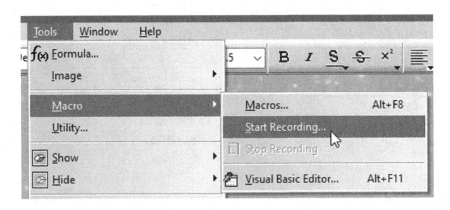

FIGURE 5.7   Launching the *Start Recording...* tool.

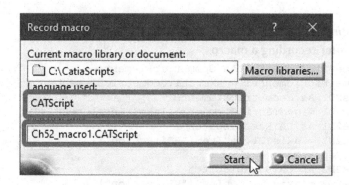

FIGURE 5.8 Setting the properties for the first recorded script.

Thus, in the directory selected for saving the script, *C:\CatiaScripts*, the file *Ch52_macro1. CATScript* can be found, the content of which is shown in Table 5.2. Although it is spread over several pages, the script code will be explained.

TABLE 5.2

Code obtained after recording a macro

```
001: Language="VBSCRIPT"
002: Sub CATMain()
003: Dim partDocument1 As Document
004: Set partDocument1 = CATIA.ActiveDocument
005: Dim part1 As Part
006: Set part1 = partDocument1.Part
007: Dim bodies1 As Bodies
008: Set bodies1 = part1.Bodies
009: Dim body1 As Body
010: Set body1 = bodies1.Item("PartBody")
011: Dim sketches1 As Sketches
012: Set sketches1 = body1.Sketches
013: Dim reference1 As Reference
014: Set reference1 = part1.CreateReferenceFromName("Selection_RSur:
(Face:(Brp:(AxisSystem.1;1);None:();Cf11:());AxisSystem.1;Z0;G4074)")
015: Dim sketch1 As Sketch
016: Set sketch1 = sketches1.Add(reference1)
017: Dim arrayOfVariantOfDouble1(8)
018: arrayOfVariantOfDouble1(0) = 0.000000
019: arrayOfVariantOfDouble1(1) = 0.000000
020: arrayOfVariantOfDouble1(2) = 0.000000
021: arrayOfVariantOfDouble1(3) = 1.000000
022: arrayOfVariantOfDouble1(4) = 0.000000
023: arrayOfVariantOfDouble1(5) = 0.000000
024: arrayOfVariantOfDouble1(6) = 0.000000
025: arrayOfVariantOfDouble1(7) = 1.000000
026: arrayOfVariantOfDouble1(8) = 0.000000
027: sketch1.SetAbsoluteAxisData arrayOfVariantOfDouble1
028: part1.InWorkObject = sketch1
029: Dim factory2D1 As Factory2D
030: Set factory2D1 = sketch1.OpenEdition()
```

(Continued)

TABLE 5.2 (*Continued*)

Code obtained after recording a macro

```
031: Dim geometricElements1 As GeometricElements
032: Set geometricElements1 = sketch1.GeometricElements
033: Dim axis2D1 As GeometricElement
034: Set axis2D1 = geometricElements1.Item("AbsoluteAxis")
035: Dim line2D1 As CATBaseDispatch
036: Set line2D1 = axis2D1.GetItem("HDirection")
037: line2D1.ReportName = 1
038: Dim line2D2 As CATBaseDispatch
039: Set line2D2 = axis2D1.GetItem("VDirection")
040: line2D2.ReportName = 2
041: Dim point2D1 As Point2D
042: Set point2D1 = factory2D1.CreatePoint(50.000000, 30.000000)
043: point2D1.ReportName = 3
044: Dim point2D2 As Point2D
045: Set point2D2 = factory2D1.CreatePoint(50.000000, -30.000000)
046: point2D2.ReportName = 4
047: Dim line2D3 As Line2D
048: Set line2D3 = factory2D1.CreateLine(50.000000, 30.000000,
50.000000, -30.000000)
049: line2D3.ReportName = 5
050: line2D3.StartPoint = point2D1
051: line2D3.EndPoint = point2D2
052: Dim point2D3 As Point2D
053: Set point2D3 = factory2D1.CreatePoint(-50.000000, -30.000000)
054: point2D3.ReportName = 6
055: Dim line2D4 As Line2D
056: Set line2D4 = factory2D1.CreateLine(50.000000, -30.000000,
-50.000000, -30.000000)
057: line2D4.ReportName = 7
058: line2D4.StartPoint = point2D2
059: line2D4.EndPoint = point2D3
060: Dim point2D4 As Point2D
061: Set point2D4 = factory2D1.CreatePoint(-50.000000, 30.000000)
062: point2D4.ReportName = 8
063: Dim line2D5 As Line2D
064: Set line2D5 = factory2D1.
CreateLine(-50.000000, -30.000000, -50.000000, 30.000000)
065: line2D5.ReportName = 9
066: line2D5.StartPoint = point2D3
067: line2D5.EndPoint = point2D4
068: Dim line2D6 As Line2D
069: Set line2D6 = factory2D1.CreateLine(-50.000000, 30.000000,
50.000000, 30.000000)
070: line2D6.ReportName = 10
071: line2D6.StartPoint = point2D4
072: line2D6.EndPoint = point2D1
073: Dim constraints1 As Constraints
074: Set constraints1 = sketch1.Constraints
075: Dim reference2 As Reference
076: Set reference2 = part1.CreateReferenceFromObject(line2D3)
077: Dim reference3 As Reference
078: Set reference3 = part1.CreateReferenceFromObject(line2D2)
```

(*Continued*)

TABLE 5.2 (*Continued*)

Code obtained after recording a macro

```
079: Dim constraint1 As Constraint
080: Set constraint1 = constraints1.AddBiEltCst(catCstTypeVerticality,
reference2, reference3)
081: constraint1.Mode = catCstModeDrivingDimension
082: Dim reference4 As Reference
083: Set reference4 = part1.CreateReferenceFromObject(line2D4)
084: Dim reference5 As Reference
085: Set reference5 = part1.CreateReferenceFromObject(line2D1)
086: Dim constraint2 As Constraint
087: Set constraint2 = constraints1.AddBiEltCst(catCstTypeHorizontality,
reference4, reference5)
088: constraint2.Mode = catCstModeDrivingDimension
089: Dim reference6 As Reference
090: Set reference6 = part1.CreateReferenceFromObject(line2D5)
091: Dim reference7 As Reference
092: Set reference7 = part1.CreateReferenceFromObject(line2D2)
093: Dim constraint3 As Constraint
094: Set constraint3 = constraints1.AddBiEltCst(catCstTypeVerticality,
reference6, reference7)
095: constraint3.Mode = catCstModeDrivingDimension
096: Dim reference8 As Reference
097: Set reference8 = part1.CreateReferenceFromObject(line2D6)
098: Dim reference9 As Reference
099: Set reference9 = part1.CreateReferenceFromObject(line2D1)
100: Dim constraint4 As Constraint
101: Set constraint4 = constraints1.AddBiEltCst(catCstTypeHorizontality,
reference8, reference9)
102: constraint4.Mode = catCstModeDrivingDimension
103: Dim reference10 As Reference
104: Set reference10 = part1.CreateReferenceFromObject(line2D3)
105: Dim reference11 As Reference
106: Set reference11 = part1.CreateReferenceFromObject(line2D5)
107: Dim point2D5 As CATBaseDispatch
108: Set point2D5 = axis2D1.GetItem("Origin")
109: Dim reference12 As Reference
110: Set reference12 = part1.CreateReferenceFromObject(point2D5)
111: Dim constraint5 As Constraint
112: Set constraint5 = constraints1.AddTriEltCst(catCstTypeEquidistance,
reference10, reference11, reference12)
113: constraint5.Mode = catCstModeDrivingDimension
114: Dim reference13 As Reference
115: Set reference13 = part1.CreateReferenceFromObject(line2D4)
116: Dim reference14 As Reference
117: Set reference14 = part1.CreateReferenceFromObject(line2D6)
118: Dim reference15 As Reference
119: Set reference15 = part1.CreateReferenceFromObject(point2D5)
120: Dim constraint6 As Constraint
121: Set constraint6 = constraints1.AddTriEltCst(catCstTypeEquidistance,
reference13, reference14, reference15)
122: constraint6.Mode = catCstModeDrivingDimension
123: Dim reference16 As Reference
124: Set reference16 = part1.CreateReferenceFromObject(line2D4)
```

(*Continued*)

TABLE 5.2 (*Continued*)

Code obtained after recording a macro

```
125: Dim constraint7 As Constraint
126: Set constraint7 = constraints1.AddMonoEltCst(catCstTypeLength,
reference16)
127: constraint7.Mode = catCstModeDrivingDimension
128: Dim length1 As Dimension
129: Set length1 = constraint7.Dimension
130: length1.Value = 100.000000
131: Dim reference17 As Reference
132: Set reference17 = part1.CreateReferenceFromObject(line2D3)
133: Dim constraint8 As Constraint
134: Set constraint8 = constraints1.AddMonoEltCst(catCstTypeLength,
reference17)
135: constraint8.Mode = catCstModeDrivingDimension
136: Dim length2 As Dimension
137: Set length2 = constraint8.Dimension
138: length2.Value = 60.000000
139: sketch1.CloseEdition
140: part1.InWorkObject = sketch1
141: part1.Update
142: Dim shapeFactory1 As Factory
143: Set shapeFactory1 = part1.ShapeFactory
144: Dim pad1 As Pad
145: Set pad1 = shapeFactory1.AddNewPad(sketch1, 20.000000)
146: part1.Update
147: Dim reference18 As Reference
148: Set reference18 = part1.CreateReferenceFromBRepName("FSur:(Face:
(Brp:(Pad.1;2);None:());Cf11:());WithTemporaryBody;WithoutBuildError;
WithInitialFeatureSupport;MonoFond;MFBRepVersion_CXR15)", pad1)
149: Dim hole1 As Hole
150: Set hole1 = shapeFactory1.AddNewHoleFromPoint(-0.198835, 3.563585,
20.000000,
reference18, 10.000000)
151: hole1.Type = catSimpleHole
152: hole1.AnchorMode = catExtremPointHoleAnchor
153: hole1.BottomType = catFlatHoleBottom
154: Dim limit1 As Limit
155: Set limit1 = hole1.BottomLimit
156: limit1.LimitMode = catOffsetLimit
157: Dim length3 As Length
158: Set length3 = hole1.Diameter
159: length3.Value = 10.000000
160: hole1.ThreadingMode = catSmoothHoleThreading
161: hole1.ThreadSide = catRightThreadSide
162: hole1.SetOrigin -7.428782, 3.138123, 20.000000
163: hole1.BottomType = catTrimmedHoleBottom
164: limit1.LimitMode = catUpToLastLimit
165: length3.Value = 30.000000
166: Dim sketch2 As Sketch
167: Set sketch2 = hole1.Sketch
168: part1.InWorkObject = sketch2
169: Dim factory2D2 As Factory2D
170: Set factory2D2 = sketch2.OpenEdition()
```

(Continued)

TABLE 5.2 (*Continued*)

Code obtained after recording a macro

```
171: Dim geometricElements2 As GeometricElements
172: Set geometricElements2 = sketch2.GeometricElements
173: Dim point2D6 As GeometricElement
174: Set point2D6 = geometricElements2.Item("Point.1")
175: point2D6.SetData 0.198835, -3.563585
176: Dim axisSystems1 As AxisSystems
177: Set axisSystems1 = part1.AxisSystems
178: Dim axisSystem1 As AxisSystem
179: Set axisSystem1 = axisSystems1.Item("Absolute Axis System")
180: Dim reference19 As Reference
181: Set reference19 = part1.CreateReferenceFromBRepName("FVertex:(Vertex:
(Neighbours:(Face:(Brp:(AxisSystem.1;2);None:();Cf11:());
Face:(Brp:(AxisSystem.1;3);None:();Cf11:());Face:(Brp:(AxisSystem.1;1);
None:();Cf11:()));Cf11:());WithPermanentBody;WithoutBuildError;
WithInitialFeatureSupport;MonoFond;MFBRepVersion_CXR15)", axisSystem1)
182: Dim geometricElements3 As GeometricElements
183: Set geometricElements3 = factory2D2.CreateProjections(reference19)
184: Dim geometry2D1 As GeometricElement
185: Set geometry2D1 = geometricElements3.Item("Mark.1")
186: geometry2D1.Construction = True
187: Dim constraints2 As Constraints
188: Set constraints2 = sketch2.Constraints
189: Dim reference20 As Reference
190: Set reference20 = part1.CreateReferenceFromObject(point2D6)
191: Dim reference21 As Reference
192: Set reference21 = part1.CreateReferenceFromObject(geometry2D1)
193: Dim constraint9 As Constraint
194: Set constraint9 = constraints2.AddBiEltCst(catCstTypeOn,
reference20, reference21)
195: constraint9.Mode = catCstModeDrivingDimension
196: sketch2.CloseEdition
197: part1.InWorkObject = hole1
198: part1.Update
199: End Sub
```

Before explaining the code, a few remarks are needed. Variables are the basis of any programming language. Basically, variables store information that can be accessed by referring to a name or '*variable*'. These variable names, also called identifiers, are very important and are subject to rules: they must be less than 255 characters long, they cannot start with a number or special character, and they must not create conflicts (two variables with the same name).

- Good examples of variable names: *sketch2, length2, constraint4*;
- Wrong examples of variable names: *2sketch, dimension one*.

Variables can be one of several types. If a user wants to store a name, he will use a *string* variable. An *integer* variable will be used to store someone's age, and for a real number value, a *double*-type variable will be used. The correct declaration of the types of variables allows the programmer to know what operations can be done with these variables. Thus, in *CATIA*, the following types of variables can be found, as in Table 5.3.

TABLE 5.3

Types of variables that can be defined in CATIA

Variable Type	Default Value	Description
String	Null	Text (letters, numbers and spaces)
Integer	0	Whole number
Double	0	Any number, positive or negative
Long	0	An *integer* represented in 4 *bytes*
Boolean	FALSE	A logical statement of the false or true type, yes or no
CATVariant	Null	*Index* of a list of objects
CATBStr	Null	*String* of a *CATIA expression*

In *VBA,* a variable is declared with name and type using the Dim. Thus, a variable can be of *primitive* type – *string, single, double* and *integer* – or of *object* type. If the variable type is not specified in the Dim statement, then *VBA* will declare the variable as *Variant* and will accept any type of variable. To reduce the code errors, it is a good idea to explicitly state the variable type. Another reason is the documentation of the code, which must be done, being a quality requirement. Such a declaration of a variable, using Dim, is called *dimming.*

Once the variable is declared by *dimming,* it must be assigned a value by the Set statement. For *object*-type variables, the Set command is used to define the object by variable. In the program, the variable is the *object,* if it has not been set to another.

An example of Dim and Set statements is as follows:

```
003 : Dim partDocument1 As Document
```

The variable partDocument1 is declared, of an *object* type, the *object* being Document; see Figure 5.5.

```
004 : Set partDocument1 = CATIA.ActiveDocument
```

The partDocument1 variable is initialized with the value CATIA.ActiveDocument; i.e., the *CATIA v5* document opened in the current session is retained by the partDocument1 variable.

The code in Table 5.2 is explained as follows:

- A *VB CATIA v5* program or *macro* consists of a subroutine called CATMain(). *CATIA v5* recognizes this feature as a starting point for a *VBA application.* Any such routine must end with the End Sub declaration, line 199. A program/*macro* code is written or generated between two Sub declarations.

```
001 : Language = "VBSCRIPT"
002 : Sub CATMain ()
...
199 : End Sub
```

- A partDocument1 variable is declared of an *object* type, the object being Document. This variable is initialized with the document active in the *CATIA v5* session.

```
003 : Dim partDocument1 As Document
004 : Set partDocument1 = CATIA.ActiveDocument
```

- The part1 variable is declared, which is initialized as part of the current document in the open *CATIA v5* session, partDocument1.

```
005 : Dim part1 As Part
006 : Set part1 = partDocument1.Part
```

- A variable, bodies1, is then declared as an *object* collection type and is initialized as part of the part1 object. Another object, body1, of *object* type, is declared, which is initialized as part of the bodies1 collection and is called PartBody.

```
007 : Dim bodies1 As Bodies
008 : Set bodies1 = part1.Bodies
009 : Dim body1 As Body
010 : Set body1 = bodies1. Item (" PartBody ")
```

- A new variable is declared, sketches1, of the *object* collection type, and is initialized as part of the body1 object.

```
011 : Dim sketches1 The Sketches
012 : Set sketches1 = body1.Sketches
```

- The reference1 variable is also declared, of *object* type, and is initialized as part of the part1 object.

```
013 : Dim reference1 As Reference
014 : Set reference1 = part1. CreateReferenceFromName ("Selection_
RSur: (Face: (Brp: (AxisSystem.1; 1); None :(); Cf11 :());
AxisSystem.1; Z0; G4074)")
```

The hierarchy of objects listed above, *Document, Part, Bodies, Body, Sketches* and *Sketch*, is shown in Figure 5.9.

- A sketch1 variable is declared, which is of *object* type, and reference1 is assigned to it. An absolute axis system is initialized, which is defined by an 8-element array variable, declared by arrayOfVariantOfDouble1 (8)

```
015 : Dim sketch1 As Sketch
016 : Set sketch1 = sketches1. Add (reference1)
017 : Dim arrayOfVariantOfDouble1 (8)
018 : arrayOfVariantOfDouble1 (0) = 0.000000
019 : arrayOfVariantOfDouble1 (1) = 0.000000
020 : arrayOfVariantOfDouble1 (2) = 0.000000
021 : arrayOfVariantOfDouble1 (3) = 1.000000
022 : arrayOfVariantOfDouble1 (4) = 0.000000
023 : arrayOfVariantOfDouble1 (5) = 0.000000
024 : arrayOfVariantOfDouble1 (6) = 0.000000
025 : arrayOfVariantOfDouble1 (7) = 1.000000
026 : arrayOfVariantOfDouble1 (8) = 0.000000
027 : sketch1.SetAbsoluteAxisData arrayOfVariantOfDouble1
```

- The sketch1 is set as the working object in the part1 object. From this moment, the user can move on to create geometric elements.

```
part1.InWorkObject = sketch1
```

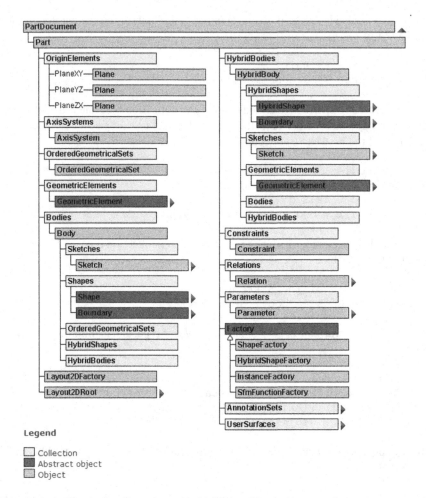

FIGURE 5.9 Displaying the *PartDocument* object's hierarchy.

- The *Sketch* object contains all the 2D geometric elements that define a sketch. Geometric elements are created using the *Factory2D* object. These elements are contained in the *GeometricElements* collection (Figure 5.10). To create *2D* elements, the user needs to *'open an edition'* for the sketch to work with, using the *OpenEdition ()* method.

```
029 : Dim factory2D1 As Factory2D
030 : Set factory2D1 = sketch1. OpenEdition ()
031 : Dim geometricElements1 As GeometricElements
032 : Set geometricElements1 = sketch1.GeometricElements
033 : Dim axis2D1 As GeometricElement
034 : Set axis2D1 = geometricElements1. Item (" AbsoluteAxis ")
035 : Dim line2D1 As CATBaseDispatch
036 : Set line2D1 = axis2D1. GetItem (" HDirection ")
037 : line2D1. ReportName = 1
038 : Dim line2D2 As CATBaseDispatch
039 : Set line2D2 = axis2D1. GetItem (" VDirection ")
040 : line2D2. ReportName = 2
```

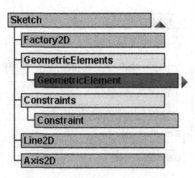

FIGURE 5.10 Displaying the *Sketch* object's hierarchy.

- The code between lines 040–127 creates a rectangle centred at the origin of the axis system.
- The code between lines 128–138 sets two dimensions, called length1 and length2, with the values 100 and 60, respectively, for the two sides of the centred rectangle, constrained by the variables constraint7 and constraint8.

```
128 : Dim length1 As Dimension
129 : Set length1 = constraint7.Dimension
130 : length1. Value = 100.000000
131 : Dim reference17 As Reference
132 : Set reference17 = part1. CreateReferenceFromObject (line2D3)
133 : Dim constraint8 As Constraint
134 : Set constraint8 = constraints1. AddMonoEltCst
(catCstTypeLength, reference17)
135 : constraint8. Mode = catCstModeDrivingDimension
136 : Dim length2 As Dimension
137 : Set length2 = constraint8.Dimension
138 : length2. Value = 60.000000
```

- Exit the sketch1, closing the *'edition'* with CloseEdition, and Update the part1 object.

```
139 : sketch1.CloseEdition
141 : part1.Update
```

- Once the sketch is created, two 3D operations follow – creating a solid using the *Pad* tool and drilling a hole on the solid's upper face, having the coordinates of the centre in the origin. Both 3D entities, *Pad* and *Hole*, are created using an object called ShapeFactory, which belongs to another Factory abstract object, as in Figure 5.11. This is how an *object*-type shapeFactory1 variable is declared. This object is used to make the *Pad*, line 145, using the AddNewPad method, which has as parameters the sketch1 and the extrusion height of 20.000000. Once the *Pad* is created, an update is run on the part1 object, at line 146.

```
142 : Dim shapeFactory1 As Factory
143 : Set shapeFactory1 = part1.ShapeFactory
144 : Dim pad1 As Pad
```

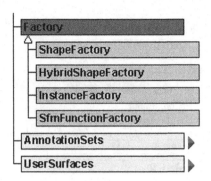

FIGURE 5.11   Displaying the *Factory* object's hierarchy.

```
145 : Set pad1 = shapeFactory1. AddNewPad (sketch1, 20.000000)
146 : part1.Update
```

- The next step, in the code, is to create a reference, by declaring a `reference18` variable, of *object* type, which represents the upper face of the `pad1` object.

```
147 : Dim reference18 As Reference
148 : Set reference18 = part1. CreateReferenceFromBRepName ("FSur:
(Face: (Brp: (Pad.1; 2); None :(); Cf11 :()); WithTemporaryBody;
WithoutBuildError; WithInitialFeatureSupport; MonoFond;
MFBRepVersion_CXR15)", pad1)
```

- Thus, using this reference, a hole is added, `line 150`, originating at an arbitrary point, with the other parameters having default values. One of the default parameters is the diameter of the hole, which has the value 10.000000, `line 159`, defined by the variable `length3`. Later, at `line   165`, this value is changed according to the sketch in Figure 5.11, to the value of 30.000000.

```
149 : Dim hole1 As Hole
150 : Set hole1 = shapeFactory1. AddNewHoleFromPoint (-0.198835,
3.563585, 20.000000, reference18, 10.000000)
151 : hole1. Type = catSimpleHole
152 : hole1. AnchorMode = catExtremPointHoleAnchor
153 : hole1. BottomType = catFlatHoleBottom
154 : Dim limit1 As Limit
155 : Set limit1 = hole1.BottomLimit
156 : limit1. LimitMode = catOffsetLimit
157 : Dim length3 As Length
158 : Set length3 = hole1.Diameter
159 : length3. Value = 10.000000
160 : hole1. ThreadingMode = catSmoothHoleThreading
161 : hole1. ThreadSide = catRightThreadSide
162 : hole1. SetOrigin -7.428782, 3.138123, 20.000000
163 : hole1. BottomType = catTrimmedHoleBottom
164 : limit1. LimitMode = catUpToLastLimit
165 : length3. Value = 30.000000
```

- Like any hole, it must be positioned. The 3D drawing in Figure 5.6 requires that this hole be in the centre of the upper face, and therefore the origin of the axis system, in the X and Y directions, respectively. An *object*-type sketch2 variable is declared, which controls the positioning of the hole. In order to position the hole, an *'edition'* is opened, line 170, in which the origin of the axis system, line 183, is projected. Two variables are declared, reference20 and reference21, which are initialized with the projected point in line 183 and, respectively, with the centre of the hole, point2D6, arbitrary point, at this moment, lines 190–192. In order to bring the hole's centre point to the projected point of the origin of the axis system, a constraint is created, lines 193–195. At the end, the *'edition'* for sketch2 is closed and the part1 object is updated.

```
166 : Dim sketch2 As Sketch
167 : Set sketch2 = hole1.Sketch
168 : part1. InWorkObject = sketch2
169 : Dim factory2D2 As Factory2D
170 : Set factory2D2 = sketch2. OpenEdition ()
171 : Dim geometricElements2 As GeometricElements
172 : Set geometricElements2 = sketch2.GeometricElements
173 : Dim point2D6 As GeometricElement
174 : Set point2D6 = geometricElements2. Item ("Point.1")
175 : point2D6.SetData 0.198835, -3.563585
176 : Dim axisSystems1 The AxisSystems
177 : Set axisSystems1 = part1.AxisSystems
178 : Dim axisSystem1 As AxisSystem
179 : Set axisSystem1 = axisSystems1. Item ("Absolute Axis System ")
180 : Dim reference19 As Reference
181 : Set reference19 = part1. CreateReferenceFromBRepName ("FVertex:
(Vertex: (Neighbors: (Face: (Brp: (AxisSystem.1; 2); None :(); Cf11
:()); Face: (Brp: (AxisSystem.1; 3); None :(); Cf11 :()); Face:
(Brp: (AxisSystem.1; 1); None :(); Cf11 :())); Cf11 :());
WithPermanentBody; WithoutBuildError; WithInitialFeatureSupport;
MonoFond; MFBRepVersion_CXR15) ", axisSystem1)
182 : Dim geometricElements3 As GeometricElements
183 : Set geometricElements3 = factory2D2. CreateProjections
(reference19)
184 : Dim geometry2D1 As GeometricElement
185 : Set geometry2D1 = geometricElements3. Item ("Mark.1")
186 : geometry2D1. Construction = false
187 : Dim constraints2 The Constraints
188 : Set constraints2 = sketch2.Constraints
189 : Dim reference20 As Reference
190 : Set reference20 = part1. CreateReferenceFromObject (point2D6)
191 : Dim reference21 As Reference
192 : Set reference21 = part1. CreateReferenceFromObject
(geometry2D1)
193 : Dim constraint9 As Constraint
194 : Set constraint9 = constraints2. AddBiEltCst (catCstTypeOn,
reference20, reference21)
195 : constraint9. Mode = catCstModeDrivingDimension
196 : sketch2.CloseEdition
197 : part1. InWorkObject = hole1
198 : part1.Update
```

It is important to observe that the first thing to do after recording a macro is to run it under the same conditions as it was recorded. Subsequently, the code can be optimized, if necessary.

5.3 DEVELOPMENT OF VBSCRIPT SCRIPTS

So far, the recording of *macros* and their running have been presented. It should be noted that since the code is run in the same process as *CATIA v5*, it is called the '*in-process*' macro. This chapter will cover the aspects of how to run a code sequence in another application, in a process other than *CATIA v5*, such as *Visual Basic*, *VBA* and *Windows Scripting Host with VBScript*.

The development and execution of a script from a process other than *CATIA v5* is called the '*out-process*' macro. In principle, the *Visual Basic editor* will be used to write code that will be executed, regardless of whether the application is open or not. The code will open *CATIA v5* and manipulate objects in the 3D design program.

A variable naming convention, called '*Hungarian notation*', will be used whenever necessary. In this convention, a letter or group of letters precedes the name of the variable, to suggest its type, for example:

- *o* or *obj* – object, e.g. *oPartDoc*
- *s* or *sel* – selection, e.g. *sSelect1*
- *str* – character string, e.g. *strFileName*
- *i* or *int* or *n* – integer, e.g. *iCount*
- *rng* – interval (range), e.g. *rngCamp*.

Visual Basic and *VBA* provide the tools needed for easy code development, with information about objects, their properties and methods that can be used with exposed objects.

To be able to write the code, the user should follow the steps below:

1. Create a new library, *VBA projects* type, as in Figure 5.12.
2. Create a *macro*, using the *MS VBA* language, as in Figure 5.13. The first script will be named *MyFirstScript*. Once created, the script will be edited by clicking the *Edit...* button.
3. The *Visual Basic editor* will open. In order to access the *CATIA v5* objects, properties and methods, from the *Tools* menu → *References*, all *CATIA v5* references must be selected (Figure 5.14).
4. By default, *CATIA v5* does not force the user to declare variables. This must be changed from the *Tools* menu → *Options* → *Require Variables Declaration*, as in Figure 5.15. This option adds as the first line of a module's code the `Option Explicit` statement, which causes *CATIA v5* to declare all variables.
5. A *Visual Basic* program, as mentioned above, consists of a 'subroutine' called `CATMain()`. *CATIA v5* only recognizes `Sub CATMain` as a starting point for any *VBA* application. Each subroutine must end with the `End Sub` declaration. The following code (without line numbers: 1, 2,...) will be entered in the editor:

```
1 : Option Explicit
2 :
3 : Under CATMain ()
4 : Dim strMessage As String
5 : strMessage = "This is my first script!"
6 : MsgBox strMessage
7 : End Sub
```

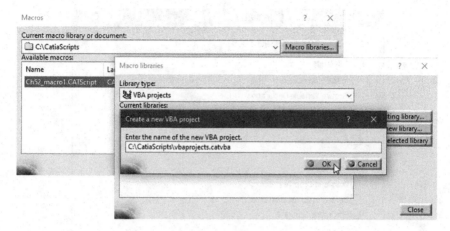

FIGURE 5.12 Displaying the *Create a new VBA project* window.

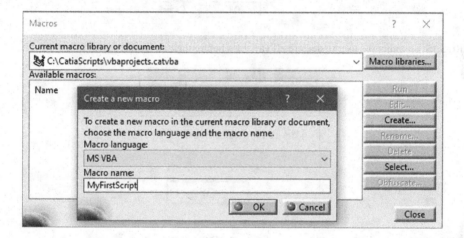

FIGURE 5.13 Setting the macro's language and name.

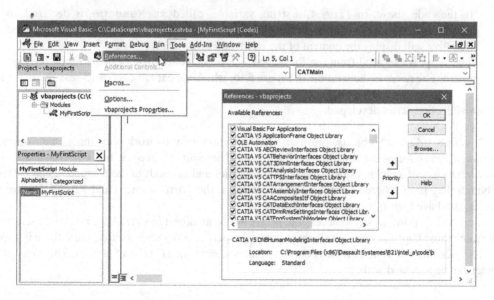

FIGURE 5.14 Enabling the *CATIA v5 References*.

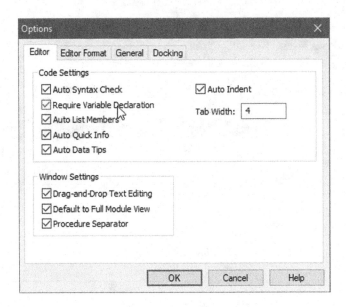

FIGURE 5.15   Forcing the user to declare all variables before their assignment.

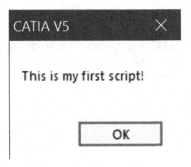

FIGURE 5.16   The result obtained after running the first *VBA* script.

In the code above, in `line 4`, a string variable, called `strMessage`, is declared, to which, in `line 5`, is assigned the value in quotation marks. In `Line 6`, a window called `MsgBox` will display the content of the `strMessage` variable.

6. The script is executed from the *Run* menu → *Run Sub/UserForm* or using *F5* on the keyboard. The result is the window in Figure 5.16, which appears in *CATIA v5* session.

This was the first script developed.

To access *CATIA v5* objects, the user needs to know how to work with three main objects: *Document, PartDocument* and *ProductDocument*. These are all classes, but *PartDocument* and *ProductDocument* are classes that inherit the properties and methods of the *Document* class. Thus, although they have inherited properties and methods, the *PartDocument* and *ProductDocument* classes also have their own methods and properties.

In normal working mode, to access the properties of an object in *CATIA v5*, it must be selected. The same thing happens when you want to access properties programmatically. Thus, there is a special object, called '*Selection*', which contains what is selected in a *CATIA v5* session. The '*Selection*' object can be accessed with:

```
1 : Dim oSel As Selection
2 : Set oSel = CATIA.ActiveDocument.Selection
```

If nothing is selected, then the selection is empty. If one or more objects are selected, then it contains those objects. The Add function is used to add an item to the selection. Thus, the code becomes:

```
1 : Dim oSel As Selection
2 : Set oSel = CATIA.ActiveDocument.Selection
3 : oSel.Clear
4 : oSel.Add (ObjectToAdd)
```

As a good practice, it is recommended that, when working with a selection, it be cleared (line 3: oSel.Clear) before and after use.

Within a selection, items can be searched by certain criteria: name, type, colour, etc., as follows:

- To search for 'Pad', use the following code:

  ```
  oSel.Search " Name = Pad. *, all "
  ```

- To search for all points in a selection, use the command:

  ```
  oSel.Search " Type = Point, all "
  ```

- To search for all items coloured in blue, use the code:

  ```
  oSel.Search " Color = Blue, all "
  ```

The result is a list of objects that must be traversed iteratively before the components can be used.

From Figure 5.5, which shows the architecture of objects hierarchy in *CATIA v5*, it can be seen that, in addition to objects, there are also collections. The main collections of an object, in *CATIA v5*, are the following:

- Documents – such as *CATPart*, *CATProduct*, *CATProcess*, *CATDrawing* and *CATAnalysis*.
- Windows – this collection contains information about how the data in the document collection are viewed in *CATIA v5* (view: *shaded, wireframe*; *zoom level, background*, etc.).

CATIA v5 documents and collections have been presented so far, but it is not known how they are called, what are the methods used to manipulate them, and what is the syntax for these methods.

CATIA v5 documentation provides a useful tool for programmers called the *Object Browser*. It can be accessed in the *Visual Basic editor* from the *View* menu → *Object Browser*, or by pressing *F2* on the keyboard. The result will be the one presented in Figure 5.17.

In the *Object Browser* (Figure 5.17), two areas can be observed, marked with *3* and *4*, respectively. In area *3*, all the classes that can be used when developing *VB* programs for *CATIA v5* are displayed, and area *4* shows the members of these classes. The *Object Browser* tool also has an area that can be used to search for members or classes by name.

For example, if the user wants to find the method that allows the creation of a *Pad*, then the name of the method, the syntax, as well as the parameters necessary for its realization must be determined. In the search area, the '*pad*' keyword is entered and the *Enter* key is pressed. The result will be the one in Figure 5.18.

In Figure 5.18, the following areas are observed, according to the numbering:

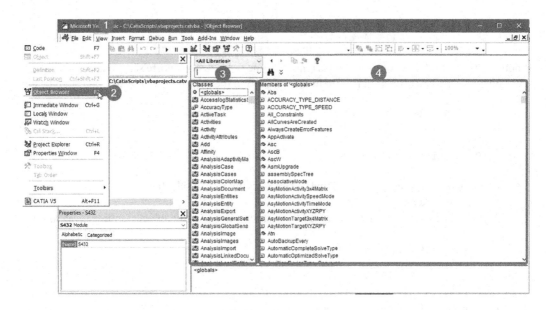

FIGURE 5.17   Displaying how to access the *Object Browser*.

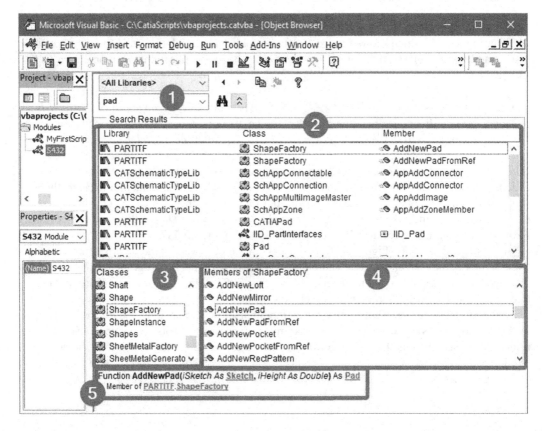

FIGURE 5.18   Displaying the '*pad*' keyword classes and members.

1. search area – where the user enters the keyword for searching;
2. search results area – these are organized by libraries – the first column, classes – the second column, and members – the third column;
3. zone 3 – which automatically selects the class from the first search result;
4. zone 4 – which displays the members of the class selected in zone 3;
5. zone 5 – which shows the function name, syntax and input parameters for the function selected in zone 4.

In this example, there is an `AddNewPad` method, which is part of the `ShapeFactory` class and supports as input parameters a sketch, `iSketch`, and a real number, `iHeight`, for the height of the *Pad*. Here are some examples.

Example 5.1

For the part in Figure 5.6, created using the registered macro *Ch52_macro1.CATScript*, it is desired to determine its mass and display the value in a *MessageBox* window. In addition to its mass value, the name of the part will be displayed in the same window. It starts from the premise that the *CATPart* file is already open.

The steps for creating the script are the following:

1. Open the *Visual Basic editor* using the combination *ALT + F11* or from the *Tools* menu → *Macro* → *Visual Basic editor*.
2. Add a new module from the *Insert* menu → *Modules*; its name should be changed to, for example, *App5_1*, as in Figure 5.19.
3. Enter the code from Figure 5.20, in which a series of variables of different types have been defined: *mass (double)*, *name (string)*, *doc1, part1, body1, objInertia* and *objSPA-Workbench* as the object type. It is worth noting, as novelty elements, the two objects, *objInertia* and *objSPAWorkbench*. The two objects are necessary to obtain the mass property for the considered part.
4. Run the script from the *Run* menu → *Run* or using *F5* on the keyboard, as in Figure 5.21.

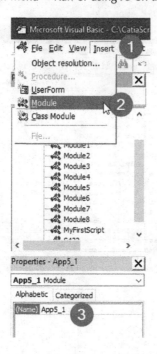

FIGURE 5.19 Displaying how to insert a new module and change its name.

```
01: Option Explicit
02:
03: Sub CATMain()
04:      Dim mass As Double
05:      Dim name As String
06:      Dim doc1 As Document
07:      Dim part1 As Part
08:      Dim body1 As Body
09:      Dim objInertia As Inertia
10:      Dim objSPAWorkbench As Workbench

11:      Set doc1 = CATIA.ActiveDocument
12:      Set part1 = doc1.Part
13:      Set body1 = part1.MainBody
14:      Set objSPAWorkbench = part1.Parent.GetWorkbench("SPAWorkbench")
15:      Set objInertia = objSPAWorkbench.Inertias.Add(body1)

16:      mass = objInertia.Mass
17:      name = part1.Name

18:      MsgBox "The name of the part is: " & name & vbNewLine &
"The mass of the part is: " & mass & " kg!"
19: End Sub
```

FIGURE 5.20   Displaying the code of Example 5.1.

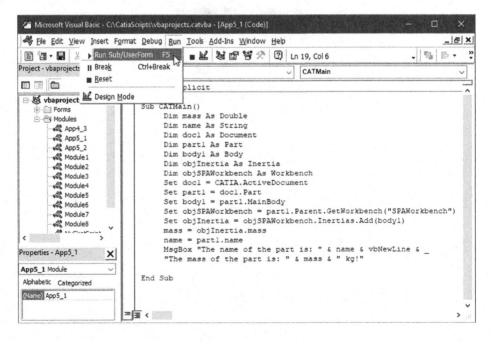

FIGURE 5.21   Displaying how to run a script.

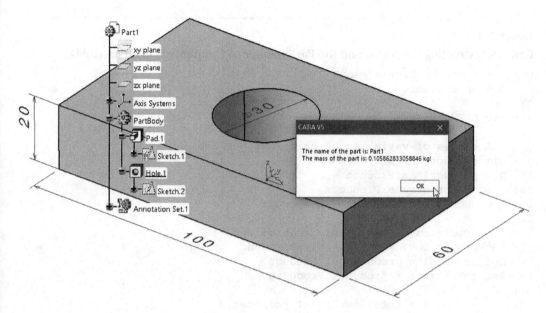

FIGURE 5.22 Displaying the result after running the script.

5. The result obtained by running the script is an information window in which the two properties are displayed: part name – *Part1* – and mass value – 0.10586 … kg, as in Figure 5.22.

Example 5.2

It is required that for a *CATProduct* file, traverse the specification tree and extract the name of each instance (Name) and component (PartNumber). The properties will be written to a text file. For ease of reading the code, it has been documented.

It is also assumed that a *CATProduct* file with at least two components is open. The steps for making the script are the following:

1. Open the *Visual Basic editor* using the combination *ALT + F11* or from the *Tools* menu → *Macro* → *Visual Basic editor.*
2. Add a new module from the *Insert* menu → *Modules,* and change its name to *App5_2.* The application will use a data type, ArrayList, which is not available in *VBA,* but an external reference will need to activate. Thus, from the *Tools* menu → *References,* the *mscorlib.dll* option will be checked.
3. Enter the code from Table 5.4.
4. Run the code from the *Run* menu → *Run* or using *F5* on the keyboard. The result of running the code will be a text file, named *export_product_parts.txt,* which will be saved, for example, in the *C:\CatiaScripts* location.
5. The code was tested for two *CATProduct* files, and the result is shown in Figure 5.23.

TABLE 5.4

Code for extracting the Name and the PartNumber of components in an assembly

```
01: Attribute VB_Name = "App5_2"
02: Option Explicit
03:
04: Sub CATMain()
05:
06: ' A series of variables of certain types are declared
07: Dim productDoc1 As Document
08: Dim product1 As Product
09: Dim components As Products
10: Dim number_of_components As Integer
11:
12: 'The declared variables are initialized
13: Set productDoc1 = CATIA.ActiveDocument
14: Set product1 = productDoc1.Product
15: Set components = product1.Products
16:
17: 'Determine the total number of components
18: number_of_components = components.Count
19:
20: 'A counter is declared to scroll through the list of components
iteratively
21: Dim contor As Integer
22:
23: 'A list is declared that will store the component names
24: Dim list_components As ArrayList
25: Set list_components = New ArrayList
26:
27: 'Scroll through the specification tree and add two values to the
list, Name and PartNumber. This makes the number of components in the
list always even, i.e. twice the number of components in the set.
28: For contor = 1 To number_of_components
29:     list_components.Add (components.Item(contor).Name)
30:     list_components.Add (components.Item(contor).PartNumber)
31: Next contor
32:
33: 'Declare a series of variables required to write to the text file.
34: Dim file_name As String
35: Dim text_components As String
36: Dim textRow As String
37: Dim fileNo As Integer
38:
39: 'A location will be set where the text file will be saved
40: file_name = "C:\CatiaScripts\export_product_parts.txt"
41:
42: 'fileNo is an identifier for open text files. It takes values between
1 and 511, and to avoid the confusion it causes, the function FreeFile
will be used.
43: fileNo = FreeFile
44:
```

(Continued)

TABLE 5.4 (*Continued*)

Code for extracting the Name and the PartNumber of components in an assembly

```
45: 'Opening file file_name for writing. It will be overwritten each time
the script is run. If no overwritten required, Output is replaced by Append
46: Open file_name For Output As #fileNo
47: 'Iterate the list up to half of its length. The counter starts at 0,
because the first item in the list has the index 0, list_components(0).
At line 49, the length of the list_components list is divided by 2 to
determine the number of components in the set. Write in the text file the
elements of the list in pairs, even/odd. For example, the list components
for a three-component set are Name1, PartNumber1, Name2, PartNumber2,
Name3 and PartNumber3. The length of the list is 6. As the index of the
first element is 0, the range 1-3 (three elements) will be moved to the
left by one unit, this becoming 0-2 (three elements). To find the
component name, the even-indexed elements will be extracted, and for
PartNumber, the odd-indexed elements will be extracted and written, at
line 50, as pairs (Name, PartNumber) in the text file, in the format set
on line 50.
48: Dim i As Integer
49: For i = 0 To list_components.Count / 2 - 1
50:     text_components = list_components(2 * i) &"
>> " & list_components(2 * i + 1)
51:     Print #fileNo, text_components
52: Next i
53:
54: 'Close the file
55: Close #fileNo
56:
57: End Sub
```

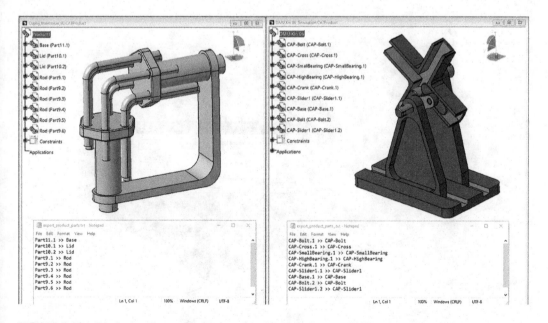

FIGURE 5.23 Displaying the result after running the script for two *CATProduct* files.

Example 5.3

For a *CATPart* file, write a script that changes the name of the main body with a user-given name. The script code is commented and presented in Table 5.5. The result of running this script is shown in Figure 5.24.

TABLE 5.5

Code for changing the Name of the main body

```
01 : Option Explicit
02 :
03 : Sub CatMain ()
04 :
05 : 'Declare 3 variables for document, part and body
06 : Dim oDoc As Document
07 : Dim oPart As Part
08 : Dim oPartBody As Body
09 :
10 : 'The current document is being read
11 : Set oDoc = CATIA.ActiveDocument
12 :
13 : 'The current part is read
14 : Set oPart = oDoc.Part
15 :
16 : 'The current body is read
17 : Set oPartBody = oPart.MainBody
18 :
19 : 'The user is asked for the new name of the main body
20 : oPartBody. Name = InputBox ("Provide a new name for the PartBody:")
21 :
22 : End Sub
```

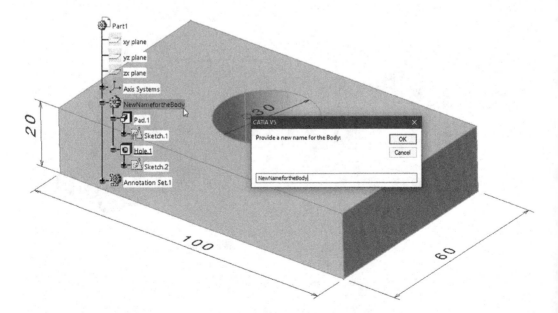

FIGURE 5.24 Displaying the result after running the script for changing the *PartBody* name.

6 Knowledge Assessment Tests and 2D Drawings of Parts Proposed for Modelling

6.1 MULTIPLE-CHOICE QUESTIONS

T1. What is the full name of the *CATIA v5* program?

T2. What file format is used to save a sketch created in the *CATIA Sketcher* workbench?

    a. *.CATPart   b. *.CATSketch   c. *.CATDwg   d. *.CATDraft

T3. Is it possible to change the display size of the working planes of an *Axis System*? If 'yes', how?

T4. Is it possible to hide the specification tree?

T5. What is the use of the *Hide/Show* option available in the context menu?

T6. When is the *Cut Part by Sketch Plane* option used?

T7. In what *CAD* formats it is possible to save a part created in the *CATIA Part Design* workbench?

    a. *.CATPart   b. *.stp   c. *.stl   d. *.igs

T8. What is the main purpose of the *Multi-Pad* tool in the *CATIA Part Design* workbench?

T9. The selection of the current working plane is made according to Figure 6.1:

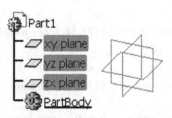

FIGURE 6.1   Top of a specification tree for a 3D model.

    a. by choosing it in the specification tree

    b. by choosing it in the orthogonal *Axis System*

    c. by selecting the specification tree that contains the planes

DOI: 10.1201/9781003281153-6

T10. Imposing the symmetry of the vertical sides of a rectangle with respect to the *V* axis (Figure 6.2) is done by:

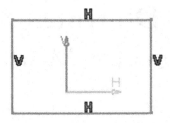

FIGURE 6.2   Sketch containing a rectangle and the coordinate system.

    a.   choosing the *Symmetry* option

    b.   choosing the *Equidistant point* option

T11. To extrude a closed profile (Figure 6.3), the *Pad* tool is used with the icon:

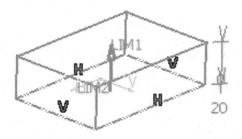

FIGURE 6.3   A simple extruded profile.

    a.   b.   c.

T12. The *Mirror* tool for creating symmetric copies is present:

    a.   according to the icon, only in the *CATIA Sketcher* workbench

    b.   according to the icon, only in the *CATIA Part Design* workbench

    c.   in both workbenches, but also in others, the icons have different representations.

T13. To pierce a solid 3D model, the user may apply the *Pocket* tool with the options:

    a.   *Up To Next*

    b.   *Up To Last*

    c.   *Dimension*, but the value of *Depth* is important

T14. To obtain a bored hole, the *Hole* tool is used with the option:

    a. *Simple*   b. *Counterbored*   c. *Threaded*   d. *Tapered*

T15. The rounded edge in Figure 6.4 is obtained with the help of the tool:

FIGURE 6.4   A rounded edge on a 3D solid model.

    a. *Edge Fillet*   b. *Chamfer*   c. *Close Surface*

T16. When using the *Pad* tool, the *Mirrored extent* option has the role of:

    a. creating two identical extrusions on either side of the plane containing the sketch with the profile to be extruded

    b. opening an extensive list of options when accessing the *Mirror* tool

    c. creating a single extrusion on one side or the other of the plane containing the sketch with the profile to be extruded

T17. What tool is used to obtain an M8 thread (Figure 6.5) at the end of a shaft created in the *CATIA Part Design* workbench?

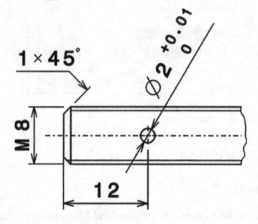

FIGURE 6.5   Representation of a M8 thread in a 2D drawing.

    a. *Groove*   b. *Thread/Tap*   c. *Plane*   d. *Hole*

    e. *Pocket*   f. *Edge Fillet*   g. *Chamfer*

T18. A support plane can only be changed (Figure 6.6) for a sketch if:

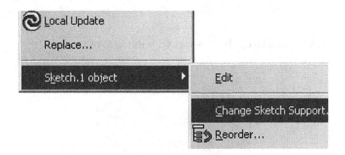

FIGURE 6.6 Option *Change Sketch Support* available in the context menu of a sketch.

a.   the support plane is parallel to the plane that contains the sketch

b.   the support plane was created before the sketch

c.   this operation is not possible

d.   the plane is placed in a geometric set

T19. After activating an icon, the dialog box in Figure 6.7 opens. What tool was used? Explain the *Offset from plane* option. Give an example.

FIGURE 6.7 The *Plane Definition* dialog box.

T20. How to define volumetric models?

T21. What is a *CAD* system?

T22. Explain the purpose of the *Construction/Standard Element* icon. Give an example.

T23. What are these tools used for? Give an example.

 *Constraint*   *Constraints Defined in Dialog Box*

T24. Explain the parent-child relation in feature-based modelling (Figure 6.8).

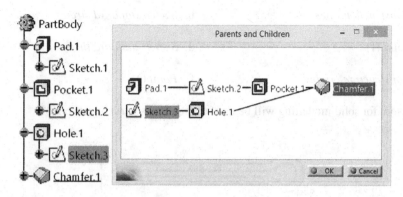

FIGURE 6.8 Relations between parents and children.

T25. Which feature categories do the items shown in Figure 6.9 belong to?

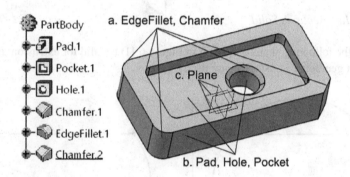

FIGURE 6.9 Features marked on a 3D solid model.

a. *Edge Fillet, Chamfer*

b. *Pad, Pocket, Hole*

c. *XY plane, YZ plane, ZX plane*

T26. Which of these *CATIA Part Design* tools remove or add volume by rotating the profile/sketch around an axis?

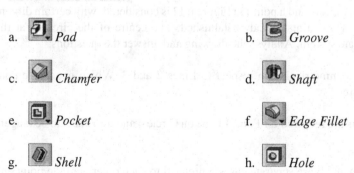

a. Pad   b. Groove

c. Chamfer   d. Shaft

e. Pocket   f. Edge Fillet

g. Shell   h. Hole

T27. Which of the following will not be visible when the user leaves the *Sketcher* workbench?

    a. *Construction lines*   b. *Sketch standard circles*

    c. *Point (+)*   d. *Sketch standard lines*

    e. *Construction circles*   f. *Point (·)*

T28. What tool for solid modelling will be used to create this part (Figure 6.10)?

FIGURE 6.10   2D drawing of a solid model.

    a.   *Multi-sections Solid*   b.   *Slot*   c.   *Rib*

T29. Which of the following sketches (a or b in Figure 6.11) is valid for extrusion using the *Pad* tool? What geometric constraints are used?

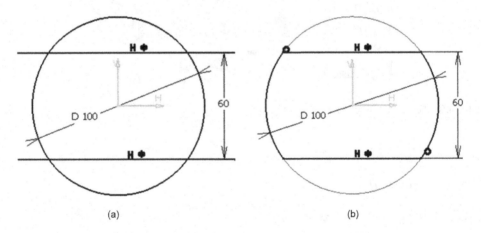

                          (a)   (b)

FIGURE 6.11   (a and b) Two sketches proposed for *Pad* extrusion.

T30. The sketch (two lines, a circle and a point) in Figure 6.12 is considered, with certain dimensional and geometric constraints already established. The centre of the circle is at the origin of the coordinate system. Analyse the drawing and answer the questions:

    a. Point 1 must be symmetrical with respect to Lines 2 and 3. Which geometric constraints are required?

    b. Which constraints are required for Point 1 to be on Circle 4 and for Line 3 to be tangent to Circle 4?

It is considered that a and b are separate cases, unrelated to each other, and the point and the lines are in their initial positions at the beginning of each question.

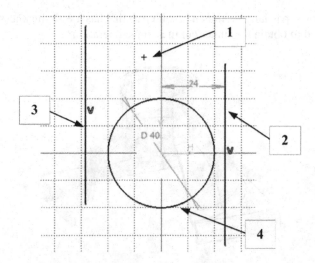

FIGURE 6.12 Sketch containing two lines, a circle and a point.

T31. Which of the following sketches (a, b, c and d in Figure 6.13) can be successfully used by the *Shaft* tool? Argue the answers.

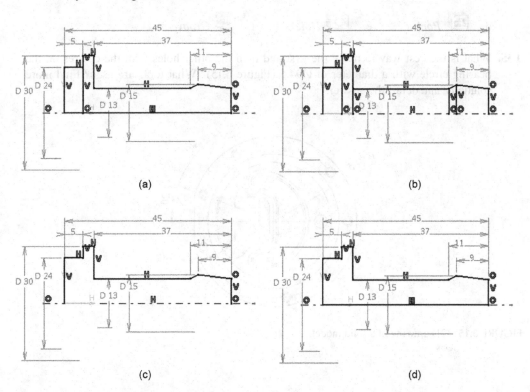

FIGURE 6.13 (a–d) Four sketches proposed to be used by the *Shaft* tool.

T32. When to use and what is the name of the tool represented by the icon ?

T33. What colour are the geometric elements fully constrained in a sketch?

T34. Figure 6.14 shows, partially sectioned, a solid part that has a wall thickness of 2 mm. What tool can be used to obtain this inner cut in a single operation?

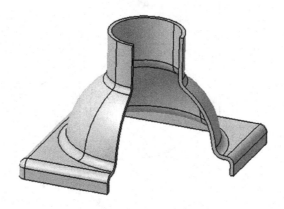

FIGURE 6.14   Representation of a 3D solid part that has a wall thickness of 2 mm.

a. *Shell*   b. *Stiffener*   c. *Mirror*

d. *Pocket*   e. *Hole*   f. *Groove*

T35. What is the best way to make the first and then the other holes with the centres on the bearing circle with a diameter of Ø94.5 (Figure 6.15)? What tools are used? Find more solutions, and detail the answer.

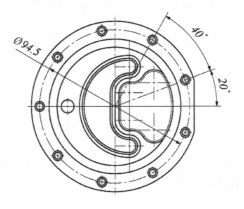

FIGURE 6.15   2D drawing of a solid model.

T36. What are the names of the entities in Figure 6.16 a and b?

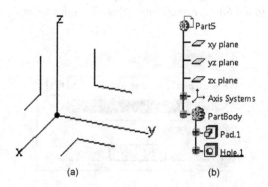

FIGURE 6.16   (a and b) Two of the *CATIA v5* interface entities.

T37. What tool is used to make the circular cut-out defined by the angle of 30⁰ and the diameter of Ø36, according to the drawing of the nut in Figure 6.17?

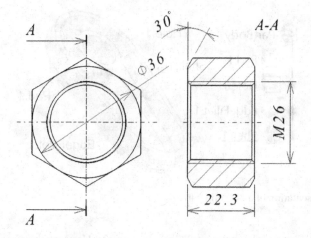

FIGURE 6.17   2D drawing of a hexagonal nut.

a. *Shaft*   b. *Groove*   c. *Pocket*

d. *Round*   e. *Edge Fillet*   f. *Chamfer*

g. *Plane*   h. *Trim*   i. *Fillet*

T38. What reference feature will be created by activating the *Plane Definition* dialog box (Figure 6.18)? Explain the *Tangent to surface* option and which features from a 3D model should be selected in the *Surface* and *Point* fields. Consider an example.

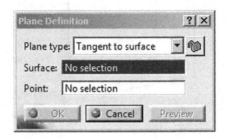

FIGURE 6.18   The *Plane Definition* dialog box.

T39. What happens if *Pad.1* is deleted from the specification tree? What if the *EdgeFillet.1* is deleted? Explain how the part in Figure 6.19 was created.

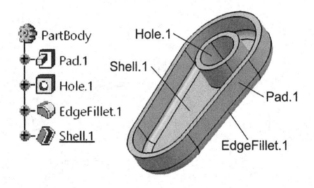

FIGURE 6.19   Representation of a 3D solid part.

T40. What are the dimensional constraints? Give an example. How are they imposed in *CATIA*?

T41. In what situations are these *CATIA Sketcher* tools used?

a.   *Project 3D Elements*   b.   *Project 3D Silhouette Edges*

T42. What is a sketch? What elements can it contain? How to check a sketch?

T43. What must be the constraints status of a sketch (*Sketch Analysis* option in the *Tools* menu, *Diagnostics* tab; Figure 6.20) to be valid and usable in modelling solids and surfaces?

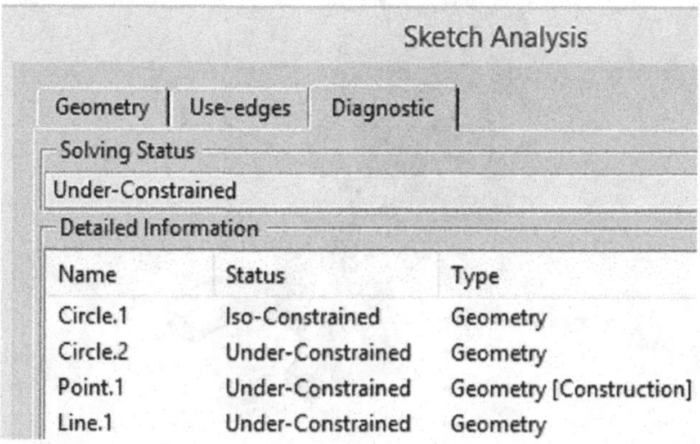

Sketch Analysis		
Geometry	Use-edges	Diagnostic

Solving Status

Under-Constrained

Detailed Information

Name	Status	Type
Circle.1	Iso-Constrained	Geometry
Circle.2	Under-Constrained	Geometry
Point.1	Under-Constrained	Geometry [Construction]
Line.1	Under-Constrained	Geometry

FIGURE 6.20 Example of constraints status from a sketch.

a. *Iso-Constrained*

b. *Under-Constrained*

c. *Over-Constrained*

d. All of the above

T44. According to Figure 6.21, on a 15 mm thick plate, a hole is positioned and created using the *Hole* tool. Which of the following options can be used for the hole to be pierced?

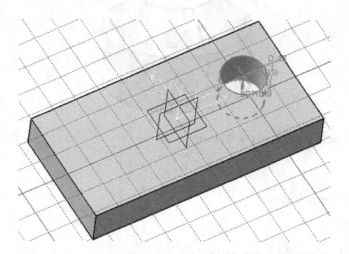

FIGURE 6.21 Representation of a pierced hole through a rectangular 3D solid part.

a. *Blind* with a depth of 5 mm

b. *Up To Last*

c. *Up To Surface* and selection of the opposite surface

d. *Up To Next*

T45. What is the name of the feature 1 in Figure 6.22? What modelling tool is used to obtain it?

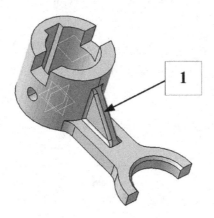

FIGURE 6.22   Representation of a 3D solid part.

a. *Shell*   b. *Stiffener*   c. *Mirror*

T46. To obtain the representation view from Figure 6.23, the user applies the tool:

FIGURE 6.23   Representation of a 3D solid part.

a.  *Apply Material*

b. *Shading with Edges and Hidden Edges*

c.  *Hide/Show*

T47. Changing the hatch step (Figure 6.24) is done by changing the value in the field:

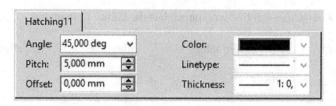

FIGURE 6.24 Options of a hatch from a 2D drawing.

    a. *Angle*   b. *Pitch*   c. *Offset*

T48. To add dimensions to a chamfer (Figure 6.25) in a 2D drawing, it is used one the following icons:

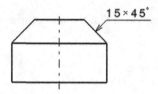

FIGURE 6.25 Adding dimensions to a chamfer.

    a.   b.   c.

T49. In the *CATIA Drafting* workbench, the transition from the projections (views and sections) space to the sheet space is done from the *Edit* menu with the option:

    a. *Replace*   b. *Sheet Background*

    c. *Repeat*   d. *Working views*

T50. Is it possible to add one or more parameters to an existing design table?

T51. What is a 2D drawing and in which workbench of *CATIA* it can be created?

T52. The *Center graph* and *Reframe on* options are placed in the context menu (Figure 6.26) of a feature in the specification tree. How and when are they used?

FIGURE 6.26 Context menu of a feature in the specification tree.

T53. How many axis-type lines can be drawn in a sketch?

T54. The *CATIA Knowledge Advisor* workbench has the ability to manage:

    a. formulas   b. design tables   c. rules   d. checks

T55. A parameter can be defined:

    a. simultaneously through a table and a formula, both of which are active

    b. simultaneously by two or more active formulas

    c. just by a table or a formula

T56. The design table can contain:

    a. any type of parameters, provided that the unit of measurement is also present, where applicable

    b. only parameters of type length

    c. any type of parameters, and the indication of the unit of measurement is not required

T57. What is the use of the *Explode* option applied to a multiplication of type pattern? Consider an example.

T58. In surface modelling, the *Extrude* tool is used to achieve:

    a. extruding only closed profiles to create surfaces

    b. extruding only open profiles to create surfaces

    c. extruding both types of profiles to create surfaces

T59. To create a solid inside a space delimited by surfaces, it is used:

    a. the *Join* tool    b. the *Fill* tool

    c. the *Sweep* tool 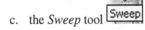   d. the *Close Surface* tool

T60. Depending on the shape and the manner they are created, the surfaces can be:

    a. of extrusion   b. of revolution

    c. through two or more sections   d. along a curve

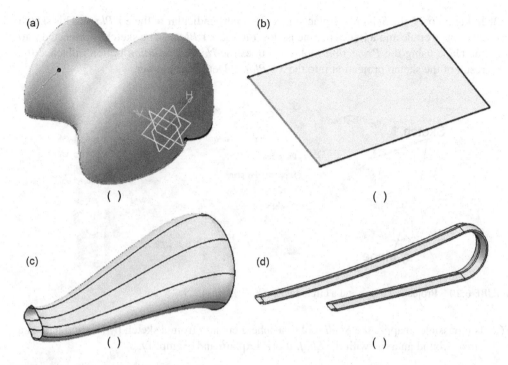

FIGURE 6.27 Four types of surfaces.

Match the surface types a, b, c and d with Figure 6.27 below and mark the letters in the parentheses:

T61. For each of the two images in Figure 6.28, specify and mark accordingly:

a. which areas of the part are obtained as surfaces of revolution

b. which areas of the part are obtained through intersections of surfaces

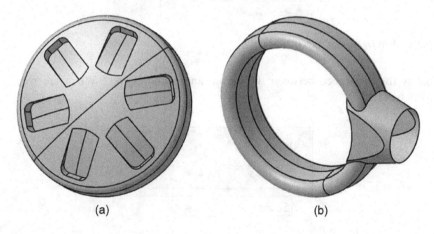

FIGURE 6.28 Representations of two parts created by surfaces.

T62. In Figure 6.29, the *Sketch.1* is placed in a plane perpendicular to the *XY Plane*. This sketch contains a circle and a line. The line is closer to the *XY Plane*. The sketch is projected into the plane using the *Projection* tool. How does the *Nearest Solution* option influence the result of the sketch projection into the *XY Plane*? Detail the answer.

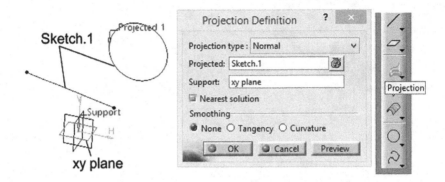

FIGURE 6.29   Projecting a line and a circle in the *XY Plane*.

T63. Is it possible to apply the *Shell* tool to a sphere created from a sketch by rotating half of a circle around an axis (with the *Shaft* tool)? Explain and exemplify.

T64. What is the role of the *Simplify the result* option in the *Join Definition* selection box? (Figure 6.30).

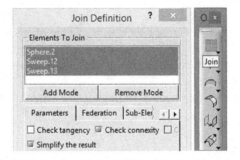

FIGURE 6.30   *Join Definition* selection box.

T65. What is the difference between the *Spline* and *Spine* curves? When are they used? (Figure 6.31).

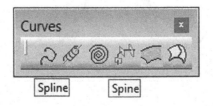

FIGURE 6.31   The *Spline* and *Spine* icons on the *Curves* toolbar.

T66. Define the following: *Blend Surface, Boundary, Construction Element, Design Table, Healing* and *Pocket*.

T67. Indicate which of the following answers are options for creating a plane:

a. *Offset from plane*

b. *Offset from point*

c. *Parallel through point*

d. *Normal to curve*

e. *Mean through points*

f. *Through point and plane*

g. *Through two lines*

h. *Through near point*

i. *Through three points*

j. *Up to next plane*

k. *Normal to surface*

l. *Through point and line*

T68. Which of the following menus and options allow the user to close *CATIA v5*?

a. *Start* and *Exit*

b. *File* and *Quit*

c. *File* and *Exit*

d. *File* and *Close*

T69. Which of the following actions lead to renaming a part (Figure 6.32)?

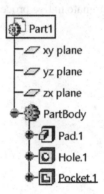

FIGURE 6.32   Name of the part in the specification tree.

a. double-clicking on its name;

b. editing the properties of the part (*Part1*) from the context menu;

c. editing the properties of the part's features (*Pad.1, Hole.1,* etc.);

d. overwriting the part's name in the specification tree;

e. using the *Rename* option in the context menu.

T70. According to Figure 6.33, the part has several threaded holes. What is the tool name that performs the analysis of these threads and what is its icon?

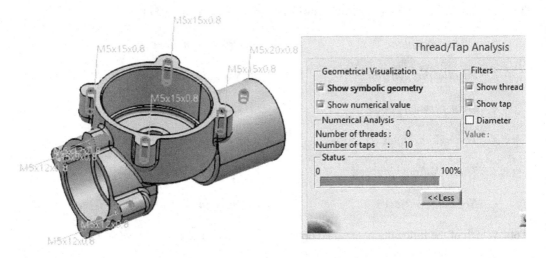

FIGURE 6.33   Threaded holes identified and highlighted on the 3D model of the part.

a.   b.   c.   d.

T71. What languages can be used to automate native processes in *CATIA v5*?

    a.   *Java*

    b.   *JavaScript*

    c.   *C++*

    d.   *VB Script/CATScript*

    e.   *Pascal*

T72. Is it possible to run scripts in other scripts?

    a.   yes, the script is called and run using the *ExecuteScript* function

    b.   no, it is not possible

    c.   yes, but the part should be previously saved

    d.   not before compilation

T73. *CATIAPart* class members are of several types (Figure 6.34).

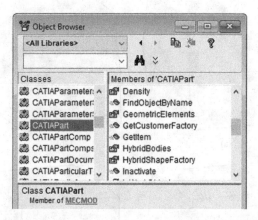

FIGURE 6.34   List of *Classes.*

The icon ⇒● represents a ....

The icon 🖾 represents a ....

T74. Figure 6.35 shows a solid part obtained by *Pad* extruding a rectangle, drawn in a sketch placed in the *XY Plane*, with the same value, on either side of the plane (using the *Mirrored extent* option). The user selects the four flat lateral faces of the part to be drafted by 5°. In the *Draft Definition* dialog box, the user chooses the *XY Plane* as the neutral element and a vertical edge in the *Pulling Direction* area, presses the *More >>* button and checks the options *Parting=Neutral* and *Draft both sides*. The other options are default and can be seen in that selection box.

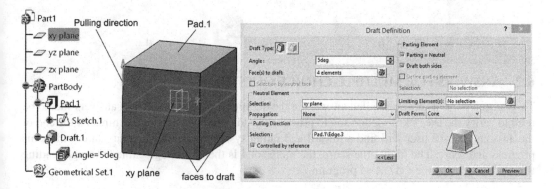

FIGURE 6.35   *Pad* extrusion of the rectangle and drafting its lateral faces.

Which of the following images is the result of applying the *Draft* tool according to the figure above and the selected options? (Figure 6.36).

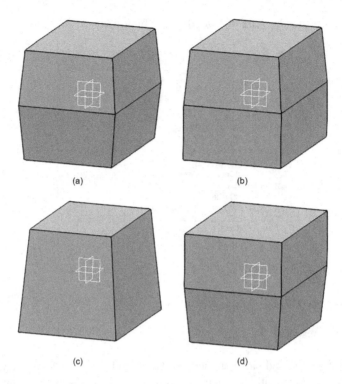

FIGURE 6.36   Four different representations and solutions of the part after applying the *Draft* tool.

T75. Which of the following icons should be used to split a curve or a surface using another curve/surface?

a.   b.   c.   d.

6.2   ANSWERS FOR MULTIPLE-CHOICE QUESTIONS

This section presents many detailed answers to the questions in Section 6.1. The answers should be used to verify the user's knowledge and to provide an additional resource of learning. At the end of the answers list, the user will find a reward and rating system through which he can calculate his progress and score. The higher the score, the more solid is the knowledge gained through reading this book and using the *CATIA v5* program.

T1. Computer-aided three-dimensional interactive application.

T2. The file format for saving a sketch made in *CATIA Sketcher* is **a.** *CATPart*.

T3. Yes, the displayed plane size is set in the menu *Tools → Options → Infrastructure → Part Infrastructure → Display*, option *Axis system display size (in mm)*.

T4. Yes, hiding the specification tree is possible by pressing the *F3* key.

T5. The *Hide* option hides certain features contained in the specification tree (parts, components, relations, analysis results, etc.), without removing them, and they still influence the document in progress. The *Show* option displays again these hidden items (Figure 6.37).

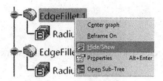

FIGURE 6.37 Context menu of a feature with the *Hide/Show* option selected.

T6. The option is important to make certain edges visible in the sketch, sectioning the solid of the part with a plane. Only the portion of this solid between the user and the sketch plane is visually removed, as shown in Figure 6.38.

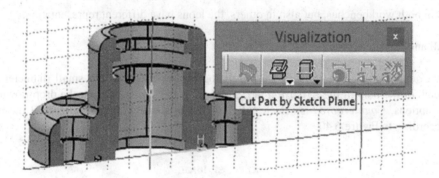

FIGURE 6.38 Representation of a 3D solid part cut by the sketch plane.

T7. All answers are correct.

T8. The *Multi-Pad* tool (Figure 6.39) allows the individual extrusion, at different distances, of several profiles drawn in the same sketch, respecting certain conditions: the profiles should be closed and not (self-) intersect. Each profile and extrusion distance may be selected in a dialog box. The tool is applicable on either side of the plane that contains the profile sketch.

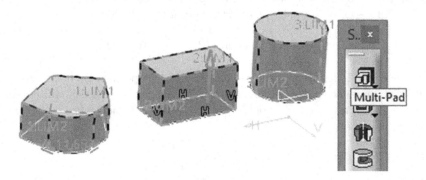

FIGURE 6.39   Representation of three extruded profiles from the same sketch.

T9. a. by choosing it in the specification tree and **b.** by choosing it in the orthogonal plane system.

T10. a. choosing the *Symmetry* option.

T11. b. icon of the *Pad* tool

T12. c. in both workbenches, but also in others, the icons have different representations.

T13. All answers are correct.

T14. b. *Counterbored*. The *Hole Definition* dialog box first defines the small diameter and depth of the hole in the *Extension* tab, and then in the *Type* tab, from the drop-down list of options, *Counterbored* is selected. Below, the user enters the parameters of the hole, according to Figure 6.40.

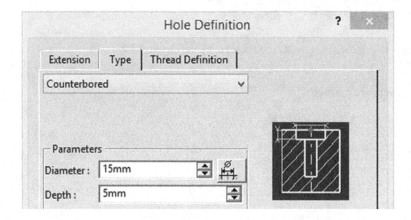

FIGURE 6.40   The *Hole Definition* dialog box with parameters for *Counterbored* holes.

T15. a. Edge Fillet.

T16. a. to create two identical extrusions on either side of the plane containing the sketch with the profile to be extruded.

T17. b. *Thread/Tap.* Depending on the selected cylindrical surface (inner or outer), one of the two available options is used: *Thread* or *Tap* for the shaft, respectively, for the cylindrical hole (Figure 6.41).

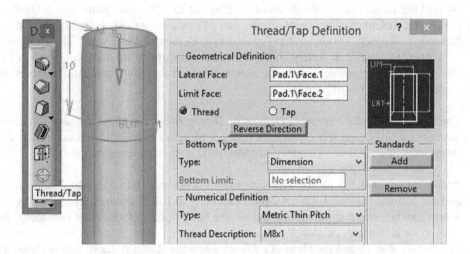

FIGURE 6.41 The *Thread/Tap Definition* dialog box with parameters to apply on a shaft.

T18. b. the support plane was created before the sketch.

T19. The *Plane* tool is used. The *Offset from plane* option allows the user to obtain a new plane, parallel to the reference plane, at a distance imposed by the *Offset* parameter. In the example in Figure 6.42, the *Plane.1* was inserted parallel and at a distance of 20 mm from the *XY Plane*.

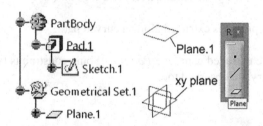

FIGURE 6.42 *Plane.1* created parallel and at a distance of 20 mm from the *XY Plane*.

T20. Volumetric models are geometric entities created and represented in 3D space that have volume properties and mass by adding a material defined by density.

T21. The *CAD* (*computer-aided design*) system is a set of software, hardware and peripheral components installed on a computer system. They allow a design engineer to create, modify and visualize geometric elements; perform calculations and simulations (finite element, NC machining, kinematics, etc.); work with complex assemblies; obtain 2D and assembly drawings; and render 3D models in images and videos.

T22. By clicking on the *Construction/Standard Element* icon, the user transforms a standard drawn object/element (line, point, circle, arc, polygon, spline curve, etc.) of the *CATIA Sketcher* workbench into an auxiliary construction element and vice versa. These elements are particularly useful in sketches, influence other standard or auxiliary construction elements, and are not visible outside the sketch in which they were created. Examples are angle bisector and bearing circle.

T23. The *Constraint* and *Constraints Defined in Dialog Box* tools allow the user to impose dimensional and geometric constraints between elements in a sketch. Examples are distance, angle, parallelism, symmetry, perpendicular, coincidence and concentricity.

T24. According to Figure 6.8, each model is composed of features (*Pad.1*, *Pocket.1*, *Hole.1*, etc.) and sub-features (*Sketch.1*, *Sketch.2* and *Sketch.3*), all called features. These are, in one way or another, interconnected. In principle, the basic feature of a model is considered the 'parent' for all features ('children'). The sketch of a basic feature is 'parent' for it (example: *Sketch.1* for *Pad.1*), and the plane (*XY Plane*) where the sketch is drawn is considered the 'parent' for *Sketch.1*. The user should note that if he edits a 'parent', that change will propagate to the 'child' as well. Also, if a 'parent' is deleted, the *Delete* dialog box will ask the user to decide if the 'child' can be deleted too.

T25. a. *Edge Fillet, Chamfer* (*Dress-Up Features*), **b.** *Pad, Pocket, Hole* (*Sketch-Based Features*) and **c.** *XY Plane, YZ Plane, ZX Plane* (*Default reference planes*).

T26. b. *Groove* and **d.** *Shaft.*

T27. a. *Construction lines*, **e.** *Construction circles* and **f.** *Point* (·).

T28. c. *Rib*, a half-circle profile is extruded along a curved path.

T29. b. the sketch can be extruded using the *Pad* tool. Some constraints of coincidence, parallelism and symmetry are observed too.

T30. a. The line on the right is dimensionally constrained with respect to the centre of the circle and, implicitly, with respect to the vertical axis *V* of the coordinate system. It is also geometrically constrained, being drawn vertically, parallel to the *V* axis. The left line is drawn and constrained vertically, but can move in the coordinate system of the sketch, because it is not fully constrained. Because Point 1 must be symmetrical about these Lines 2 and 3, an equidistance constraint is imposed as follows: the user selects the right line, holds down the *Ctrl* key, selects the left line and then, holding down the *Ctrl* key, clicks on the point. Thus, three geometric elements are selected. The user clicks the *Constraints Defined in Dialog Box* icon and chooses *Equidistant point* from the list of options that appears (Figure 6.43).

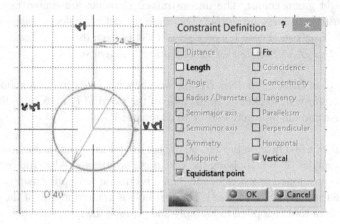

FIGURE 6.43   Use of the *Equidistant point* constraint.

b. In order for Point 1 to be on Circle 4, a coincidence constraint is established between the two geometric elements: the *Constraint* icon is pressed, the circle is selected and then the point; between them, the program proposes a linear distance (offset). The user right-clicks and chooses *Coincidence* from the context menu. In order for Line 3 to be positioned tangent to Circle 4, a tangent constraint is established between the two geometric elements: the user activates the *Constraint* icon, clicks on the circle, then on the line, presses the right button on the mouse and, from the context menu, chooses *Tangency* (Figure 6.44).

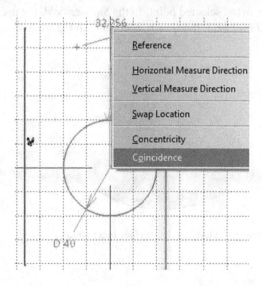

FIGURE 6.44   Use of the *Coincidence* constraint.

T31. c. and **d.** only these profiles can be rotated around the horizontal axis to create a solid of revolution using the *Shaft* tool. Situations **a.** and **b.** are not possible because one or more vertical lines appear inside each profile in each sketch.

T32. The tool is called *User pattern*, and it is used when the user wants to multiply a feature (pocket cut, hole, etc.) according to a user-defined pattern other than rectangular or circular. There are other specific tools and options for these.

T33. In a sketch, the geometric elements completely and correctly constrained are highlighted using the light green colour. The unconstrained elements are drawn using white, and the over-constrained ones are purple. These are the default colours imposed by the program, but the user can choose his own set of colours from the menu *Tools → Options → Mechanical Design → Sketcher → Visualization of diagnosis*.

T34. The *Shell* tool in the *CATIA Part Design* workbench is used to cut a solid body inside, keeping a certain wall thickness. In the case of the part in Figure 6.14, both planar faces (from the base and the top) were chosen to be removed.

T35. The user makes a first hole on the left, on the horizontal axis, and then multiplies it in nine instances, at angles of 40°, applying the *Circular Pattern* tool. Also, the centre (*Point*) of a hole can be positioned in a sketch, and then it is circular multiplied using the *Rotate* tool (another eight instances are created). The first hole is centred at one of these points, and the *User Pattern* tool is applied to multiply and position the rest of the holes in the sketch points. However, the first solution is faster and easier to apply.

T36. a. Axis system, coordinate system and planes system; **b.** Specification tree.

T37. b. *Groove*.

T38. The dialog box opens after activating the *Plane* icon. It is considered a truncated cone solid, which has an inclined surface, obtained using the *Chamfer* tool. A point has been inserted in *Sketch.2* located on the upper flat surface of the body. The point is coincident with the top circular edge. Using the *Tangent to surface* option, the user will get the *Plane.1*, which is tangent to surface in the selected point. In the *Surface* field, the user chooses the conical surface and, in the *Point* field, the sketch, according to Figure 6.45.

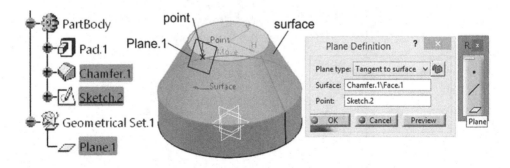

FIGURE 6.45 Placing a plane to be tangent to a surface in a point.

T39. The *Pad.1* feature cannot be deleted from the specification tree and from the part structure because it is built on it, and the other features (*Hole.1*, *EdgeFillet.1* and *Shell.1*) edit the initial solid, which was obtained by the extrusion of a closed profile. By removing the *EdgeFillet.1* feature, all the outer and inner connecting radii, created by *Shell.1*, disappear from the 3D model. The inner radius is given by the difference between the value of the *EdgeFillet.1* outer radius and the wall thickness of the part (Figure 6.46).

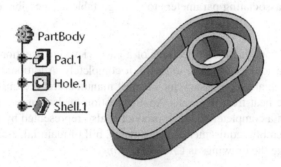

FIGURE 6.46 Representation of the part without the *EdgeFillet.1* feature.

T40. *Dimensional Constraints* are restrictions imposed on geometric elements in a sketch. They can be defined by numerical values and formulas, being applied to linear and angular dimensions. Examples are linear distance (offset), length, radius, diameter and formulas established between different parameters. Dimensional constraints are added to the sketch using the *Constraint* and *Constraints Defined in Dialog Box* icons.

T41. **a.** The *Project 3D Elements* tool projects in the current sketch edges belonging to some flat faces. **b.** The *Project 3D Silhouette Edges* tool projects in the current sketch edges belonging to some bodies of revolution.

T42. The sketch is the basic element for creating solid parts and/or surfaces, and it is defined in a plane or on a flat face of an existing 3D model. Sketches contain 2D geometric elements. For the correct creation of the sketch, certain relations defined by geometric and dimensional constraints are established. The sketch can contain various geometric entities (lines, arcs, spline curves, points, etc.), in standard or of auxiliary construction. To model solids and surfaces, the user requires closed and open sketches. As a recommendation, the user should draw sketches as simple and clear as possible, with the definition of all parameters, and then verify them with the *Sketch Analysis* option in the *Tools* menu.

T43. Profiles belonging to a sketch can be fully or partially constrained (**a.** *Iso-Constrained* and **b.** *Under-Constrained*). Sketches that are **c.** *Over-Constrained* are not valid.

T44. The plate is 15 mm thick, so the first option, **a.** *Blind*, with a depth of 5 mm, does not pierce the part. All other options **b.** *Up To Last*, **c.** *Up to surface* and **d.** *Up To Next* are possible to be applied in this case.

T45. A stiffener is provided between the cylindrical and the flat-shaped elements. It has the role of stiffening the part. The tool **b.** *Stiffener* is used to create it.

T46. **b.** *Shading with Edges and Hidden Edges*.

T47. The hatch step/pitch is the distance between two successive hatch lines. The field that defines this parameter is **b.** *Pitch*.

T48. c. *Chamfer dimensions*.

T49. b. *Sheet Background*.

T50. Yes, adding and associating parameters to a design table are possible using the *Associate* button.

T51. The 2D drawing contains the projections of a part: views and sections with dimensions, tolerances and various conditions, to define it completely. Additionally, the 2D drawing contains the information necessary for the part manufacture: the material, the technical requirements, the heat treatment, etc. Another notion of drawing refers to the assembly drawing in which a completely defined product is also represented by views and sections, as well as by assembly requirements, adjustments, bill of material, etc. The workbench to create and manage the drawings is *CATIA Drafting*.

T52. The *Center graph* option applied to a feature scrolls the specification tree until the feature is positioned in the middle of the working screen. *Reframe on* zooms in on the selected feature placing it in the centre of the screen.

T53. A sketch can only contain a single axis-type line. Although the user has the possibility to draw several such lines, only the last one will be considered as the axis and the others are automatically converted into auxiliary construction elements.

T54. All answers are correct.

T55. At some moment, a parameter can be defined **c.** just by a table or a formula. If the parameter is defined by one formula and the user adds another, the first one is deactivated. Of course, a parameter cannot take different values from two or more formulas or from design tables.

T56. A parameterization table can contain **a.** any type of parameters if their units of measurement are indicated (mm, kg, N/mm^2, mm^3, etc.).

T57. Using the *Explode* option in the context menu of a pattern multiplication, individual copies of the multiplied features are obtained. Then, they can be edited separately. For example, considering the model in Figure 6.47, it is observed that the *Hole.1* was multiplied by the *Rectangular Pattern* in two more holes, these being stored in the *RectPattern.1* feature. To have access to the sketch of each hole, to its diameter, depth, etc., the user 'exploded' the multiplication in individual holes: *Hole.2* and *Hole.3*.

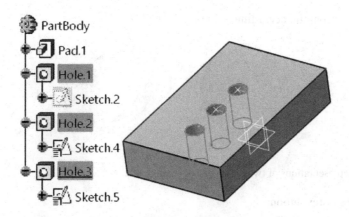

FIGURE 6.47   Rectangular pattern of a hole exploded in individual holes.

T58. In surface modelling, the *Extrude* tool is used to perform **c.** extrusion of both types of profiles, closed and open.

T59. The *Close Surface* tool is used to create a solid inside a space bounded and closed by surfaces.

T60. Correspondence between the types of surfaces **a.** of extrusion, **b.** of revolution, **c.** through two or more sections and **d.** along a curve (Figure 6.48):

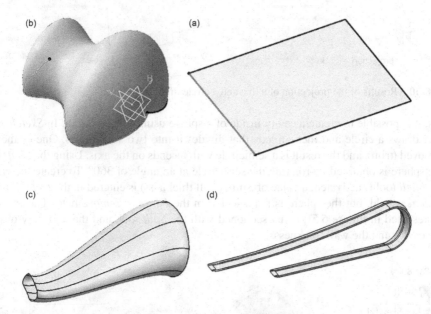

FIGURE 6.48   Correspondence between surfaces and types of surfaces.

T61. The two parts contain, according to Figure 6.49:

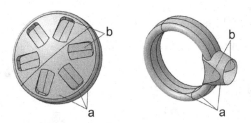

FIGURE 6.49   Representations of two parts created by surfaces.

a.   surfaces of revolution;

b.   surfaces obtained by intersections of other surfaces.

T62. The sketch contains two profiles, the circle and the line (Figure 6.50). The line is closer to the *XY Plane*, which will contain the projection. If **a.** the *Nearest solution* option is checked, then only the line will be projected in the plane because it is the nearest profile. If **b.** the *Nearest solution* option is not checked, then the whole sketch will be projected in the plane. In both cases, the projection is a line, but in the case **b**, the line is longer.

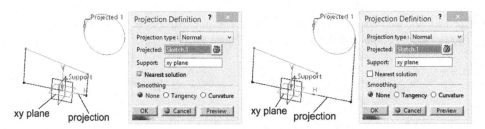

FIGURE 6.50   Results of the projection of a line and a circle in the *XY Plane*.

T63. Yes, it is possible to create a cavity inside of a sphere using the *Shell* tool. In *Sketch.1*, the user draws a circle and then an axis that divides it into two semicircles. One of these is removed (trim), and the result is a semicircle with its ends on the axis. Using the *Shaft* tool, the sphere is obtained by rotating the semicircle at an angle of 360°. To create the cavity, the *Shell* tool is activated, a value of 3 mm (wall thickness) is entered in the *Default inside thickness* field, but the sphere is not selected in the *Faces to remove* field. The sphere is represented in Figure 6.51, being sectioned with the *Split* tool, and the user may observe the cavity and the wall thickness.

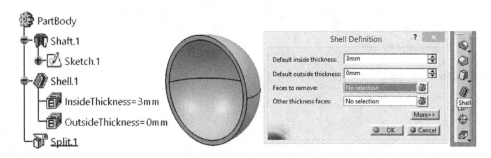

FIGURE 6.51   Cavity created inside a sphere using the *Shell* tool.

T64. When joining two or more complex surfaces, the user checks the *Simplify the result* option to allow the program to automatically reduce (if possible) the number of faces and edges required to create the *Join* surface. This may, however, affect the surface resolution.

T65. The user creates a *Spline* curve through several points in the 3D space (example: as a path for a *Sweep* extrusion). The *Spline* curve shows a tangent continuity at the points through which it passes. The *Spine* curve passes through a point belonging to one plane and then continues through other planes, always being perpendicular to them.

T66. *Blend Surface*: A surface created between and connecting two other existing surfaces.

   Boundary: the topological limit of a geometric element.

   Construction Element: An auxiliary construction element belongs to a sketch and can only be created, edited and viewed in that sketch. The element is used as a positioning reference for the other elements of the sketch, those of standard construction, which are the basis of 3D solids and/or surfaces.

   Design Table: This table allows the user to create families of parts. Generally, the parts have the same shape, but differ in size and the presence/absence of elements such as fillets, chamfers and threads. The design tables contain many parameters of type dimensional, Boolean, numerical, etc., and the very important is the presence of the unit of measurement along with the value in the table.

   Healing: It is a *Join*-like tool for joining several adjacent surfaces, but it is applied when there are some small gaps between them that needs to be closed.

   Pocket: It makes cuts in the volumes of solids using a sketch that contains a closed and non-self-intersecting profile.

T67. According to Figure 6.52, the following options can be used by the *Plane* tool to create a plane:

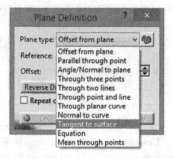

FIGURE 6.52   List of the available options to create planes.

a. *Offset from plane*   c. *Parallel through point*

d. *Normal to curve*   e. *Mean through points*

g. *Through two lines*   i. *Through three points*

l. *Through point and line*

T68. The options for closing the *CATIA v5* program are: **a.** *Start and Exit* and **c.** *File and Exit*. Option **b.** *File and Quit* does not exist, and option **d.** *File and Close* closes the current working document. The user should note that the program allows working in parallel in several workspaces (parts, assemblies, 2D drawings, finite element analyses, surface modelling, etc.).

T69. To change the name of a part, option **b.** is used. Thus, to edit the properties of the part from the context menu (the user right-clicks on *Part1*, chooses the *Properties* option and opens a selection box with the same name, selects the *Product* tab and changes the part name in the *Part Number* field).

T70. The tool for analysing the internal and external threads is called *Tap-Thread Analysis*, and its icon is **c.**

T71. To automate native processes in *CATIA v5*, option **d.** *VB Script/CATScript* can be used.

T72. **a.** yes, the script is called and run using the *ExecuteScript* function.

T73. The icon ⮑ represents a method, and the icon 🖻 represents a property.

T74. The result of applying the *Draft* tool to the part using the options presented is in Figure 6.36 **b.**

T75. The icon of the *Split* tool is **b.**
    Correct answers: ...... Incomplete answers: .... Wrong answers: ....
    The maximum score that can be obtained is **90 points**.
    The questions have different scores as follows: tests **T6, T24, T30, T31, T35, T39, T57, T64** and **T66** are worth **2 points** each, tests **T25, T62** and **T74** are worth **3 points** each, and the rest of the tests, if the explanations are given and the user considered the examples, have **1 point** each.
    For no explanation and/or no example, but still with a correct answer, each test has **0.5 points**.
    Sum up the scores and mark the total value below:

Total score:	.... points
Score **1–30**:	rating: **Unsatisfactory**; we would suggest **Need to study more!**
Score **31–50**:	rating: **Satisfactory**; we would suggest **So close! Keep on learning!**
Score **51–70**:	rating: **Good!**
Score **71–80**:	rating: **Very Good!**
Score **81–90**:	rating: **EXCELLENT**. This means **passion! Congratulations!**

6.3 2D DRAWINGS OF THE PARTS TO PRACTISE MODELLING

The part is the simplest constructive element in the configuration of mechanisms, machine tools, mechanical devices, complex tools, subassemblies and assemblies, various machines, machinery and equipment and, generally, in all technical systems.

In the majority of cases, in the process of designing the part, the design engineer establishes its basic shape and dimensions, determined by the functional role in the assembly which it belongs to, thus: to support or join other parts, to transmit motion by acting on components and to represent an element of safety, protection and ornament.

The final shape of the part and its dimensions are, of course, also influenced by the complex factors such as the loads to which it is subjected (forces, moments, accelerations, temperatures, pressures), technology and manufacturing costs, machine tools available in the company, the material of the part, and design and ergonomic requirements.

All these are established mainly on the basis of the assembly/product design from which the functional role, the connection with the other parts, the degree of technological difficulty, the type of assembly, technical conditions that are required for manufacturing, assembly and disassembly, maintenance plans, repairs, decommissioning, recycling, etc., result.

In Section 6.3, there are 40 various practical parts with very different functional roles, dimensions and constructive shapes proposed for modelling.

The diversity and complexity of the parts are determined by the continuous increase in the quality and precision characteristics of the devices, machines, installations, equipment and performance indicators (functionality, resistance, reliability and durability).

The proposed exercises identify different types of parts, and each one is represented by 2D drawing and one or two isometric views necessary for a better understanding of its geometry.

The parts proposed can be modelled in 3D by several methods, depending on the user experience, gradually acquired using the *CATIA v5* program. *Sketcher, Part Design, Generative Shape Design* and *Generative Sheetmetal Design* workbenches can be used to create the 3D models of these interesting parts.

The 2D drawings of the parts have, in general, a didactic character, with different degrees of difficulty. Many parts have shapes that can only be obtained with surface features. There are some hard parts with complex geometry, that are not very accessible to the beginners, but, to help the user in the modelling process and, at the same time, to offer a correct solution, below the most of the 2D drawings, the authors provided links on the YouTube channel pointed to video tutorial. It is recommended to access and view these tutorials only after the user has tried at least to understand the geometry of the part. The user is also encouraged and advised to find another modelling solution for the respective part and then post it in the comments section of the video tutorial he followed. This will help the community with a growing number of modelling solutions.

Through these proposed tutorials, the authors want to provide a good video support for this book, to accustom users to the e-learning system and resources, and to establish a method for each one to individually verify their knowledge gained using the *CATIA v5* program.

The users should also note that some of drawings can be created using mainly *Part Design* workbench, but for most of them, a hybrid design is inevitable.

Drawing exercise 1 (Figure 6.53).

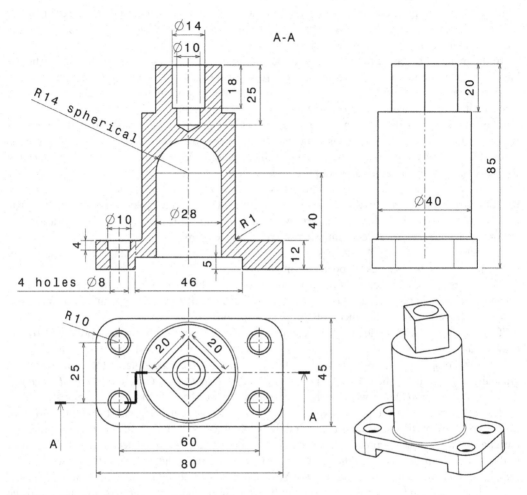

FIGURE 6.53 Drawing of the practising part.

Drawing exercise 2 (Figure 6.54).

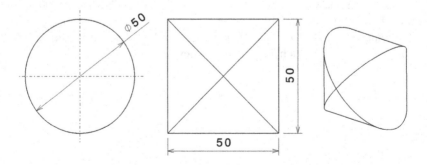

FIGURE 6.54 Drawing of the practising part. (Modelling solutions: https://youtu.be/8t-Q2LQHU-0 and https://youtu.be/y8U7cg_KqtU.)

Drawing exercise 3 (Figure 6.55).

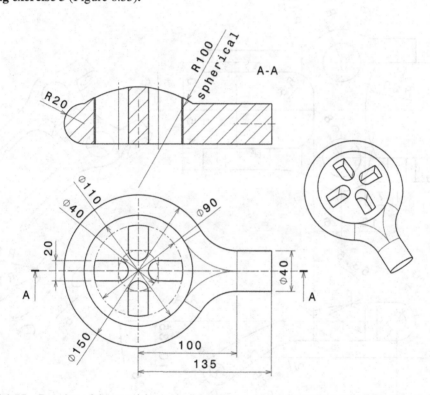

FIGURE 6.55 Drawing of the practising part. (Modelling solution: https://youtu.be/hEp_bK_e6c8.)

Drawing exercise 4 (Figure 6.56).

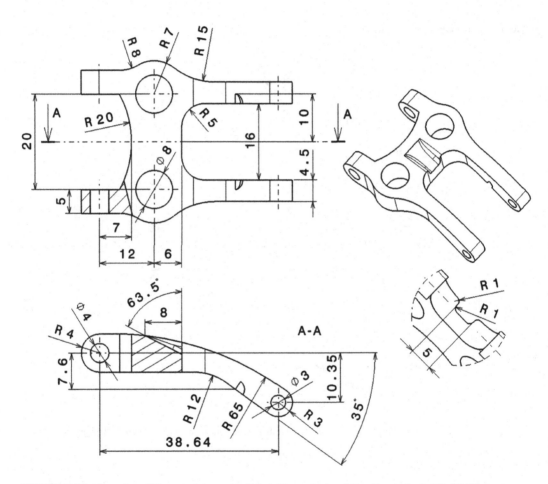

FIGURE 6.56   Drawing of the practising part. (Modelling solution: https://youtu.be/p-PltvhLiUg.)

Drawing exercise 5 (Figure 6.57).

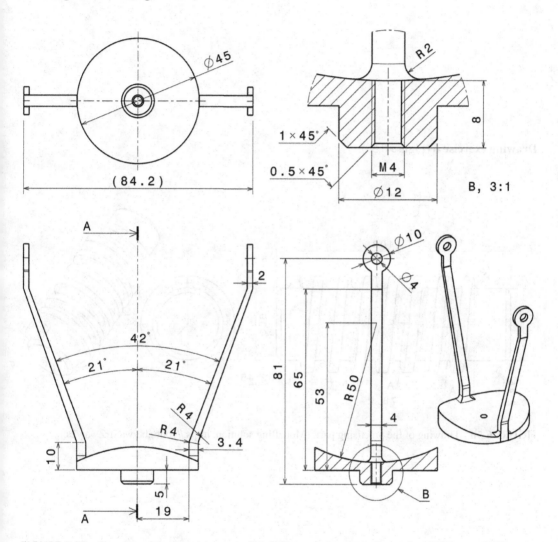

FIGURE 6.57 Drawing of the practising part. (Modelling solution: https://youtu.be/pMbTw_pwOU4.)

Drawing exercise 6 (Figure 6.58).

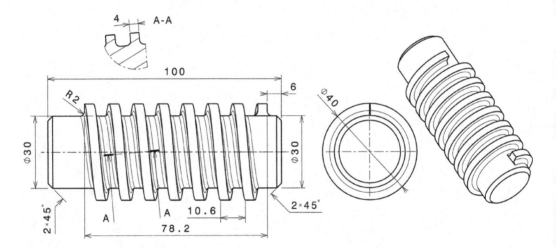

FIGURE 6.58   Drawing of the practising part. (Modelling solution: https://youtu.be/waSfrZwow2s.)

Drawing exercise 7 (Figure 6.59).

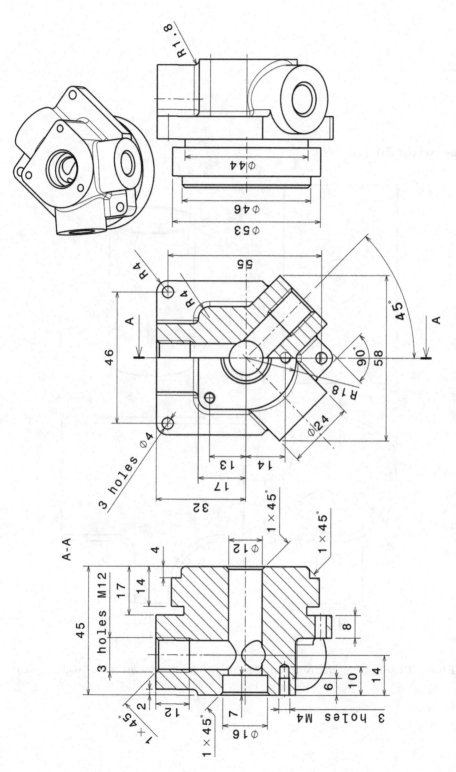

FIGURE 6.59 Drawing of the practising part. (Modelling solution: https://youtu.be/QpY0B_nMnJQ.)

Drawing exercise 8 (Figure 6.60).

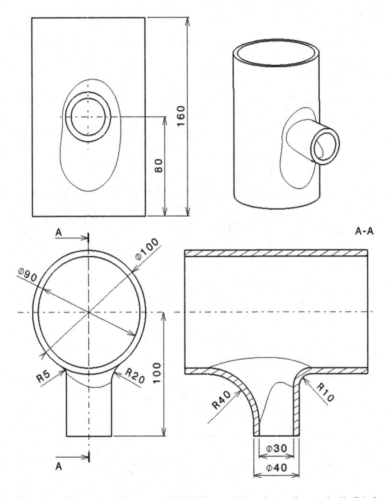

FIGURE 6.60   Drawing of the practising part. (Modelling solution: https://youtu.be/1sGdx8wq-Ik.)

Drawing exercise 9 (Figure 6.61).

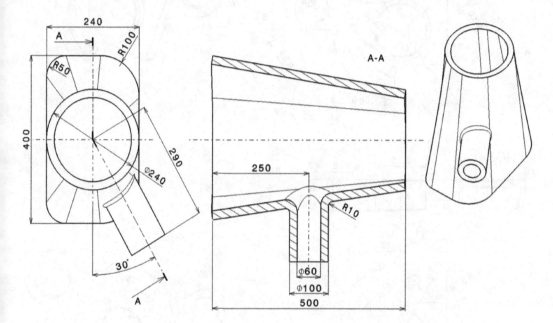

FIGURE 6.61   Drawing of the practising part. (Modelling solution: https://youtu.be/_-ddF1lXSuM.)

Drawing exercise 10 (Figure 6.62).

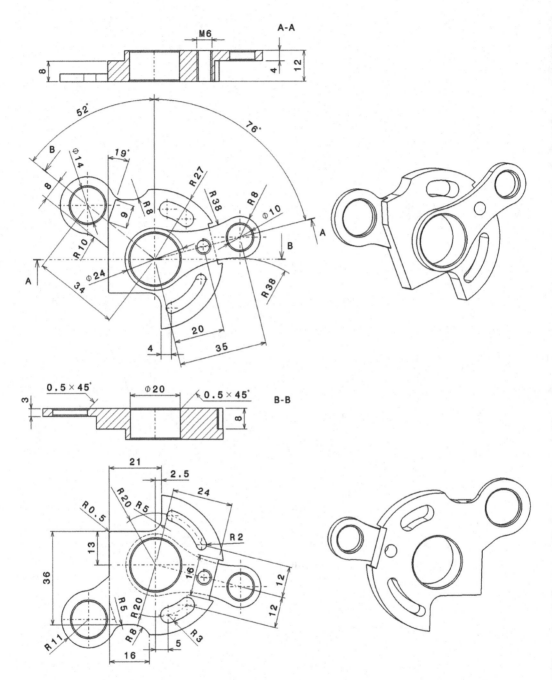

FIGURE 6.62   Drawing of the practising part. (Modelling solution: https://youtu.be/RJoHkDlwSh8.)

Drawing exercise 11 (Figure 6.63).

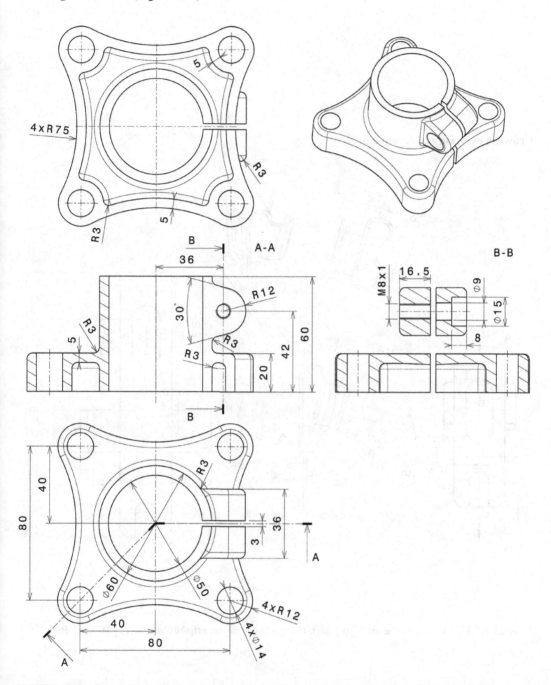

FIGURE 6.63 Drawing of the practising part. (Modelling solution: https://youtu.be/aGYiwVcZTNA.)

Drawing exercise 12 (Figure 6.64).

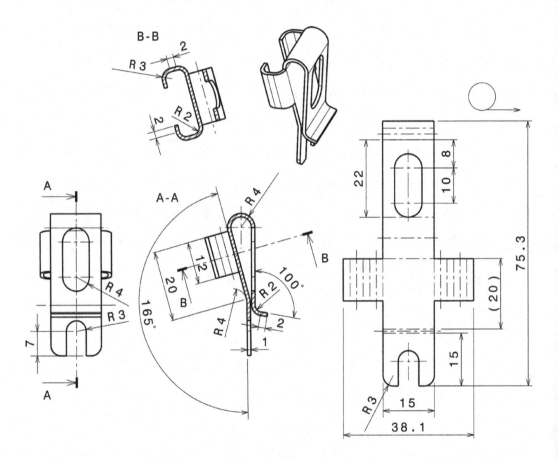

FIGURE 6.64 Drawing of the practising part. (Modelling solution: https://youtu.be/PcpdKXW1-Rw.)

Drawing exercise 13. Model the part by transforming it from phase 1 to phase 2 (Figure 6.65).

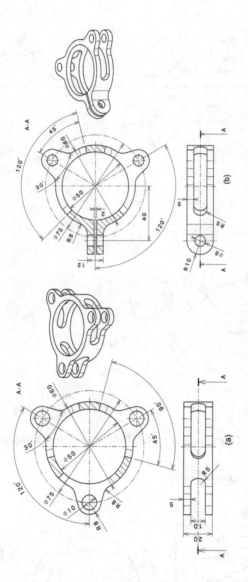

FIGURE 6.65 (a) Drawing of the practising part in phase 1. (b) Drawing of the practising part in phase 2. (Modelling solution: https://youtu.be/Mxzp899bIXM.)

Drawing exercise 14 (Figure 6.66).

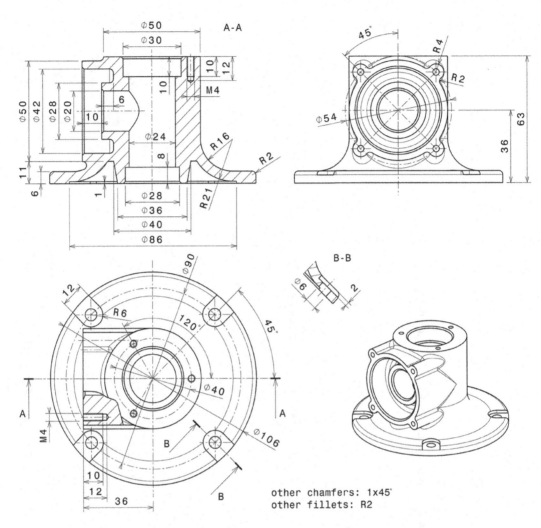

FIGURE 6.66   Drawing of the practising part. (Modelling solution: https://youtu.be/47QYuwrSR20.)

Drawing exercise 15 (Figure 6.67).

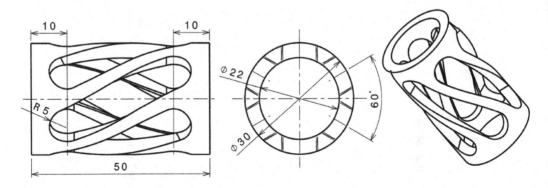

FIGURE 6.67   Drawing of the practising part. (Modelling solution: https://youtu.be/V-MNtOCSiyk.)

Drawing exercise 16 (Figure 6.68).

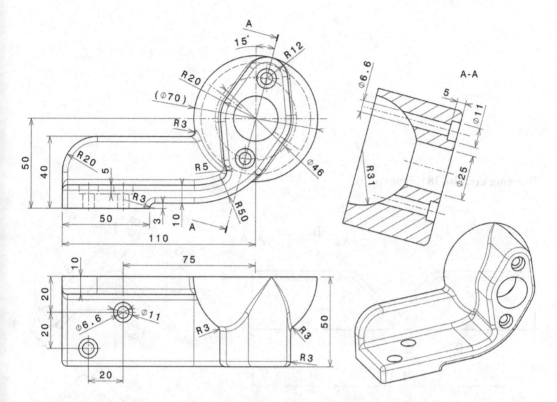

FIGURE 6.68 Drawing of the practising part. (Modelling solution: https://youtu.be/V8Z66e-i4qc.)

Drawing exercise 17 (Figure 6.69).

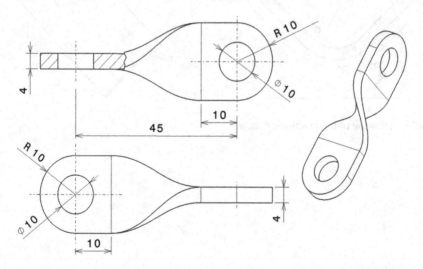

FIGURE 6.69 Drawing of the practising part. (Modelling solutions: https://youtu.be/zAWMkHAsCVo and https://youtu.be/JEkoZ25Hi0g.)

Drawing exercise 18 (Figure 6.70).

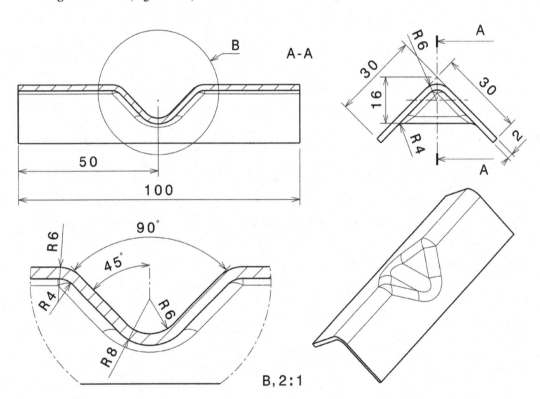

FIGURE 6.70 Drawing of the practising part.

Drawing exercise 19 (Figure 6.71).

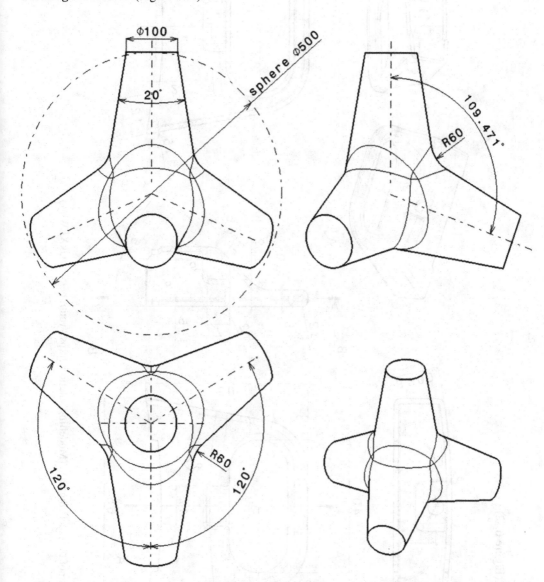

FIGURE 6.71 Drawing of the practising part. (Modelling solution: https://youtu.be/dmaLLjZS36A.)

Drawing exercise 20 (Figure 6.72).

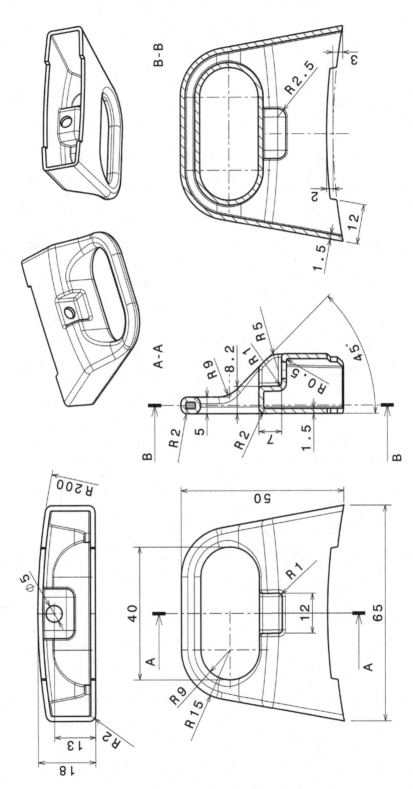

FIGURE 6.72   Drawing of the practising part. (Modelling solution: https://youtu.be/5Z1zGkXLWno.)

Drawing exercise 21 (Figure 6.73).

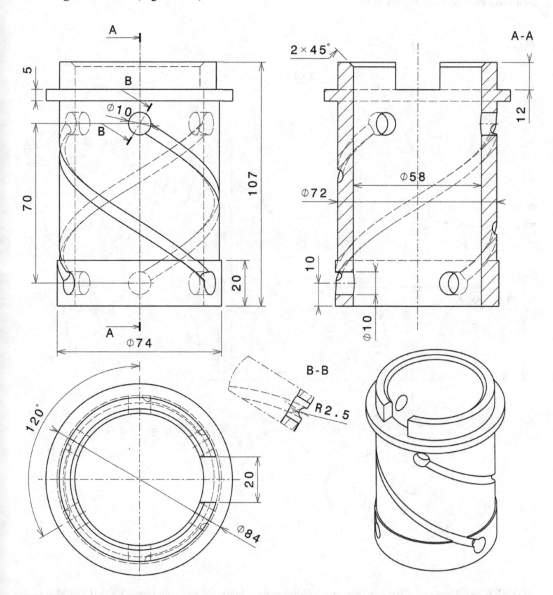

FIGURE 6.73 Drawing of the practising part. (Modelling solution: https://youtu.be/w_EZpVque74.)

Drawing exercise 22 (Figure 6.74).

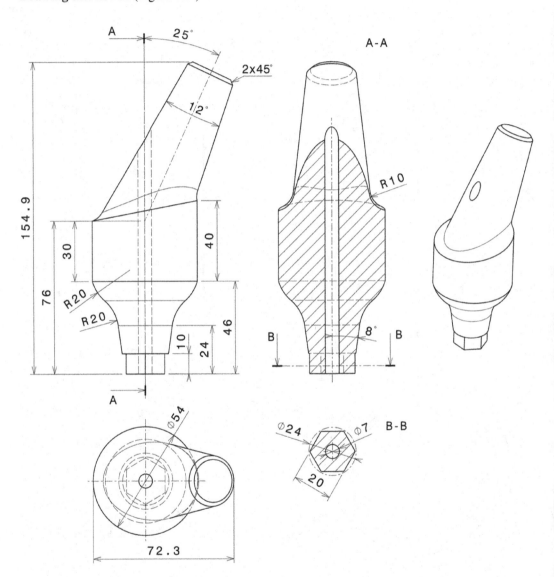

FIGURE 6.74 Drawing of the practising part. (Modelling solution: https://youtu.be/R50OT1LUcPM.)

Drawing exercise 23 (Figure 6.75).

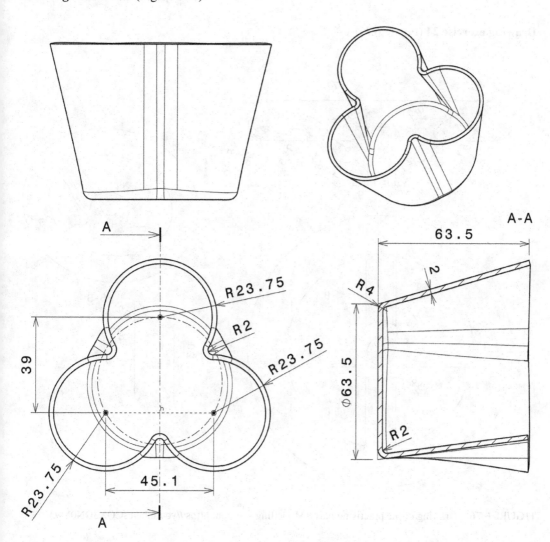

FIGURE 6.75   Drawing of the practising part. (Modelling solution: https://youtu.be/byRE-Iv03Io.)

Drawing exercise 24 (Figure 6.76).

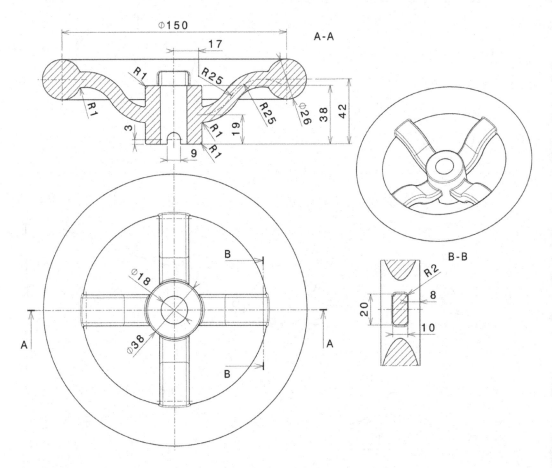

FIGURE 6.76   Drawing of the practising part. (Modelling solution: https://youtu.be/csCQX3DN0Yw.)

Drawing exercise 25 (Figure 6.77).

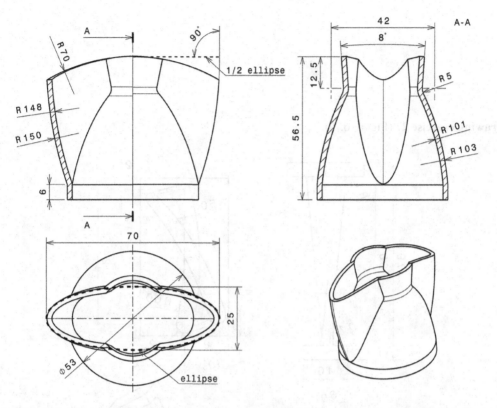

FIGURE 6.77 Drawing of the practising part. (Modelling solution: https://youtu.be/hH4EwGRz6rw.)

Drawing exercise 26 (Figure 6.78).

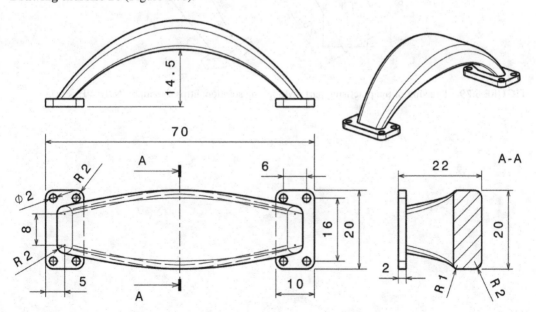

FIGURE 6.78 Drawing of the practising part. (Modelling solution: https://youtu.be/2R353x_hByw.)

Drawing exercise 27 (Figure 6.79).

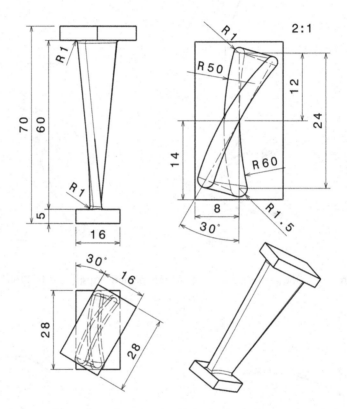

FIGURE 6.79   Drawing of the practising part. (Modelling solution: https://youtu.be/NaIFWExslhU.)

Drawing exercise 28 (Figure 6.80).

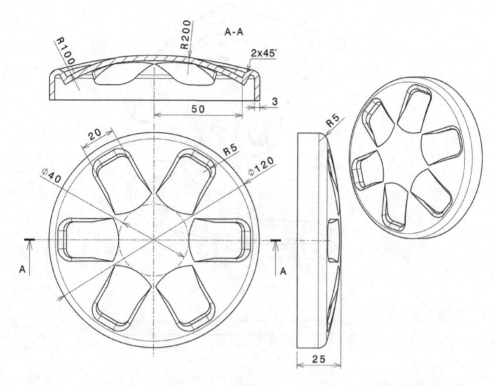

FIGURE 6.80 Drawing of the practising part. (Modelling solution: https://youtu.be/qPk0F5cjCWk.)

Drawing exercise 29 (Figure 6.81).

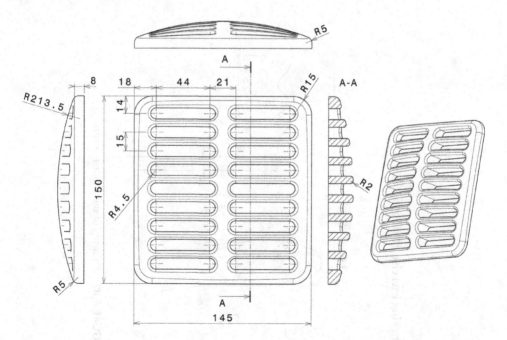

FIGURE 6.81 Drawing of the practising part. (Modelling solution: https://youtu.be/m4cEViMt2VY.)

Drawing exercise 30 (Figure 6.82).

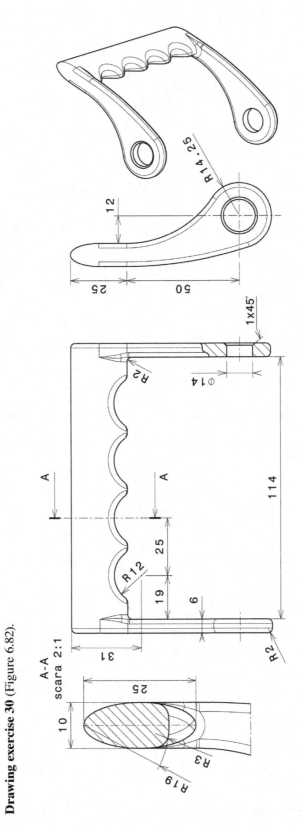

FIGURE 6.82   Drawing of the practising part.

Drawing exercise 31 (Figure 6.83).

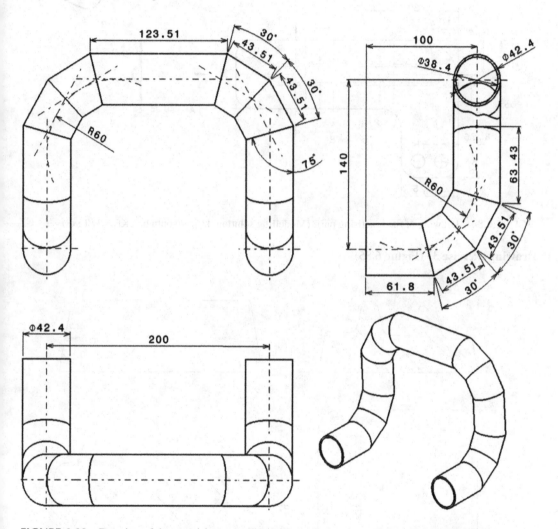

FIGURE 6.83 Drawing of the practising part. (Modelling solution: https://youtu.be/ltXsO2i2Fkg)

Drawing exercise 32 (Figure 6.84).

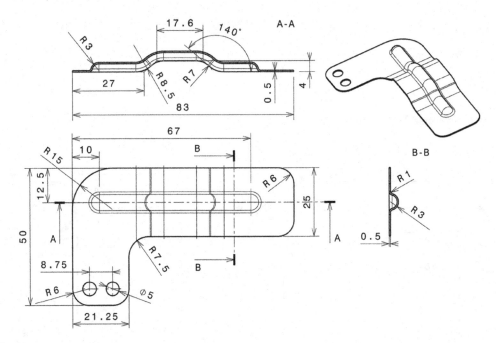

FIGURE 6.84 Drawing of the practising part. (Modelling solution: https://youtu.be/cKt_K2wT1x8)

Drawing exercise 33 (Figure 6.85).

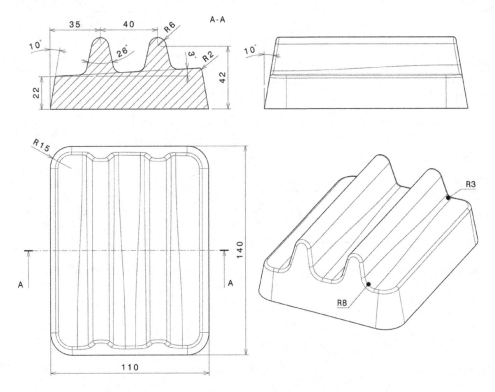

FIGURE 6.85 Drawing of the practising part. (Modelling solution: https://youtu.be/NdlKNszr_JQ.)

Drawing exercise 34 (Figure 6.86).

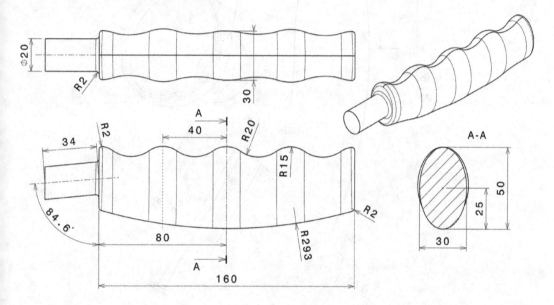

FIGURE 6.86 Drawing of the practising part. (Modelling solution: https://youtu.be/84ckdPsPRjI.)

Drawing exercise 35 (Figure 6.87).

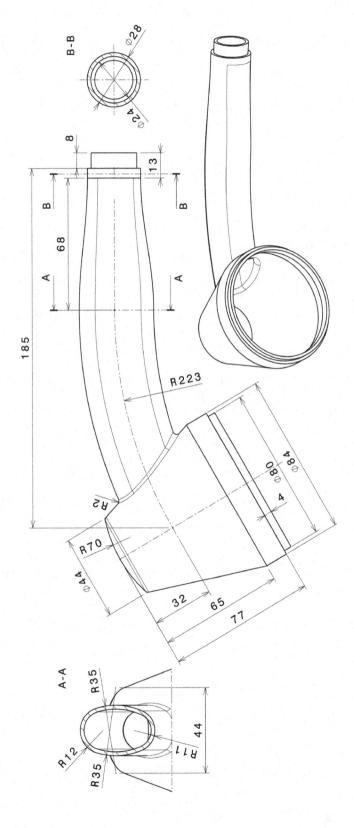

FIGURE 6.87 Drawing of the practising part.

Drawing exercise 36 (Figure 6.88).

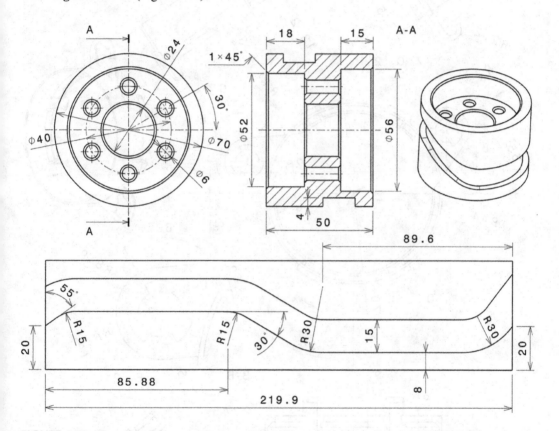

FIGURE 6.88 Drawing of the practising part. (Modelling solution: https://youtu.be/HLiXeLGOqmc.)

Drawing exercise 37 (Figure 6.89).

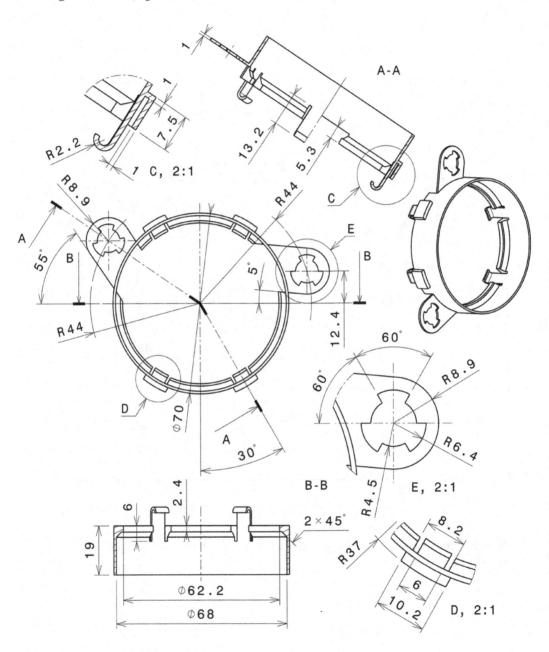

FIGURE 6.89   Drawing of the practising part.

Drawing exercise 38 (Figure 6.90).

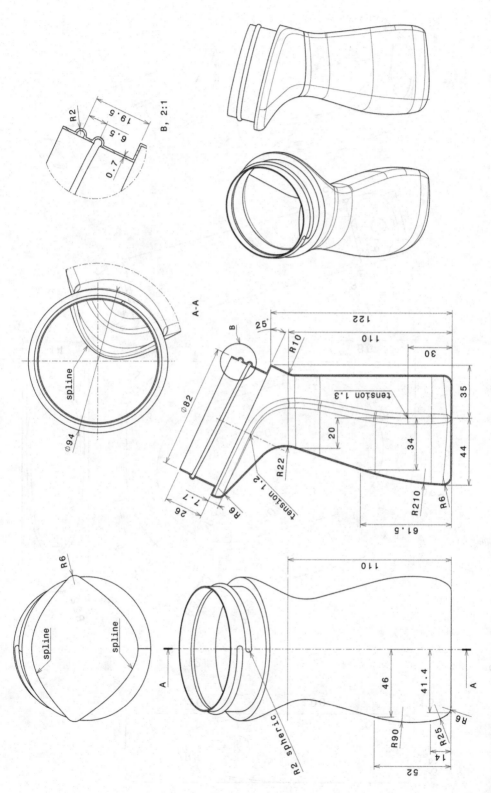

FIGURE 6.90 Drawing of the practising part. (Modelling solution: https://youtu.be/cPFGBQtOpDY.)

Drawing exercise 39 (Figure 6.91).

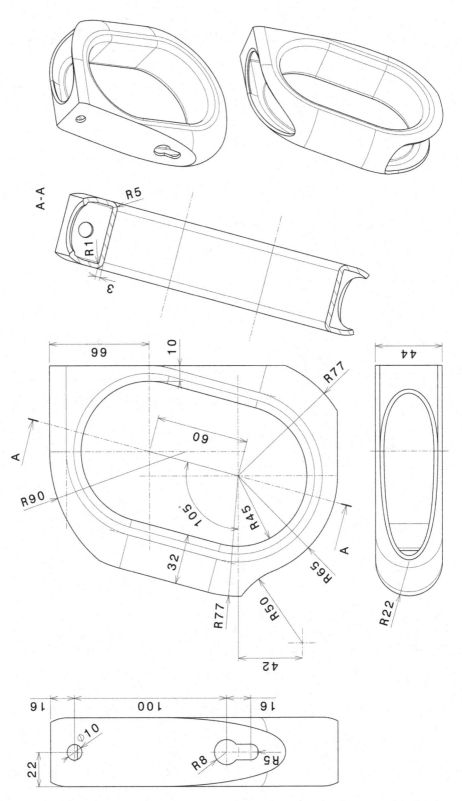

FIGURE 6.91   Drawing of the practising part. (Modelling solution: https://youtu.be/KTkK8anm4hM.)

Drawing exercise 40 (Figure 6.92).

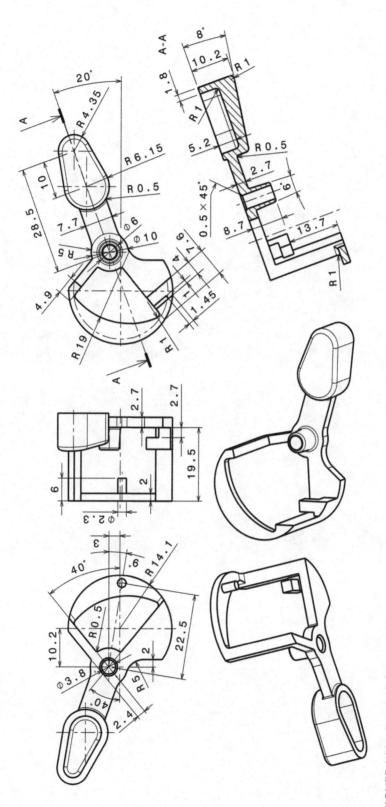

FIGURE 6.92 Drawing of the practising part.

Annexes

A1 Additional Online Resources: User Communities, Forums and Video Tutorials

A1.1 CATIA USER COMMUNITY

Dear *CATIA* Lovers, Welcome to your place to be! You are now part of the community which shapes the world we live in.

You will discover a world of expertise, of achievements with the world leading Design & Engineering Brand: *CATIA*! Even more you will get the possibility to experience interactive services which will develop and strengthen your skills and expertise. Be social, and do not hesitate to interact with *CATIA* team and me, to grow together!

Welcome to a World of Creativity and Innovation – Welcome to the *CATIA* Community!

Olivier Sappin, *CATIA CEO Dassault Systèmes*

https://r1132100503382-eu1-3dswym.3dexperience.3ds.com/#community:6

 CATIA User Community
Public

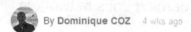 Shape Healing app is very useful in...

 By **Dominique COZ** 4 wks ago

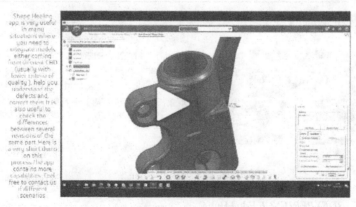

Some hint on healing Parts

👍 3   💬 0

FIGURE A1.1 *CATIA* User Community.

FIGURE A1.2 QR code to access the *CATIA* User Community, login is required for all users.

A1.2 3DEXPERIENCE EDU | HUB

Discover *3DEXPERIENCE* Edu Learning Labs Network are learning labs created by teachers, champions of the *3DEXPERIENCE* platform, who are convinced that project-based learning, which we rather prefer to call experience-based learning, is the best way for students to practice their skills and get new skills such as project management and collaboration.

https://r1132100503382-eu1-3dswym.3dexperience.3ds.com/#community:HNbVc2T1Qp2OBY9BVYYngA

 3DEXPERIENCE Edu | Hub
🌐 Public

 3DEXPERIENCE Edu Hub: our Learning Labs network in Video!

 By **Natacha BECARD**   2020-06-19

   Discover our **3DEXPERIENCE Edu Learning Labs Network**, who are learning labs created by teachers, champions of the 3DEXPERIENCE platform and convinced that project based learning, that we rather prefer to call experience based learning, is the best way for student to practice their skills and get new skills like project management, collaboration...

👍 20   💬 12   ↪

FIGURE A1.3 *3DEXPERIENCE* Edu I Hub.

FIGURE A1.4   QR code to access the *3DEXPERIENCE* Edu | Hub, a login is required for all users.

A1.3   LEARN ONLINE

From quick videos to the very complete user training for industry, hundreds of self-paced educational materials are available for students and educators, whatever is their preferred learning style.

  https://edu.3ds.com/en/learn-online

FIGURE A1.5   *3DEXPERIENCE* Edu Learn Online.

FIGURE A1.6   QR code to access the *3DEXPERIENCE* Edu Learn Online.

A1.4   DOCUMENTATION

Discover all the documentation you need to install, get started with and effectively use your Dassault Systèmes products.

https://www.3ds.com/support/documentation

A1.5   CERTIFICATION FOR STUDENTS & EDUCATORS

Get certified right now and leverage your expertise on Dassault Systèmes solutions with the 3DS Certification Program. Certification brings credibility to your CV, leading to significant opportunities for career growth. By getting certified, you demonstrate your expertise and prove your capability to differentiate yourself from the others in today's increasingly competitive job market.

Figures are based on a survey we conducted with 7,000 3DS Certified Engineers:

- Increased employability and better job opportunities: 41% said they found a better job (73% for students).
- Better-paid jobs and salary increase: 10% said they received a salary increase.
- Recognition within the company and among peers: 47% said they received more recognition in their company.

https://edu.3ds.com/en/be-recognized/academic-certification-program

CERTIFICATION FOR STUDENTS & EDUCATORS

Get certified right now and leverage your expertise on Dassault Systèmes solutions with the 3DS Certification Program.

FIGURE A1.7   Certification for students and educators.

FIGURE A1.8   QR code to access the certification for students and educators.

A1.6   ENG-TIPS.COM

Professional forum and technical support for engineers of Dassault: *CATIA* products includes problem-solving collaboration tools.

https://www.eng-tips.com/threadminder.cfm?pid=560

FIGURE A1.9   Professional forum and technical support for engineers.

FIGURE A1.10   QR code to access the professional forum and technical support for engineers.

A1.7   COE

Community of Experts (COE) of Dassault Systèmes solutions. COE helps users and their company leverage their Dassault Systèmes solutions through education, training, networking opportunities, product influence and best practices available exclusively to COE members.

The primary objectives of COE are to provide a forum for the interchange of knowledge, experience and technical information relating to the application of the Dassault Systemes family of solutions and the environment in which they operate, and to communicate with Dassault Systemes regarding the current and future capabilities and the use of these products.

http://www.coe.org

FIGURE A1.11   Community of Experts of Dassault Systèmes solutions.

FIGURE A1.12   QR code to access the Community of Experts of Dassault Systèmes solutions.

A1.8   COMPLETE VIDEO TUTORIALS LIST

This continuing growing list contains various exercises for:

- sketches
- solid parts
- surfaces parts
- assemblies
- 2D drawings
- parameterization
- tips & tricks about using the *CATIA v5* program.

https://qrgo.page.link/kLqJm

How to create a mechanical part using Generative Shape Design and CATIA Part Design 99
workbenchstuff

 How to create a Cover model using Generative Shape Design and CATIA Part Design 98
workbenchstuff

 How to create a Knot model using Generative Shape Design and CATIA Part Design 97
workbenchstuff

 How to create a mechanical part using Generative Shape Design and CATIA Part Design 96
workbenchstuff

 How to create a mechanical part using Generative Shape Design and CATIA Part Design 95
workbenchstuff

 How to create a mechanical part using Generative Shape Design and CATIA Part Design 94
workbenchstuff

FIGURE A1.13 Video tutorials list.

FIGURE A1.14 QR code to access the video tutorials list.

A1.9 QUESTIONS TAGGED [CATIA]

Macros, scripting and programming in *CATIA v5*
   https://stackoverflow.com/questions/tagged/catia

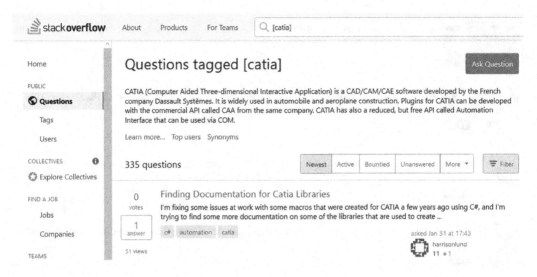

FIGURE A1.15 Macros, scripting and programming in *CATIA v5*.

A2 Video Tutorials to Support the Presented Written Tutorials

3.1. **Methods for creating the working planes**
      https://youtu.be/yCg_aoW34DE
3.2. **Modelling of a nut-type part**
      https://youtu.be/MqcIIMy5XDk
3.5. **Modelling of a hinge part**
      https://youtu.be/Hhl_R7UF24o
3.6. **Modelling of a complex spring**
      https://youtu.be/vPqYgyX-C6Q
3.7. **Modelling of a complex spiral ornament part for wrought iron fence**
      https://youtu.be/IexKFnGl1ZI
3.9. **Modelling of a 3D Knot**
      https://youtu.be/VrxsD0cZI5A
3.15. **Modelling of a reinforced key button part**
      https://youtu.be/1VMgEUaQQ64
3.16. **Modelling of a complex fitting part**
      https://youtu.be/LyILHjZYbEA
3.17. **Modelling and transformation of a part into two constructive solutions**
      https://youtu.be/AeC-J1kxNhQ
3.18. **Editing and reconstruction of solids using surfaces – twisted area**
      https://youtu.be/EcE1R7gEZGI

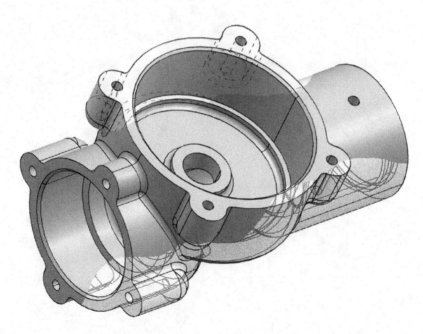

FIGURE A2.1 Representation of a 3D solid part.

3.19. Editing and reconstruction of solids using surfaces – connected surfaces
https://youtu.be/L3Neywur0og

3.20. Modelling of a gearbox shifter knob
https://youtu.be/oH47PD9rp-g

3.21. Modelling of a complex plastic cover
https://youtu.be/_K5vTErh040

3.22. Modelling of a window crank handle
https://youtu.be/YvNi3Q-uHjk

3.23. Modelling of a shield using laws
https://youtu.be/oLus-3yBhMQ

3.25. Modelling of an ornament panel
https://youtu.be/HMQyIJgMgJ8

3.26. Modelling of parametric bellows in different constructive solutions
https://youtu.be/uXrFScT8oJU

4.3. Parametric modelling of a connector cover and optimizations of the part
https://youtu.be/EcHTF466_L8
https://youtu.be/am2fbtPqyrA

4.6. Modelling of a sheet metal cover
https://youtu.be/I6z_SoMb2e8

4.7. Modelling of a sheet metal closing element
https://youtu.be/chvGIuMlseg

All the video tutorials are free, do not require a registered *YouTube* user account and were last accessed on 15 July 2022. These videos are not identical to the applications explained in this book, so the authors recommend viewing the video tutorials after reading the written material.

Bibliography

Barlier, C., Poulet, B., (1999) - *Mémotech. Génie mécanique, productique mécanique*. Deuxième édition. Editions Casteilla, ISBN 2-7135-2063-0, Paris.

Bondrea, I., Frăţilă, M., (2005) - *Proiectarea asistată de calculator utilizând CATIA v5 (Computer Aided Design using CATIA v5)*. Editura Alma Mater, Sibiu, 247 p.

Bondrea, I., Pîrvu, B., Goia, A., Şerbana, O., Paşa, A., Arieşan, I., Anghel, C., (2010) - *Reingineria prin CATIA v5 între teorie şi aplicaţii (CATIA v5 Reengineering between Theory and Applications)*. Editura Universităţii Lucian Blaga din Sibiu, ISBN 978-973-739-948-9, Sibiu, 487 p.

Brkić, D., Stajić, Z., (2021) - Excel VBA-based user defined functions for highly precise colebrook's pipe flow friction approximations: A comparative overview. *Facta Universitatis, Series Mechanical Engineering*, vol. 19, no. 2, University of Niš, ISSN: 0354-2025 (Print), ISSN: 2335-0164 (Online), Serbia.

Cather, H., Monius, R., Philip, M., Rose, C., (2001) - *Design Engineering*. Heinemmann Publishing House, ISBN 978-0-7506-5211-7, Butterworth.

CATIA v5, (2015) - *Official documentation*. Dassault Systemes.

Ćuković, S., Devedžić, G., Pankratz, F., Ghionea, I., Subburaj, K., (2015) - *Praktikum za CAD/CAM Augmented Reality*, University of Kragujevac, Faculty of Engineering, CIRPIS Center, ISBN 978-86-6335-020-5, Kragujevac, Serbia, 200 p.

Ćuković, S., Devedžić, G., Fiorentino, M., Ghionea, I., (2017) - A comparative study of CAD data exchange based on the STEP standard. *Scientific Bulletin of the University Politehnica of Bucharest, Series D*, vol. 79, no. 4, ISSN 1454-2358, Bucharest, pp. 187–198.

Dăscălescu, A., (2005) - *Desen tehnic industrial. Reprezentările, cotarea, notarea şi înscrierea desenului tehnic. Aplicaţii (Industrial Technical Drawing. Representations, Dimensions and Annotations of the Technical Drawing. Applications)*. Editura Risoprint, ISBN 973-751-080-1, Cluj-Napoca.

Devedžić, G., Ćuković, S., Petrović, S., Maksić, J., (2016) -*3D Product Modeling - Methodical Problems Collection*, Second edition. University of Kragujevac, Faculty of Engineering, CIRPIS Center, ISBN 978-86-6335-023-6, Kragujevac, Serbia, 432 p.

Dobre, G., (2003) - *Organe de maşini (Machine Parts)*, vol. I. Editura Bren, ISBN 973-8143-99-2, Bucureşti.

Dubbel, H., (1998) - *Manualul inginerului mecanic. Fundamente (Mechanical Engineer's Manual. Fundamentals)*. Editura Tehnică, ISBN 973-31-1271-2, Bucureşti.

Ghionea, I., (2004) - *Module de proiectare asistată în CATIA v5 cu aplicaţii în construcţia de maşini (CATIA v5 Assisted Design Workbenches with Applications in Machine Manufacturing)*. Editura Bren, ISBN 973-648-317-7, Bucureşti, 226 p.

Ghionea, I., (2007) - *Proiectare asistată în CATIA v5. Elemente teoretice şi aplicaţii (Assisted Design in CATIA v5. Theoretical Elements and Applications)*. Editura Bren, ISBN 978-973-648-654-8. doi: 10.13140/RG.2.1.1077.0642, Bucureşti, 462 p.

Ghionea, I., (2009) - *CATIA v5. Aplicaţii în inginerie mecanică (CATIA v5. Applications in Mechanical Engineering)*. Editura Bren, ISBN 978-973-648-843-6. doi: 10.13140/ RG.2.1.2387.7848, Bucureşti, 258 p.

Ghionea, I., Devedžić, G., Ćuković, S., (2015) Parametric modeling of surfaces using CATIA v5 environment. *7th International Conference on Advanced Manufacturing Technologies – ICAMaT 2014*, Advanced Technologies in Designing and Progressive Development of Manufacturing Systems, volume Applied Mechanics and Materials, ISSN 1660-9336, Trans Tech Publications Ltd., Bucharest, pp. 93–98.

Ghionea, I., Tarbă, C., Ćuković, S., (2021) - *CATIA v5. Aplicaţii de proiectare parametrică şi programare (CATIA v5. Parametric Design and Programming Applications)*, Editura Printech, ISBN 978-606-23-1264-0, August 2021, Bucureşti, 532 p.

Giesecke, F. E., Mitchell, A., Spencer, H. C., Hill, I. L., Dygdon, J. T. (1986) - *Technical Drawing*. Eighth edition. Macmillan Publishing Company, ISBN 0-02-342600-4, New York.

Ispas, C., Ghionea, I., (2003) - *The Management of Informations and CAD in the Conception and Development Phases of a Product*. Scientific Bulletin. Serie C, volume XVII, Fascicle Mechanics, Tribology, Technology of Machine Manufacturing, International Multidisciplinary Conference, 5th edition, North University of Baia Mare, ISSN 1224–3264, Baia Mare.

Ionescu, N., Vişan, A., Rohan, R., (2016) - *Toleranţe. Aplicaţii (Tolerances. Applications)*, Editura Politehnica Press, ISBN 978-606-515-693-7, Bucureşti, 200 p.

Kumar, K., Ranjan, C., Davim, J., P., (2021) - *Understanding CATIA. A Tutorial Approach*, CRC Press, Taylor & Francis Group, Science, Technology and Management Series, ISBN: 978-0-367-48794-2 (hbk), ISBN: 978-1-003-12165-7 (ebk), 261 p.

Madsen, D. A., Madsen, D. P., (2012) - *Engineering Drawing & Design*. 5th edition. Delmar Cengage Learning, ISBN 978-1-111-30957-2, New York.

Marin, D., (2007) - *Desen tehnic. Elemente de proiectare (Technical Drawing. Design Elements)*, Editura Bren, ISBN 978-973-648-633-3, București, 179 p.

Marinescu, A., Alupei, O., (2003) - *Toleranţe şi ajustaje pentru piese în construcţia de maşini (Tolerances and Adjustments for Parts in Machine Construction)*, Editura Bren, ISBN 973-648-222-7, București.

Miller, N., (2022), *Video tutorial: Facades Generative Parametric Modelling in CATIA on the Cloud*. 3DExperience CATIA Buildings & Infrastructure.

Orlov, P., (1977) - *Fundamentals of Machine Design*, vols. 1–4. Mir Publishers, Moscow.

Popescu, D., Zapciu, A., Tarbă, C., Laptoiu, D., (2020) - Fast production of customized three-dimensional-printed hand splints. *Rapid Prototyping Journal*, vol. 26, no. 1, ISSN 1355–2546. https://doi.org/10.1108/RPJ-01-2019-0009, pp. 134–144.

Spasić, Ž., Jovanović, M., Bogdanović-Jovanović, J., Milanović, S., (2020) - Numerical investigation of the influence of the doubly curved blade profiles on the reversible axial fan characteristics. *Facta Universitatis, series Mechanical Engineering*, vol. 18, no. 1, University of Niš, ISSN: 0354–2025 (Print), ISSN: 2335-0164 (Online), Serbia, pp. 57–68.

SR EN ISO 5455:1997, *Desene tehnice. Scări (Technical drawings. Scales)*.

SR EN ISO 128-3:2020, *Desene tehnice. Principii generale de reprezentare. Partea 3: Vederi, secţiuni şi rupturi (Technical drawings. General principles of representation. Part 3: Views, sections and breakings)*.

SR ISO 129-1:2014, *Desene tehnice. Indicarea cotelor şi toleranţelor. Partea 1: Principii generale (Technical drawings. Indication of dimensions and tolerances. Part 1: General principles)*.

SR EN ISO 1101:2017, *Specificaţii geometrice pentru produse. Tolerare geometrică. Toleranţe de formă, de orientare, de poziţie (Geometric specifications for the products. Geometric tolerance. Tolerances of shape, orientation, position)*.

Stăncescu, C., Manolache, D. S., Pârvu, C., Ghionea, I., Tarbă, C., (2012) - *Proiectare asistată cu Autodesk Inventor. Îndrumar de laborator (Assisted Design with Autodesk Inventor. Practical Applications)*. Ediţia a II-a, Editura FAST, ISBN 978-973-86798-7-0, București, 372 p.

Stăncescu, C., (2014) - *Modelare parametrică şi adaptivă cu Inventor (Parametric and Adaptive Modelling with Inventor)*. Editura FAST, ISBN 978-973-86798-8-7, București, 657 p.

Stăncescu, C., (2016) – *Album cu 100 piese mecanice (Album with 100 Mechanical Parts)*. Editura Din Condei, ISBN 978-606-8707-23-5, București, 207 p.

Tarbă, C., (2017) - Development of a web based instrument on higher education structures of industrial engineering. *MATEC Web Conference*, vol. 112, article number 08013, eISSN: 2261-236X, https://doi.org/10.1051/matecconf/201711208013, 6 p.

Vișan, A., Ionescu, N., (2006) - *Toleranţe. Bazele proiectării şi prescrierii preciziei produselor (Tolerances. Basics of Product Design and Accuracy Prescribing)*, Ediţia a II-a, Editura Bren, ISBN 978-973-648-280-4, București.

Vlase, A., (1996) - *Tehnologia construcţiilor de maşini (Machine manufacturing technology)*, Editura Tehnică, ISBN 973-31-0777-8, București.

Wächter, K., (1987) - *Konstruktionslehre für Maschineningenieure. Grundlagen, Konstruktions-und Antriebselemente*. VEB Verlag Technik, ISBN 3-341-0045-3, Berlin.

Ziethen, D. R., (2013) - *CATIA v5: Macro Programming with Visual Basic Script*. McGraw-Hill Education, ISBN 978-0-07-180002-0, New York, 560 p.

Index

Add Formula 369, 371
Advanced Surfaces 354
Affinity 67
Angle 41, 93, 149, 181, 201, 297, 423
Angle/Normal to Plane 34
Angular Spacing 43, 101, 143, 162, 294
Assemble 162, 282, 324
Associate 396, 406
Automation 13, 435, 438
Axis 34, 40, 121, 166
Axis System 18, 50, 451, 463, 483

Bi-Tangent Line 164, 166
Blend 132, 155, 214, 479, 493
Boolean 30, 146, 269, 282, 363, 388, 446
Boundary 79, 98, 133, 168, 213, 318, 479

Catalogue 13, 22, 409–419
CATDrawing 5, 455
CATIA (Computer Aided Three dimensional Interactive
　　　Applications) 1, 3
CATIA Assembly Design 2, 12, 409
CATIA DMU Kinematics 2
CATIA Drafting 2, 19, 48, 475, 490
CATIA Generative Shape Design (GSD) 2, 79, 101, 147
CATIA Generative Sheetmetal Design 2, 420, 426, 495
CATIA Generative Structural Analysis 2
CATIA Knowledge Advisor 2, 15, 361, 373, 390, 476
CATIA Part Design 2, 5, 29, 100, 148, 348, 467, 488
CATIA Prismatic Machining 2
CATIA Sketcher 2, 19, 256, 463, 483, 486
CATIA Wireframe and Surface Design 37
CATMain 438, 441, 446, 452, 460
CATPart 3, 5, 8, 385, 405, 409–415, 437, 483
CATProduct 3, 6, 437, 455, 459
CATScript 13, 435–441, 457, 480, 494
Center curve 96, 105, 120, 253
Chamfer 48, 172, 196, 423, 465, 488
Change Sketch Support 329, 466
Check connexity 87, 99, 216, 230, 249, 353
Circle 40, 71, 96, 114, 190, 206, 238, 253, 300
Circle Limitations 238, 300, 308
Circle Type 238, 300, 307
Circular Pattern 43, 100, 294, 367, 424
CloseSurface 100, 120, 161, 231, 491
Closing Point 174–176, 216, 276, 302–303
Coincidence 25, 152, 201, 486–487
Combine 181, 186
Compass 3, 27–28, 38–39
Complementary Circle 308
Complete Crown 43, 100–101, 138, 308, 367
Computer Aided Design (CAD) 1, 3, 33, 361, 486
Connect Curve 83–84
Constraints 17–19, 25, 132, 190, 243, 385–388, 468
Constraints Defined in Dialog Box 90, 466, 486–487
Construction/Standard Element 106, 112, 150, 350,
　　　466, 486
Continuity 79–80, 99, 133, 147, 212–215

Corner 75, 90, 106, 110–111, 174, 187, 433
Counterbored 200–202, 400, 465, 484
Counterclockwise 93, 266, 285
Coupling 175–176
Ctrl key 8, 12, 25, 72, 119, 188, 220, 328, 487
Curvature 79–80, 253, 356
Curve 83–88, 96–99, 103–110, 120, 186, 229, 253, 318,
　　　331–341
Cutout 422–425, 428–431
Cutting/Stamping 422, 427–428, 433
Cutting elements 82, 98, 141, 225, 235, 259, 323

Default Bend Radius 421–422, 426
Define In Work Object 145, 160, 203, 296, 318, 339, 348
Depth 41, 48, 76, 127, 369, 391, 464, 484
Design Table 364, 394–419, 475–476, 493
Dim 437–438, 441–462
Dimension 17, 41, 76, 102, 153, 177, 361, 464, 489
Dimming 446
Disassemble 221–222, 331
Draft 58, 61–69, 180–182, 201–202, 288, 481–482
Draft Direction 288
Dress-Up Features 46–48, 56–58, 135–136, 163, 180, 486

Edge Fillet 46, 138–140, 178–179, 201, 220, 261–262, 304
Elements To Join 87, 120, 134, 160, 216, 249, 353
Elongated Hole 256, 282, 321, 426
Exit workbench 41, 91, 351
Explicit 95, 132, 228, 329, 351, 357
Extension 48, 484
Extract 147, 154, 212–216, 239–245, 272–274, 287
Extrapolate 79, 82, 215
Extrude 37, 101–103, 140–145, 153–155, 352–358, 476

Face-Face Fillet 311–312
Faces to keep 146, 181–182
Faces to remove 146, 182, 211, 492
Federation 249
Fill 81–82, 157, 160–162, 168, 229–230, 278, 341, 476
First Angle 41, 122, 131
Flange 187, 422–424
Flanged Hole 423
Fold/Unfold 426, 428–429
Formula 12–15, 193–197, 342–347, 362–402, 476, 489
Formula Editor 342, 369, 371

Geometrical Set 92, 149, 189, 295–308, 335, 342–346
Guiding curve 216, 243, 260, 270, 288

Height 93, 424, 427, 457
Helix 79–83, 92–96
Hem 433–434
Hide/Show 11–12, 48, 81, 119, 135, 175, 213, 249, 298, 329
Hole 56–57, 108, 111, 172, 187–189, 282, 418–419, 450, 484

Insert 12, 29, 50, 140, 144, 216, 247, 296, 404
Intersection 96, 102, 114, 154–155, 272–273, 329, 338
Invert Orientation 330–331

Join 87, 99, 120, 160–161, 216, 230,
    249, 293

Keep Angle 105
Knowledge 15, 30, 193, 342, 362–363, 369, 390,
    396, 402

Law 287–290
Length 14, 79, 85, 180, 193, 342, 369–370, 432, 444
Line 35, 52, 80, 94, 164, 174, 224, 239, 273, 334

Macro 13, 435–452
Mean through points 39, 479, 493
Measure Between 105, 115, 118
Measure Item 183, 202, 231, 235, 429, 434
Measure toolbar 105, 382, 416, 429
Metric Thick Pitch 48
Microsoft VBScript 13, 435, 441, 452, 494
Mirror 75, 106–108, 204, 391, 424–425, 464, 470
Mirrored extent 41, 51, 76, 102, 177, 199, 334, 465
Multi-Sections Surface 174, 215, 243, 276, 301, 308

Nearest solution 116, 186, 224, 257, 478, 492
Next solution 308
No propagation 72, 213, 318, 352
Normal to Curve 37, 80, 83, 93, 109, 337, 356

Object(s) to fillet 46, 138, 163, 201
Object Browser 438, 455
Offset 33, 72–73, 106, 190, 371–373
Offset from plane 33–34, 124, 150, 200, 270
Operation toolbar 46, 72, 87, 98, 133, 178, 312
Options 24, 193, 231, 342, 362–364, 452, 488
Orientation 50, 93
Other side 98, 141, 158, 165, 224, 318

Pad 43, 63, 137, 404
Pan 8, 24
Parallel through point 34, 83, 104, 114, 207
Parameters 13, 193, 342, 362, 393, 396–398
Parameters and Measure 15, 193, 342, 363
PartBody 100, 145–147, 181, 324, 369–374
PartDocument 438, 441, 446, 454
Pitch 79, 93, 475
Plane 33–39, 80–86, 109–110, 241, 335
Pocket 43, 200, 205, 369, 464
Point Definition 84, 228, 238, 333
Point-Direction 94, 115, 240, 334
Point or Curve Mapping 427
Positioned Sketch 50, 56, 306–307
Preselection navigator 328
ProductDocument 454
Profile 40, 91, 102, 132, 307
Profile type 95, 132, 253, 351
Projection 116, 168, 223, 252, 333
Project 3D Elements 46, 128, 259, 472, 489
Project 3D Silhouette Edges 141, 472, 489
Propagation 72, 99, 212, 318
Properties 7, 24, 194, 331, 338
Pulling direction 58, 95, 201, 351
Punch & Die 423

Quick Trim 40, 55, 90, 131–132, 235

Radial Alignment of Instance(s) 424
Ratio 70, 332–333, 371
RectPattern 209, 490
Reference Elements 33–34, 136, 200, 206, 431
Reference point 85, 103, 273, 431–432
Relations 13, 17, 193, 197, 342
Relimitations 73
Render Style 10
Replace Closing Point 175
Reverse Direction 44, 85, 103, 270, 276, 347
Reverse Side 44, 122, 204, 378
Revolution axis 91, 125, 139, 167, 315
Revolution Surface 91, 128, 139, 158, 164–167, 255–256
Revolve 91, 158, 285, 315
Rib 79, 88, 105, 186, 468
Rolled Wall 426, 433
Rotate 8–9, 190, 296–305, 309, 488
Rule Editor 373–374, 390

Semimajor / semiminor 270
Sew Surface 216
Shaft 41, 131, 149, 278, 280
Shape Fillet 304
Sheet Metal Parameters 421, 426
Shell 136, 210–212, 378, 467
Shift key 24, 335, 344
Sketch 2, 19, 25, 29, 50–52
Sketch Analysis 73, 106, 235, 473, 489
Sketch-Based Features 29, 41, 180, 486
Sketch Tools 71, 80, 106
Specification Tree 3, 24–25, 58, 472, 475
Sphere 41, 92, 96–98, 238, 318–322
Spline 37, 142, 271, 274, 298, 307–308
Split 98–99, 141, 182, 224–225, 235, 258–260
Starting Angle 93
Starting point 93
Start recording 13, 440
Surface-Based Features 100, 120, 160, 216, 323
Surface Stamp 427–428
Swap 50, 314
Swap visible space, 12, 96
Swept Walls 432
Sweep 95–96, 110, 132–133, 260–262, 288–290
Sweep Surface 95
Symmetry 90, 132–133, 168, 243, 247, 276

Tangent Tension 340
Tangent to Surface, 37, 472, 488
Taper Angle 93
Tear Drop 431–434
Thick Surface 144, 170, 179, 263, 294, 323
Thread 48, 282, 465
Three Point Arc 132, 150
Through Planar Curve 35, 37
Through Point and Line 35, 37, 113, 479
Through three points 34, 165, 479
Through two Lines 35, 240, 334
Tools 5, 12, 15, 73, 193
Total Angle 43, 101
Transformation Features 43, 100, 138, 296, 307
Translate 298, 350
Trim 71, 90, 110, 122, 133
Trim elements 85, 158, 318

Union Trim 146, 181
Update 6, 18, 385, 449
Update Diagnosis 189, 205
Up to element 79
Up to Last 48, 56, 65, 381, 464
Up to Next 43, 56, 74, 108, 124, 171
Up to plane 106
User Pattern 187, 488

Variable Fillet 312
VBScript 13, 435, 446, 452

Visual Basic Application 364
Visual Basic Editor 452, 455, 457, 459

Walls 421, 426, 432–433
Wireframe 10, 37, 79, 110, 269
With pulling direction 95, 351
Workbench 2–3, 29, 37, 41, 79
Wrap Surface 354–359
Wrap type 354–359

Zoom 8–9, 24, 188, 437, 455

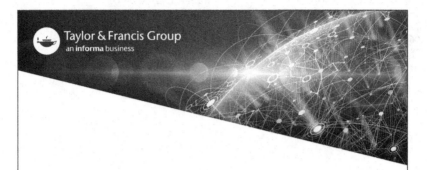

Taylor & Francis eBooks

www.taylorfrancis.com

A single destination for eBooks from Taylor & Francis
with increased functionality and an improved user
experience to meet the needs of our customers.

90,000+ eBooks of award-winning academic content in
Humanities, Social Science, Science, Technology, Engineering,
and Medical written by a global network of editors and authors.

TAYLOR & FRANCIS EBOOKS OFFERS:

A streamlined
experience for
our library
customers

A single point
of discovery
for all of our
eBook content

Improved
search and
discovery of
content at both
book and
chapter level

REQUEST A FREE TRIAL
support@taylorfrancis.com

 Routledge
Taylor & Francis Group

 CRC Press
Taylor & Francis Group

Printed in the United States
by Baker & Taylor Publisher Services

Printed in the United States
by Baker & Taylor Publisher Services